Textbook Series of University of Chinese Academy of Sciences
（Graduate Level）

Condition Monitoring of Insulation Gases and Fault Diagnosis of Electrical Apparatus

Editor in Chief ZHANG Guoqiang
Participate in editing HAN Dong LI Kang
QIU Zongjia ZHANG Shen

中国电力出版社
CHINA ELECTRIC POWER PRESS

This book is inspired by the Gas Insulation and Electrical Apparatus course taught at the University of Chinese Academy of Sciences (UCAS). It is structured into two main sections: Insulation Gases and Fault Diagnosis of Electrical Apparatus. The Insulation Gases section focuses on sulfur hexafluoride (SF_6) and the environmentally friendly alternative C_4F_7N, incorporating the latest research by scientists and engineers worldwide. Key topics include insulation performance, partial discharge, decomposition characteristics, and gas-solid compatibility. The Fault Diagnosis section explores innovative condition monitoring and diagnosis technologies for high-voltage electrical apparatus, leveraging optics, acoustics, and chemistry. These advancements equip high-voltage apparatus with sensory capabilities akin to sight, hearing, smell, taste, and touch. By integrating multidisciplinary perspectives, this book examines environmental sustainability, sensing, and detection in smart grids. A standout feature is its systematic approach to practical challenges in optical spectrum analysis, quantum chemistry calculations, and molecular design/synthesis, making it especially valuable for readers seeking real-world solutions in smart grid applications.

Serving as both a graduate-level textbook and a key academic reference, this book highlights cutting-edge research and technological advances in advanced sensors and optical instruments for the power industry. It is an essential resource for researchers, engineers, and graduate students in electrical power engineering, optical/acoustic engineering, and chemical engineering.

图书在版编目（CIP）数据

绝缘气体状态监测与电气设备故障诊断 = Condition Monitoring of Insulation Gases and Fault Diagnosis of Electrical Apparatus——Textbook Series of University of Chinese Academy of Sciences (Graduate Level) / 张国强主编. --北京：中国电力出版社，2025. 6. --（中国科学院大学研究生教材系列）.
ISBN 978-7-5239-0055-0

Ⅰ. TM853；TM07

中国国家版本馆 CIP 数据核字第 2025RK4909 号

出版发行：中国电力出版社
地　　址：北京市东城区北京站西街 19 号（邮政编码 100005）
网　　址：http://www.cepp.sgcc.com.cn
策划编辑：周　娟
责任编辑：杨淑玲（010-63412602）
责任校对：黄　蓓　郝军燕　李　楠　于　维
装帧设计：王红柳
责任印制：杨晓东

印　　刷：北京雁林吉兆印刷有限公司
版　　次：2025 年 6 月第一版
印　　次：2025 年 6 月北京第一次印刷
开　　本：787 毫米×1092 毫米　16 开本
印　　张：26.75
字　　数：612 千字
定　　价：138.00 元

Preface

This book draws its inspiration from the course, "Gas Insulation and Electrical Apparatus" taught at the University of Chinese Academy of Sciences (UCAS), Beijing, China, in the spring term of 2021, 2022, and 2023. However, its origins date back to courses I developed from 2016 to 2019 titled "Contemporary Electric Measurement, Condition Monitoring and Fault Diagnosis of Electrical Equipment". Together with a number of colleagues, especially Dr. Dong HAN, Dr. Kang LI, and Dr. Zongjia QIU, we offered this forty-hour course multiple times at UCAS. Roughly half of the content in this book stems from these lecture notes.

Special thanks to Dr. Shen ZHANG for his editing contributions to the final chapter, his diligent revisions and proofreading, as well as his pursuit of the latest technological breakthroughs and significant efforts in LATEX typesetting.

The main contents of this book include two units: Condition Monitoring of Insulation Gases and Faults Diagnosis of Electrical Apparatus.

SF_6 gas-insulated electrical apparatus, including gas-insulated switchgears, gas-insulated power transformers, gas-insulated pipelines, etc., because of its high reliability, small installation clearance, compact transmission corridor, and other advantages, has been widely used in the electric power industry. However, SF_6 gases can accelerate global warming, which is currently avoided mainly by preventing harmful insulating gases from being emitted directly into the atmosphere. At present, the best solution which is recognized by domestic and overseas experts is to find new environmentally friendly insulation gases. The National Ministry of Science and Technology of China launched two key projects in the smart grid special program *Environmental Insulation Gas* research direction in July 2017 and December 2021 respectively. The authors of this book had the honor to participate in the research work of the above projects, and fully appreciate the rapid outcomes and technological progress of scientific research in the field of *Environmentally Friendly Insulation Gas*. The content of SF_6 *gas insulation* and *Environmental friendly gas* in this book brings together the latest research progress by domestic and overseas scientists and engineers to keep up with the pace of the times.

In the module on *Electrical Apparatus Fault Diagnosis*, the content of this book focuses on smart grids, smart electrical equipment, advanced sensors, and new detection technologies. Smart electrical equipment serves as the foundation of the smart grid. The integration of advanced sensors and detection instruments on a large scale enhances the equipment's intelligence, enabling it to perceive its health conditions through sensory capabilities similar to smell, vision, hearing, taste, and touch. At present, the concept of operation and maintenance engineers in the electric power industry is to discover potential defects of electrical apparatus as early as possible without interrupting the power supply and realize online, live, on-site, continuous, and real-time monitoring of electrical equipment. Therefore, optical, chemical, and acoustic technology and any other non-electromagnetic technology, due to their innate immunity to background electromagnetic interference, have become the technical means that have attracted wide attention at present, and become the research hotspot in this field. The authors of this book are honored to undertake the key project of *Basic Scientific Research Conditions and Major Scientific Instruments and Equipment Research and Development* of the Ministry of Science and Technology of China in 2022, in addition, the scientific research achievements in the research and development of largescale scientific instruments were also rated as the top ten excellent projects of the State Grid Corporation of China 2022 headquarters. This book "Electrical Apparatus Fault Diagnosis" unit summarizes and refines the latest research results in this field, which has certain reference values for operation and maintenance engineers and scientific researchers in the electric power industry and can also become a necessary study reference book for graduate students in related majors.

The book is structured as follows:

Chapter 1 - Electrical equipment is fundamental components of the power system. Insulation defects of electrical equipment are difficult to detect during daily operation and maintenance, often leading to serious accidents. Sensors monitor various physical quantities that reflect the state of equipment, such as electricity, heat, mechanical force, chemistry, optics, acoustics, and so on. They are the first and most important step in state monitoring and fault diagnosis.

Chapter 2 - The continuous growth in electric power demand, coupled with increasing environmental awareness, has led to the urgent development of high-voltage electrical equipment with large capacity and compact structures. Heptafluorisobutyronitrile (C_4F_7N), as a new environmental protection insulating medium, the research on its physical and chemical properties attracts widespread attention. After introducing SF_6 and its alternatives, this chapter focuses on the insulation, decomposition, and compatibility characteristics of C_4F_7N, proving a theoretical and experimental basis for promoting the process of environmental insulation gases.

Chapter 3 - Atomic Emission Spectrometry (AES) and its diverse applications. In power transmission lines, it has been instrumental in detecting high-voltage insulators tainted by dust and salty fog, and in gauging the coating thickness of room-temperature vulcanized silicone rubber (RTV). Beyond this, Laser-Induced Breakdown Spectroscopy (LIBS) is one kind of Atomic Emission Spectroscopy, it was employed in thermal power plants to supervise the structure and composition of coal. In nuclear power plants, the composition of reactor control rods and other highly radioactive materials can be remotely analyzed. In metal alloy industries, LIBS can be leveraged to determine the elemental percentages during melting stages or recycling clarifications. Moreover, in space explorations, LIBS serves as a tool to identify water or rare metals on celestial bodies like Mars or the Moon.

Chapter 4 - Gas Chromatography and Photoacoustic Spectroscopy in Smart Electric

Apparatus. While gas chromatographs are now ubiquitously used to assess trace dissolved gases in power transformer oil, the burgeoning technology of photoacoustic spectroscopy has captured significant attention in the power transformation and distribution sectors.

Chapter 5 - Tunable Diode Laser Absorption Spectroscopy and its environmental and industrial applications. This chapter introduces tunable diode laser absorption spectroscopy, a pivotal technique in determining concentrations of specific gases like methane and water vapor, especially in the environmental protection sector, such as within coal-fired power plants. Leveraging cavity enhancement techniques, this method has been extended to detect decomposition gases in SF_6 gas-insulated switchgears.

Chapter 6 - Infrared Spectroscopy and Thermal Imaging. This chapter elucidates how infrared radiation, though imperceptible to the human eye, can be translated into a visual spectrum through infrared cameras, showcasing thermal variations. This capability extends to detecting specific gas leaks, given unique radiation absorption wavelengths. For instance, infrared hyperspectral imaging can detect SO_2 emissions from as far as 1.5 km using a specific system.

Chapter 7 - Ultraviolet Spectroscopy & UV Camera and their diagnostic applications. Ultraviolet spectroscopy and UV cameras offer crucial diagnostic methods for partial discharges in air-insulated apparatus. As electrons discharge, they emit ultraviolet light. The number of emitted photons can thus be counted to determine the intensity of the discharge.

Chapter 8 - Electrochemical Gas Sensors for GIS and their specifications. This chapter elaborates on the comprehensive technical specifications and multifaceted properties of SF_6. As it decomposes into varied compounds and ions during discharges, electrochemical sensors have become the go-to tools for detection, especially for routine GIS inspections.

Chapter 9 - Chemiluminescence and UV Absorption Sensors and their applications. Highlighting chemiluminescence and UV absorption sensors, this chapter delves into the attributes of air as an insulation medium. The decomposition of air during discharge processes is a focal discussion point, emphasizing its applications in hydro-generators and medium voltage switchgears.

Chapter 10 - Electro-optic and Magneto-optic Sensors and their practical implications. This chapter is dedicated to Electro-optic and Magneto-optic sensors, underscoring the Kerr, Pockels, and Faraday rotation effects. These effects are pivotal for gauging variations in voltage, electric fields, currents, and magnetic fields. Real-world applications of these sensors are also showcased in this chapter.

Chapter 11 - Michelson and Fabry-Perot Interferometers and their precision measurement applications. The intricacies of Michelson and Fabry-Perot interferometers are explored in this chapter. Renowned for their superlative precision, these tools can measure minuscule deformations or pressures, sometimes as small as a few nanometers or picometers.

Chapter 12 - Acoustic and Ultrasonic imaging techniques. In this chapter, readers are introduced to how acoustic or ultrasound cameras discern noises birthed by mechanical vibrations (frequency $\leqslant 20$ kHz) or partial discharges within electric apparatus (frequency range from 20 kHz to 500 kHz).

This book serves as both a graduate-level textbook and a significant academic resource, shedding light on the recent research and application milestones in advanced sensors and optical instruments within the electrical power industry's smart apparatus segment.

I am extremely grateful that this book is funded by the Textbook Publishing Center of University of Chinese Academy of Sciences.

Beijing, China   Guoqiang ZHANG

Table of Contents

Preface

1

Introduction

1.1 Overview of Electrical Equipment

A substation is an intermediate connecting power plants and users, serving as a connection point for power grid lines. It plays a role in transforming voltage, controlling current flow, and distributing power. According to voltage level, substations can fall into step-up or step-down substations. According to the scale, it can be divided into hub, liaison, or terminal substations.

Various electrical equipment of different voltage levels are installed in the substation according to strict design requirements to ensure a safe, reliable, and economical power supply. The equipment that directly produces, exchanges, transmits, and uses electrical energy is called primary equipment. The equipment for monitoring, measuring, controlling, regulating, and protecting primary equipment is called secondary equipment.

1.1.1 Primary Equipment

1.1.1.1 Equipment for Transforming Voltage or Energy

Power transformers are the leading equipment for converting electrical energy in substations, and their function is to change voltage for power transmission. A step-up power transformer boosts the voltage to reduce line losses, improve the power transmission economy, and achieve the goal of long-distance transmission. A step-down power transformer can convert high voltage into the required voltage levels for meeting the usage needs.

1.1.1.2 Switch

- Circuit breaker

A circuit breaker has arc extinguishing and current breaking capabilities, used to cut off the no-load current and load current in high-voltage circuits. In addition, when a system malfunction occurs, the circuit breaker, in conjunction with protection and automatic devices, can quickly cut off the fault current and prevent the accident from expanding. According to the arc extinguishing medium, circuit breakers can be divided into oil immersed (more oil, less oil), SF_6, vacuum, and compressed air circuit breakers.

- Isolating switch

The isolating switch is the most used device in the high-voltage switchgears, playing an isolation role in the circuit. Compared with the circuit breaker, the isolating switch has no arc extinguishing capability, and only opens and closes the circuit under the condition of no load current. The structure and working principle of the isolating switch are simple. However, due to its large usage and high-reliability requirements, the isolating switch has a great impact on the design, establishment, and safe operation of substations and power plants.

- Switchgear cabinet

A switchgear cabinet integrates circuit breakers, isolation switches, grounding switches, and current transformers into a metal cabinet. It has the advantages of space-saving and convenient installation. Unlike open-type equipment, the internal structure of the switchgear is closed, and the component status observation inside the cabinet is not intuitive enough, which is more likely to cause misoperation.

1.1.1.3 Gas-Insulated Switchgear (GIS)

GIS is the optimized integration of primary equipment in a substation, excluding transformers. A GIS mainly includes circuit breakers, isolation switches, grounding switches, voltage transformers, current transformers, lightning arresters, busbars, cable terminals, and inlet and outlet bushings. HGIS (hybrid gas insulated switchgear)'s structure is similar to GIS, but the busbar of HGIS adopts an open type. HGIS is mainly used for the renovation of 500 kV substations or old stations.

A GIS usually uses SF_6 gas as insulation and arc extinguishing medium, and the charged parts are completely enclosed in SF_6 gas. As a result, a GIS has the advantages of miniaturization, high reliability and safety, and less maintenance or even no maintenance.

1.1.1.4 Reactive Power Compensation Devices

The power loads, such as motors and transformers, are mostly inductive loads that need to absorb corresponding reactive power during operation. Reactive power compensation devices can provide reactive power consumed by inductive loads, mainly including power capacitors, shunt reactors, and arc suppression coils.

- Power capacitors

There are two forms of power capacitor compensation - parallel and series compensation. Parallel compensation refers to the parallel connection between power capacitors and electrical equipment. The reactive power generated by the capacitors is supplied to the local area, avoiding long-distance transmission of reactive power and reducing line energy loss and voltage drop. Series compensation connects power capacitors in series with transmission lines to counteract part of inductance and improve the voltage level of the system.

- Shunt reactors

Shunt reactors are generally connected between the end of ultra-high voltage transmission lines and the ground to absorb excess reactive power. Thus, the voltage and reactive power distribution of the system and line can be improved, thereby improving transmission efficiency.

- Arc suppression coils

The arc suppression coil compensates for the single-phase grounding capacitive current of a small current grounding system to prevent excessive capacitive current.

1.1.1.5 Current/Voltage Transformers

Current/voltage transformers are sensors that enable secondary equipment such as measuring instruments and relay protection devices to obtain electrical primary circuit current/voltage information.

A current transformer can proportionally convert high current into low current, playing a role in current conversion and electrical isolation. The primary side of the current transformer is connected to the primary system, while the secondary side is connected to measuring instruments and relay protection devices.

Similar to power transformers, a voltage transformer is used to change the voltage. However, the purpose of a power transformer with a large capacity is to transmit electrical energy, usually measured in kV·A or MV·A. A voltage transformer is to convert high AC voltage into low AC voltage, providing power for measuring instruments and relay protection devices. The capacity of a voltage transformer is small, usually only a few or tens of volt-amperes.

1.1.1.6 Lighting Protection Devices

Lightning protection devices, such as lighting arresters and lighting rods, release lightning or overvoltage energy during power system operation. The lightning protection device can protect power equipment from the harm of instantaneous overvoltage, and can also cut off the continuous current to avoid system grounding short circuit. It is usually connected between wires and ground, in parallel with the protected equipment.

The working principle of a lightning arrester is as follows: when the overvoltage value reaches the specified operating voltage, the lightning arrester immediately acts to limit the overvoltage amplitude and protect equipment insulation; after the voltage value is normal, the lightning arrester quickly returns to its original state, ensuring the normal power supply of the system.

1.1.2 Secondary Equipment

1.1.2.1 Monitoring Devices

Monitoring devices, such as the primary system operation monitoring machine and fault recorder, supervise the primary equipment in real-time online and timely detect abnormal operation and fault situations of the equipment.

1.1.2.2 Measuring instruments

Substation measurement equipment monitors and measures the data that affect the operation of the substation's primary equipment, such as current, voltage, power, frequency, and temperature.

1.1.2.3 Relay Protection and Automatic Control Devices

Relay protection devices act on the circuit breaker trip and automatically cut out the faulty components when the system or equipment fails. In addition, relay protection devices will send a signal when the system is abnormal. Automatic control devices mainly refer to automatic input devices, low-frequency load reduction devices, and automatic reclosing devices, which can automatically complete the necessary control measures in the case of anomalies or accidents.

1.1.2.4 DC Power Supply Equipment

DC power supply equipment provides reliable DC power supply for control systems, signals, relays, automatic devices, and emergency lights in the substation. The reliability of DC power supply equipment plays an important crucial in the safe operation of substations.

1.2 Insulation Design Requirements[1, 2]

1.2.1 Electrical Requirements

Figure 1.1 shows the electrical requirements. The electrical test voltage/operating voltage ratio varies according to the system's basic insulation level, type of equipment and its location. The value is specified by the user, guided by the data in IEC 60071-1. Depending on the particular circumstances, partial discharge, radio interference voltage and dielectric loss levels must be below specified values as manufactured. To ensure these requirements are met, appropriate tests are applied as included on the diagram. Special arrangements may be required for in-service monitoring of, for example, PDs and DDF.

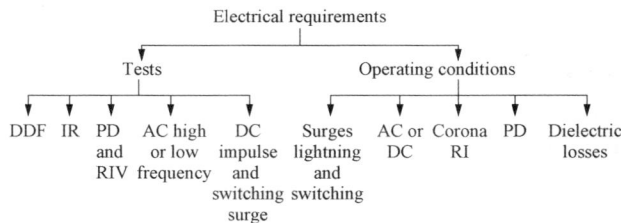

Figure 1.1   Electrical requirements[1]

1.2.2 Physical Limitations

Awareness of the physical limitations imposed by non-insulation factors is important, as shown in Figure 1.2. There are many occasions when such restrictions exist and only minor modifications are possible to improve the overall dielectric performance. Some of the physical changes to the initial overall design proposals that might be considered are indicated below.

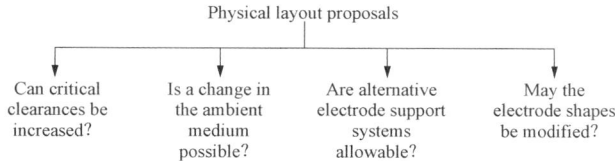

Physical layout proposals

| Can critical clearances be increased? | Is a change in the ambient medium possible? | Are alternative electrode support systems allowable? | May the electrode shapes be modified? |

Figure 1.2 Pysical limitations[1]

1.2.3 Working Environment

Figure 1.3 shown the working enviroment. No matter how well-designed the structure and how satisfactory the test laboratory results, these measurements are of little value unless attempts have been made to simulate and/or allow for possible insulation deterioration due to poor working conditions. If the environment results in control equipment becoming covered with dust in a damp atmosphere, or a rotating machine contaminated with grease and dirt, or an overhead line insulator polluted with salt and industrial fumes, then these factors must be considered at the design stage. A major deteriorative effect is due to the trapping of air or the build-up of gas leading to partial discharges and ultimate failure. This can occur in oil-insulated transformers due to poor impregnation, failure to 'top' up correctly, or not following specified procedures when installing a bushing. The ingress of moisture over a long period can increase the dielectric losses in many materials, again resulting in breakdown. This applies to solid materials such as epoxy resins and XLPE as well as oil-impregnated paper, pressboard, and wood. Acceptance and monitoring measurements have been devised to cover the various conditions.

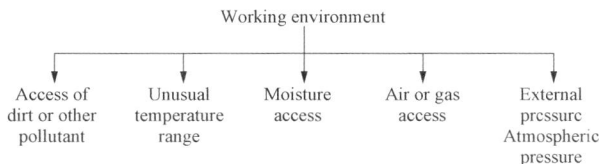

Working environment

| Access of dirt or other pollutant | Unusual temperature range | Moisture access | Air or gas access | External pressure Atmospheric pressure |

Figure 1.3 Pysical limitations[1]

1.2.4 Mechanical Requirements

As in much of electrical engineering, it is important to appreciate that allowance for the mechanical requirements in the application of insulation is essential, as shown in Figure 1.4. This is often the prime factor in ancillary low-voltage systems including coaxial cables, wiring, plugs and sockets, and some windings. At high voltages, in such items as rotating machines, suspension insulators, switchgear, transformers, compressed gas or vacuum equipment, and cables the mechanical performance of the insulation is of major consequence.

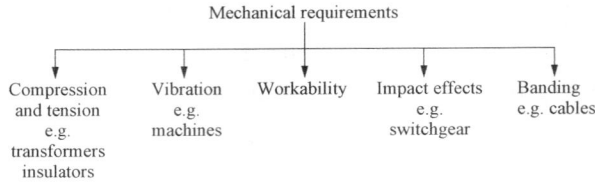

Mechanical requirements

Compression and tension e.g. transformers insulators

Vibration e.g. machines

Workability

Impact effects e.g. switchgear

Banding e.g. cables

Figure 1.4   Mechanical requirements[1]

1.2.5 Thermal Conditions

In addition to knowledge of the thermal performance of the chosen insulation, the design must allow for the imposed thermal conditions and any built-in cooling systems that exist or might be added, as shown in Figrue 1.5. Thermal conditions in some cases limit the application of large amounts of insulation and structures have to be devised to allow the heat from an inner winding or conductor, and the losses in the dielectric itself, to be dissipated satisfactorily in order to avoid overheating of the insulating materials. An outstanding example of such a problem is a high-voltage power cable using oil-impregnated paper as its insulation where a 'runaway' condition is possible. Such conditions have also occurred in power and current transformers in which the insulating materials were not correctly dried. The use of 'internal' cooling of conductors has eased the thermal problem considerably in certain cases, as demonstrated in large generators and in a number of highly rated cables.

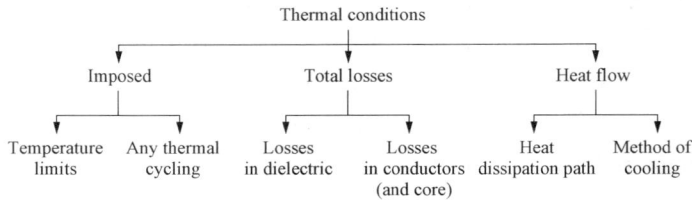

Thermal conditions

Imposed

Total losses

Heat flow

Temperature limits

Any thermal cycling

Losses in dielectric

Losses in conductors (and core)

Heat dissipation path

Method of cooling

Figure 1.5   Thermal conditions[1]

1.2.6 Processing

Figure 1.6 shows the processing limitations. The behavior and expected life of the insulation structures depend critically on the processing procedures adopted in the factory. These include large vacuum drying ovens through to small resin vacuum-pressure-impregnation (VPI) systems. In all cases, the requirements and possible limitations must be clearly defined. An insulation change from traditional material to a new plastic or resin may appear attractive technologically but prove uneconomic because of the cost of investment in additional manufacturing plants. For example, such decisions were necessary in the cable industry when introducing plastic systems, in transformer manufacture during the development of cast-resin-type distribution units, in overhead line insulator developments, replacing conventional assembly methods with VPI processes for some larger rotating machines, and replacement of oil with SF_6 in a number of plant items. The consequence of manufacturing changes or applications of new techniques must be carefully assessed following an unexpected insulation problem.

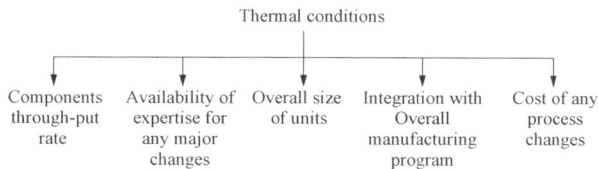

Thermal conditions

| Components through-put rate | Availability of expertise for any major changes | Overall size of units | Integration with Overall manufacturing program | Cost of any process changes |

Figure 1.6　Processing limitations[1]

1.2.7　Reliability

The above factors influence, directly or indirectly, the ultimate reliability of the insulation structures and, therefore, of the high-voltage equipment, components, and associated systems. As the reliability of possibly better than 99 percent-based on average times of outages over the whole system-is being claimed by some utilities, and is expected by customers, periodic maintenance and monitoring of the insulation is becoming of major importance. As shown in Figure 1.7, 'stress' includes electrical, thermal, and mechanical conditions. An attempt is made to show how reliability might be related to the condition (age), cost, and the level of monitoring that might appear necessary.

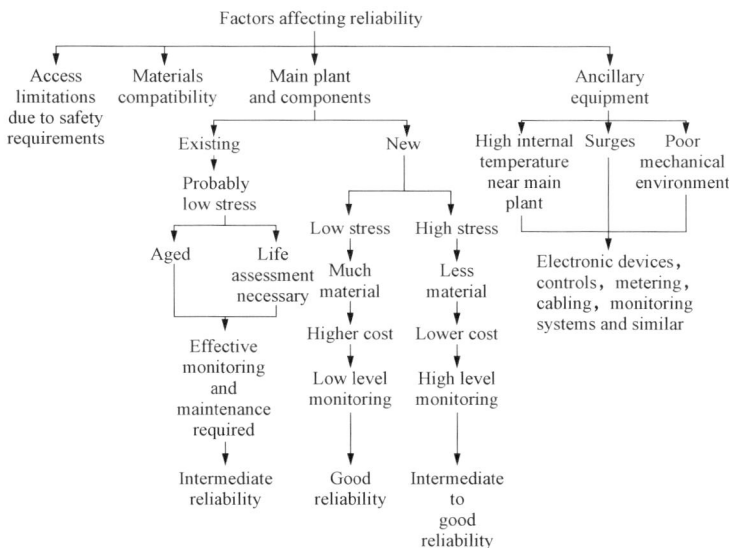

Factors affecting reliability

Access limitations due to safety requirements　Materials compatibility　Main plant and components　Ancillary equipment

Existing — Probably low stress

New — Low stress / High stress

High internal temperature near main plant　Surges　Poor mechanical environment

Aged — Life assessment necessary

Much material — Higher cost — Low level monitoring

Less material — Lower cost — High level monitoring

Electronic devices, controls, metering, cabling, monitoring systems and similar

Effective monitoring and maintenance required

Intermediate reliability　Good reliability　Intermediate to good reliability

Figure 1.7　Factors affecting reliability[1]

1.3 Insulation Failures of Electrical Equipment

1.3.1 Insulation Failures and Their Hazards[3]

Electrical equipment is fundamental components of the power system. Whether it is large key equipment such as generators, transformers, and GIS, or small equipment such as power capacitors and insulators, local or even overall power outages will occur if they fail. On April 7, 2018, the SF_6 casing at a converter station of ± 800 kV ultra-high voltage direct current in Xinjiang exploded, causing the entire converter station to burn down, resulting in a direct economic loss of one billion yuan.

Data investigation shows that insulation degradation is one of the main reasons for electric power equipment failures. The factors that cause insulation degradation include electrical, mechanical, humidity, moisture, and a lack of good management and maintenance. According to statistics, State Grid Corporation of China experienced a total of 19,939 defects in switchgear above 72.5 kV in 2015, including 1194 critical defects. Taking GIS as an example, from 2006 to 2015, the installation amount increased significantly. However, there were more and more defects and faults, which could not reflect the maintenance-free advantage of GIS. According to the statistics of GIS faults by type and cause, foreign objects cause the most discharge times and have the greatest impact, accounting for 29%; 22% of insulation component faults, such as pot insulators, insulation rods, and support insulators; faults from body cracking and damage account for 15%; flow heating faults account for 7%. That is to say, the occurrence rate of insulation faults is as high as 73% (quoted from the "Special Analysis of GIS Faults in State Grid Corporation of China in 2015"). From 2002 to 2005, the statistical analysis of 110 kV and above power transformer accidents in the State Grid Corporation of China system shows that the insulation of windings, main insulation, and leads were the main parts of transformer accidents. The proportion of longitudinal insulation and main insulation accidents at various voltage levels reached 78.6% of the total accidents.

The statistical results abroad are also similar. For example, insulation failures in the North American power system have caused explosions of 230 kV current transformers in at least three power bureaus. A statistical analysis of the failure of capacitors in a 4.8 kV distribution system in the United States from 1980 to 1989 shows that 92% of them were caused by insulation degradation. In the statistical results of Japan's Nippon Corporation on faulty transformers, insulation faults account for 45%. The North American power system blackout that occurred on August 14, 2003, affected 8 states in the United States and 1 province in Canada, with an estimated total loss of 4 billion to 10 billion in the United States. However, Canada's gross domestic product decreased by 0.7% in August. In order to study the causes of power outages and improve measures, the United States Canada Power System

Power Outage Task Force was established. The final analysis report of the task force pointed out that the main causes of power outages were computer failures at the Ohio Regional Power Bureau and short circuit accidents to the ground caused by the discharge of trees growing too fast on several key 345 kV transmission lines.

The failure of power equipment, especially large equipment, can cause huge economic losses. On November 22, 2019, a 1,000 kV main transformer in an ultra-high voltage substation in Jinan, Shandong Province exploded, resulting in one death and two serious injuries.

Insulation defects are difficult to detect during daily operation and maintenance, often leading to serious accident consequences. Accidents are often caused by insulation faults such as partial discharge, surface discharge, and abnormal arc. Insulation faults will cause discharge, leading to deterioration of insulation material performance, rapid changes in internal air pressure and temperature of the equipment, and ultimately explosion.

1.3.2 Characterization of Insulation Condition[1]

Several measurable parameters can be employed directly or indirectly-to characterize the condition of insulating materials when built into system equipment. The predominant electrical characteristics are the values of permittivity and capacitance, resistivity and insulation resistance, insulation time constants, dielectric dissipation factor, and partial discharge status. In an oilinsulated apparatus, the levels of moisture content, dissolved gas volumes, and various chemical quantities are indicative of the oil/solid condition.

1.3.2.1 Permittivity and Capacitance

The relative permittivity, ε_r, is the ratio between capacitances of identical electrode systems with and without a dielectric present.

The relationship is the capacitance.

$$C = k\varepsilon_r\varepsilon_0$$

k is a constant representing the geometrical structure of the system, and $\varepsilon_0 = 8.854\times10^{-12}$. The values of ε_r range from 1 for air through to 5.5 or so for porcelain.

Capacitance values range from a few pF for cap-and-pin insulators through to μF for cables, generators, and power capacitors. At power frequencies, little variation occurs in the value of ε_r with applied stress or temperature for the materials used in power system practice. The presence of moisture ($\varepsilon_r \approx 80$) or trapped gas might be expected to increase or decrease respectively the measured values.

1.3.2.2 Resistivity and Insulation Resistance

Volume resistivity, ρ, is defined as the resistance between opposite faces of a 1-meter cube of insulation.

The resistance $R = \rho L/A$, where L is the spacing between electrodes and A is the area of the electrodes. The unit of volume resistivity is the ohm-meter, the absolute values of which range from 10^9 to 10^{10} for oils through to the order of 10^{14} for materials such as XLPE.

Surface resistivity may be defined as the resistance between two opposite edges of a unit square of material. It is an important quantity for assessing the state of surfaces used in air and gas systems.

For a given material the values vary widely depending on impurity content (especially moisture for hygroscopic dielectrics) and temperature. In addition, the initial insulation resistance will change with the time of application for a constant direct voltage stress. This property is related to the time constant. The relative values of ρ are important in the design of direct-voltage multiple-dielectric systems such as transformers, bushings, and cable termina-tions/joints. The difference between direct-voltage electric stress distributions and those with alternating voltages can be very significant during the testing of equipment.

1.3.2.3 Time Constants

Another parameter that can be related to some aspects of the insulation quality is the time constant, as measured by applying a direct voltage pulse to the system, measuring the rise time required to reach a predetermined value, discharging the sample (or equipment) for a known time and measuring the recovery voltage. Such characteristics are influenced by the complex impedance and frequency response of the equivalent capacitance and resistance network. The effect was studied in the 1950s, being designated the dispersion effect, and was applied to the determination of moisture content in insulating materials in particular power transf-ormers.

The time constant changes also become apparent when measuring the insulation resist-ance.

1.3.2.4 Dielectric Dissipation Factor

An ideal dielectric or insulating material may be considered as a non-conductor of electricity acting as a pure capacitance when built into equipment. However, in practice conduction occurs and finite losses are produced that may be of great importance during test and operating conditions.

Angle φ represents the usual power factor value, $I_R/I = \cos\varphi$, while angle δ is designated the loss angle. I_R represents the losses and I_C the charging current. The ratio of $I_R/I_C = \tan\delta$ and is defined as the dielectric dissipation factor (DDF).

For good insulation, $\tan \delta$ is of the order of $1-3\times10^{-3}$, even for dry oil-impregnated paper at room temperature. The value for XLPE is much lower, perhaps 0.3×10^{-4}. Within this range of values, $\tan \delta \approx \delta$, enabling the DDF magnitude to be conveniently quoted in milliradians (mR) as measured. The cosine of the dielectric phase angle (or the sine of the dielectric loss angle) may be designated as the dielectric power factor by some authorities.

1.3.2.5 Partial Discharge

Partial discharges are small electrical "micro-sparks" that can occur in insulation systems operating with high electric fields. According to IEC Publication 60270, a partial discharge may be defined as a localized electrical discharge that only partially bridges the insulation system between conductors, and that may or may not occur adjacent to a conductor. The Standard states that PDs are usually due to local electrical stress concentrations in the insulation or on its surface, the pulse discharges having durations of much less than 1 μs in most cases. The 'pulse-less' discharges in gases are of a more continuous form and would not be detected by the techniques described in IEC 60270. The latter also notes that corona PD disturbances are associated with conductors in a gaseous medium away from solid or liquid insulants. The term is not to be applied to other forms of PDs. It is observed that sound, light, heat, and chemical reactions are often produced by partial discharges.

In electrical terms, PDs manifest themselves at the terminals of samples and equipment as very small-magnitude pulses (μV to volts) of rise times in the ns to μs range. The intensities of the PDs are usually represented by the charge level in picocoulombs (pC) or nanocoulombs (nC) but other parameters are also of significance in estimating possible damaging effects.

Acceptable magnitudes producing minimal deterioration in service range from a few pCs (10^{-12} C) in XLPE and GIS through to tens of pCs in oil-impregnated paper and resins, to several thousands in mica-based configurations. Higher values may be allowed in short-duration overvoltage tests.

Under alternating voltage conditions the PD sources are usually associated with trapped gas or the ambient medium. Such components of a combined insulation system have lower permittivities and breakdown strengths than the solid materials and therefore tend to be the initial location of the local partial discharges. Possible situations where electrical overstressing might occur resulting in the initiation of partial discharges include internal voids or cavities in solid and laminated insulation, relatively high stressing in ambient media at electrode or support interfaces, sharp edges at surfaces and in free space (gas/air or liquid) and conditions that a potential difference might be created due to a broken or poor connection. It should be noted that PDs can exist on earth as well as at high voltage. In air-insulated systems, such as overhead lines, discharges are usually designated corona or even radio interference, and include those from hardware components in, for example, substations.

Under direct-voltage conditions, the potential distributions across insulating materials are governed by the relative values of the resistivities that in oil-impregnated systems can result in the overstressing of the solid insulants and not the liquid. The complex charging states with direct voltages may produce partial discharges that need to be monitored.

1.3.2.6 Physical and Chemical Changes

Many deteriorative processes take place during the operation of insulation. Some of these result in by-products, which may be used for identification and monitoring of degradation. This is especially the case for oil-immersed systems, where the detection of combustible gases and furans in the oil has resulted in the development of specialized techniques for assessing the condition of the materials. The methods utilize DGA and the application of liquid chromatography (HLPC).

An important practical insulation condition indicator associated with oil-immersed equipment is the analysis of the gases and their relative concentrations as produced by particular types of faults in the structure. Possible diagnoses are quantified in the revised IEC Publication 60599.

The physical aging of paper as represented by the reduction in the degree of polymerization (DP) has been related to the amount of furanic compounds produced at hot-spot temperatures and absorbed in the oil of transformers in service. A reduction in the magnitude of the DP from a value of $>1,000$ (2-Furaldehyde 0-0.1 μL/L) when new to 250 (2-Furaldehyde $>10\ \mu$L/L) is considered to be the end of life for oil-impregnated cellulose by some authorities. Much work is being carried out in this field, including studies for assessing the characteristics of naturally aged oil-impregnated materials aimed at estimating probable lifetimes. The formation of water under certain conditions, such as in sealed units, may be of considerable importance in the prediction of the rate of aging in oil-paper systems.

Conventionally, oil acidity, interfacial tension, and sludging tendency are used as physical and chemical guides for monitoring the condition of transformer oil.

Determination of the physical changes associated with the various forms of treeing and tracking in XLPE and resins has been the aim of much research. The understanding of the mechanism of water treeing in XLPE is of considerable practical significance and remains an important research topic. This work requires the removal of samples from equipment suspected of deterioration and no longer suitable for normal service and therefore is only an indirect monitoring technique.

1.3.2.7 Modes of Deterioration and Failure of Practical Insulating Materials

The concept of insulating materials having an intrinsic breakdown strength was postulated many years ago. It was found that, for thin specimens of certain pure dielectrics and by minimization of edge effects in the test configuration, the impulse breakdown stresses could be increased to perhaps ten times that achievable with normal materials. However, in practice, impurities and manufacturing variations prevent exploitation at a large scale of these purely electronic processes.

The major deteriorated and failure modes associated with power equipment insulating materials, which often comprise combinations of solid and gas/air or solid and liquid, are listed below. This is followed by consideration of some of the processes involved. In the case of the physical and chemical changes, the present treatment is limited to mention of those aspects considered helpful in understanding the condition monitoring requirements.

(a) Dielectric losses causing thermal instability, or runaway, in the bulk of the solid material.

The failure of insulating materials due to high dielectric losses is a well-known phenomenon. In practice, the effect is due to the use of, for example, unsuitable resins and often to the presence of moisture. The latter may be produced by chemical aging, in some cases because of poor equipment sealing and, in others, by incorrect processing procedures in the factory. Additionally, the summation of the effects of energy dissipated by any high-value partial discharges can contribute significantly to the overall dielectric losses.

(b) Partial discharges representative

Partial breakdown in voids and gaps enclosed within the solid insulation producing local erosion of adjacent material; Partial breakdown in the ambient medium (oil, gas, or air) at an interface between an electrode to dielectric or a dielectric to dielectric, thereby initiating flashover or creep and, perhaps, localized puncture through an adjacent surface.

PD activity can directly lead to insulation degradation and equipment failure. PD is also sometimes a symptom of poor manufacturing and/or aging of the insulation due to high temperature, mechanical forces, contamination, etc. PD might not directly lead to failure but may indicate that insulation aging due to other mechanisms is occurring and maintenance may be needed. Thus, by measuring PD activity, equipment manufacturers can often determine that the insulation system on the equipment was properly made, and equipment owners can determine if aging is occurring that could lead to failure.

(c) Ageing due to thermal, electrical, and mechanical stressing, including the effects in (a) and (b).

(d) Long-term chemical changes produced by incompatibility between materials, resulting in the creation of dangerous by-products. The changes are also related to (c).

(e) Deterioration of surface material by external pollutants, leading to a reduction in tracking strength.

1.4 Advanced Sensors and Measurement Instruments in Electric Power Industry

Sensors monitor various physical quantities that reflect the state of equipment, such as electricity, heat, mechanical force, chemistry, and other energy forms. They are the first and most important step in state monitoring and fault diagnosis, directly affecting the success or failure of monitoring and diagnosis. Because electrical signals are most easily processed, regardless of whether the physical quantity is electrical or nonelectrical, it is generally converted into electrical signals by sensors and sent to subsequent units.

The basic requirements for sensors are (1) being able to detect signals that reflect the characteristics of equipment status, and having good static and dynamic characteristics. The former includes sensitivity, linearity, resolution, accuracy, stability, and hysteresis; The latter refers to frequency response characteristics. (2) It has no impact on the tested equipment, absorbs minimal energy from the tested system, and can match well with subsequent units. (3) Good work reliability and long service life.

1.4.1 Electrical and Magnetic Sensors

Each partial discharge is accompanied by a current pulse. These current pulses can be detected by various types of sensors and measurement instruments. A portion of the current pulse can be measured at the terminals of the HV equipment using a suitable sensor usually a measuring impedance (quadripole) in series with a coupling capacitor that is connected in parallel to the high-voltage terminals with the equipment under test. Alternatively, a high-frequency current transformer (HFCT) installed on a ground lead in series with the HV equipment can measure the discharge current pulse. When measured by an instrument operating at less than 1 MHz, the current pulse is integrated to yield an indication of the apparent charge in the discharge pulse. This is the "conventional" way of measuring PD and is often called electrical or charge-based detection. Charge-based PD measurements are overwhelmingly used for factory PD testing.

In addition to the current frequencies below 1 MHz, the current may contain frequencies up to the hundreds of MHz range, which can also be detected with coupling capacitors or HFCTs. There is also a radiated EM signal from each discharge pulse, which tends to predominantly occur at frequencies above 1 MHz. When the conducted or radiated signals associated with a discharge are measured at frequencies above 3 MHz, this is commonly referred to as electromagnetic detection and sometimes referred to as "unconventional" PD detection. EM detection of PD is most commonly performed in onsite/offline PD tests and online PD tests.

Electrical and electromagnetic detection provides a wealth of quantitative information about the PD activity, including:

- the PD magnitude (in units such as pC or mV).
- the number of PD pulses per second.
- the phase position of each PD pulse with respect to the 50/60 Hz AC cycle.
- usually, the relative predominance of positive vs. negative polarity PD.
- total energy expended by the PD in one AC cycle or one second, or other PD quantities that indicate the total or average PD activity.
- phase- resolved PD (PRPD) plots that display in a convenient form the PD magnitude,

polarity, and pulse count, as well as where the PD is occurring with respect to the 50/60 Hz AC cycle.

Unfortunately, there are limits to electrical/electromagnetic PD measurements, including:

- Such PD measurements generally do not precisely locate where the PD may be occurring within the HV apparatus. For many types of test objects, an electrical measurement can only locate the PD to one phase.
- Sometimes the PD signals are so overwhelmed by electrical interference that the electrical detection methods can produce a high rate of false-positive indications. If the rate is too high, users will not trust the results.
- Some end users do not wish to pay for the cost of offline electrical/EM PD tests, or the cost of installing PD sensors for online PD testing.

In addition, even if an electrical/EM PD test is performed and high PD readings are measured, most end users of critical HV equipment will want some type of independent confirmation that there is a problem with the insulation. This is because repairing the insulation may involve significant costs and extended shutdowns of the HV equipment.

For the above reasons, nonelectrical methods are available to measure PD. These methods may often not be possible to implement depending on the circumstances, but when they can be applied, they are very useful. The remainder of this chapter reviews the main nonelectrical/EM PD detection methods in use today.

1.4.2 Optical, Chemical, Acoustic, and Mechanical Detection

1.4.2.1 Optical Detection[4,5]

That discharges emit light is well known. Two common examples are lightning and static discharges. The optical emissions are primarily the result of collisions between electrons in the discharge and gas molecules that do not result in ionization, as well as electron-ion recombination. When the molecule returns to its ground state, a photon is emitted. The energy (and thus the wavelength) of the photon depends on the initial energy of the collision, which excites the molecule. Another process that emits light occurs when a positive ion recombines with an electron. PD is easily visible to the naked eye in a low-light level environment. The spectra will change depending on the type of gas. The human eye is most sensitive to light in the wavelength range of 400-780 nm. Ultraviolet (UV) light is defined to occur in the 100-400 nm range.

In the past, the human eye was most often used to detect surface PD and corona at night or in darkened rooms. In the 1960s, "night vision" cameras, also called image intensifiers, were commercialized that intensified visible light and thus made the eye more sensitive to low-level PD. Corona on transmission lines and in substations at night is easily visible to the human eye and has become acclimatized to darkness, or more quickly with the aid of a night vision camera. It was also common to detect surface PD on stator windings and air-insulated switchgear when the area was dark. PD tests in the dark using the human eye are called a "blackout" or "lightsout" test. The great advantage of the blackout test was that the locations of surface PD could be easily determined (of course PD within the insulation is not visible unless the insulation is transparent). The problem with doing PD tests on energized high-voltage equipment in the dark is that it is more likely that an observer may inadvertently contact the high voltage in an offline test, a clear safety hazard.

Researchers sometimes use vacuum photomultiplier tubes (PMTs) to detect very low-intensity optical emissions from PD. PMTs use a special surface that emits an electron when an optical photon hits it. This electron is then directed to a series of metal plates (dynodes) at progressively higher voltages that accelerate the initial electron such that when it hits a dynode, the energy of the electron impact causes multiple electrons to be emitted (secondary emission). After several stages, a relatively high current pulse is created for each incident photon. These current pulses can be measured to indicate the intensity of the light from each discharge. The number of discharge events are also counted. PMTs can only be used in the dark.

In the 1990s, the UV imaging camera was invented. The first commercial device to detect PD was developed by Dr. Kieth Forsyth of Forsyth Electrooptics. It displayed the UV light from PD while suppressing sensitivity to visible light. By filtering out the visible light, it became possible to perform a "blackout" test without turning off the lights or waiting until the sun had set, greatly improving the safety of the test. In addition, by adding a still-image camera or video camera, the PD images could be kept for later comparison and precise location assessment. This made the optical detection of surface PD much more objective.

Later developments allowed the strong visible and UV light from the sun to be filtered out while still being sensitive to the UV frequencies associated with PD and corona. This enables modern UV cameras to work outdoors in the daylight. Helicopter- and drone-mounted UV cameras have revolutionized the inspection of transmission lines and insulator strings, with inspections done in normal daylight.

A UV camera commonly used for PD and corona detection is the Daycor line made by OFIL Systems. Note that most glass is opaque to UV light, so UV cameras cannot be used to detect PD through normal windows. Also, UV imaging cannot detect PD that is not on a surface or where a direct line-of-sight to that surface cannot be obtained. Thus, UV imaging and blackout tests may miss many important sources of PD.

Another critical advance with UV imaging occurred when Dr. Claude Hudon and his colleagues at Hydro Québec discovered that a UV camera could be "calibrated" against the human eye. He found that at least some commercial UV cameras had about the same sensitivity to surface PD as the human eye, when people had been in a very dark room for about 15 minutes to acclimate their night vision. That is, the acclimatized human eye would detect almost the same PDIV as many UV cameras. This led to the development of IEEE 1799 to detect PD on the surface of stator coils and windings, a calibration procedure, and set voltage levels below which surface (both phase-to-ground and phase-to-phase) PD should not be present.

1.4.2.2 Chemical Detection[5-7]

Partial discharge can change the chemical composition of the surrounding medium, which sometimes allows the detection of PD using chemical analysis. The changes that occur depend on the surrounding medium: air, SF_6, or oil. The most widely used chemical tests are ozone detection for PD in air; dissolved gas analysis (DGA) for PD in oil - commonly used in power transformers; and SF_6 decomposition products in the case of GIS or GIL.

1. Ozone in Air

When PD occurs in air, a gas called ozone - O_3 - is created. A discharge creates many energetic electrons. An excited electron may have inelastic collisions with an oxygen molecule (O_2) in air and have sufficient energy to break the molecular bond, creating two oxygen atoms ($O + O$). Oxygen atoms are very chemically reactive, and if an oxygen atom is near an oxygen molecule, they will combine, creating O_3. Ozone itself is also very chemically reactive with many other molecules, often rusting metals and degrading some organic materials. Thus, any ozone in the air tends to disappear in a few minutes. However, if the PD sources are continuously active, new ozone is created to replace the depleted ozone. Eventually, a steady-state level of ozone gas concentration occurs which is the balance between the creation and depletion rates. Aside from ultraviolet radiation, a few other processes create ozone. Thus, the presence of ozone near HV equipment is a good indicator that PD is occurring in the air or on the surface of solid insulation in the air. Continuous exposure to ozone above 0.1 μL/L is deemed by many health and safety regulators to be hazardous to human beings.

Ozone can be created by PD in many types of HV equipment:

- corona from transmission lines and outdoor substation components.
- PD from surface discharges in metal-clad air-insulated switchgear.
- PD on the surface of high-voltage stator coils/bars, both in the stator slot and in the end winding.

Generally, PD within a void enclosed within solid insulation does not produce ozone in the surrounding environment.

The amount of ozone produced by PD depends on both the number of discharge sites and the magnitude of the discharges. That is, the ozone concentration is an indicator of the total PD activity that is occurring in the air. The ozone concentration in the air depends not only on the steady-state creation/depletion rate of the ozone but also on how open the air is to the surrounding environment. For example, if a lot of air is moving near the PD sites (due to high winds if outdoors, or from a fan blowing external air over an open-ventilated stator winding), the ozone concentration will be low. The PD activity may be lower in enclosed HV equipment such as totally enclosed rotating machines and metal-clad air-insulated switchgear, yet show a much higher ozone concentration since little air may be circulating.

There are two main methods for measuring the ozone concentration in air: chemical reaction tubes and electronic instruments.

The ozone chemical reaction tubes are similar to the litmus test used to check acidity. If one breaks the tube open in the presence of ozone, the reagent changes color, depending on the ozone concentration. Draeger is a common supplier of such tubes.

Electronic instruments are a more convenient way to measure the ozone concentration. These devices may contain a special MOSFET sensor that changes its resistance in the presence of adsorbed ozone. Another type of sensor uses the principle that UV absorption is higher with higher ozone concentrations. The associated instruments can take spot measurements or can continuously monitor the ozone concentration.

The ozone concentration for PD detection is usually measured in parts per million ($\mu L/L$). The typical ozone concentration in the air is usually $<0.1\ \mu L/L$ and is caused by thunderstorms, UV radiation, and/or certain types of industrial pollution. The human nose has a detection sensitivity of about $0.1\ \mu L/L$. In open-ventilated HV equipment, readings of about $1\ \mu L/L$ indicate widespread surface PD activity. In enclosed HV equipment where there is little air circulation, ozone concentrations up to $25\ \mu L/L$ have been measured. As mentioned above, both cases may indicate the same amount of PD activity. In spite of this limitation, ozone monitoring is an excellent way to confirm high PD activity detected by electrical methods.

2. Dissolved Gas Analysis (DGA)

DGA has been used since the 1960s to determine if PD, arcing, or thermal aging of the insulation is occurring in oil-filled transformers. When PD occurs in mineral oil, hydrogen (H_2) and various hydrocarbons are produced. These gases are normally dissolved in the oil and can be detected either by oil sampling together with chemical lab analysis, or by using specialized continuous monitors that detect these gases in the oil. Overheated organic insulation tends to produce ethylene, PD produces hydrogen, and arcing (including surface tracking) produces acetylene. Thus, DGA can detect both PD and arcing, although the response time is much slower than with electrical PD detection. In fact, since PD in transformers is sometimes intermittent, and the gases in the oil can have a relatively long lifetime, it can be argued that DGA may be a more reliable method of detecting discharges than conventional electrical detection, especially when there is a high level of electrical disturbances.

There are thousands of papers on DGA by sampling or by monitoring at the transformer, and it is well beyond the scope of this book to discuss this technology in detail. The technology for online monitoring of the gas in oil has advanced tremendously since it was introduced by Duval and others in the mid-1970s for detecting hydrogen alone. Now commercial monitors for nine or more gases dissolved in oil have been commercially introduced.

3. SF$_6$ Decomposition Products in Gas Insulted Equipment

The SF$_6$ used in gas insulated equipment is delivered exceptionally pure and dry. However, there are usually some impurities such as H$_2$O and O$_2$ present. When PD and arcing are occurring, adjacent SF$_6$ will decompose, and these decomposition products then combine with H$_2$O and O$_2$ to create characteristic gases. Research has shown that the main decomposition products produced by PD and arcing include CF$_4$, CO$_2$, SOF$_2$, SO$_2$F$_2$, SO$_2$, and H$_2$S. Although it is still not used widely, these decomposition products can be identified by sampling the gas in GIS and analyzing certain gas ratios using a gas chromatograph. As with ozone monitoring and DGA, the sensitivity of the method depends on the number of discharge sites and their magnitude. There will be a delayed response to the PD.

1.4.2.3 Acoustic Detection

The acoustic techniques described in the previous section located PD sites for HV equipment operated in the air, using detectors remote from the HV apparatus for reasons of safety. The acoustic signals from the PD were transmitted through the air. These methods are not effective for HV apparatus such as power transformers, power cables, gas-insulated switchgear, and HV capacitors where the electrical insulation, and any associated PD, is completely enclosed by metal. Acoustic PD signals within an enclosure are reflected back by the enclosure (and its high acoustic impedance). Furthermore, the signal that is transmitted through the enclosure is not efficiently coupled into the air for reception by a remote microphone.

However, acoustic microphones or vibration sensors [acoustic emission (AE) or accelerometers] can be installed within the enclosure or on the enclosure's outside surface to detect internal PD. This is common for liquid-filled transformers, GIS, and HV capacitors, all of which have grounded enclosures. Acoustics was first applied to power transformers as far back as 1939. Using acoustic methods to detect PD within such apparatus may sometimes be the primary PD detection method for the equipment if UHF sensors are not installed, since LF and HF electrical methods may be severely compromised by high levels of disturbances, and optical methods are not possible since the enclosure blocks the light emission.

Piezoelectric sensors are most common for acoustic detection of PD signals in enclosed equipment; these usually employ specialized crystals that produce an electrical output signal caused by mechanical pressure transient arising from the acoustic PD signal. These devices can detect signals from a few Hz to 5 MHz, depending on the design. The sensor enclosure is metallic and grounded via the sensor output cable, therefore the sensor cannot be used in or near high electric or magnetic fields. Piezoelectric sensors are usually mounted on the grounded enclosure of the HV equipment. There are many manufacturers of piezoelectric sensors with a wide variation of design and application.

In the past 20 years, fiber-optic acoustic sensors have been developed, which use either no or only microscopic amounts of metal and thus can be installed within HV equipment such as power transformers and switchgear, and can even be attached to components operating at high voltage. These sensors use a variety of principles for operation: Fiber Bragg Grating (FBG), Fabry-Perot interferometry, and resonant vibrating reads or diaphragms. Fiber-optic sensors tend to be more expensive than piezoelectric sensors and thus are not yet widely used in practice for acoustic detection in HV equipment.

1.4.2.4 Mechanical Detection[5, 8]

Noise and vibrations are produced by all motors. The health of the motor can be estimated by analyzing these noises and vibrations. Vibration monitoring is a very important content, including mechanical vibration of rotating motors, and vibration caused by electrostatic or electromagnetic forces. For example, the impact of charged particles on the shell under the action of an electric field in GIS, and the weak vibration caused by partial discharge inside the transformer. The range of vibration intensity is wide, and there are three parameters for measuring vibration, namely displacement, velocity, and acceleration. The measurement can be determined based on the vibration frequency. As the speed of vibration increases, displacement decreases while acceleration increases. Therefore, as the frequency increases, displacement sensors, velocity sensors, acceleration sensors, and acoustic emission sensors can be selected separately. The general method of operation is that the measurement affects some mechanical features of the sensor and a secondary sensor using capacitance, piezoresistance, or, to a lesser extent piezoelectricity, provides the electrical signal.

1. Displacement Sensors

Displacement sensors are the most effective in the lowfrequency region and are widely used to measure the vibration and eccentricity of heavy-duty motor frames. The high-frequency power supply generates an electromagnetic field on the sensor probe, and when there is relative displacement between the surface of the measured object and the probe, the energy on the system changes to measure the relative displacement. The sensitivity of the displacement sensor can reach 10 mV/μm.

2. Velocity Sensors

The most effective speed sensor for vibration within the range of 10 Hz to 1 kHz is commonly used to measure the total mean square value of various motor vibrations. Its basic structure is to place a permanent magnet inside the coil, firmly attach the coil to the sensor housing, and then install the sensor and probe together on the surface of the measured object. Once vibration occurs, there will be relative displacement between the sensor shell and coil and the magnetic iron block and an induced electromotive force will be generated in the coil. The speed of vibration is measured by the magnitude of the electromotive force. The characteristic of a speed sensor is its large output signal, but its disadvantage is that it is not sturdy enough.

3. Acceleration Sensors

Acceleration sensors are commonly used to measure vibrations with higher frequencies, especially those with frequencies exceeding 1 kHz. Since acceleration is the second derivative of displacement, the acceleration sensor has the highest sensitivity among the three types of sensors. Usually, piezoelectric sensors are used, consisting of magnetic seats, mass blocks, and piezoelectric crystals. The entire sensor is tightly attached to the surface of the device to be tested, and the acceleration a generates a force $F = ma$ through the mass block m. The force is transmitted to the piezoelectric film, generating an electric charge. After being amplified by a charge amplifier, the amplitude of the output signal is proportional to the acceleration. The characteristics of piezoelectric acceleration sensors are good rigidity, high sensitivity and stability, good linearity, and more convenient use after built-in amplifiers.

4. Acoustic Emission Sensors

When monitoring higher frequency signals, acoustic emission sensors are required. The coverage frequency of acoustic emission is very wide, ranging from infrasound below 20 Hz, audible sound at 20 Hz$\sim$20 kHz, to high-frequency sound at 100 MHz. Acceleration sensors can be used for detection below 20 kHz, ultrasonic sensors can be used for 20 kHz to 60 kHz, and acoustic emission sensors can be used for 60 kHz to 100 MHz. The acoustic emission sensor uses piezoelectric chips as energy exchange components. Compared with piezoelectric acceleration sensors, the difference lies in utilizing the resonant characteristics of the piezoelectric chips themselves to work. There are two types of acoustic emission sensors: narrowband and broadband: the former has a bandwidth of only 200 kHz; The latter is 700 kHz, but its sensitivity is low. Narrowband sensors are generally used in online monitoring.

A number of new materials are finding applications in mechanical micro-sensors. Silicon has been used for several decades and is attractive because the manufacturing technology is welldeveloped to produce large, extremely pure, single crystals. These can be doped, etched, or machined as desired. The mechanical properties of silicon are also very attractive to sensor design engineers. It has:

- high tensile strength.
- good fatigue strength.
- high Young's modulus, similar to steel.
- good strength/weight.
- good hardness.
- good resistance to deforming plastically so creep and hysteresis effects are minimal.

1.5 Online Monitoring and Condition-Based Maintenance[1, 3]

Since the 1970s, with the rapid growth of installed capacity in the world, the demand for power supply reliability has become increasingly high. Considering the limitations of the original preventive maintenance system, a new concept of predictive maintenance or condition-based maintenance is proposed to reduce power outages and maintenance costs. Its specific content is to conduct live detection or continuous online monitoring (also known as state monitoring) of the insulation status of electrical equipment in operation, and to obtain information that can reflect changes in insulation status at any time; After analysis and processing, diagnose the insulation condition of the equipment; Arrange necessary repairs based on the diagnosis, that is, carry out repairs with a targeted approach. Therefore, condition-based maintenance should include three steps, namely obtaining equipment status data, evaluating and analyzing the status, diagnosing it, and conducting predictive maintenance.

Condition-based maintenance has the following advantages:

(1) More effective use of equipment, improving equipment utilization. (2) Reduced the inventory of spare parts and the cost of replacing parts and maintaining them. (3) Targeted maintenance can improve the maintenance level and make equipment operation safer and more reliable. (4) It is possible to systematically feedback equipment quality information to the equipment manufacturing department to improve product reliability.

1.5.1 The Main Problems with Offline Monitoring

Offline condition assessment of the insulation in HV equipment is applied extensively in order to minimize the possibility of failure in service. However, the required testing and measurement procedures are sometimes impractical, costly, and not indicative of operating conditions. This chapter considers the alternative of online monitoring by means of which more continuous assessment is possible under operating conditions. Some of the developments during the past decade are discussed, including a number of new techniques aimed at overcoming many difficulties in implementing in-service measurements. The cost, reliability, and convenience of the new systems need to be balanced against the savings effected by a reduction in outages and the extension of the life of the insulating materials.

Although offline insulation tests, either destructive or non-destructive, are valuable for the assessment of HV equipment conditions, there exist several disadvantages.

(1) The equipment has to be taken out of service, which may cause an unnecessary outage or reduction in electricity supply. For example, a large steam turbine generator has to run continuously during the maintenance interval of 3-4 years. A power transformer at a critical location in a power system may need 3-6 months' preparation for loads to be transferred before it can be taken out of service.

(2) The equipment cannot be continuously monitored during operation. A fault may occur between planned offline tests. Such periodic measurements cannot guarantee the detection of all developing defects of significance.

(3) The equipment's electrical, mechanical, and chemical stresses may be impossible to duplicate when testing offline. Some insulation defects cannot be activated and detected under

offline conditions. For example, during insulation tests on a standstill generator, mechanical vibration does not exist and the voltage and temperature distributions can be very different from those present during operation.

(4) The cost involved in offline tests can be expensive and time-consuming. Since the equipment has to be taken out of service and all necessary safety issues need to be addressed before the testing voltage source and instruments can be connected, long preparation and measurement times are common.

The above factors indicate that it is preferable to monitor the insulation condition online. However, there are additional difficulties when testing online, such as the noise problem during PD measurements. Also, online condition monitoring results may be less accurate, although they provide a meaningful continuous record, and the instruments are more expensive. Nevertheless, there is an increase in the development of online condition-monitoring techniques and this is expected to lead to the wide application of centralized (integrated) online condition-monitoring systems.

1.5.2 Online Monitoring

Online monitoring is one of the important bases for conducting status assessments. Of course, establishing an online monitoring system for equipment also requires investment. Therefore, when analyzing whether it is necessary to establish an online monitoring system for certain electrical equipment, economic accounting should be conducted and decisions should be made based on its economic benefits.

In the 1960s, the United States was the first to develop monitoring and diagnostic technology, established a large research institution for fault diagnosis, and held 1-2 academic exchange conferences annually. In the early 1960s, the United States had already used Total Combustible Gas (TCG) detection devices to measure the free gas on the oil surface of transformer oil storage tanks to determine the insulation status of transformers. However, during the latent fault stage, most of the decomposed gas is dissolved in oil, so this device cannot detect latent faults.

In the stage of sexual failure, most of the decomposed gas is dissolved in oil, so this device cannot detect latent faults. In response to this limitation, countries such as Japan have studied the use of gas chromatography to analyze free gases while also analyzing dissolved gases in oil, in order to detect early faults. Its disadvantage is that oil samples need to be taken and analyzed in the laboratory, which takes a long time and cannot be continuously monitored online. In the mid-1970s, the invention and application of polymer plastic permeable membranes that could separate gases from oil solved the problem of online continuous monitoring. Since the late 1970s, Japan has developed oil gas monitoring devices for H_2, three component gases (H_2, CO, CH_4), and six component gases (H_2, CO, CH_4, C_2H_2, C_2H_4, C_2H_6) in oil. Canada successfully developed an online monitoring device for gas analysis in oil in 1975, which was subsequently developed by Syprotec as an official product called the transformer early fault monitor.

In recent years, China has also developed an online monitoring system for transformers that can simultaneously monitor seven gas components, including H_2, CO, CH_4, C_2H_2, C_2H_4, C_2H_6, and CO_2, as well as micro water content. It has been installed in multiple substations and has gained valuable operating experience. This system has played an important role in the online monitoring and condition evaluation of transformers.

Gas chromatography analysis technology has become increasingly mature and has been proven to be an effective monitoring and diagnostic technology through long-term practice. It has been widely used in the monitoring of various oil-filled electrical equipment. Its limitation is that the generation of gas undergoes a developmental process, so it is not sensitive to sudden faults, which requires the use of partial discharge monitoring.

The online monitoring of partial discharge is difficult, and its development has been limited for decades. With the development of sensor technology, signal processing technology, electronic and optoelectronic technology, and computer technology, its monitoring sensitivity and anti-interference level have improved. For example, in the past 20 years, the improvement of the sensitivity of piezoelectric components and the application of low-noise integrated amplifiers have greatly improved the signal-to-noise ratio and monitoring sensitivity of ultrasonic sensors, making them widely used for online monitoring of partial discharge. By the 1980s, the monitoring technology of partial discharge had developed significantly. The Ontario Water and Electricity Administration of Canada has developed a partial discharge analyzer (PDA) for generators and has successfully applied it to hydroelectric generators in countries such as Canada. This device was equipped with over 500 units between 1981 and 1991. The Quebec Institute of Water and Electricity (IREQ) has developed a multi-parameter monitoring system (AIM), which can not only monitor the partial discharge of 735 kV transformers, but also analyze dissolved gas components in oil and line overvoltage, and has a preliminary automatic diagnosis function. In the 1980s, Tokyo Electric Power Company of Japan developed an automatic partial discharge monitoring instrument for transformers, which uses optical fiber to transmit signals, uses a combination of sound and electricity to monitor and suppress interference, and locates the fault point of the discharge source.

DMS (Diagnostic Monitoring System) developed the world's first partial discharge online monitoring system based on ultra-high frequency signal detection in 1993. It is a commonly used partial discharge detection technology for gas-insulated switchgear (GIS) equipment internationally, with a detection signal frequency range of 100 MHz and 1,500 MHz. The high-frequency partial discharge live detection device developed by the Italian Techimp company uses a sampling rate of 100 MSa/s to obtain a partial discharge.

The original waveform of electrical signals, which uses equivalent time-frequency separation technology to separate signals from noise or different types of discharge signals, has been widely used in live detection and online monitoring of power cables.

In recent years, the research and application of smart grids in China have been vigorously developed, and multiple achievements with world-advanced levels have been achieved. In the transformation process of the smart grid, it is proposed to achieve the intelligence of high-voltage equipment; In terms of information access, a complete solution has been provided. A large amount of research and practice has been conducted on online monitoring and diagnostic evaluation of power transformers, circuit breakers, lightning arresters, transformers, GIS, and other equipment, as well as dielectric material aging detection and fault mechanism analysis. Infrared temperature measurement, multi-component oil chromatography online monitoring, GIS ultra-high frequency partial discharge online monitoring, and other technologies have been widely applied, greatly improving monitoring technology and methods.

From the overall development situation both domestically and internationally, it can be seen that online monitoring and diagnosis systems for power equipment are moving towards remoteness, intelligence, and comprehensiveness.

Currently, China has successfully developed comprehensive monitoring systems for transformers, circuit breakers, and capacitive equipment, and these monitoring systems have gradually been applied. Taking the comprehensive online dynamic monitoring and fault diagnosis system for transformers as an example, it consists of two main parts: online monitoring and fault diagnosis. The online monitoring section includes basic monitoring units such as dissolved gases and micro water in oil, dielectric loss factor of casing, and partial discharge. It can also expand the online monitoring of iron core grounding, winding deformation, temperature load, and switch input interfaces (cooling fans, oil pumps, gas relays). The fault diagnosis section includes a module for determining the presence or absence of faults, a module for qualitative and localized diagnosis of faults, a module for analyzing the severity and development trend of faults, a module for assessing the harm of faults, and a module for maintenance strategies.

On the basis of the monitoring system for a certain type of equipment mentioned above, a comprehensive online monitoring system for substations and a comprehensive monitoring system for transmission lines have been developed. For example, the substation monitoring system integrates multiple monitoring units of high-voltage equipment in the substation, such as transformers, GIS, capacitive devices, lightning arresters, and switchgear. It collects various operating data in real-time and multi-channel, achieving comprehensive data analysis and diagnosis of the status of high-voltage equipment in the substation. The system can detect latent faults within the equipment at the first time, and estimate the operating characteristics and life loss of high-voltage equipment based on the analysis results of comprehensive monitoring data, providing a reliable basis for the safe operation of the equipment.

1.6   Technical Requirements for Online Monitoring Systems[3]

Online monitoring systems generally include the following basic units.

(1) Signal Transmission

Usually completed by corresponding sensors. It monitors physical quantities that reflect the equipment status from electrical equipment, such as current, voltage, temperature, pressure, and gas composition, and converts them into appropriate electrical signals for transmission to subsequent units.

(2) Signal Preprocessing

Its function is to preprocess the signal transmitted by the sensor appropriately and adjust the signal amplitude to the appropriate level.

Hardware circuits such as filters and polarity discriminators are used to suppress the interference caused by aliasing, in order to improve the signal-to-noise ratio of the system.

(3) Data Collection

Convert the preprocessed signal into a digital quantity and store it.

(4) Signal Transmission

Send the monitoring results to the monitoring platform in a unified format, usually using fiber optic Ethernet for data transmission. For fixed monitoring systems, special signal transmission units need to be configured due to the distance of the data processing unit from the site. For portable live detection or monitoring devices, only on-site display, recording, or remote data transmission through General Packet Radio Service (GPRS) technology and other means are required.

(5) Data Processing

Process and analyze the collected data, for example, perform time-domain and frequency-domain analysis of the obtained digital information, and further process the signal using software filtering, averaging, and other techniques to improve signal-to-noise ratio. Obtain feature values that reflect the device status, providing effective data and information for diagnosis.

(6) Diagnosis

After comparing and analyzing the processed data with historical data, criteria, and other information, diagnose the status or fault location of the equipment. If necessary, further measures should be taken, such as arranging maintenance plans and whether it is necessary to exit operation, usually completed in the background of the monitoring system.

The technical requirements for online monitoring systems can be summarized as follows:

1) The input and use of the system should not alter or affect the normal operation of primary electrical equipment.

2) Capable of automatically and continuously monitoring, data processing, and storage.

3) Equipped with self-check and alarm functions.

4) Have good anti-interference ability and reasonable monitoring sensitivity.

5) The monitoring results should have good reliability, repeatability, and reasonable accuracy.

6) It has the function of online calibration of its monitoring sensitivity.

7) It has the diagnostic function for electrical equipment faults, including fault location, judgment of fault nature and degree, and Prediction of insulation life, etc.

8) It has a unified communication interface and data remote transmission function.

Bibliography

[1] R.E. James, Q. Su. Codition assessment of high voltage insulation in power system equipment. IET Power and Energy Seires 53, 2008.

[2] IEC 60071-1: Insulation co-ordination-part 1: Definiations, principles and rules. IEC International Standard, 2019.

[3] Gao Shengyou, Wang Changchang, Li Fuqi. On-line monitoring and diagnosis for power equipment. Tsinghua University Press, Beijing China, 2018.

[4] Greg C. Stone, Andrea Cavallini, Glenn Behrmann, Claudio Angelo Serafino. Practical Partial Discharge Measurement on Electrical Equipment. IEEE Press Series on Power and Energy Systems, Wiley, 2023.

[5] R.G. Jackson. Series in Sensors: Novel Sensors and Sensing. Institute of Physics Publishing, bristol and Philadelphia, 2004.

[6] Sivaji Chakravorti, Debangshu Dey, Biswendu Chatterjee. Recent trends in the condition monitoring of transformers: Theory, Impementation and Analysis. Springer, 2013.

[7] Sebastian Coenen. Measurement of partial discharges in power transformers using electromagnetic signals. University of Stuttgart, Germany, 2012.

[8] Hermann J. Koch. Gas insulated substations [2nd Edition]. John Wiley&Sons Ltd., New Jersey, 2022.

SF₆ Gas and Its Environmental-Friendly Alternatives

2.1 Review of Insulating Materials[1]

The materials are reviewed under the general headings of gases, liquids, and solids; the last of these includes composites and conditions where an impregnant is necessary to obtain the required dielectric strength. [1] To assist in the choice of appropriate materials for specific temperatures, a classification guide was introduced within the IEC standards system in 1957. A number of Guidelines and Standards have been issued since that time covering the determination of thermal endurance properties, identification of insulation systems for particular temperature conditions, associated aging mechanisms and diagnostics, appropriate statistical methods and functional tests for evaluating the expected performance in service. In IEC 60085 the insulation thermal classifications are tabulated according to the recommended operating temperatures. Also included is the earlier method of alphabetical classification as this is still used in the power industry.

2.1.1 Gases

In addition to air, several other gases serve as insulating agents in power systems. Noteworthy among them are sulfur hexafluoride (SF_6), nitrogen, and Freon variants such as C_2F_6 and C_2F_5Cl. These gases find applications in various equipment, including switchgear, cables, transformers, and large turbine generators, where they contribute to effective insulation. Hydrogen is also utilized in substantial turbine-generator systems.

The electrical characteristics of air are extensively documented, given its prominent role as the primary insulator in numerous components of overhead power systems. In certain scenarios, the known breakdown strength and self-restoring properties of air are harnessed in protective devices like rod gaps and gap-type surge arresters. Particularly at elevated voltages, and correspondingly longer gaps, it has been observed that switching surges can induce flashovers at relatively low voltages, especially when the electrode under higher stress is positively charged. This consideration holds significant implications for the design of transmission lines and substations, particularly in the context of higher system voltages.

A profound understanding of the characteristics of air in the presence of moisture is of utmost importance, primarily because, unlike most insulating materials, it must consistently perform well in adverse climatic conditions. When subject to testing in high-voltage laboratories, adjustments to flashover voltage levels are meticulously made with consideration for humidity values.

The insulation properties of air at interfaces, particularly those associated with porcelain and synthetic polymer surfaces, hold significant practical importance in various equipment. The behavior of SF_6 has undergone comprehensive scrutiny, with early investigations dating back to the 1950s. These studies encompass breakdown tests conducted at pressures up to 0.8 MPa (8 bar), involving a diverse range of electrode configurations relevant to switchgear, current transformers, power transformers, substation hardware, and cables. Additionally, extensive examinations of the chemical structures and long-term stability of this electronegative gas have been conducted. To circumvent moisture-related issues, it becomes imperative to achieve dew points of 20 ℃ or better at working pressures. This precautionary measure is particularly crucial in ensuring the optimal performance and longevity of the equipment.

In practical applications, the effect of particles within the SF_6 gas of gas-insulated systems (GIS) can be very significant. Following a flashover within GIS equipment or disconnector operation dangerous surges with rise times of the order of tens of ns may be injected into the local power system. The volt-time characteristics tend to be flat after about 10 μs.

Nitrogen is used at pressures up to 1.0 MPa in standard capacitors and in some forms of cables, while the low density of hydrogen is exploited in large water-cooled turbo-generators. The breakdown strength of hydrogen at atmospheric pressure is about half that of air. Operating pressures are in the region of 0.4/0.5 MPa with moisture contents corresponding to dew points of the order of 20 ℃.

2.1.2 Vacumn

In its pristine state, a high vacuum stands out as an ideal dielectric over short distances due to the absence of electron multiplication. However, in real-world applications like high-voltage circuit breakers, the presence of contaminants from metallic and insulation surfaces, coupled with residual oil and gases, imposes limitations on the attainable voltage stresses. Through meticulous design and the use of suitable electrode materials, vacuum circuit breakers have now found application in circuits up to and including 36 kV.

2.1.3 Liquids

Oil serves as a commonly employed insulant, either independently or as an impregnant, imparting favorable properties to laminated or porous 'solid' materials. This is particularly evident in transformers and certain high-voltage cable designs, where oil functions as a heat-transfer medium between active conductors and water or air coolers.

The choice of oil, both in type and quality, is contingent upon the specific application. Specifi-cations span from standard hydrocarbon oils utilized in switchgear and transformers to specialized variants designed for use in cables and capacitors. When employed as impregnants, these liquids undergo meticulous processes, including thorough drying, degassing, and filtration, to generate structures with high dielectric strength.

The motivation to substitute "paraffinic"-based oils for "naphthenic" types appears to have diminished, with extensive research primarily focused on the former. This research delved into aspects such as viscosity at low temperatures and aging characteristics. While the qualities of various oils are evaluated through specified tests, the results may not fully reflect their behavior when integrated into a complex structure over an extended period.

Recent concerns have surfaced regarding the impact of particles on the strength of oil used in high-voltage power transformers. Previous experimental findings revealed that an increase in the density of suspended particles (5 μm) from 2,000/100 cm^3 to 12,000/100 cm^3 led to a significant reduction, approximately 40%, in the breakdown voltage of large oil volumes.[2] The test configuration involved concentric cylinders as electrodes, with an oil volume of approximately 4×10^5 cm^3. A subsequent report by CIGRE compiled data from 15 laboratories, offering insights into the particle concentrations anticipated in practical transformers.[3] While this data informs design considerations, additional information is likely needed, particularly concerning partialdischarge inception stresses, especially under surge voltages.

As the gas-absorbing characteristics of oils can vary, some difficulties may arise in the interpretation of gas-in-oil analyses. These gas-absorbing oils are now used in transformers as well as cables and capacitors. In the latter cases, they help minimize the formation of bubbles in the tightly packed insulation structures. A new test incorporating a point-sphere electrode system for checking the partial discharge characteristics of oils has been developed by CIGRE (SC 15) and is now an IEC Standard.

A persistently intriguing oil characteristic involves the electrostatic charging effects induced by flow rates, typically around 1.5 meters/second, within the insulation configuration of specific high-voltage power transformers. In-depth studies on this phenomenon have been conducted in Japan, the USA, and Europe. The findings suggest that localized charges can accumulate on insulation surfaces, potentially leading to hazardous partial discharges and even flashovers at the interface or within the bulk oil. Techniques have been developed to evaluate the electrostatic charging tendencies (ECT) of different oils, whether new or aged, either independently or in conjunction with pressboard surfaces.

To address fire hazards linked to hydrocarbon oils, synthetic liquids were introduced for use in distribution transformers several years ago. One common liquid variant predominantly comprised polychlorobiphenyl (PCB), a substance now deemed unacceptable due to environmental and health concerns. Existing equipment is subject to stringent regulations, allowing extremely low concentrations, for instance, 0.5 μL/L.

The disposal of PCB liquids has necessitated the development of specialized collection and waste-disposal techniques, incurring significant costs for the electrical industry. One notable advantage of PCB liquids was their high permittivity when employed in capacitors. To address the environmental concerns associated with PCBs, replacement oils, such as silicone liquids, have become available for use in smaller transformers. Additionally, a diverse range of individually designed synthetic oils has been developed for power capacitors.

Among the well-established synthetic liquids is dodecylbenzene (DDB), widely used for impregnating wrapped insulation in high-voltage cables. It is asserted that DDB exhibits superior aging and gas-absorption characteristics compared to natural oils. In the realm of low-temperature cables, if deemed commercially viable, there is an exploration of using liquid nitrogen and/or helium as potential fluids for impregnating lapped plastic dielectrics.

2.1.4 Solids

Certainly, the structures of all power system equipment necessitate the incorporation of solid insulating materials proficient in efficiently supporting and isolating conductors at varying potentials. These materials must exhibit adequate puncture and creep/tracking strengths while withstanding anticipated thermal, mechanical, and chemical conditions. Furthermore, they must maintain electrical stresses to enable economically viable and technically acceptable designs. The materials under scrutiny encompass a select group of the most crucial synthetic polymers employed in power system engineering. While the industrial development of these materials is ongoing, insulation engineers must exercise caution when evaluating a new material for a specific application. In certain cases, opting for well-established materials may prove to be the optimal solution, particularly when considering long-service performance. The emphasis lies not only on innovation but also on reliability and proven durability in the demanding operational contexts of power systems.

2.1.4.1 Wood

Wood is one of the oldest insulations used by electrical engineers and despite limitations in its natural form, it is widely applied. The outstanding application is in overhead line systems, where its relative cheapness and insulation properties are attractive. It is also utilized in transformers, some older switchgear, and generators in a laminated form suitably dried and glued/impregnated with resins to give a high mechanical strength and acceptable electrical properties.

2.1.4.2 Porcelain

For numerous years, porcelain, and to a lesser extent, glass, stood unrivaled as insulation for overhead line insulators. Its resilience to weathering, even in the face of moderate pollution, coupled with commendable flashover characteristics, made it a preferred choice. Methods have been developed to meet stringent mechanical requirements, although a drawback lies in its density during handling and vulnerability to cracking or fracturing when subjected to abrupt physical shocks.

Despite its relatively low puncture strength, porcelain finds extensive use in bushings, highvoltage instrument transformers, standoff insulators, and similar components. In these applications, the material's resistance to atmospheric conditions and its capability to withstand potential flashovers without catastrophic failure are the key insulating properties being leveraged. The enduring popularity of porcelain in certain applications underscores its ability to fulfill essential electrical and mechanical requirements in diverse power system components.

2.1.4.3　Glass

In addition to its application for sheds of overhead line insulators, glass is used in the power industry in fiber form to produce insulation components, including tapes, tubes, boards, and tie rods in composite insulators.

2.1.4.4　Rigid Laminates

A diverse category of insulating materials falls under the classification of "laminates." This encompasses boards made of paper or cloth bonded with resin, specialized plywood that is either fully or partially impregnated with resin, layers of bonded pressboard designed for oil impregnation, and high-quality materials composed of glass fiber layers impregnated with formulations such as silicone, epoxy, or polyester.

In outdoor applications and certain dry-type equipment, the material selection often hinges on tracking properties and resistance to degradation caused by moisture and dirt. In situations demanding robust mechanical strength or support in high electric fields, such as in power transformers, the presence of voids between layers of laminates can lead to partial discharges. Additionally, the use of unsuitable resins may result in excessive local dielectric heating, and impurities within layers parallel to the electric field may compromise breakdown strength. The intricacies of material selection are crucial, with considerations extending beyond just electrical properties to encompass mechanical strength, environmental resilience, and overall performance in specific applications.

2.1.4.5　Sheet Composites and Tapes

Polymers are used in the production of sheet composites, which may be formed as required and cured in situ, e.g. for machine slot insulation and interwinding wraps in dry-type transformers. The application of Melinex/Mylar (polyethylene terephthalate film) with polyester fiber mat produces a material suitable for Class 130 (Class B) temperatures. At higher temperatures, Nomex (polyamide paper) for Class 155 and Kapton (polyamide film) for Class 180 (Class H) and above (Class 200) may be used for specific applications. Nomex is a sheet material also applied in dry-type transformers impregnated with resin, usually under vacuum, operating in the Class 130 (Class B) range.

These types of materials may be used in tape form, especially for machines. Research continues in order to develop tapes for insulating conductors and leads suitable for operating at higher temperatures and more onerous conditions. Mica-based resin systems in formed or tape configurations are widely used in high-voltage rotating machines-both motors and generators. The techniques are specialized but are of considerable interest in power station operations. Studies continue concerning the partial discharge characteristics. The action of the mica-resin system in resisting high PD levels is not fully understood. Thousands of pico-coulombs may be withstood for many years of operation.

2.1.4.6　Epoxy Resins

The utilization of cast resins in power engineering has become firmly established. Manu-facturers provide an extensive range of components where designs, originally incorporating traditional materials, have been modified to harness the advantages offered by cast resins, specifically thermoset epoxy resin systems. The ultimate characteristics of these components are influenced by the type and quantity of fillers, such as glass fiber, silica, or other inorganic materials, with dielectric constants typically falling within the range of 3.0 to 4.0, depending on the epoxy filler.

Despite challenges related to thermal differences between the resin and conductors, a significant development hurdle has been the elimination of partial discharges in voids. It was discovered that the resins were susceptible to low-value partial discharges, possibly in the range of tens of pico Coulombs, which could lead to service failures. Effective quality control, especially the monitoring of partial discharges, is crucial in addressing this concern.

Originally restricted to indoor use, cast resins saw a breakthrough with the introduction of cycloaliphatic epoxy resins, accompanied by suitable fillers, for outdoor applications. Several current transformers with these resins have been in service for many years, and long-term tests have been conducted on online insulators. However, newer polymeric composites appear to outperform in the latter application, highlighting the continuous evolution and improvement in materials for diverse power engineering requirements.

2.1.4.7　Elastomers

Elastomers are a group of polymers that have rubber-like mechanical properties. They can be molded into such components as insulator sheds and extruded to form medium-voltage HV cables. In all applications, a wide range of fillers and chemical modifications are required to obtain satisfactory performance characteristics. This is especially important for outdoor insulators and shells. There are two major systems in use in the supply industry.

Ethylene propylene rubber (EPR). For medium-voltage cables extruded EPR is widely applied. It is flexible, does not form water trees, is resistant to internal partial discharges, and has good electrical properties. Ethylene propylene hexadiene monomer (EPDM) is a preferred form for outdoor insulators because it is more resistant to UV radiation and corona than EPR, although its long-term behavior under pollution is still being monitored.

Silicone rubbers (HTV, LSR, RTV). The high-temperature vulcanized form of silicone rubber HTV is used for outdoor insulators as it has good resistance to corona, ozone, UV, and many forms of pollution. The material is hydrophobic, which results in surface water forming as droplets, and consequently a high-resistance path for leakage currents, thus minimizing the possibility of dry-band arcing. Silicone is probably the only polymer that has this very desirable property.

Liquid silicone rubber (LSR) appears to be advantageous where complex shapes are required and avoidance of voids is essential. Room-temperature-vulcanized silicone rubber (RTV) has been applied to porcelain insulators to improve their wet and pollution performance. A comparison of the properties of EPDM and HTV silicone rubber is given in Reference 1. Dielectric constant values are of the order of 2.5-3.5 and 3.3-4.0 respectively.

2.1.4.8 Heat-Shrinkable Materials

The advent of heat-shrinkable polymeric materials marked a significant development, prompting alterations in the methodologies employed for 11 kV (and higher) cable terminations at switchgear and similar installations. Extensive testing has been conducted both in laboratory settings and outdoor test sites. Evaluation processes encompassed the study of the shrunken material's behavior under thermal cycling conditions, mimicking those experienced by a cable.

A critical consideration in these assessments is the prevention of air gaps between the heatshrink sleeve and the cable insulation, whether it is plastic or oil-impregnated. The presence of such gaps could lead to partial discharges, posing a risk of subsequent failure. Ensuring a seamless interface between the heat-shrinkable material and the cable insulation is paramount for maintaining the reliability and performance of cable terminations in high-voltage environments. The rigorous testing conducted aims to validate the effectiveness and safety of these heat-shrinkable polymeric materials in real-world applications.

2.1.4.9 Polyethylene (including XLPE)

The utilization of polyethylene as an insulating material is appealing due to its low losses and moderately high electric strength. However, its initial drawback was thermal instability at the temperatures required in power engineering. The breakthrough came with the introduction of cross-linked polyethylene (XLPE), which led to its widespread application.

XLPE, a material that can be extruded, was found suitable for cable manufacture once several challenges were addressed. Critical among these was the development of methods for curing and cooling long lengths, simultaneously eliminating voids where partial discharges (PDs) might initiate. Although XLPE is susceptible to discharges of the order of tens of pico

Coulombs, it is now widely applied as the primary insulation in cables across a broad voltage range, including some as high as 500 kV. The dielectric constant of XLPE is approximately 2.3. This evolution from conventional polyethylene to XLPE has significantly enhanced the thermal stability and overall performance of insulating materials in power engineering applications.

2. 1. 4. 10   Polyviny Chloride

Polyvinyl chloride (PVC) finds extensive use in the insulation of wires and as a sleeve material for low-voltage applications. The integrity of secondary wiring is crucial in many power situations, and PVC serves this purpose effectively. PVC typically has a thermal rating of 105 ℃, and this can be increased with suitable formulations. The dielectric constant of PVC falls in the range of 3.0-4.0, depending on the specific form and composition.

While PVC exhibits resistance to a variety of liquids, it is susceptible to attack by certain substances, particularly aromatic hydrocarbons. This vulnerability may be associated with the migration of plasticizers from the PVC material. Careful consideration of the chemical environment and potential exposure is essential when selecting PVC for specific applications to ensure its long-term reliability and performance.

2. 1. 4. 11   Polytetrafluoroethylene

Polytetrafluoroethylene (PTFE) is employed in extrusions, molds, and films in applications where demanding requirements for dielectric, mechanical, chemical, and thermal conditions exist. Specialist applications include insulators, cables, wires, and windings, justified in cases where a high cost can be warranted. However, PTFE is susceptible to corona discharge and radiation. Its dielectric constant is 2.0.

Fluorinated ethylene propylene copolymer (FEP), commonly known as Teflon, shares similar properties to PTFE. However, it is not as tough and has a more limited temperature range. FEP can be processed using conventional extrusion and molding methods. Both PTFE and FEP are chosen for their exceptional chemical resistance, thermal stability, and dielectric properties, making them suitable for applications in challenging environments where other materials might not perform as effectively.

2. 1. 4. 12   Polypropylene

Polypropylene (PE) film serves as a dielectric material in power capacitors, characterized by high electric strength and low losses. It boasts a low dielectric constant, and the appropriate grade must be carefully chosen to minimize swelling when exposed to certain dielectric liquids used in capacitors.

In its bulk form, polypropylene is utilized for molding components and extruded as insulation for cables operating at less than 5 kV. Polypropylene tapes laminated with paper (PPL) have been successfully employed for high-voltage cable insulation for a decade or more. For instance, the National Grid in the UK has adopted this technology. The use of polypropylene in these applications is driven by its favorable electrical properties and suitability for specific voltage ranges, making it a reliable choice in the realm of power engineering.

2.2 SF₆ and Its Alternatives

2.2.1 SF₆

2.2.1.1 History and Applications

Atmospheric air, while economically advantageous, exhibits poor insulating properties, leading to larger geometric dimensions and extremely non-uniform electrical fields in insulation systems. With increasing transmission voltages, the need for larger structures like towers, lines, and substations to maintain clearances has not only compromised aesthetics but also posed economic challenges due to substantial land and material requirements. The performance of electrical equipment in the open air is heavily influenced by atmospheric conditions and environmental pollution, necessitating a shift toward more compact, friendly, and economically viable high-voltage electrical installations.

To address these challenges, engineers sought to produce electrical installations with smaller dimensions, transitioning from extremely non-uniform fields in atmospheric air insulation systems to weakly non-uniform fields in GIS, also known as "Metalclad" systems. Weakly non-uniform fields in GIS enable better utilization of dielectric properties, resulting in reduced dimensions for equipment at a given rated voltage.

Atmospheric air, even under high pressure, has relatively poor dielectric strength. The search for gaseous dielectrics with improved properties has led to the adoption of sulfur hexafluoride (SF₆) as an alternative to air and nitrogen since the 1960s. In 1976, seven years after the introduction of GIS in Japan, the first 550 kV full GIS system was installed. By 1986, approximately 40% of newly built substations and replacements in Japan featured GIS. Gas-insulated switchgear with rated voltages up to 765 kV has been in service for over a quarter of a century, with Toshiba in Japan pioneering the development of the world's first complete SF₆ gas-insulated substation up to the highest rated voltage of 1,100 kV by the year 2,000.

SF₆ gas was initially employed for high-voltage circuit breakers, leading to the abbreviation "Gas Insulated Switchgears". As complete metal-clad substations utilizing SF₆ gas were developed, many countries adopted the same abbreviation for "Gas Insulated Substations." Both terms are widely used but may vary across regions. Additionally, SF₆ gas-insulated coaxial cables are referred to as "Compressed Gas Insulated Transmission Lines" (CGITL) by some. Interestingly, the abbreviation GIS has become more familiar than its full form. To streamline the terminology, some authors have adopted the more general expression "Gas-insulated Systems" for SF_6 gas-insulated substations in recent literature.

In an SF_6 molecule, six fluorine atoms uniformly arrange themselves in an octahedral structure around a central sulfur atom. The excited sulfur atom can form six stable covalent bonds with the strongly electronegative fluorine atoms by sharing electron pairs. Both fluorine and sulfur have high coefficients of electronegativity, around 4 and 2.5, respectively, making them highly attractive to electrons and conducive to forming dipole bonds.

The densities of gases increase with the relative molecular mass. As the molecular mass of SF_6 gas is quite high, it has a high density. Because of high density, the charge carriers have a short mean free path. This property, along with the properties of electron attachment, that is, electro-negativity and high ionization energy, result in the high dielectric strength of SF_6 gas, as shown in Table 2.1.

Table 2.1   **Physical properties of SF_6**

Property	Physical conditions	Symbol	Unit	Value
Relative permittivity	0.1 MPa, 25 ℃ −51 ℃(liquid)	ε_r	—	1.002, 1.81±0.02
Dielectric loss tangent	0.1 MPa −51 ℃(liquid)	$\tan\delta$	—	$<5\times10^{-6}$ $<1\times10^{-6}$
Critical temperature	—	θ_{cr}	℃	−50.8
Triple point	$P = 0.22$ MPa	θ_r	℃	−63.8
Sublimation point	—	θ_s	℃	
Specific heat capacity	at 10 ℃ and constant $p = 0.1$ MPa at 10 ℃ and const. volume	c_p, c_v	J/(mol·K)	5.13 4.06
Heat conductivity	30 ℃	—	J/(cm·s·K)	0.82×10^{-5}
Heat transition No.	—	—	J/(cm²·s·K)	0.44×10^{-5}

The rigid symmetrical structure, small binding distance, and high binding energy in SF_6 gas molecules provide stability, rendering its properties close to rare gases at relatively low temperatures. Thermal dissociation in highly purified SF_6 gas begins at extremely high temperatures (above 1,000 K), typically occurring only in electrical arcs. Even at continuous temperatures up to about 500 K, no reports exist of thermal decomposition or chemical reactions of SF_6 gas with other materials. SF_6 is non-toxic, colorless, odorless, non-flammable, non-explosive, chemically inert, and thermally stable, making it suitable for use in power system equipment.

2.2.1.2 Situation and Development of Environmentally Friendly Insulating Gas

The continuous growth in electric power demand, coupled with increasing environmental awareness, has led to the urgent development of high-voltage electrical equipment with large capacity and compact structures. Researchers are actively seeking non-flammable and non-aging insulating materials. Gas insulation, in comparison with other methods, occupies a smaller area, which is crucial in crowded cities, and has lower maintenance costs. However, the environmental impact of the insulating gas has raised concerns.[4, 5]

In the past century, significant changes have occurred in the Earth's climate, with global warming being a key feature. This warming is attributed to climate fluctuations and human activities. There is a pressing need to reduce greenhouse gas emissions and mitigate global climate change for sustainable development. The world is under increasing pressure to address and limit greenhouse gas emissions.

The greenhouse effect involves gases such as carbon dioxide allowing solar radiation to penetrate the atmosphere, warming the Earth's surface. Simultaneously, the atmosphere blocks long-wave radiation from the Earth's surface, leading to higher atmospheric temperatures a phenomenon known as the "greenhouse effect." Gases like carbon dioxide are termed "greenhouse gases," as they absorb and reflect infrared radiation emitted from the ground, contributing to heat retention. Greenhouse gases have strong infrared absorption characteristics in the 7-13 μm wavelength range.

Life on Earth depends on a normal greenhouse effect to maintain suitable temperatures for growth and sustenance. However, increased emissions of greenhouse gases enhance the greenhouse effect, disrupting the normal balance and leading to changes in the radiation balance. This imbalance contributes to climate change and is a significant driver behind the global efforts to reduce greenhouse gas emissions.

Carbon dioxide (CO_2), methane (CH_4), nitrous oxide (N_2O), hydrofluorocarbon compounds (HFCs), perfluorocarbons (PFCs), and SF_6 are the six main greenhouse gases released by human activities. Among these, carbon dioxide has the largest impact, accounting for 60%, while the contribution of SF_6 is relatively minimal at 0.1%. However, SF_6's greenhouse effect poses potential hazards due to its high global warming potential (GWP), with an impact 25,000 times greater than that of CO_2 molecules (GWP of 23,900). Additionally, SF_6 gas in the atmosphere has a long lifespan of about 3,200 years, leading to cumulative effects on global warming.

Although SF_6 emissions are significantly lower than CO_2 emissions (annual CO_2 emissions into the atmosphere are about 21 billion tons), the impact of SF_6 is substantial when considering its GWP. The annual emissions of SF_6 into the atmosphere are equivalent to 125 million tons of CO_2. Therefore, while SF_6 contributes a relatively small proportion to overall greenhouse gas emissions, its potent warming effect and long atmospheric lifetime make it an environmentally significant concern. Efforts to reduce SF$_6$ emissions are crucial for mitigating its impact on climate change.

The global warming caused by the greenhouse effect can cause a serious threat to the environment and may have disastrous consequences. Global warming has become one of the three greatest international environmental problems (ozone layer destruction, global warming, and the drastic decrease in biological species). In recent years, the international community has launched a wideranging global cooperation. In December 1997 in Kyoto, Japan, the 5th Meeting of States Parties of the United Nations Framework Convention on climate change was held, and the Kyoto Protocol was signed at that meeting. The protocol confirmed the impact of greenhouse gases on global climate change; in this protocol, CO_2, CH_4, N_2O, SF_6, PFC (perfluorinated hydrocarbons), CFC (chlorofluorocarbons), HCFC (hydrogen fluoride hydrocarbons) and HFC (hydrofluorocarbons) are all greenhouse gases. The meeting also required developed countries to control greenhouse gas emissions at the level of the 1990s, then during 2008-2012 to reduce greenhouse gas emissions by 5.2% based on that level, and by 2020 prohibit the use of SF$_6$ gas.

The world's annual production of SF$_6$ is about 8,500 t and more than half is used in the power industry. In the power industry, high-voltage switchgear consumption accounts for more than 80% of SF$_6$ use, and medium-voltage switchgear consumption accounts for about 1/10. SF$_6$ is mainly used at 126-252 kV and 330-800 kV, especially in the 126 kV, 252 kV, and 550 kV circuit breaker (GCB), SF$_6$ GIS, gas-insulated cabinet (C-GIS) and SF$_6$ GIL. Therefore, SF$_6$ gas should be properly used.

Efforts to address the environmental concerns associated with pure SF$_6$ gas have led to research on SF$_6$ gas mixtures, exploring the possibility of using binary gas mixtures to replace pure SF$_6$. Studies indicate that incorporating gases like N_2, CO_2, or air into SF$_6$ can offer several advantages. For instance, at the same gas pressure, the liquefaction temperature of gas mixtures is lower than that of pure SF$_6$, making SF$_6$ gas mixtures suitable for use in circuit breakers in alpine regions to prevent gas liquefaction at low temperatures.

The addition of certain gases to SF$_6$ can also help reduce the roughness effect of electrodes, resulting in a smaller sensitivity of the electric field strength compared to pure SF$_6$. This improvement in design can significantly enhance the positive breakdown voltage. Additionally, SF$_6$/N_2, SF$_6$/CO_2, or SF$_6$/air binary mixtures can potentially reduce the overall cost of the gas, while also decreasing the Global Warming Potential (GWP).

However, research on the arc-quenching performance of SF$_6$/N_2 mixtures has indicated that a 25% N_2 content has a similar performance to pure SF$_6$, but the performance deteriorates with 50% N_2 content. Therefore, for high-voltage circuit breakers, SF$_6$ gas mixtures may not be suitable for applications where superior breaking performance is required. The challenge lies in finding the right balance of gas composition to maintain optimal performance while addressing environmental concerns.

2.2.2 Situation and Development of Environmentally Friendly Insulating Gas

In recent years, there has been research on electronegative gases containing fluorine atoms that exhibit similar electronegative performance to SF_6 but have less impact on the greenhouse effect. Octafluorocyclobutane (c-C_4F_8), perfluoropropane (C_3F_8), and hexafluoroethane (C_2F_6) are among the gases studied, showing excellent performance in various research endeavors. For example, Kyoto University conducted a study on the feasibility of using c-C_4F_8 as an insulating medium in high-voltage equipment. Experimental results indicated that c-C_4F_8 mixtures have performance like SF_6/N_2 gas mixtures, leading researchers to propose c-C_4F_8 as an alternative insulating gas to replace SF_6. German scholars conducted experiments to examine the relationship between the electron drift velocity and effective ionization coefficient of c-C_4F_8 under different pressures using a pulse Townsend experiment. Similarly, researchers at Keio University in Japan tested the electron drift velocity and electronic vertical diffusion coefficient of c-C_4F_8 and its gas mixtures. Some results suggested that elastic collisions between electrons and c-C_4F_8 molecules are relatively strong.

The potential application of c-C_4F_8 gas mixtures as an insulating medium has garnered attention. In 1997, the US National Institute of Standards and Technology conference identified c-C_4F_8 gas mixtures as a potential insulating gas for further study. In 2001, the Tokyo Electric Power Industrial Center and Tokyo University presented the application of c-C_4F_8 gas mixtures as an insulation medium. However, its Global Warming Potential (GWP) remains high, posing challenges in addressing the greenhouse effect. Moreover, c-C_4F_8 can generate carbon particles under discharge conditions, posing a risk to stable power equipment operation.

The development of gases like C_4F_7N and $C_5F_{10}O$ has garnered attention due to their promising environmental friendliness and insulation properties. However, the high liquefaction temperature of $C_5F_{10}O$ limits its application in high-pressure fields, making it more suitable for medium to low-pressure applications. Studies have shown that under certain conditions, the electrical strength of SF_6 under a mixture of 20% C_4F_7N and 80% CO_2 is comparable.[6] Adjusting air pressure and C_4F_7N content in the range of 4% to 10% can achieve insulation levels of over 85% of SF_6.

Although C_4F_7N has low toxicity, its toxicity is further reduced when mixed with buffer gas, making it considered non-toxic. In 2014, Alstom (now part of GE) developed g^3 gas, a mixture of C_4F_7N/CO_2, successfully creating prototypes of 420 kV GIL and 145 kV GIS using g^3 gas as the insulation medium.

The development of prototypes for C_4F_7N gas-insulated equipment, including circuit breakers, GIL, and switchgear, suggests that C_4F_7N mixed gas has promising insulation performance and potential as a substitute for SF_6 gas. The physical and chemical parameters of SF_6 and its alternative mediums are compared in Table 2.2.

Table 2.2   **Physical and chemical parameters of SF_6 and its potential alternative mediums**

Physical and chemical properties	SF_6	C_4F_7N	$C_5F_{10}O$	$C_6F_{12}O$	CF_3I	$c\text{-}C_4C_8$
GWP_{100yr}	23,500	2,090	1	1	1	8,700
ODP	0	0	0	0	0	0
Atmospheric lifetime	3,200 years	22 years	15 days	5 days	<2 days	—
LC50 (4h rat)	—	10,000~15,000	>20,000	>100,000	160,000	—
Liquefaction temperature	−63	−4.7	29	49	−8	−22.5
Molecular weight/(g/mol)	146	195	266	316	196	200
Relative permittivity	1	>2.2	2	2.7	1.2	1.3

Note: "−" represents that no relevant data was found.

2.3   Insulating and Extinguishing Performance of C_4F_7N

2.3.1   Power Frequency and DC Breakdown Characteristics

The research on the insulation performance of C_4F_7N/CO_2 mixed gas under various conditions has provided valuable insights. In medium and low voltage applications, it has been demonstrated that the insulation performance of C_4F_7N/CO_2 mixed gas, especially with 15% and 20% C_4F_7N content at 0.1-0.2 MPa, is comparable to that of SF_6. The critical breakdown field strengths for these mixtures are reported to be equivalent to SF_6. The synergistic effect between CO_2 and C_4F_7N is noted to be superior to N_2, making C_4F_7N/CO_2 mixtures more promising for application compared to C_4F_7N/N_2.

Under certain pressures, the power frequency breakdown voltage of C_4F_7N/N_2 mixture gas is reported to be lower than that of C_4F_7N/CO_2 and C_4F_7N/air mixtures. The insulation performance of a 3.7% C_4F_7N/96.3% CO_2 gas mixture is claimed to reach SF_6 levels at specific pressures. Regarding DC breakdown, the negative DC breakdown voltage of C_4F_7N/CO_2 gas mixtures with 4% and 8% C_4F_7N content is reported to reach a percentage of SF_6 breakdown voltage at certain pressures.

While the insulation performance of C_4F_7N is noted to be more than twice that of SF_6, considerations for high-voltage equipment operating under minimum temperature limits indicate that the insulation performance of C_4F_7N/CO_2 mixed gas may not reach the level of pure SF_6 under the same conditions. The study suggests the need to increase the buffer gas to improve the insulation performance of the mixed gas while meeting the minimum liquefaction temperature of C_4F_7N. For medium and low voltage applications, when the content of C_4F_7N in the mixed gas exceeds 15%, its insulation performance relative to SF_6 is reported to be better. In addition, O_2 also has a certain improvement effect on the insulation performance of the C_4F_7N-CO_2 mixed gas. The power frequency breakdown voltage of the 15% C_4F_7N-79% CO_2-6% O_2 mixed gas under a slightly uneven electric field is 7.7% higher than that of the 15% C_4F_7N-85% CO_2 mixed gas.[7]

In conclusion, the research on the power frequency and DC breakdown characteristics of C_4F_7N mixed gas has confirmed its excellent insulation resistance performance. The available data can serve as crucial references for equipment design and optimization. Future considerations for high-voltage gas insulation equipment should include adjustments to operating pressure, volume, and design temperature to meet insulation-type test requirements. Attention should also be given to avoiding the introduction of uneven electric field environments during equipment design and manufacturing processes to ensure the reliability of equipment operation.

2.3.2　Lighting Impulse and Surface Flashover Characteristics

The U50 (50% breakdown voltage) of the 3.7% C_4F_7N-96.3% CO_2 mixture gas is reported to increase at a lower rate with gas pressure compared to SF_6. In terms of lightning impulse characteristics, the 3.7% C_4F_7N-96.3% CO_2 mixed gas is claimed to reach SF_6 levels at specific pressures in a quasi-uniform electric field environment. The breakdown voltage of the mixed gas under negative polarity lightning impulse is noted to increase faster with gas pressure than that of positive polarity.

Under extremely non-uniform electric field conditions (when the electric field non-uniformity is greater than 5.56), it is reported that the negative lightning impulse voltage of the 3.7% C_4F_7N-96.3% CO_2 mixed gas cannot reach SF_6 levels at certain pressures.

The lightning impulse flashover voltage along the surface of the mixed gas with varying C_4F_7N content (5%, 9%, and 13%) is claimed to reach certain percentages of SF_6 levels under the same conditions.[8] Power frequency surface flashover characteristics of the C_4F_7N-CO_2 mixture gas and the 252 kV basin insulator surface flashover characteristics tests indicate that insulator flashover mostly occurs on the concave side of the basin insulator under mixed gas environments. At specific pressures, the surface flashover voltage of the 9% C_4F_7N-91% CO_2 mixture gas is claimed to reach the level of 0.5 MPa SF_6.

Concerning the influence of non-uniformity, the lightning impulse voltage of the C_4F_7N mixed gas in a slightly non-uniform electric field environment reportedly shows a saturation growth trend with the mixing ratio and a quasi-linear growth trend with air pressure. As the non-uniformity of the electric field increases, both positive and negative lightning impulse breakdown voltages of the mixed gas significantly decrease, with the positive polarity being much lower than the negative polarity. The mixed gas is described as exhibiting high sensitivity to the non-uniformity of the electric field.

2.3.3 Arc Extinguishing Characteristics

The 4% C_4F_7N -96% CO_2 mixture gas was charged into the SF_6 isolation switch with a rated voltage of 420 kV (0.55 MPa), and 1,000 C/O tests were conducted (1,600 A, 20 V).[9] The comparison of the arcing time between the 4% C_4F_7N-96% CO_2 mixture gas and the SF_6 C/O operation is shown in Figure 2.1. The experiment found that the arc discharge time of the C_4F_7N mixed gas was relatively stable in 100 operations, with an average arc discharge time of about 12 ms (typical value of 15 ms for SF_6), and the electrical wear of the arc contacts was similar to that of SF_6.

In tests based on the IEC 62271-102 standard, a 5% C_4F_7N-95% CO_2 mixed gas in a 245 kV GIS equipment achieved bus transfer current (BTC) disconnection with an isolation switch at 0.65 MPa, demonstrating a disconnection time between 2-8 ms.[10] However, after the power frequency breakdown and lightning flashover voltage tests of the mixed gas, its withstand performance was reported to decrease by more than 20%. Additionally, a significant number of solid substances precipitated on the surface of the insulator in GIS. The surface flashover and smoke spot tracks on the insulator after the 5% C_4F_7N-95% CO_2 gas BTC test are shown in Figure 2.2. In addition, the content of C_4F_7N in the mixed gas decreased to 3.6% after the experiment, indicating that C_4F_7N was consumed during the disconnection process.

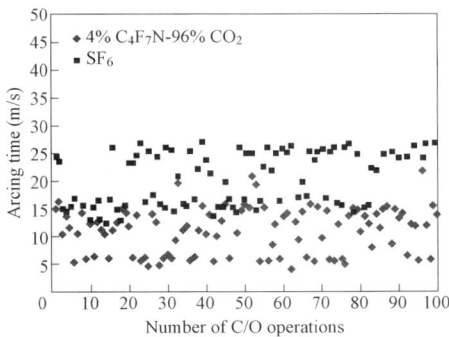

Figure 2.1   Arcing time versus C/O operation number on a 420 kV disconnector for 4% C_4F_7N-96% CO_2 mixture and SF_6 negative impact

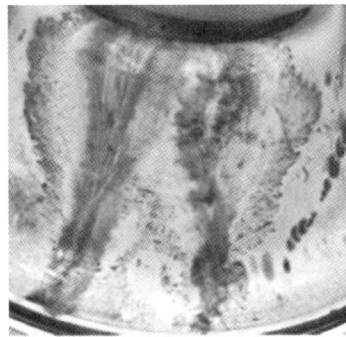

Figure 2.2   Tracks of surface flashover and soot spots on an insulator after BTC tests of 5% C_4F_7N-95% CO_2

The breaking performance of a 9.5% C_4F_7N-9.5% O_2-81% CO_2 gas mixture shows that the di/dt of the mixed gas was 9 A/s. Same as 92% CO_2-8% O_2 mixed gas. In addition, after the experiment, the air pressure in the chamber increased by 150 kPa (the temperature rise effect is about 40 kPa), which is related to the decomposition of C_4F_7N to produce multiple components. In a test of a 145 kV GIS and high-voltage tank circuit breaker with 0.8 MPa C_4F_7N-CO_2-O_2 mixed gas, it was found that the mixed gas could successfully break the 63 kA short circuit current generated by terminal fault (TF) and short line fault (SLF), and the circuit breaker's design life could reach 450 kA^2 Xs.

The relative molecular weight of C_4F_7N is large and its molecular structure is relatively complex. When used as an arc extinguishing medium, its performance differs significantly from that of pure SF_6. In addition, the transient energy of a high-energy arc can lead to the continuous decomposition and consumption of C_4F_7N during interruption, and a large number of dissociated particles cannot fully composite into C_4F_7N, resulting in many small molecule products with low insulation performance and also triggering the precipitation of solid carbon fluorine decomposition products. Therefore, further exploration is needed for the application research of C_4F_7N mixed gas as an arc extinguishing medium, In the future, it is necessary to focus on the optimal formulation of gas components, gas decomposition products, and solid precipitation suppression schemes in arc extinguishing scenarios. At the same time, it is necessary to optimize or improve the circuit breaker structure, contact materials, auxiliary arc extinguishing methods, etc. to meet relevant technical standards and service life requirements.

2.4 Decomposition Characteristics of C_4F_7N

C_4F_7N has a liquefaction temperature of -4.7 ℃ and cannot be used independently; it requires mixing with a buffer gas. Commonly used buffer gases include air, N_2, and CO_2. The electrical performance of C_4F_7N mixed with N_2, CO_2, or air is comparable. Studies on the decomposition characteristics of SF_6 and c-C_4F_8 mixed gases have shown that the type of buffer gas influences the types and concentrations of decomposition products in the insulating medium. Additionally, the buffer gas significantly affects the formation of solid by-products. Given the comparable electrical performance, further research is warranted to systematically investigate the selection of buffer gases from a decomposition perspective.

Compared to other fault forms, corona discharge exhibits good stability and is a typical form of partial discharge. Therefore, combining the decomposition and product formation mechanisms of C_4F_7N, this chapter employs GC/MS detection methods to continuously measure the types and concentrations of decomposition products in binary mixtures of C_4F_7N with CO_2, N_2, or dry air under corona discharge conditions. The study quantitatively compares the content of three types of C_4F_7N mixed gas decomposition products under different mixing ratios. Additionally, X-ray energy-dispersive spectroscopy is used for semi-quantitative analysis of the elemental composition and content on the electrode surface after discharge. Finally, considering the types, concentrations, chemical properties, toxicity, and solid deposits of decomposition components, CO_2 is selected as the buffer gas from a decomposition perspective. Based on this, the decomposition characteristics of C_4F_7N/CO_2 mixed gas under discharge and overheating conditions are analyzed.

2.4.1　Introduction of Quantum Chemistry Theory

Theoretical research on reaction mechanisms involves the use of computer simulations to model the process of chemical reactions. From a theoretical perspective, the goal is to identify all elementary reactions and potential intermediates and transition states that may arise during the complex chemical reaction. By calculating reaction energies, the theoretical approach aims to determine the lowest energy reaction pathway.

Quantum chemistry is a crucial component of theoretical chemistry, developed as a computational method based on the principles of quantum mechanics to address various chemical problems. Typically, quantum chemistry calculations refer to solving the Schrödinger equation, under the Born-Oppenheimer approximation, for the electronic motion within atoms or molecules. This process yields the total electronic energy that varies with changes in molecular structure, allowing the description of important information such as molecular structures and reaction pathways in chemistry. In solving the Schrödinger equation, quantum chemistry calculations do not rely on any empirical parameters but start from fundamental physical constants, earning them the designation of "ab initio" methods.

Various quantum chemistry calculation methods have been developed, including early semiempirical methods like AM1, PM3, PM7, Hartree-Fock methods, and various electron-correlation methods such as configuration interaction (CI), perturbation theory (RS, MP), and coupledcluster (CC) theory. With the rapid advancement of computer technology, the precision and speed of quantum chemistry calculations have significantly improved. Currently, they can achieve what is known as "chemical accuracy," and for certain typical chemical molecular systems, theoretical results can even surpass experimental precision.

In summary, quantum chemistry calculations possess powerful explanatory and predictive capabilities. They have formed a comprehensive, systematic, and self-consistent theoretical framework and have been successfully applied across various branches of chemistry. These include materials chemistry, chemical biology, medicinal chemistry, nanotechnology, environmental chem-istry, analytical chemistry, inorganic chemistry, organic chemistry, physical chemistry, colloid and interface chemistry, and macromolecular biochemistry, where the influence of quantum chemistry calculations can be found.

In 1929, Pople, a graduate student in the mathematics department at the University of Cambridge, created the renowned Gaussian program, providing modern computer technology support for solving the stationary equations of quantum chemistry theory. The Gaussian computational program is utilized to solve the Schrödinger equation for complex chemical systems like molecules, which is the fundamental problem in quantum chemistry. The Gaussian program supports various quantum chemistry computational theories and methods, including transition state energy and structure, bond and reaction energies, molecular orbitals, atomic charges and potentials, vibrational frequencies, infrared and Raman spectra, nuclear magnetic properties, polarizabilities and hyperpolarizabilities, thermodynamic properties, reaction pathways, and more. The calculations can be performed for the system's ground state or excited states, and they can predict the energy, structure, and molecular orbitals of periodic systems.

Therefore, Gaussian is regarded as a powerful tool for researching various topics in the field of chemistry, such as the impact of substituents, chemical reaction mechanisms, potential energy surfaces, and excited states.

Density Functional Theory (DFT) simplifies the wave function from ab initio calculations into a three-dimensional particle density. It utilizes the electron density to describe various properties of the system, resulting in a significant reduction in computational complexity compared to ab initio methods. Its widespread application is achieved through the implementation of the Kohn-Sham method. Currently, DFT is extensively used in electronic structure calculations for both molecules and condensed matter systems, encompassing studies on ionization potentials, spectra, catalysis, reaction mechanisms, and other related issues.

2.4.2 Decomposition and By-product Formation Mechanism

2.4.2.1 Initial Isomerization and Dissociation Processes of C_4F_7N

Figure 2.3 presents potential energy surfaces for the isomerization and dissociation of C_4F_7N, calculated at the CCSD(T)/cc-pVTZ//M06-2X/6-311G(d,p) level. The decomposition of C_4F_7N primarily involves two isomerization reaction channels and four unimolecular dissociation channels.

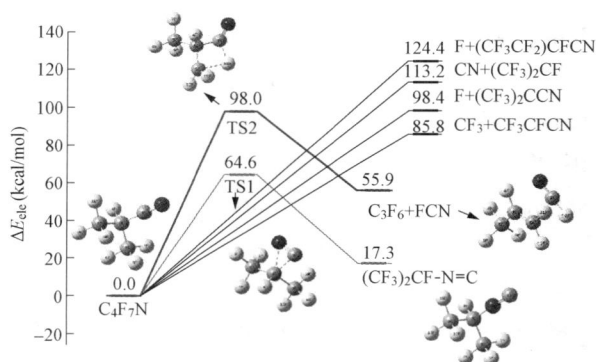

Figure 2.3   The potential energy surface of C_4F_7N isomerization and dissociation of based on the relative energies at CCSD(T)/cc-pVTZ//M06-2X/6-311G(d,p) level

The two isomerization reaction channels are as follows:

(1) C_4F_7N undergoes a C—CN bond rotation via the transition state TS1 to isomerize into $(CF_3)_2CF$—N=C.

The electron energy barrier is 64.6 kcal/mol, and the energy of the product $(CF_3)_2CF$—N ═C is 17.3 kcal/mol higher than that of the reactant C_4F_7N. IRC calculations confirm the connection between the reactant C_4F_7N and the product $(CF_3)_2CF$—N═C through the transition state TS1.

(2) C_4F_7N undergoes an isomerization through F atom migration on the CF_3 group, resulting in C_3F_6+FCN.

The free energy barrier is 98.0 kcal/mol, and the energy of the product C_3F_6+FCN is 55.9 kcal/mol higher than that of C_4F_7N. IRC calculations confirm the connection between C_4F_7N and the complex C_3F_6+FCN through the transition state TS2.

The four unimolecular dissociation channels are as follows.[11, 12] C_4F_7N can dissociate into CF_3, CF_3CFCN, F, $(CF_3)_2C(CN)$, $(CF_3)_2CF$, CN, and $(CF_3CF_2)CFCN$ radicals through the cleavage of covalent bonds in C—CF_3, C—F, C—CN, or C—F bonds in the CF_3 group. The energies of the products CF_3+CF_3CFCN, F+$(CF_3)_2C(CN)$, CN+$(CF_3)_2CF$, and F+$(CF_3CF_2)CFCN$ are higher than that of C_4F_7N by 85.8 kcal/mol, 98.4 kcal/mol, 113.2 kcal/mol, and 124.4 kcal/mol, respectively. Among the four dissociation pathways, the one leading to C_4F_7N⟶ CF_3CFCN+CF_3 has the lowest free energy change, making it the dominant dissociation channel. The dissociation channel C_4F_7N⟶F+$(CF_3CF_2)CFCN$ is not favorable due to the higher bond energy of the C—F bond in the CF_3 group and can be neglected.

2.4.2.2 Electronic Structure during the Isomerization Process

The localized orbital locator (LOL) coloring maps and Mayer bond order analysis are effective tools for understanding changes in electronic structure during the isomerization process of C_4F_7N. This section provides a detailed discussion of the changes in electronic structure during the isomerization reaction C_4F_7N ⟶ $(CF_3)_2CF$—N═C through LOL coloring maps and Mayer bond order analysis.

(1) Isomerization Reaction C_4F_7N ⟶ $(CF_3)_2CF$—N═C.

Figure 2.4 depicts the LOL coloring map for the isomerization process of C_4F_7N to $(CF_3)_2CF$—N═C. Higher numerical values in the map indicate stronger localization in that region.

In Figure 2.4(a), it can be observed that the regions between C3—C1 and C3—C5 exhibit high localization, indicating strong localized covalent bonds corresponding to C3—C1 and C3—C5 in the C_4F_7N molecule. The region between C1—N2 also demonstrates high localization, suggesting the presence of a strongly localized covalent bond between C1 and N2. Additionally, there is a large numerical region pointing outward from N(2), corresponding to the lone pair electrons on N(2), indicating the capability of N(2) to form coordinate bonds. Furthermore, the C1—N2 bond's localized region appears flat and concave, displaying clear characteristics of multiple bonds.

In Figure 2.4(b), the localization between C3—C1 weakens, indicating a decrease in the covalent interaction between C3 and C1. In the optimized transition state TS1, the distance between C(1) and C(3) gradually elongates from 1.471 Å to 1.752 Å as the CN group rotates. N(2) approaches C(3), and C(1) moves away from C(3). The localized region of lone pair electrons on N(2) weakens, and the lone pair electrons gradually disappear. A region with strong localization appears outward from C(1), indicating the presence of lone pair electrons on C(1).

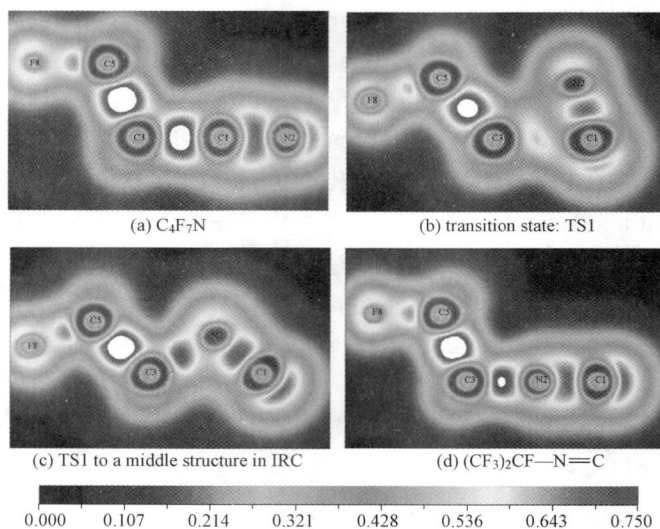

(a) C_4F_7N (b) transition state: TS1

(c) TS1 to a middle structure in IRC   (d) $(CF_3)_2CF{-}N{=}C$

| 0.000 | 0.107 | 0.214 | 0.321 | 0.428 | 0.536 | 0.643 | 0.750 |

Figure 2.4   The localized orbital locator (LOL) color-filled maps of the reaction $C_4F_7N \longrightarrow (CF_3)_2CF{-}N{=}C$ in the C(3)—C(1)—N(2) plane

As shown in Figure 2.4(c), with the rotation of the CN group, the lone pair electrons on N(2) completely disappear, and the localized region between C3 and N2 becomes connected. The localization in the region between C3 and N2 strengthens, indicating the formation of a covalent bond between C3 and N2. The connected region between C3 and C1 disappears entirely, indicating the complete rupture of the covalent bond between C3 and C1. The localized region of lone pair electrons on C(1) increases, displaying distinct characteristics of lone pair electrons. The rotation of the CN group concludes, leading to the final product $(CF_3)_2CF{-}N{=}C$.

In Figure 2.4(d), the region between C3 and N2 is fully connected, indicating the formation of a strong covalent bond between C3 and N2, completing the entire isomerization process. The resulting configuration of the product $(CF_3)_2CF{-}N{=}C$ also displays a flat and concave localized region in the C1—N2 bond, showing clear characteristics of multiple bonds.

Due to significant changes in Mayer bond orders for the C1—C3, C1—N2, and N2—C3 covalent bonds during the isomerization process of C_4F_7N, the Mayer bond orders were calculated for these three bonds and plotted the variations with respect to the Intrinsic

Reaction Coordinate (IRC), as shown in Figure 2.5. In the C_4F_7N molecule, the bond between C(1) and C(3) forms a single bond with a Mayer bond order of approximately 1.0, the bond between C(1) and N(2) forms a covalent triple bond with a Mayer bond order of approximately 3.0, and there is no covalent bond formed between N(2) and C(3), resulting in a Mayer bond order of 0.0.

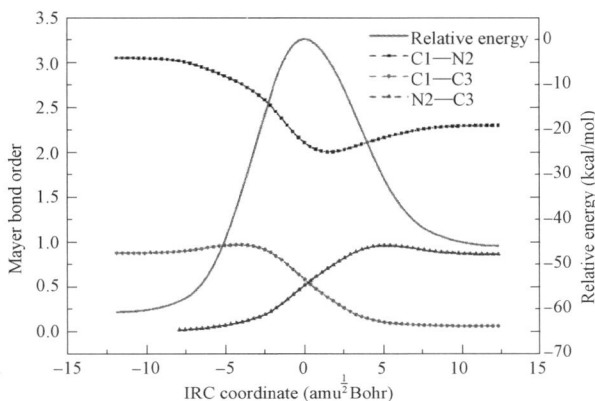

Figure 2.5   The localized orbital locator (LOL) color-filled maps of the reaction $C_4F_7N \longrightarrow$
$(CF_3)_2CF—N\!=\!C$ in the C(3)—C(1)—N(2) plane

Before the reaction proceeds to $s=-3(amu^{1/2}$ Bohr), the bond orders for C1—C3 and N2—C3 remain nearly constant, while the bond order for C1—N2 decreases from the original 3.0 to around 2.3. This suggests that, with the rotation of the —C≡N group, the triple bond between C(1) and N(2) gradually transforms into a double bond. After the reaction passes $s=-3(amu^{1/2}$ Bohr), the bond order for C1—C3 starts to decrease, and the bond order for N2—C3 begins to increase, indicating that, as the isomerization reaction proceeds, C(1) starts moving away from C(3), and N(2) starts approaching C(3). When $s=5(amu^{1/2}$ Bohr), the covalent single bond between N2 and C3 has formed, while the covalent single bond between C1 and C3 is completely broken. The carbon-nitrogen triple bond (C≡N) in the C_4F_7N molecule transforms into a carbon-nitrogen double bond (N=C), and thereafter, the bond orders for N2—C3, C1—C3, and C1—N2 essentially remain constant. Additionally, it can be observed that the Mayer bond orders change very littleat the beginning and end of the reaction but undergo drastic changes during the formation of new covalent bonds and the rupture of old covalent bonds. The variations in Mayer bond orders, along with LOL coloring maps, clearly illustrate the changes in electronic structure during the isomerization of C_4F_7N to $(CF_3)_2CF—$ $N\!=\!C$, and the two approaches align well.

(2) Isomerization reaction C_3F_6+FCN of C_4F_7N.

The LOL coloring map for the isomerization reaction of C_4F_7N to C_3F_6+FCN is presented in Figure 2.6, where higher numerical values indicate stronger localization in that region. In Figure 2.6(a), it can be observed that the regions between C3—C6, C1—C3, and N2—C1 exhibit high localization, indicating strong localized covalent bonds corresponding to C3—C6, C1—C3, and N2—C1 in the C_4F_7N molecule. Additionally, the localized region of the C1—N2 bond appears flat and concave, displaying distinct characteristics of multiple bonds.

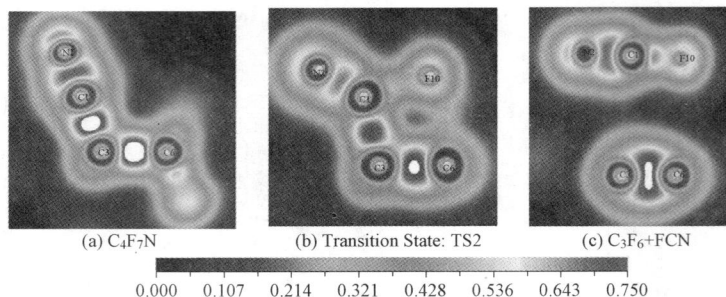

| (a) C_4F_7N | (b) Transition State: TS2 | (c) C_3F_6+FCN |

0.000   0.107   0.214   0.321   0.428   0.536   0.643   0.750

Figure 2.6   The localized orbital locator (LOL) color-filled maps of the reaction $C_4F_7N \longrightarrow$
C_3F_6+FCN in the C(6)—C(3)—C(1) plane

In Figure 2.6(b), the F(10) atom moves away from C(6) and approaches C(1). In the transition state TS2, the localized region between F(10) and C(1) is connected, but with weak localization. The localization between C1—C3 slightly decreases, primarily due to the elongation of the distance between C(1) and C(3) from the original 1.471 Å to 1.590 Å in transition state TS2. As shown in Figure 2.6(c), the localized region between C(1) and F(10) is fully connected, indicating the formation of a strong covalent bond between C1 and F10. The connected region between C1 and C3 has disappeared entirely, indicating the complete rupture of the covalent bond between C1 and C3, completing the isomerization process. In the product, the localized region of the C3—C6 bond appears flat and concave, displaying clear characteristics of multiple bonds.

In the process of C_4F_7N isomerizing to C_3F_6+FCN, significant changes are observed in the Mayer bond orders for the C1—C3, C1—F10, C3—C6, and C6—F10 covalent bonds. The Mayer bond orders were calculated for these four bonds and plotted their variations with respect to the Intrinsic Reaction Coordinate (IRC), as shown in Figure 2.7.

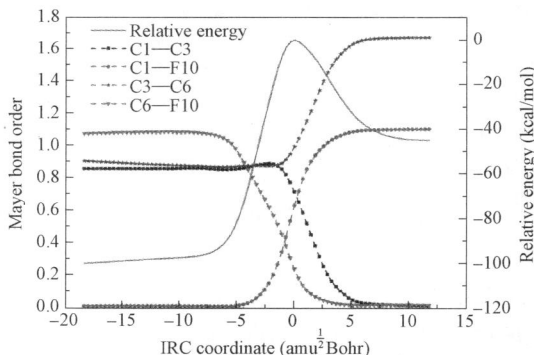

Figure 2.7   Mayer bond order Variation of key bonds along IRC path in the
isomerization reaction of $C_4F_7N \longrightarrow C_3F_6$+FCN

In the C_4F_7N molecule, single bonds with Mayer bond orders of approximately 1.0 are formed between C(1) and C(3), C(3) and C(6), and C(6) and F(10). However, there is no covalent bond formed between C(1) and F(10), resulting in a Mayer bond order of 0.0.

Before the reaction proceeds to $s=0$ (amu$^{1/2}$ Bohr), the bond orders for C1—C3 and C3—C6 remain almost constant, while the bond order for C6—F10 drastically decreases from the original 1.1 to around 0.2. The bond order for C1—F10 increases from zero to approximately 0.4. This indicates that the F atom on the CF_3 group has moved away from the C(6) atom on CF_3 and is gradually approaching the C(1) atom. At this point, the covalent interaction between C(6) and F(10) is very weak, and a covalent interaction begins to form between C(1) and F(10).

After the reaction proceeds beyond $s=0$ (amu$^{1/2}$ Bohr), the bond order for C1—C3 starts to decrease, while the bond order for C1—F10 continues to increase. This suggests that, as the reaction progresses, the C(1) atom in the C_4F_7N molecule begins to move away from C(3) and gradually approaches F(10). The bond order for C3—C6 starts to increase, indicating a shorter distance between C(3) and C(6) and an enhancement of covalent interaction. When $s=5$(amu$^{1/2}$ Bohr), the covalent single bonds for C1—C3 and C6—F10 are completely broken, the covalent single bond for C1—F10 has formed, and the C3—C6 bond transitions from a C—C single bond to a C=C double bond. After this point, the bond orders essentially remain constant, indicating the completion of the isomerization process. Additionally, it can be observed that Mayer bond orders change very little at the beginning and end of the reaction but undergo drastic changes during the formation of new covalent bonds and the rupture of old covalent bonds. The variations in Mayer bond orders, along with LOL coloring maps, clearly illustrate the changes in electronic structure during the isomerization of C_4F_7N to C_3F_6+FCN, and the two approaches align well.

2.4.2.3 Isomerization and Dissociation of CF₃CFCN

Figure 2.8 depicts the potential energy surface for the isomerization and dissociation reactions of the CF_3CFCN radical obtained at the CCSD(T)/cc-pVTZ//M06-2X/6-311G(d,p) level. The CF_3CFCN radical undergoes a fluorine atom transfer reaction, transforming through the transition state TS3 to produce the CF_2CF_2CN radical. This process requires overcoming an electronic barrier of 57.6 kcal/mol. Transition state TS3 belongs to the C1 point group, representing a first-order saddle point with a virtual frequency of 334.83i. The confirmation of TS3 connecting CF_3CFCN and CF_2CF_2CN is further verified through IRC calculations. The energy of the CF_2CF_2CN radical is 18.0 kcal/mol higher than that of CF_3CFCN, indicating its instability. Consequently, the CF_2CF_2CN radical further dissociates directly by breaking the C—CF_2 and C—CN covalent bonds to generate the products CF_2CN, CF_2, CN, and C_2F_4. Among these products, the energy of CF_2CN+CF_2 is 49.6 kcal/mol higher than that of CF_2CF_2CN radical, and the energy of CN+C_2F_4 is 60.4 kcal/mol higher than that of CF_2CF_2CN radical.

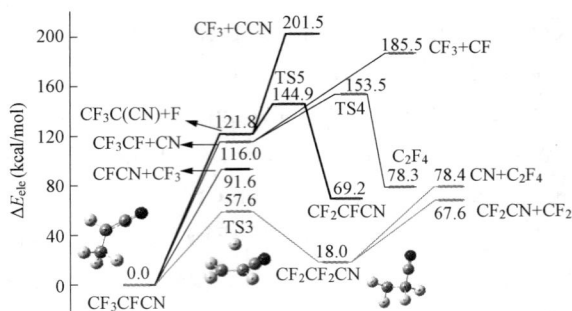

Figure 2.8 The potential energy surface of the isomerization and dissociation of CF$_3$CFCN radical based on the relative energies at CCSD(T)/cc-pVTZ//M06-2X/6-311G(d,p) level

In addition, the CF$_3$CFCN radical has three other fragmentation channels: CF$_3$CFCN$\longrightarrow$ CF$_3$ + CFCN, CF$_3$CFCN$\longrightarrow$CF$_3$CF + CN, and CF$_3$CFCN$\longrightarrow$F + CF$_3$CCN, with energy barriers of 91.6 kcal/mol, 116.0 kcal/mol, and 121.8 kcal/mol, respectively. The dissociation products CF$_3$CF and CF$_3$CCN further undergo isomerization or dissociation. From Figure 2.6, it can be observed that the CF$_3$CF carbene can overcome a barrier of 37.5 kcal/mol via transition state TS4 to generate the product C$_2$F$_4$ or undergo direct fragmentation into CF$_3$ + CF without a barrier. Transition state TS4 is a first-order saddle point with a virtual frequency of 501.10i. The energy of the product C$_2$F$_4$ is 37.7 kcal/mol lower than that of CF$_3$CF, while the energy of the product CF$_3$ + CF is 69.5 kcal/mol higher than that of CF$_3$CF. Since the energy required for the fragmentation reaction CF$_3$CF$\longrightarrow$CF$_3$ + CF is higher than the barrier for isomerization to C$_2$F$_4$, the isomerization reaction CF$_3$CF$\longrightarrow$C$_2$F$_4$ is the primary pathway in the decomposition of CF$_3$CF.

Similarly, the carbene CF$_3$CCN, generated through fragmentation, has an important isomerization pathway. It proceeds through transition state TS5 to form CF$_2$CFCN with a barrier height of 23.1 kcal/mol. Transition state TS5 is a first-order saddle point with a virtual frequency of 430.77i. The energy of the product CF$_2$CFCN is 52.6 kcal/mol lower than that of the reactant CF$_3$CCN, making this reaction thermodynamically feasible. IRC calculations confirm that TS5 indeed connects CF$_3$CCN and CF$_2$CFCN.

2.4.2.4 Isomerization and Dissociation of (CF$_3$)$_2$CCN

Figure 2.9 illustrates the potential energy surface obtained at the CCSD(T)/cc-pVTZ// M06-2X/6-311G(d,p) level for the isomerization and dissociation reactions of the (CF$_3$)$_2$C(CN) radical. The (CF$_3$)$_2$C(CN) radical can undergo direct cleavage of the C—CN bond and the C—CF$_3$ bond, leading to the fragments (CF$_3$)$_2$C + CN and CF$_3$CCN + CF$_3$, with energy barriers 142.4 kcal/mol and 109.2 kcal/mol higher than (CF$_3$)$_2$C(CN), respectively. The fragmentation products (CF$_3$)$_2$C and CF$_3$CCN further undergo isomerization or dissociation. The isomerization of (CF$_3$)$_2$C can overcome a barrier of 14.1 kcal/mol, forming the intermediate IM1 (C$_3$F$_6$)

through transition state TS6 or undergoing barrierless direct cleavage into $CCF_3 + CF_3$. The energy of the intermediate IM1 (C_3F_6) is 62.5 kcal/mol lower than that of $(CF_3)_2C$, while the energy of the fragmentation products $CCF_3 + CF_3$ is 84.2 kcal/mol higher than that of $(CF_3)_2C$. Overall, the isomerization reaction $(CF_3)_2C \longrightarrow$ IM1 (C_3F_6) is thermodynamically more favorable. Additionally, the low barrier (14.1 kcal/mol) for isomerization makes the carbene $(CF_3)_2C$'s isomerization to produce IM1 (C_3F_6) kinetically feasible.

From the intermediate IM1 (C_3F_6), there are four reaction pathways. The first and second pathways involve direct cleavage of the C—C bond, leading to the products $CF_3 + CF_2CF$ and $CF_2 + CF_3C$, with energy barriers 112.0 kcal/mol and 107.3 kcal/mol higher than IM1 (C_3F_6), respectively.

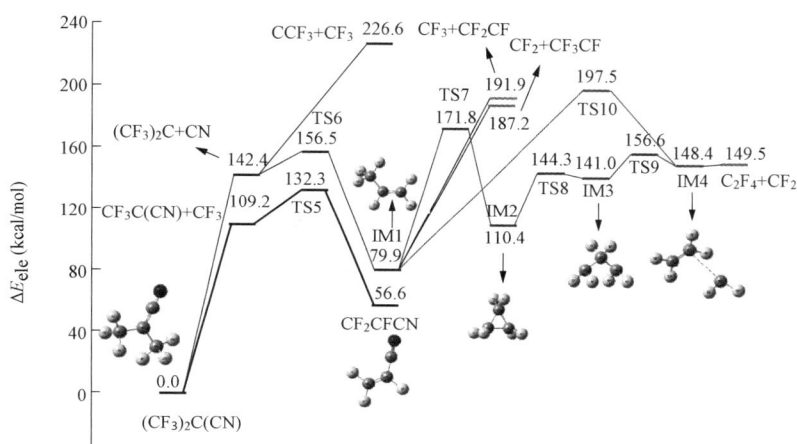

Figure 2.9   The potential energy surface of the isomerization and dissociation of $(CF_3)_2C(CN)$ radical based on the relative energies at CCSD(T)/cc-pVTZ//M06-2X/6-311G(d,p) level

The third pathway involves several transition states and intermediates that isomerize to the complex $C_2F_4 + CF_2$. Firstly, IM1 (C_3F_6) undergoes isomerization to IM2 (c-C_3F_6) through transition state TS7, overcoming a barrier of 91.9 kcal/mol. In this process, the F atom on the CF_3 group in IM1 (C_3F_6) migrates toward the central C atom, reducing the $\angle$C—C—C angle from the original 126.90° to 60.00°, bringing the two C atoms on each side closer until IM2 (c-C_3F_6) is formed in a cyclic structure. The energy of the product IM2 (c-C_3F_6) is 30.5 kcal/mol higher than that of IM1 (C_3F_6). Secondly, IM2 (c-C_3F_6) can overcome a barrier of 33.9 kcal/mol, isomerizing through transition state TS8 to form IM3 ($CF_2CF_2CF_2$ biradical). The energy of the product IM3 is 30.6 kcal/mol higher than that of IM2. The stability of the biradical $CF_2CF_2CF_2$ was confirmed by optimizing the geometry through a symmetry-breaking approach at the UM06-2X/6-311G(d,p) level and subsequently conducting stability tests using the unrestricted openshell wavefunction. The isomerization of IM2 (c-C_3F_6) to IM3 ($CF_2CF_2CF_2$ biradical) involves breaking the three-membered ring, with the two end C atoms gradually moving apart, as shown in Figure 2.2. The angle formed by the three C atoms in transition state TS8 increases from the original $60.00°$ to $111.25°$, and the distance between C(2) and C(3) increases from 1.514 Å to 2.490 Å. Transition state TS8 belongs to the C2v point group, and is a first-order saddle point with a virtual frequency of $-327.53i$, and IRC calculations confirm its connection between intermediates IM2 and IM3. Thirdly, the intermediate IM3 ($CF_2CF_2CF_2$ biradical) can overcome a barrier of 15.6 kcal/mol, isomerizing through transition state TS9 to form IM4 (complex $C_2F_4 + CF_2$). The energy of the product IM4 (complex $C_2F_4 + CF_2$) is 7.4 kcal/mol higher than that of IM3. Transition state TS9, as shown in Figure 2.9, features a decrease in the angle formed by the three C atoms from $111.25°$ to $96.45°$, indicating the impending rupture of the C—C bond, with the length of the C—C bond stretching from 1.508 Å to 1.932 Å. Transition state TS9 belongs to the Cs point group, with a virtual frequency of $393.01i$. The generated IM4 (complex $C_2F_4 + CF_2$) can dissociate without a barrier into C_2F_4 molecule and CF_2 radical, with the energy of monomers C_2F_4 and CF_2 being 1.1 kcal/mol higher than the complex $C_2F_4 + CF_2$.

2.4.2.5 Isomerization and Dissociation of $(CF_3)_2CF$

Figure 2.10 depicts the potential energy surface for the isomerization and dissociation reactions of the $(CF_3)_2CF$ radical at the CCSD(T)/cc-pVTZ//M06-2X/6-311G(d,p) level. The decomposition channels of the $(CF_3)_2CF$ radical include one isomerization channel and two cleavage channels.

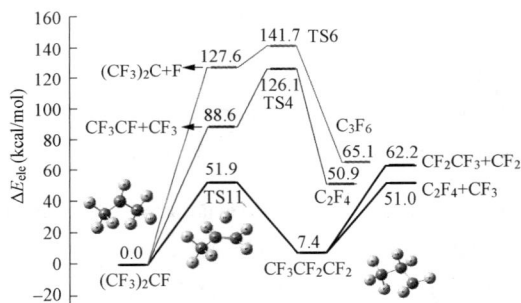

Figure 2.10   The potential energy surface of the isomerization and dissociation of $(CF_3)_2CF$ radical based on the relative energies at CCSD(T)/cc-pVTZ//M06-2X/6-311G(d,p) level

The $(CF_3)_2CF$ radical can undergo isomerization to form the $CF_3CF_2CF_2$ radical via transition state TS11 with a barrier of 51.9 kcal/mol. The energy of the $CF_3CF_2CF_2$ radical is 7.4 kcal/mol higher than that of $(CF_3)_2CF$. The $CF_3CF_2CF_2$ radical can further cleave into $C_2F_4 + CF_3$ and $CF_3CF_2 + CF_2$, with energy barriers 43.6 kcal/mol and 54.8 kcal/mol higher than the $CF_3CF_2CF_2$ radical, respectively.

Alternatively, the $(CF_3)_2CF$ radical can directly cleave through C—C and C—F bond breaking into $CF_3CF + CF_3$ and $(CF_3)_2C + F$, with energy barriers 88.6 kcal/mol and 127.6 kcal/mol higher than $(CF_3)_2CF$, respectively. Compared to the isomerization channel $(CF_3)_2CF \longrightarrow$

CF$_3$CF$_2$CF$_2$, these two cleavage channels have higher energy barriers and are not advantageous kinetically.

2.4.2.6　By-Products Formation of C$_4$F$_7$N

Under the influence of an electric field or at high temperatures, the initial breakdown of C$_4$F$_7$N leads to the formation of free radicals such as CF$_3$, F, CN, CF$_3$CFCN, (CF$_3$)$_2$CF, and (CF$_3$)$_2$C(CN). These free radicals can interact with each other to form stable compounds. Alternatively, they may undergo further fragmentation to generate small molecular free radicals or intermediates. Subsequently, these radicals and intermediates resulting from deep fragmentation can interact with each other to form stable products. The stable products arising from the decomposition of C$_4$F$_7$N may include CF$_4$, C$_2$F$_6$, C$_3$F$_8$, C$_3$F$_6$, i-C$_4$F$_{10}$, CF$_3$CN, CNCN, C$_2$F$_5$CN, CF$_2$CFCN, 2-C$_4$F$_8$, and 2-C$_4$F$_6$.

Figure 2.11 provides a theoretically analyzed pathway diagram for the formation of stable products. In the figure, black arrows represent decomposition reactions, while red arrows denote pathways leading to product formation. Intermediate free radicals or intermediates produced during these processes are labeled in black, and the resulting products are labeled in red.

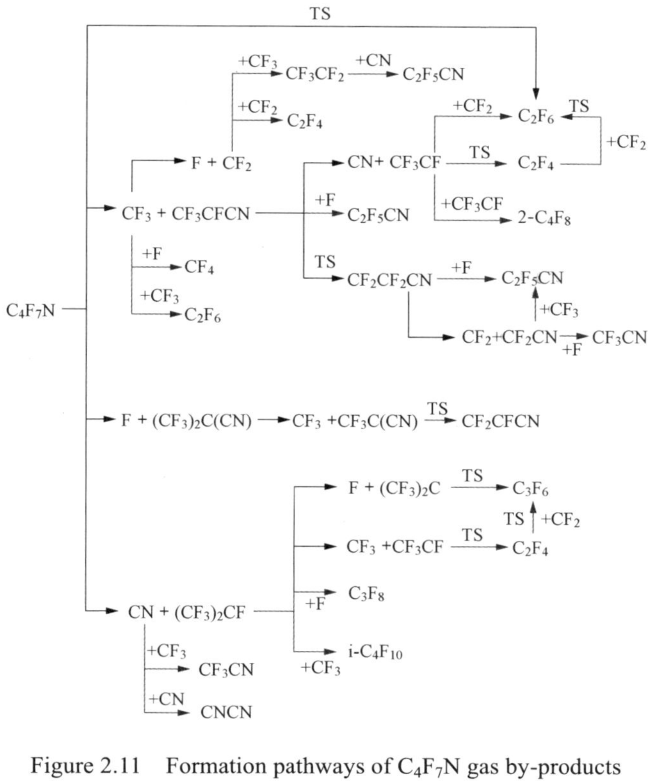

Figure 2.11　Formation pathways of C$_4$F$_7$N gas by-products

Table 2.3 presents the enthalpy changes and free energy changes for these reaction pathways under standard conditions, as calculated at the CCSD(T)/cc-pVTZ//M06-2X/6-311G(d,p)

level. It can be observed that the free energy changes for these free radical coupling reactions are all negative, indicating that these reactions can occur spontaneously. Additionally, the enthalpy changes for these reactions are also negative, indicating that the processes leading to the formation of these products are exothermic. CF_3 and F free radicals can combine to produce CF_4, releasing 128.4 kcal/mol of heat to the surroundings. The combination of CF_3 free radicals produces the product C_2F_6, with an exothermic heat release of 97.1 kcal/mol. The $(CF_3)_2CF$ free radical can either combine with F free radicals to form the stable product C_3F_8, releasing 115.5 kcal/mol of heat to the surroundings, or combine with CF_3 free radicals to generate i-C_4F_{10}, releasing 90.2 kcal/mol of heat to the surroundings. The combination of CN free radicals directly forms CNCN, releasing 137.2 kcal/mol of heat to the surroundings.

Table 2.3   Gibbs free energy and enthalpy changes of decomposition by-product formation process at CCSD (T)/cc-pVTZ level (kcal/mol)

Reactions	$\Delta G_r^{298\,K}$	$\Delta H_r^{298\,K}$
$CF_3 + F \longrightarrow CF_4$	−117.5	−128.4
$CF_3 + CF_3 \longrightarrow C_2F_6$	−83.2	−97.1
$(CF_3)_2 CF + F \longrightarrow C_3F_8$	−104.8	−115.5
$(CF_3)_2 CF + CF_3 \longrightarrow$ i-C_4F_{10}	−74.6	−90.2
$CN + CN \longrightarrow CNCN$	−125.6	−137.2
$CF_3 + CN \longrightarrow CF_3CN$	−101.0	−112.9
$CF_2CN + F \longrightarrow CF_3CN$	−106.2	−117.1
$CF_3CFCN + F \longrightarrow C_2F_5CN$	−95.1	−105.7
$CF_2CFCN_2 + F \longrightarrow C_2F_5CN$	−113.5	−123.8
$CF_3CFCN_2 + CN_2F_5CN$	−98.8	−111.0
$CF_2CN + CF_3 \longrightarrow C_2F_5CN$	−74.9	−88.7
$CF_2 + CF \longrightarrow C_2F_4$	−52.5	−65.3
$CF_3CF + CF_2 \longrightarrow C_3F_6$	−90.1	−103.3

The combination of CF_3CFCN and F free radicals produced in the initial cleavage of C_4F_7N results in the formation of C_2F_5CN, releasing 105.7 kcal/mol of heat. The CF_3CFCN free radicals first isomerize to CF_2CF_2CN, and then CF_2CF_2CN can combine with F free radicals to also generate C_2F_5CN, releasing 123.8 kcal/mol of heat. As shown in Figure 2.6, CF_3CFCN undergoes isomerization to produce CF_2CF_2CN, which further decomposes to generate CF_2CN. CF_2CN can combine with CF_3 free radicals to produce C_2F_5CN. However, the generation of CF_2CN involves multiple steps of secondary reactions and requires high energy. In contrast, the direct combination of CF_3CFCN and F free radicals from the cleavage of C_4F_7N can easily form C_2F_5CN, and the energy required for this process is the lowest among the four cleavage reaction channels. Therefore, the free radical coupling reaction

CF₃CFCN + F——→C₂F₅CN is the main pathway for C₂F₅CN production.

There are four main pathways for the formation of C₂F₄. (1) CF₃ free radicals further cleave under high temperature or electric field to produce CF₂ and F free radicals, and the generated CF₂ radicals combine to form C₂F₄, with an enthalpy change of −65.3 kcal/mol. (2) C₄F₇N initially cleaves to generate CF₃CFCN free radicals, which then undergo further cleavage to produce CF₃CF carbene, ultimately isomerizing to form C₂F₄. (3) CF₃CFCN free radicals isomerize to CF₂CF₂N, which then decomposes to generate C₂F₄. (4) C₃F₆ can overcome a relatively high energy barrier and undergo multiple transition states and intermediate isomerization to form C₂F₄. However, due to the high energy barrier of the isomerization reaction and the higher energy of the product compared to the reactant, the thermodynamically stable existence of the C₂F₄ + CF₂ complex is unlikely, and it is highly likely to undergo a reversible reaction to produce C₃F₆. Therefore, the possibility of the C₃F₆ isomerization reaction to generate C₂F₄ is relatively low.

There are four main pathways for the generation of C₃F₆: (1) C₄F₇N molecules overcome a 98.0 kcal/mol energy barrier to directly isomerize into C₃F₆, which is the primary pathway for C₄F₇N to generate C₃F₆. (2) The CF₃CF carbene generated in the cleavage path of C₄F₇N combines with CF₂ carbene to produce C₃F₆. (3) As shown in Figure 2.7, the unstable (CF₃)₂C carbene generated in the cleavage path of C₄F₇N can overcome a lower energy barrier (14.1 kcal/mol) to isomerize into C₃F₆. (4) C₃F₆ can undergo isomerization to generate C₂F₄ and CF₂ carbene. However, due to the high energy barrier and the relatively high energy of the products C₂F₄ + CF₂, a reversible reaction is likely to occur. Therefore, C₃F₆ can be generated by the isomerization reaction of C₂F₄ and CF₂ carbene. We calculated the reaction potential energy curve for the generation of C₃F₆ from C₂F₄ molecules and CF₂ carbene, considering zero-point energy correction with a correction factor of 0.97, as shown in Figure 2.12.

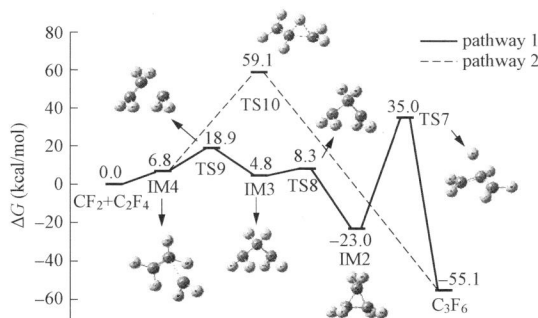

Figure 2.12 The potential energy surface of the reaction C₂F₄+CF₂——→C₃F₆ based on the Gibbs free energy at CCSD(T)/cc-pVTZ//M06-2X/6-311G(d,p) level

C₂F₄ and CF₂ carbene absorb a certain amount of heat to first form the complex C₂F₄ + CF₂, accompanied by an energy rise of 6.8 kcal/mol. Then, through multiple transition states and intermediate formations, C₃F₆ is generated. From Figure 2.12, it can be seen that there are two pathways for the isomerization of C_2F_4 and CF_2 into C_3F_6:

Pathway 1: $C_2F_4 + CF_2 \longrightarrow IM4 \longrightarrow TS9 \longrightarrow IM3 \longrightarrow TS8 \longrightarrow IM2 \longrightarrow TS7 \longrightarrow C_3F_6$

Pathway 2: $C_2F_4 + CF_2 \longrightarrow IM4 \longrightarrow TS10 \longrightarrow C_3F_6$

The critical step in Pathway 1 is the process from IM2 (c-C_3F_6) to TS7, and the free energy barrier for the critical step is 58.0 kcal/mol. The free energy barrier for the critical step in Pathway 2 is 52.3 kcal/mol. Clearly, the reaction pathway in which the complex $C_2F_4 + CF_2$ generates C_3F_6 through transition state TS10 has a lower energy barrier, making this pathway more favorable kinetically. Moreover, C_3F_6 has lower energy than $C_2F_4 + CF_2$ by 55.1 kcal/mol, making this isomerization reaction thermodynamically feasible. However, the free energy change for the formation of IM4 (complex $C_2F_4 + CF_2$) from C_2F_4 and CF_2 carbene is greater than zero, indicating that the combination of the two is not easy, and a certain amount of energy is required from the surroundings to form the complex.

2.4.3 Decomposition By-product Measurement for C_4F_7N

2.4.3.1 Decomposition By-Product Measurement Based on GC-MS

Prior to the experiment, insulation material C_4F_7N was analyzed using GC/MS, revealing the presence of an unknown component, as illustrated in Figure 2.13 for the total ion chromatogram.

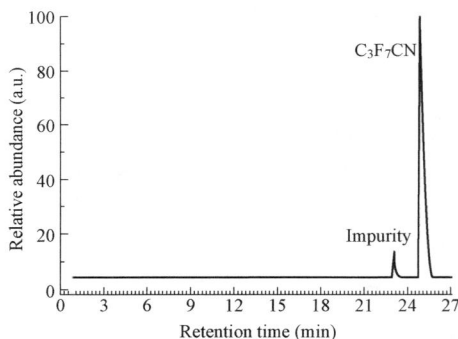

Figure 2.13 Total ion chromatogram of C_4F_7N before experiment

Under the action of the EI ion source in mass spectrometry, compounds generate characteristic fragment ions. The production of these fragment ions follows a pattern, closely related to the functional groups. After GC/MS analysis, it was found that the impurity with a retention time of 22.76 min contained fragment ions and their relative abundances: $m/z=69$ (CF_3^+, 100%), $m/z=151$ ($C_3HF_6^+$, 22%), $m/z=82$ (CF_2CHF^+, 18%), $m/z=51$ (CHF_2^+, 12%), $m/z=31$(CF^+, 4%), and $m/z=101$ (CF_3CHF^+, 2%). These characteristic fragment ions and their relative abundances match those of the standard gas heptafluoropropane CF_3CHFCF_3 (C_3HF_7), and the impurity has the same retention time as C_3HF_7 under the same chromatographic detection conditions. Therefore, we speculate that the impurity in C_4F_7N is heptafluoropropane C_3HF_7 (HFC-227ea). During mass spectrometric detection, the molecular ion peak $m/z=170$ ($C_3HF_7^+$) of heptafluoropropane C_3HF_7 was not detected, possibly due to complete fragmentation of C_3HF_7 under the 70 eV EI ion source bombardment. Through external standard quantitative calibration, it was determined that the volume fraction of the impurity C_3HF_7 in C_4F_7N is approximately 1,000 μL/L.

Through a series of corona discharge experiments, we observed that when the mixing ratio $k(C_4F_7N)$ is higher, the content of discharge decomposition components is also relatively high. However, the types of decomposition components are independent of the mixing ratio of C_4F_7N. Therefore, to better analyze the types of discharge decomposition components, we chose a mixing ratio of $k(C_4F_7N) = 15\%$ for the experiment, applying a voltage of 35 kV with a needle-plate spacing of 10 mm.

Figure 2.14 shows the chromatographic detection results for 15% C_4F_7N/85% CO_2, 15% C_4F_7N/85% N_2, and 15% C_4F_7N/85% air after corona discharge.[13] It can be observed that the chromatographic peaks exhibit good separation. By comparing the peak retention times of decomposition components under the same chromatographic detection method with those of standard gas chromatography peaks and consulting the NIST mass spectrometry database, we identified the decomposition products as CO, CO_2, CF_4, C_2F_6, C_3F_6, and C_3F_8. For decomposition components without standard gases, we obtained corresponding matching substances from the Total Ion Chromatogram (TIC) obtained under the full scan mode, cross-referenced with the NIST mass spectrometry database, and determined them based on the matching degree and peak patterns.

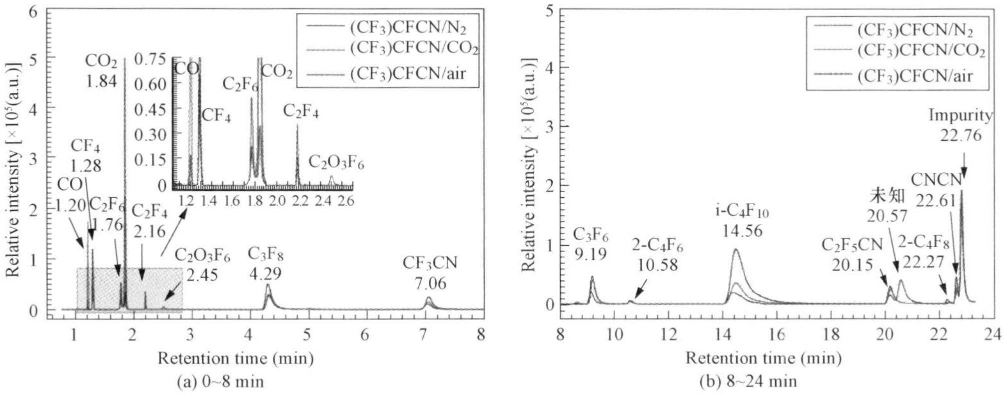

Figure 2.14　Total ion chromatogram of decomposition by-products of
C_4F_7N mixtures under corona discharge

Table 2.4 presents the mass spectrometry data for decomposition components without standard gases. After the discharge of C_4F_7N mixed gas, the predominant decomposition components are perfluorocarbons (PFCs). Under electron impact ionization (EI source bombardment), these PFCs typically yield fragment ions at $m/z=69$ (CF_3^+), and there is often a repetition of other fragment ions in the decomposition process. To identify these decomposition components of perfluorocarbons, we employed two methods: first, comparison with the NIST mass spectral database, and second, reference to the elution order of homologous compounds in chromatography, which is often based on the number of carbon atoms.

Table 2.4 **The mass spectrum fragmentations and their relative abundance of by-products without standard gases**

Retention time /min	Mass-to-charge ratio and relative abundance	Compounds
2.16	31(CF^+, 100%), 81 ($C_2F_3^+$, 73%), 100($C_2F_4^+$, 43%), 50(CF_2^+, 30%)	C_2F_4
2.45	69(CF_3^+, 100%), 47(COF^+, 14%), 66(COF_2^+, 7%), 50(CF_2^+, 3%)	$C_2O_3F_6$
7.06	69(CF_3^+, 100%), 76(CF_2CN^+, 46%), 50(CF_2^+, 25%), 31(CF^+, 20%)	CF_3CN
10.58	93($C_3F_3^+$, 100%), 143($C_4F_5^+$, 77%), 162($C_4F_6^+$, 42%), 31(CF^+, 40%), 69(CF_3^+, 39%)	$CF_3C\equiv CCF_3$
14.56	69(CF_3^+, 100%), 131($C_3F_5^+$, 16%), 219($C_4F_9^+$, 8%), 100($C_2F_4^+$, 7%), 150($C_3F_6^+$, 6%), 31(CF^+, 3%)	$(CF_3)_3 CF$
20.15	69(CF_3^+, 100%), 76(CF_2CN^+, 40%), 126($C_3F_4N^+$, 20%), 31(CF^+, 16%), 50(CF_2^+, 6%), 119($C_2F_5^+$, 3%), 100($C_2F_4^+$, 2%)	C_2F_5CN
20.57	69(CF_3^+, 100%), 169($C_3F_7^+$, 13%), 97(CF_2CFO^+, 4%), 31(CF^+, 3%), 50(CF_2^+, 2%), 47(COF^+, 2%), 119($C_2F_5^+$, 1%)	Unknown
22.27	131($C_3F_5^+$, 100%), 69(CF_3^+, 45%), 181($C_4F_7^+$, 43%) 200($C_4F_8^+$, 25%)	$CF_3CF = CFCF_3$
22.61	52($CNCN^+$, 100%), 26(CN^+, 7%)	CNCN

In Table 2.4, the substance at a retention time of 2.16 minutes has a mass spectrum that matches 93% with C_2F_4 in the NIST mass spectral library. Therefore, this substance is identified as tetrafluoroethylene (C_2F_4), as illustrated in Figure 2.15(a). The substance at a retention time of 2.45 minutes has a mass spectrum that matches 87% with $C_2O_3F_6$ in the NIST mass spectral library. Consequently, it is identified as $C_2O_3F_6$, and its mass spectrum is shown in Figure 2.15(b). The substance at a retention time of 7.06 minutes has a mass spectrum that matches 97% with CF_3CN in the NIST mass spectral library. Hence, it is identified as CF_3CN, and its mass spectrum is presented in Figure 2.15(c). The substance at a retention time of 10.58 minutes has a mass spectrum that matches 87% with hexafluoro-2-butyne ($CF_3C\equiv CCF_3$) in the NIST mass spectral library. Therefore, it is identified as $CF_3C\equiv CCF_3$, referred to as $2\text{-}C_4F_6$, and its mass spectrum is depicted in Figure 2.15(d). The substance at a retention time of 22.27 minutes has a mass spectrum detailed in Table 2.4, where $m/z=200$ is the molecular ion peak of this substance, all of which are characteristic fragment ions generated by the EI source bombardment of C_4F_8. Compared with the NIST mass spectral library, it was found that the match with $CF_3CF\equiv CFCF_3$ is 85%. Consequently, this substance is determined to be perfluoro-2-butene ($CF_3CF\equiv CFCF_3$), abbreviated as $2\text{-}C_4F_8$, and its mass spectrum is shown in Figure 2.15(e). The substance at a retention time of 22.61

minutes has a mass spectrum with a 93% match to CNCN in the NIST mass spectral library. Thus, it is identified as CNCN, and its mass spectrum is presented in Figure 2.15(f).

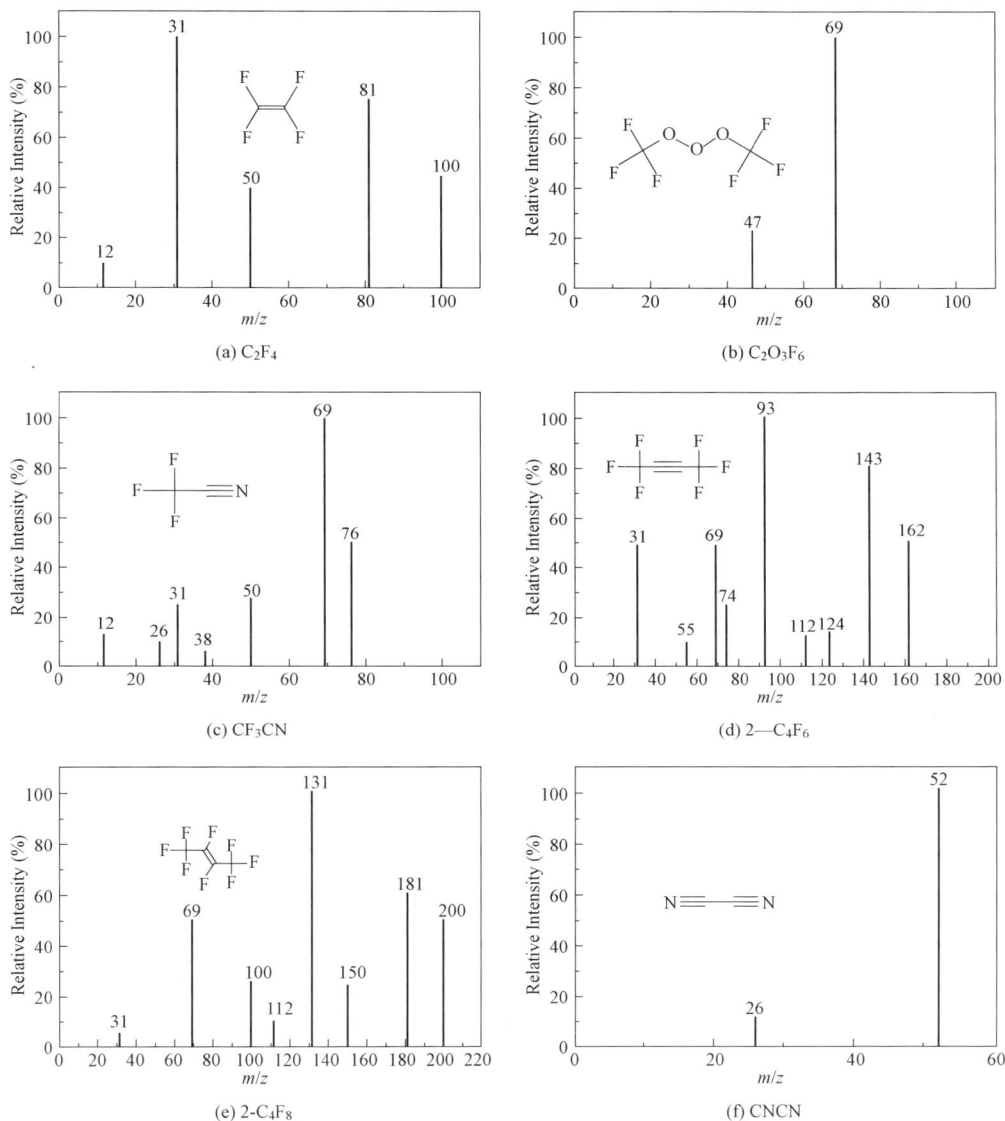

Figure 2.15　Mass spectra of decomposition by-products of C_4F_7N mixtures

The unknown substance at a retention time of 14.56 minutes has a mass spectrum with peaks at m/z=69(CF_3^+, 100%), 131($C_3F_5^+$, 16%), 219($C_4F_9^+$, 8%), 100($C_2F_4^+$, 7%), 150($C_3F_6^+$, 6%), and 31(CF^+, 3%). Although the mass spectrum of this compound is not found in the NIST mass spectral library, the mass spectra of $CF_3CF_2CF_2CF_3$ (n-C_4F_{10}) are available. Based on the fragmentation pattern of this compound under EI source bombardment, these fragments are indicative of the characteristic fragment ions produced by n-C_4F_{10} at 70 eV. However, the mass spectrum of this substance differs significantly from that of n-C_4F_{10}, as the unknown substance at a retention time of 14.56 minutes does not produce the m/z=119 ($C_2F_5^+$) fragment observed in the spectrum of n-C_4F_{10}. Considering these fragment ions produced after electron impact and applying the principles of mass spectral resolution, we hypothesize that the substance at a retention time of 14.56 minutes is an isomer of n-C_4F_{10}, denoted as $(CF_3)_3CF$.

The unknown substance at a retention time of 20.15 minutes has a mass spectrum that is not found in the NIST mass spectral library. Based on the acquired mass spectral data and the principles of mass spectral resolution, this substance has fragment ions at m/z=69, 76, 126, 31, 50, 119, and 100. Among these, 69, 31, 50, and 119 are typical fragment ions of perfluorocarbons, namely CF^+, CF_2^+, CF_3^+, and $C_2F_5^+$. Additionally, according to the decomposition pattern of C_4F_7N, m/z=76 and 126 are identified as CF_2CN^+ and $C_3F_4N^+$ fragment ions, respectively. These fragments are typical characteristic ions formed by C_3F_5N after electron impact. Therefore, it is inferred that the chemical formula of this substance is C_3F_5N. However, the mass spectrum of this substance differs significantly from that of 2,3-difluoro-2-trifluoromethyl-2H-azirine in the NIST mass spectral library. The substance at a retention time of 20.15 minutes lacks the m/z=145 fragment ion, and the peaks at 69 (CF_3^+, 100%), 76 (CF_2CN^+, 40%), and 126 ($C_3F_4N^+$, 20%) have relatively high relative abundances. Conversely, in the NIST mass spectral library, 2,3-difluoro-2-trifluoromethyl-2H-azirine exhibits high relative abundances in the m/z=69(CF_3^+), 31(CF^+), and 119($CF_3CF_2^+$) peaks, which do not align with the mass spectrum of the substance at a retention time of 20.15 minutes. Therefore, it is determined that this substance is not 2,3- difluoro-2-trifluoromethyl-2H-azirine. Based on the decomposition mechanism of C_4F_7N, it is concluded that this substance is pentafluoropropionitrile (CF_3CF_2CN).

After querying the NIST mass spectral database, no spectrum corresponding to the compound with a retention time of 22.57 minutes was found. Based on the obtained mass spectral data, a preliminary identification of the unknown substance was conducted. The mass spectrum of this substance includes peaks at m/z=69(CF_3^+, 100%), 169($C_3F_7^+$, 13%), 97(CF_2CFO^+, 4%), 31(CF^+, 3%), 50(CF_2^+, 2%), 47(COF^+, 2%), and 119($C_2F_5^+$, 1%). As this substance is exclusive to the decomposition components of C_4F_7N/air mixed gas during discharge and its mass spectrum features peaks at m/z=47 and 97, we infer the presence of oxygen in this substance.

From Figure 2.14, it is evident that the decomposition components of C_4F_7N /air mixed gas are the most diverse. Compounds such as $C_2O_3F_6$ and the unknown substance at 22.57 minutes are exclusively present in the decomposition components of C_4F_7N /air discharge, indicating that the presence of O_2 significantly increases the probability of oxygen-containing compound formation.

In summary, the discharge of C_4F_7N mixed gas primarily produces CO, CO_2, CF_4, C_2F_6, C_3F_8, C_3F_6, i-C_4F_{10}, 2-C_4F_8, 2-C_4F_6, CF_3CN, C_2F_5CN, and CNCN. The majority of these decomposition components are associated with the decomposition of C_4F_7N.

2.4.3.2 Correlation between Buffer Gases and Decomposition By-Products

Three types of C_4F_7N binary gas mixtures exhibit two main categories of gas decomposition products after corona discharge: first, substances strongly correlated with C_4F_7N, where the free radicals CF_3, F, and CN generated from the decomposition of C_4F_7N undergo coupling reactions to form perfluorinated hydrocarbons and nitrile compounds; second, substances strongly correlated with the buffer gas, such as CO and CO_2 gases. By analyzing the changes in the content of perfluorinated hydrocarbons, nitrile gases, CO, and CO_2 generated from the three types of C_4F_7N binary gas mixtures, the correlation between decomposition components and buffer gases is explored. In this context, PFCs represent perfluorinated hydrocarbon gases CF_4, C_2F_4, C_2F_6, C_3F_6, C_3F_8, 2-C_4F_6, 2-C_4F_8, and i-C_4F_{10}; R-CN represents nitrile gases CNCN, CF_3CN, and C_2F_5CN, where R represents CN, CF_3, and C_2F_5 groups.

Results from the study of SF_6 decomposition characteristics indicate that the content of most decomposition products decreases with increasing pressure. To better quantify the content of gas components, corona discharge experiments were conducted with C_4F_7N mixed with CO_2, N_2, and air binary gases under an absolute pressure of 0.2 MPa. The applied voltage was 35 kV, the needle electrode was made of stainless steel, the plate electrode was made of brass, and the needle-plate electrode spacing was 10 mm. The C_4F_7N mixing ratios were 10%, 15%, and 20%. Each mixing ratio was repeated three times, and the content was averaged.

The external standard method was used to determine the content of six decomposition components: CO, CO_2, CF_4, C_2F_6, C_3F_6, and C_3F_8. Additionally, due to the positive correlation between gas content and chromatographic peak area, the changes in the content of decomposition components C_2F_4, CF_3CN, 2-C_4F_6, 2-C_4F_8, i-C_4F_{10}, C_2F_5CN, and CNCN were analyzed, for which standard gases could not be configured, based on the peak area of specific fragment ions, with respective mass-to-charge ratios of m/z = 81, 69, 93, 131, 69, 76, 52. The content of 2-C_4F_6 and 2-C_4F_8 gases was relatively low, with a large integration error, and these substances are not analyzed in this chapter.

(1) Perfluorinated Hydrocarbons (PFCs) and Nitrile Gases

The peak areas of PFCs and R-CN gases were normalized to facilitate the discussion of the changes in the content of different decomposition components under various C_4F_7N mixing ratios. The chromatographic peak areas of PFCs and R-CN gases after discharging 10% C_4F_7N/90% N_2 were set as the baseline values, and the relative content of decomposition components under different buffer gases and mixing ratios was expressed as the ratio of the peak area of each component to the baseline value.

Figure 2.16 shows the trend of the total relative content of PFCs and R-CN gases in the decomposition components of the three C_4F_7N gas mixtures with varying C_4F_7N mixing ratios. It can be observed that the content of PFCs and R-CN gases in the decomposition components of C_4F_7N/CO_2 is the lowest, while that in C_4F_7N/N_2 gas mixture is the highest. As the proportion of C_4F_7N increases, the content of PFCs and R-CN gases in the decomposition components of C_4F_7N/CO_2 and C_4F_7N/air increases. This is mainly due to the increase in the proportion of C_4F_7N, leading to an increase in the number density of C_4F_7N particles in the chamber. Consequently, the probability of electron collision-induced decomposition reactions increases, resulting in a gradual increase in the content of PFCs and R-CN gases with an increasing proportion of C_4F_7N.

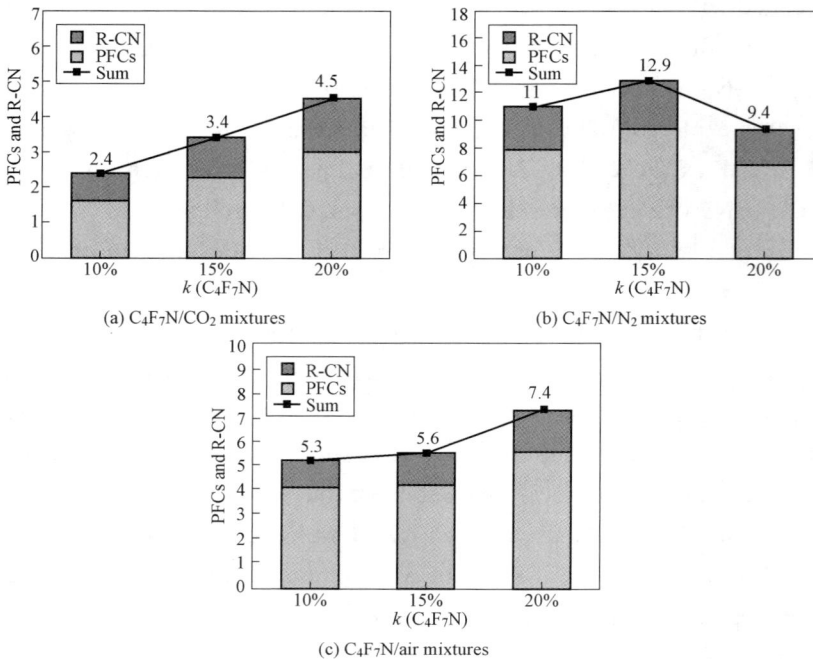

(a) C_4F_7N/CO_2 mixtures

(b) C_4F_7N/N_2 mixtures

(c) C_4F_7N/air mixtures

Figure 2.16   Total content of PFCs and R-CN under different mixing ratios

The electronegativity of the three buffer gases is in the order $O_2 > CO_2 > N_2$. Therefore, CO_2 and O_2 to some extent weaken the electron collision ionization process, reducing the degree of C_4F_7N molecular decomposition and subsequently lowering the probability of generating free radicals such as F, CN, and CF_3. As a result, the content of PFCs and R-CN gases in the decomposition components of C_4F_7N /CO_2 and C_4F_7N/air decreases with the increasing proportion of C_4F_7N. In addition, dry air contains a significant amount of O_2 molecules, which consume free radicals like CF_3 in the chamber and generate $C_2O_3F_6$. In contrast, C_4F_7N/CO_2 and C_4F_7N/N_2 decomposition components contain almost no such gas. This leads to a lower total content of PFCs and R-CN substances in the decomposition components of C_4F_7N/air compared to the C_4F_7N/N_2 gas mixture.

(2) CO_2 and CO

In the C_4F_7N/CO_2 gas mixture, CO_2 serves as a buffer gas. However, when detecting CO_2, saturation phenomena may occur due to its high content. In the decomposition components of C$_4$F$_7$N/N$_2$ gas mixture, the CO gas content is trace when gas impurities (H$_2$O, O$_2$) are present at very low levels. Therefore, in this section, we temporarily refrain from quantifying the CO$_2$ gas in the decomposition components of C$_4$F$_7$N/CO$_2$ gas mixture and the CO gas in the C$_4$F$_7$N/N$_2$ gas mixture, given the potential saturation issues and trace amounts, respectively.

Figure 2.17 illustrates the content of CO and CO$_2$ in the decomposition components after discharge for different mixing ratios of C$_4$F$_7$N/CO$_2$, C$_4$F$_7$N/N$_2$, and C$_4$F$_7$N/air. The content of CO and CO$_2$ gases in the decomposition components of C$_4$F$_7$N/air mixture is relatively high.

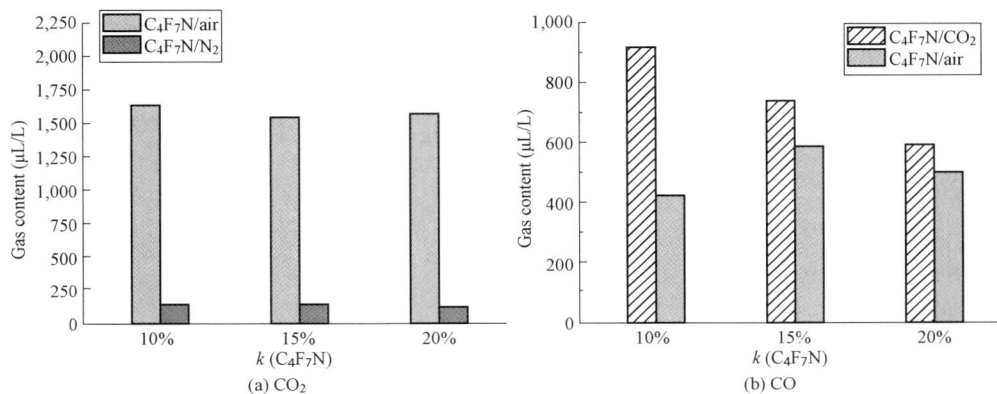

Figure 2.17   Total content of CO and CO$_2$ under different mixing ratios

This is mainly because in the buffer gas air, O$_2$, and impurity water in the chamber ionize to generate active free radicals O, H, and OH. These free radicals possess strong oxidizing properties and react with fluorocarbon gases or fluorocarbon radicals. CO is mainly derived from the oxidation reaction between fluorocarbon gases or methyl and methylene radicals and active free radicals O, H, and OH. CO$_2$ mainly originates from the oxidation reaction of CO or the hydrolysis reaction of the deep reaction product COF$_2$, as shown in Equations (2.1-2.9).

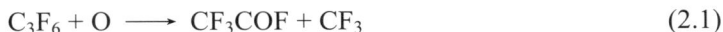

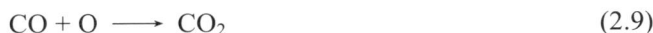

$$C_3F_6 + O \longrightarrow CF_3COF + CF_3 \tag{2.1}$$

$$CF_3COF + H \longrightarrow CF_3CO + HF \tag{2.2}$$

$$CF_3CO \longrightarrow CF_3 + CO \tag{2.3}$$

$$CF_3 + O \longrightarrow COF_2 + F \tag{2.4}$$

$$CF_2 + O \longrightarrow COF + F \tag{2.5}$$

$$COF + H \longrightarrow CO + HF \tag{2.6}$$

$$COF + OH \longrightarrow CO_2 + HF \tag{2.7}$$

$$COF_2 + H_2O \longrightarrow CO_2 + 2HF \tag{2.8}$$

$$CO + O \longrightarrow CO_2 \tag{2.9}$$

A small amount of CO$_2$ gas is generated after the discharge of C$_4$F$_7$N/N$_2$, but its content is less than one-tenth of that in C$_4$F$_7$N/air. Despite repeated gas washing and vacuum pumping operations before the experiments, trace amounts of H_2O and O_2 are inevitably present in the chamber, leading to the production of a small amount of CO_2 gas during the discharge of C_4F_7N/N_2.

In the case of C_4F_7N/CO_2 mixture, the content of CO gas produced during discharge is relatively high and gradually decreases with the increase of $k(C_4F_7N)$. CO gas mainly originates from the decomposition of the buffer gas CO_2. As the number density of C_4F_7N particles in the chamber increases with the increase of $k(C_4F_7N)$, the probability of electron collision ionization with CO_2 molecules decreases, resulting in a reduction in the content of CO gas generated during decomposition.

(3) The Total Content Comparison of Key Decomposition Components

Figure 2.18 illustrates the total content of six key decomposition components, namely CO, CO_2, CF_4, C_2F_6, C_3F_6, and C_3F_8, after discharge of C_4F_7N/CO_2, C_4F_7N/N_2, and C_4F_7N/air mixtures under different $k(C_4F_7N)$ conditions.

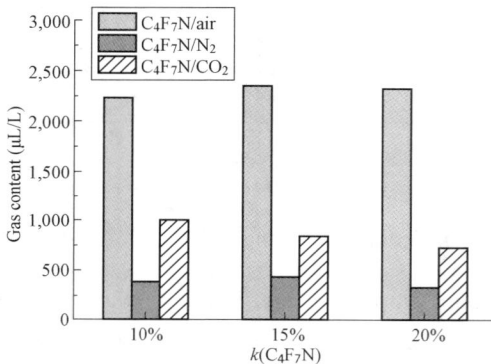

Figure 2.18   Total quantitative content of six decomposition by-products under different mixing ratios

After the discharge of C_4F_7N/air mixed gas, the content of key decomposition components is the highest, while C_4F_7N/N_2 mixed gas has the least. This is primarily because the decomposition components with the highest content in C_4F_7N mixed gas are CO and CO_2 gases, with the highest content in the decomposition products of C_4F_7N/air and the least in CO and CO_2 gases produced after C_4F_7N/N_2 discharge. As CO_2 serves as a buffer gas, when calculating the total content of decomposition components in C_4F_7N/CO_2 mixed gas discharge, CO_2 gas is not considered. The CO gas generated in the discharge of C_4F_7N/CO_2 has the largest proportion in its decomposition products, resulting in a higher content of key decomposition components compared to C_4F_7N/N_2.

The type of buffer gas has a significant impact on the content of decomposition products in C_4F_7N mixed gas.

- N_2: Although the total amount of decomposition products generated by C_4F_7N/N_2 after discharge is the least among the three C_4F_7N mixed gases, the content of perfluoroalkanes and nitrile substances produced is the highest, nearly twice that of C_4F_7N/air and 3-4 times that of C_4F_7N/CO_2. As shown in Table 3.3, nitrile gases CNCN, CF_3CN, and C_2F_5CN all have strong toxicity.
- Air: The decomposition products of C_4F_7N/air are the most diverse, producing more perfluoroalkanes and nitrile substances than C_4F_7N/CO_2, and the content of the toxic gas CO in the products is also high.

- CO$_2$: The C$_4$F$_7$N/CO$_2$ mixed gas produces the least perfluoroalkanes and nitrile substances, but the content of CO gas produced should not be ignored. The decomposition components of C$_4$F$_7$N mixed gas after discharge are mostly toxic, corrosive, and irritating. Protective measures should be taken during the operation, and good ventilation should be maintained as much as possible.

The electrical performance of C$_4$F$_7$N in binary mixed gases with CO$_2$, N$_2$, and air is similar, with CO$_2$ having the best arc extinguishing performance. Meanwhile, considering the content of decomposition products after discharge, the toxicity of decomposition products, and the content of deposits on the electrode surface, from the perspective of decomposition analysis, CO$_2$ is more suitable as a buffer gas. Compared to C$_4$F$_7$N/air and C$_4$F$_7$N/N$_2$, C$_4$F$_7$N/CO$_2$ discharge produces the least perfluoroalkanes and nitrile gases, and the toxicity of decomposition components mainly comes from CO gas. If measures are taken to reduce the CO content, the toxicity of C$_4$F$_7$N/CO$_2$ decomposition products will be significantly reduced.

2.4.4 Fault Decomposition Characteristics of SF$_6$ and C$_4$F$_7$N/CO$_2$

2.4.4.1 Decomposition Characteristics under Corona Discharge

The experiment was conducted at an absolute pressure of 200 kPa with an applied voltage of 35 kV. During the experiment, the discharge amount was approximately 300 pC, and the discharge duration was 5 hours. The ambient temperature was maintained at 25 ℃, with relative humidity ranging from 25% to 30%. Each mixing ratio was repeated three times, and the content of each decomposition component was measured and averaged. External standard quantification was employed to determine the volume fractions of CO, CF$_4$, C$_2$F$_6$, C$_3$F$_8$, and C$_3$F$_6$. For decomposition components without standard gas, normalization was performed on the chromatographic peak areas of characteristic ions. Specifically, the peak areas of each decomposition component at k(C$_4$F$_7$N) = 10% were set as the reference value of 1, and the relative content of decomposition components at other mixing ratios was obtained. Subsequently, the analysis focused on the variation in the relative content of decomposition products with mixing ratio.

Figure 2.19 illustrates the variation in the content of decomposition components in C$_4$F$_7$N/CO$_2$ mixed gas under corona discharge conditions at different mixing ratios. The content of CO gas is negatively correlated with k(C$_4$F$_7$N), while the content of other decomposition components is positively correlated with k(C$_4$F$_7$N).

A higher volume fraction of C$_4$F$_7$N in the mixed gas results in a greater number density of C$_4$F$_7$N molecules in the chamber. This, in turn, increases the probability of electron collision with C$_4$F$_7$N molecules, leading to an elevated production of free radicals such as CF$_3$, CN, F, CF$_3$CFCN, and others through free radical coupling reactions. Simultaneously, an increased volume fraction of C$_4$F$_7$N implies a reduced number density of CO$_2$ molecules in the chamber, decreasing the probability of electron collision ionization with CO$_2$ molecules and subsequently reducing the content of CO gas as the mixing ratio increases.

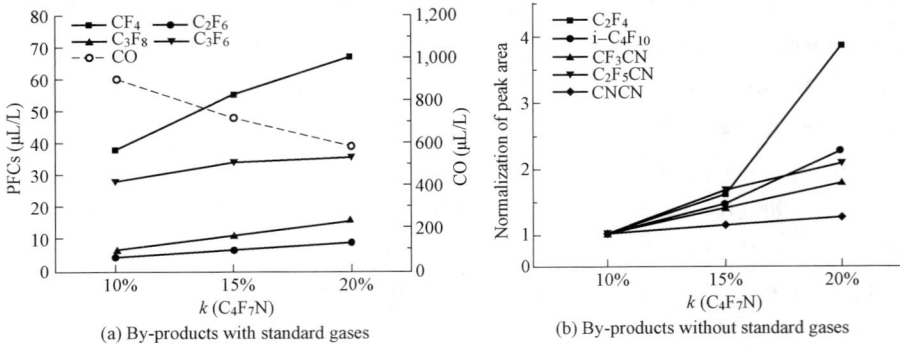

(a) By-products with standard gases

(b) By-products without standard gases

Figure 2.19   Decomposition by-product variations with mixing ratios under corona discharge

Figure 2.20 depicts the trend of decomposition component content with voltage under corona discharge experiments using pure aluminum and stainless-steel needle electrodes. As the applied voltage increases, the content of each decomposition component continues to rise. With a higher applied voltage, the corona discharge glow region and the electric field intensity E both increase, resulting in an elevated probability of electron collision ionization with C_4F_7N and CO_2 molecules. This, in turn, increases the production of free radicals such as CF_3, CN, CF_3CFCN, F, and others through free radical coupling reactions, leading to a higher content of stable products. Consequently, the content of decomposition components is positively correlated with the applied voltage.

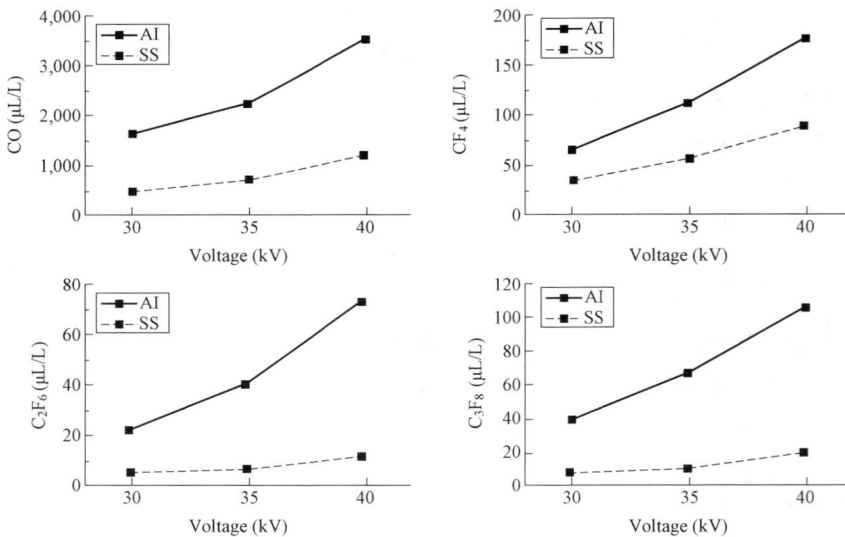

Figure 2.20   Decomposition by-product content variations with applied voltages
and needle electrode materials under corona discharge (1)

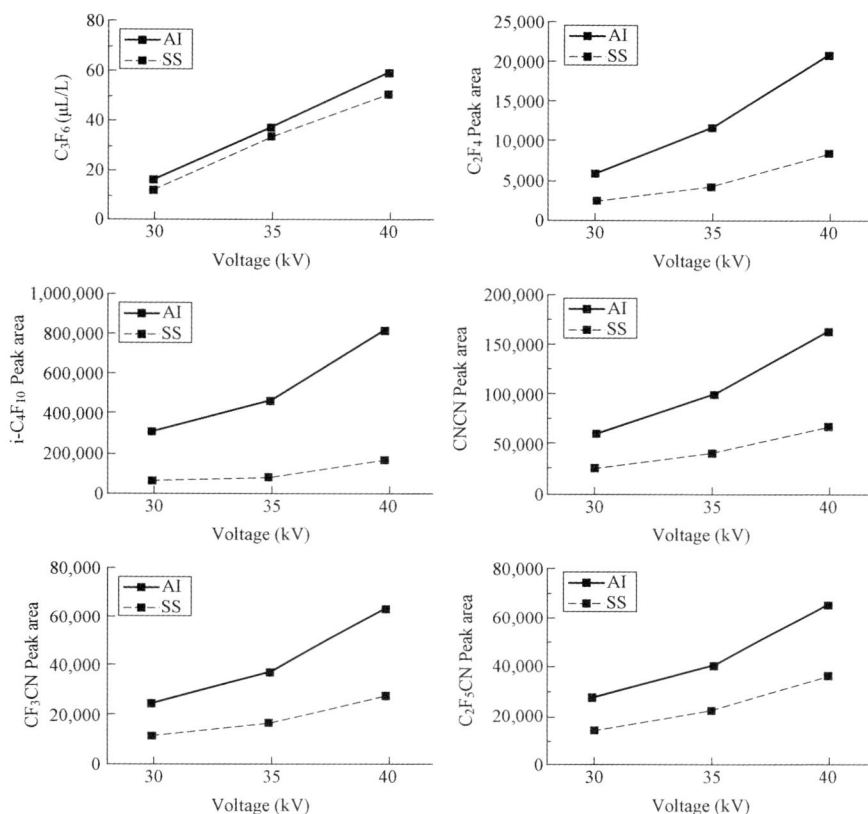

Figure 2.20   Decomposition by-product content variations with applied voltages
and needle electrode materials under corona discharge (2)

The electrode material significantly influences the generation of decomposition compo-nents. The content of decomposition components (excluding C_3F_6 gas) after corona discharge experiments with a pure aluminum needle electrode is approximately twice that after experi-ments with a stainless-steel electrode, while the C_3F_6 gas content is relatively similar between the two. This indicates that the metal material has a minimal impact on C_3F_6 gas under disch-arge conditions, whereas its influence on other products is more pronounced. After experiments with a stainless-steel electrode, the four quantifiable perfluoroalkanes are ranked in decreasing order of content as follows: $c(CF_4) > c(C_3F_6) > c(C_3F_8) > c(C_2F_6)$. On the other hand, after experiments with a pure aluminum electrode, the content of these gases is ranked as follows: $c(CF_4) > c(C_3F_8) > c(C_2F_6) > c(C_3F_6)$. According to related studies on SF_6 decomposition char-acteristics, the chemical reactivity of the metal electrode material can affect the discharge dec-omposition characteristics of SF_6 gas. The needle electrode materials used in our experiments are 304 stainless steel and pure aluminum. Stainless steel undergoes passivation treatment during processing to enhance its corrosion resistance, significantly reducing its chemical reactivity and resulting in lower content of decomposition components. In comparison, pure aluminum exhibits higher chemical reactivity, leading to a higher content of decomposition components generated during discharge.

2.4.4.2 Decomposition Characteristics under Spark Discharge

The association between various decomposed components and mixing ratios was investigated after 450 spark discharges of C_4F_7N/CO_2 gas at an absolute pressure of 200 kPa, with $k(C_4F_7N)$ values of 5%, 10%, 15%, and 20%. The experiments were conducted three times for each mixing ratio, and the content of each decomposed component was measured and averaged. For decomposed components without standard gas, such as C_2F_4, CF_3CN, CNCN, the peak areas were normalized. The peak area at $k(C_4F_7N) = 5\%$ was set as the reference value, and the relative content of decomposed components at other mixing ratios was obtained. The variations in decomposed components with mixing ratios under spark discharge conditions for C_4F_7N/CO_2 gas are shown in Figure 2.21. Under spark discharge conditions, the types of decomposed components were similar to those under corona discharge conditions. The content of CO gas decreased with increasing $k(C_4F_7N)$, while the content of perfluorocarbons (PFCs) and nitrile compounds increased with increasing $k(C_4F_7N)$, gradually reaching saturation.

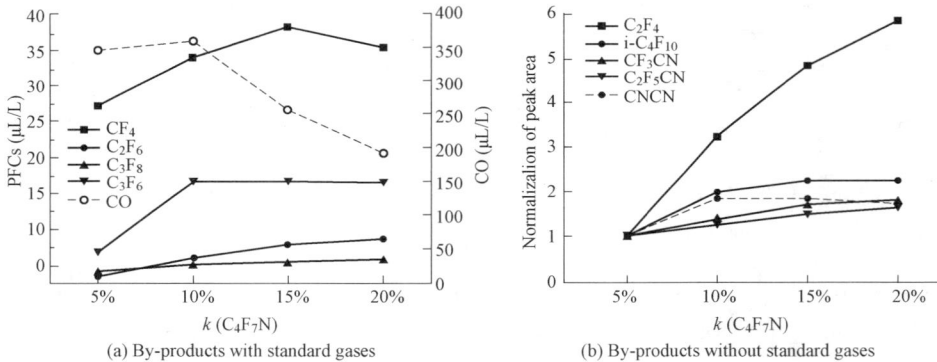

(a) By-products with standard gases   (b) By-products without standard gases

Figure 2.21   Decomposition by-product content variations with mixing ratios under spark discharge

The content of C_2F_4 gas was significantly higher than that of other decomposed components at the same mixing ratio, compared to corona discharge. Figure 2.22 illustrates the trend of the relative content of C_4F_7N with respect to CF_4 and C_2F_6 under spark discharge and corona discharge conditions. The relative content ratios, C_2F_4/CF_4 and C_2F_4/C_2F_6, were higher under spark discharge than under corona discharge.

The main reason is that, compared to corona discharge, the higher energy of a single spark discharge makes it easier for CF_3 radicals to undergo decomposition and produce CF_2 carbene, resulting in a higher concentration of CF_2 in the chamber and, consequently, a higher yield of C_2F_4. Additionally, the probability of the intermediate CF_2CF_2CN radical, produced during the C_4F_7N decomposition, undergoing decomposition to generate C_2F_4 is increased under spark discharge conditions. Therefore, the macroscopic result is that the content of C_2F_4

after spark discharge is higher than that after corona discharge.

2.4.4.3 Decomposition Characteristics under Surface Discharge

A cylindrical bisphenol A type epoxy resin block was used as the insulating material, with 13 copper wires attached to its side to simulate metallic debris adhering to the surface of the insulator. The epoxy resin block had a thickness of 35 mm and a diameter of 45 mm, while the copper wires were 25 mm in length and had a diameter of 0.2 mm. The electrodes employed a brass plate-plate configuration, with the dimensions of the plate electrode as shown in Figure 2.23. To prevent air gap discharge, a 2 μm-thick layer of pure copper powder was deposited on both planes of the epoxy resin block.

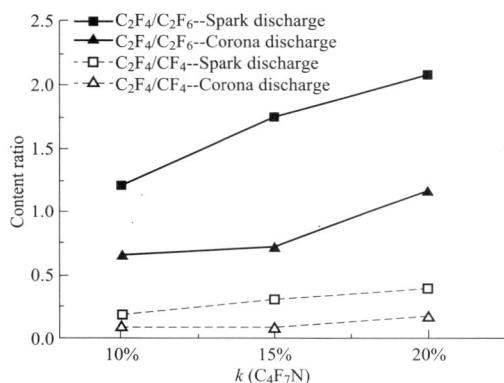

Figure 2.22 Relative content of C_2F_4, CF_4 and C_2F_6 under spark and corona discharges

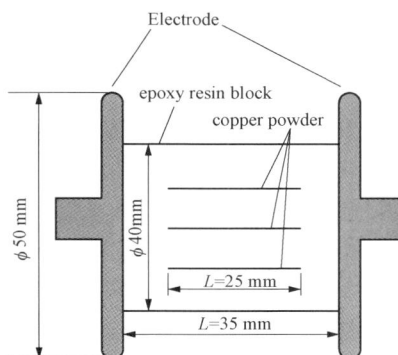

Figure 2.23 Schematic diagram of surface discharge electrode in the form of partial discharge

Gas analysis through GC/MS after discharge revealed that the types of decomposition components under surface discharge defects were the same as those under corona discharge defects, and no new gases were generated.

The variation in the decomposition components with the mixture ratio is illustrated in Figure 2.24. For the decomposition components without standard gases, the chromatographic peak areas of these components were normalized. The peak areas of each decomposition gas under the condition of $k(C_4F_7N) = 5\%$ were taken as the reference values, obtaining the relative content of decomposition components under other mixture ratios.

With the increase of $k(C_4F_7N)$, the content of perfluoroalkanes and nitrile gases showed an increasing trend, while the content of CO gas showed a decreasing trend. This trend is similar to the variation pattern of decomposition products with the mixture ratio observed under corona discharge defects and spark discharge defects. However, under surface discharge defects, the content of C_3F_6 gas is the lowest among the four quantifiable perfluoroalkanes, which is markedly different from the results under corona discharge (stainless steel needle electrode) and spark discharge defect conditions.

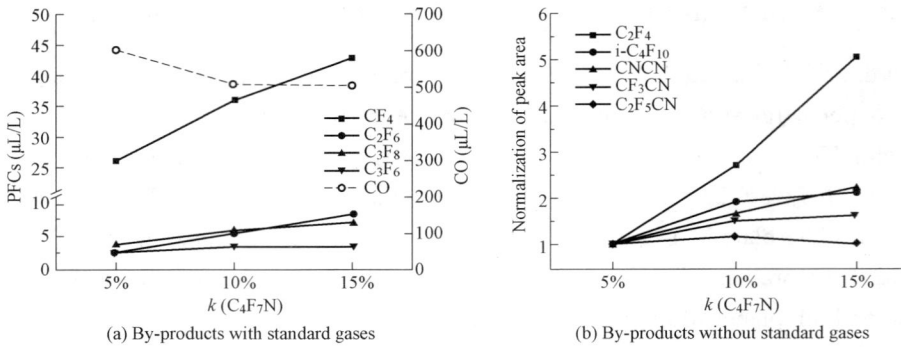

(a) By-products with standard gases   (b) By-products without standard gases

Figure 2.24   Decomposition by-product content variations with mixing ratios under surface discharge

The variation in decomposition component content under different applied voltages for surface discharge is shown in Figure 2.25.

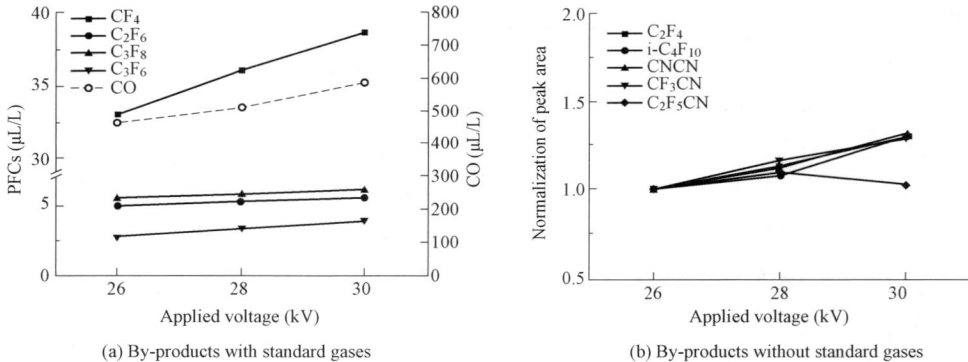

(a) By-products with standard gases   (b) By-products without standard gases

Figure 2.25   Decomposition by-product content variations with applied
voltages under surface discharge

With the increase in applied voltage or discharge quantity, the content of each decomposition component continues to increase, similar to the decomposition characteristics observed under corona discharge defects. However, as the voltage increases, C_3F_6 consistently remains the lowest among the four quantifiable perfluoroalkanes, which is markedly different from the results under corona discharge (stainless steel needle electrode) and spark discharge conditions. The content of CF_4, C_2F_6, C_3F_8, and C_3F_6 gases in a 15% C_4F_7N/85% CO_2 mixed gas was compared under corona discharge (stainless steel needle electrode) and surface discharge conditions at 200 kPa, where the applied voltage was 30 kV, and the discharge quantity was around 200 pC, as shown in Figure 2.26.

Under both discharge forms, C_3F_6 content differed significantly, while the content of the other three gases showed minor differences. Under surface discharge conditions, C_3F_6 content was the lowest, whereas under corona discharge, C_3F_6 content was second only to CF_4 gas. Combining the results of corona discharge experiments with stainless steel and pure aluminum needle electrodes, we speculate that this difference may be attributed to the influence of metal materials on decomposition components. The corona discharge experiment used a stainlesssteel needle electrode

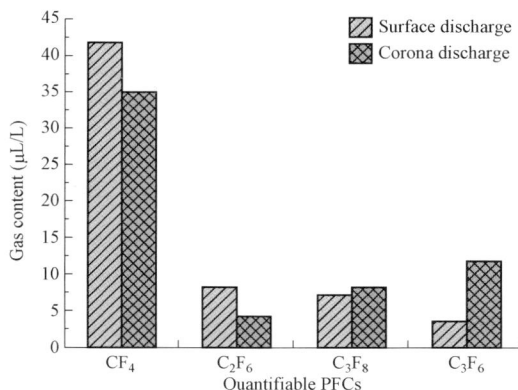

Figure 2.26   Comparison of PFCs between surface and corona discharges under the same conditions

to simulate tip discharge, while the surface discharge experiment used copper wires to simulate metal debris.

2.4.4.4 Decomposition Characteristics under Overheating

A heating rod is utilized to simulate the heat source in local overheating faults in electrical equipment, and the experimental platform for simulating local overheating is shown in Figure 2.27. The heating rod's shell is made of austenitic chromium-nickel stainless steel (2,520 stainless steel), exhibiting excellent high-temperature resistance and corrosion resistance, allowing it to operate for an extended period at temperatures up to 1,250 ℃. The heating rod has a length of 40 mm, a diameter of 14 mm, and an approximate heat exchange area of 15 cm². During the experiment, the voltage applied to the heating rod was around 25.4-26.0 V, and the current was approximately 5.1-5.3 A. We used a K-type thermocouple to measure temperature, with a temperature measurement range of 0-1,300 ℃, and an intelligent digital temperature control instrument displayed the temperature on the heating rod's surface. The material of the overheating experimental device is stainless steel, with a volume of approximately 24 L. Throughout the experiment, the environmental temperature was maintained at 25 ℃, with a relative humidity of 20%-30%, an absolute pressure of 200 kPa, and the overheating lasted for 6 hours. Before the experiment, the metal surface of the heating rod was cleaned sequentially with anhydrous ethanol and deionized water, followed by drying in a drying oven.

Figure 2.28 shows the TIC (Total Ion Chromatogram) detection results after the local overheating experiment at 650 ℃ with 10% C_4F_7N/90% CO_2 mixed gas. It is evident that after the overheating experiment, the decomposition of C_4F_7N/CO_2 produced CO, CF_4, C_2F_6, C_2F_4, C_3F_6, C_3F_8, i-C_4F_{10}, CF_3CN, CNCN, and C_2F_5CN.[14]

The experiment involved using an HF gas detection tube to assess the gas. Upon contact with the tested gas, the HF gas detection tube rapidly reached its detection limit. The particles inside the tube changed from a light yellow color to pink, indicating a high concentration of HF gas in the decomposition components. Figure 2.29 displays the physical appearance of the HF gas detection tube before and after the detection.

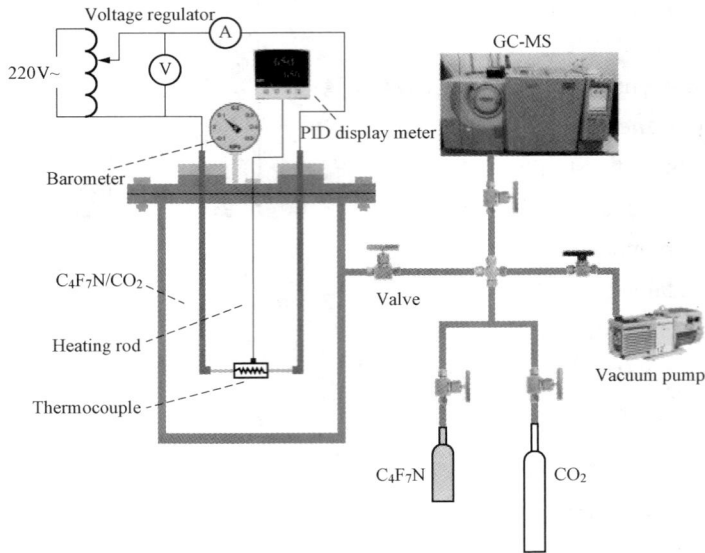

Figure 2.27　The simulation experiment platform of partial overheat

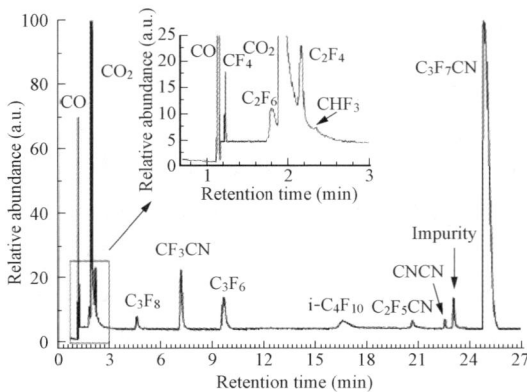

Figure 2.28　Total ion chromatogram of C_4F_7N/CO_2 mixture after 650 ℃ overheating

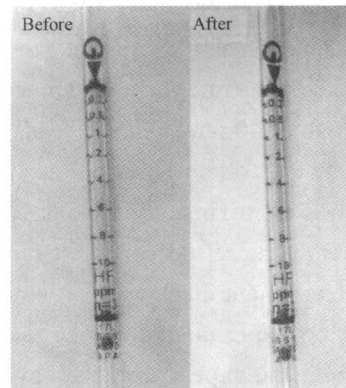

Figure 2.29　Color change of HF gas detection tube before and after 650 ℃ overheating

Table 2.5 presents the types of decomposition components at different temperatures. At 200 ℃, the C_4F_7N/CO_2 mixed gas has undergone decomposition, resulting in gaseous decompositionproducts. Within the temperature range of 200 to 500 ℃, the C_4F_7N/CO_2 mixed gas undergoes thermal decomposition, producing CO, C_3F_6, and HF gases. As the temperature increases, decomposition reactions become more favorable. At 550 ℃, the variety of decomposition components becomes more diverse, with the generation of C_2F_4, CNCN, and CF_3CN.

It is only at 650 ℃ that CF₄, C₂F₆, C₃F₈, i-C₄F₁₀, and C₂F₅CN are detected.

Table 2.5　　　　Types of decomposition by-products at different temperatures

Temperature（℃）	200	350	450	500	550	600	650	700
CO	●	●	●	●	●	●	●	●
C_3F_6	●	●	●	●	●	●	●	●
HF	●	●	●	●	●	●	●	●
C_3F_8	—	—	—	—	—	—	●	●
C_2F_6	—	—	—	—	—	—	●	●
CF_4	—	—	—	—	—	—	●	●
C_2F_4	—	—	—	—	●	●	●	●
CNCN	—	—	—	—	●	●	●	●
C_2F_5CN	—	—	—	—	●	●	●	●
i-C_4F_{10}	—	—	—	—	—	—	●	—

Note: "●" represents detected substance, "—" represents undetected substance.

The variation of different decomposition components at various temperatures after 1-hour overheating experiment of 10% C_4F_7N/90% CO_2 mixed gas is shown in Figure 2.30. The generated CO and C_3F_6 content are relatively high. In the temperature range of 200-700 ℃, the content of CO and C_3F_6 gases is much higher than other gases, exceeding the content of CF_4, C_2F_6, and C_3F_8 gases by more than an order of magnitude. With the increase in temperature, the degree of decomposition of C_4F_7N molecules increases, and a strong decomposition occurs at 700 ℃, leading to a sharp increase in the content of decomposition components.

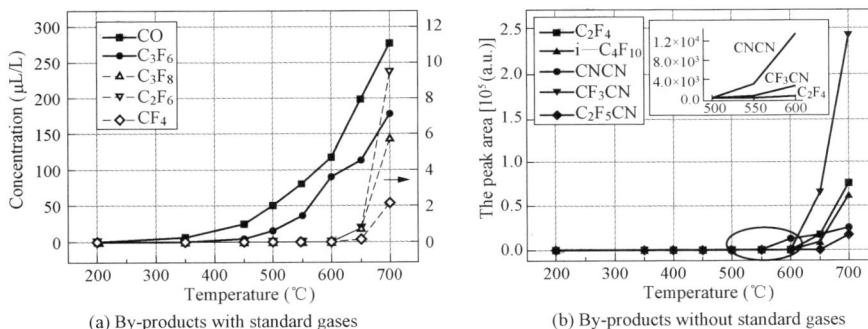

(a) By-products with standard gases　　　(b) By-products without standard gases

Figure 2.30　Relationship between decomposition by-products and temperatures

Subsequently, a 6-hour overheating experiment was conducted at 650 ℃ for 10% C_4F_7N/90% CO_2 mixed gas, and the content of its decomposition components was measured every hour. Figure 2.31 shows the variation trend of decomposition component content at 650 ℃ with increasing overheating time. It can be observed that the content of CO gas is the highest, followed by C_3F_6 gas. The content of these five quantifiable decomposition components from high to low is $c(CO)>c(C_3F_6)>c(C_2F_6)>c(C_3F_8)>c(CF_4)$. The content of these decomposition components gradually increases with the extension of overheating time, and after 6 hours, the content of CO and C_3F_6 is 768 µL/L and 367 µL/L, respectively.

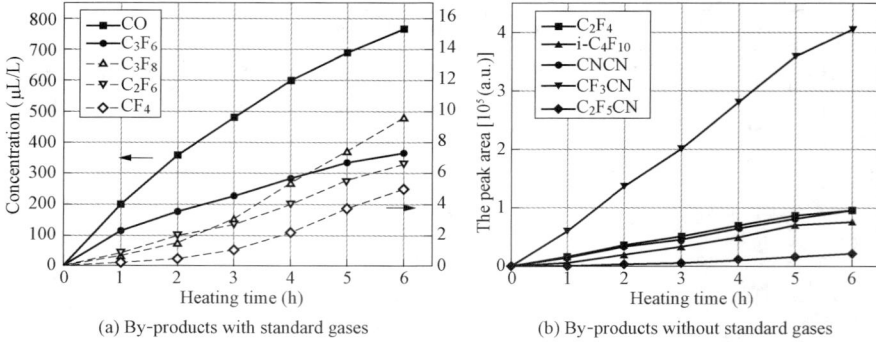

(a) By-products with standard gases   (b) By-products without standard gases

Figure 2.31   Relationship between decomposition by-product content
and overheating time at 650 ℃ overheating

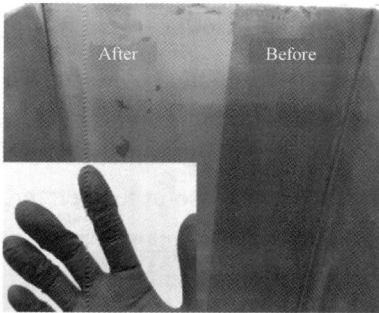

Figure 2.32   Relationship between decomposition by-product content and overheating time at 650 ℃ overheating

After the 6-hour overheating experiment at a temperature of 700 ℃, upon opening the chamber, a dense yellow-brown solid powder was observed adhering to the inner wall, as shown in Figure 2.32.

The elemental composition and phase analysis of the solid powder were conducted using EDS and XRD techniques, and the results are presented in Figure 2.33.

The EDS results indicated that the solid powder primarily contains F and Fe elements, with weight percentages (wt.%) of 48.86% and 40.90%, respectively. The XRD spectrum revealed distinct diffraction peaks in the ranges of $2\theta = 26°$-$30°$, $32°$-$36°$, and $50°$-$55°$. By comparing with JCPDS cards, these peaks corresponded to FeF_2

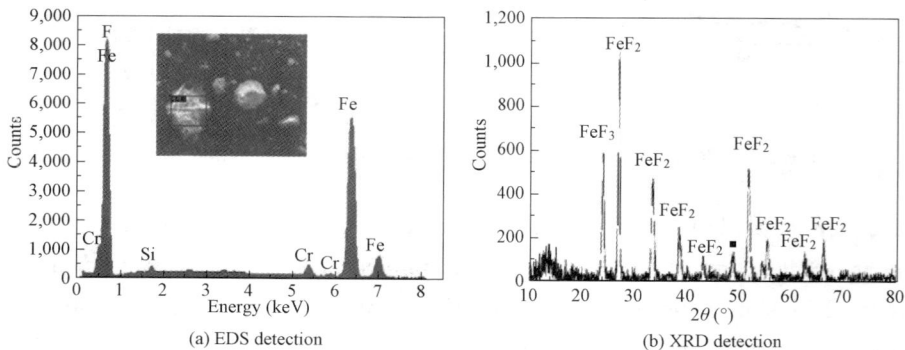

(a) EDS detection   (b) XRD detection

Figure 2.33   EDS and XRD results of solid powder

(JCPDS 76-0651, 81-2272) within the range of 22°-26° and a sharp diffraction peak within the range of 47°-50° (marked with a blue square) that could not be definitively assigned to a specific phase.

Stainless steel, when exposed to air for an extended period, forms a thin layer of oxide film (oxides of Fe and Cr) on its surface. The presence of this oxide film to some extent inhibits reactions between substances such as O_2, acids, and bases with the stainless-steel matrix. Combining the detected results of overheated decomposition components with EDS and XRD, we speculate that the source of the solid powder may be a reaction between HF gas and the oxides of Fe in the oxide layer on the inner wall of stainless steel. A large amount of HF gas generated from the overheating decomposition of C_4F_7N/CO_2 dissolves in the residual trace moisture on the stainless-steel inner wall, forming hydrofluoric acid liquid, which adheres to the stainless-steel inner wall. The hydrofluoric acid reacts with the FeO and Fe_2O_3 in the oxide film on the stainless-steel inner wall to generate FeF_2 and FeF_3, ultimately forming a uniform and dense layer of solid powder attached to the inner wall, as described in Reactions (2.10-2.11).

$$FeO + 2HF \longrightarrow FeF_2 + H_2O \qquad (2.10)$$
$$Fe_2O_3 + 6HF \longrightarrow 2 FeF_3 + 3 H_2O \qquad (2.11)$$

After each local overheating experiment at various temperatures, HF gas was detected, with the HF gas content increasing notably as the temperature rose. HF gas is highly corrosive and can react with stainless steel, potentially reducing the operational lifespan of gas-insulated electrical equipment. Therefore, it is recommended to apply anti-corrosion coatings to the inner walls of equipment using C_4F_7N/CO_2 gas insulation or to place absorbents capable of adsorbing corrosive gases like HF in appropriate quantities. Additionally, the HF gas in the products may also result from the heating decomposition of residual moisture in the chamber into H radicals, which then combine with F radicals generated from the decomposition of C_4F_7N. During equipment operation, it is essential to monitor and control the internal moisture content rigorously to prevent the generation of high concentrations of HF gas, reducing the equipment's lifespan or posing a threat to its stable operation.

2.4.4.5 Formation Mechanism of Decomposition By-Products under Overheating

During the local overheating experiment at temperatures ranging from 200 ℃ to 500 ℃, only CO, C_3F_6, and HF gases were produced, with the C_3F_6 content in the decomposition components being notably high, second only to CO gas. The decomposition mechanism of C_4F_7N indicates that the generation of C_3F_6 from C_4F_7N requires overcoming an energy barrier of 98.0 kcal/mol, while the energy change for the cleavage of C_4F_7N into CF_3 and CF_3CFCN is only 85.8 kcal/mol. Therefore, theoretically, C_4F_7N should not preferentially generate C_3F_6 and should only produce C_3F_6 gas. The GC/MS results indicate that C_4F_7N contains impurities of heptafluoropropane (C_3HF_7). It is speculated that, under overheating conditions, C_3F_6 mainly originates from the thermal decomposition of C_3HF_7. This section provides a theoretical calculation and analysis of the decomposition mechanism of C_3HF_7.

Figure 2.34 illustrates the decomposition pathways of C_3HF_7, involving six main pathways. Figure 2.35 shows the equilibrium configurations of the geometric parameters of C_3HF_7 isomerization and dissociation reaction pathways, optimized at the M06-2X/6-311G (d,p) level, where bond lengths are given in Ångströms, and bond angles are given in degrees.

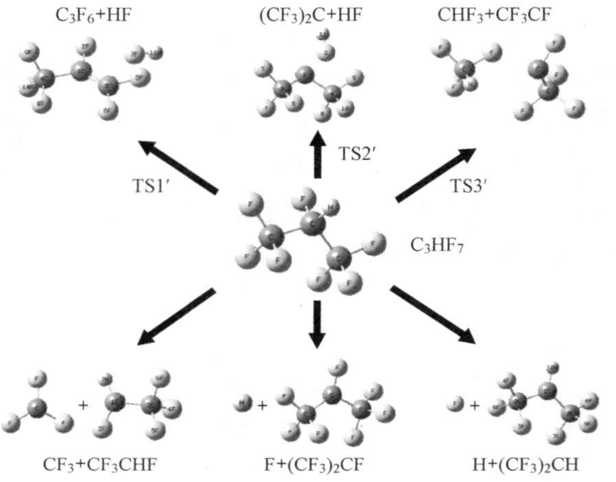

Figure 2.34   Decomposition pathways of C_3HF_7 molecule

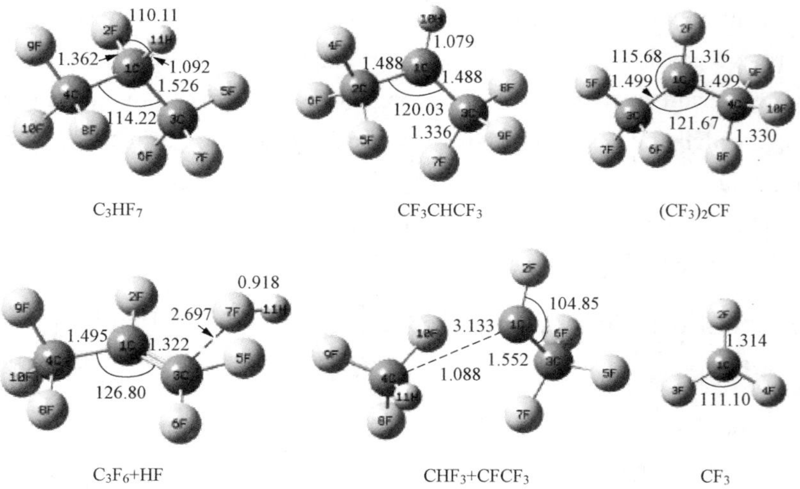

Figure 2.35   Geometrical structure parameters of reactants, intermediates, and transition states in decomposition pathways of C_3HF_7 molecule optimized at M06-2X/6-311G (d,p) level

(Bond lengths are in angstroms and angles are in degree) (1)

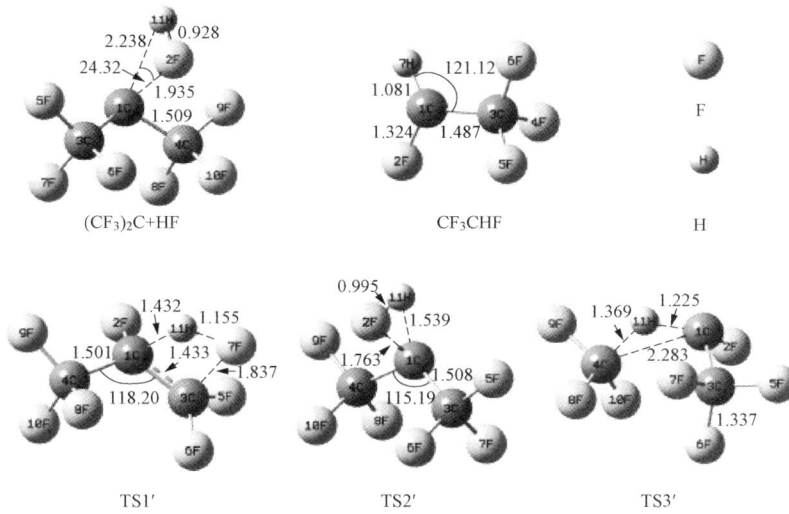

Figure 2.35   Geometrical structure parameters of reactants, intermediates, and transition states in decomposition pathways of C_3HF_7 molecule optimized at M06-2X/6-311G (d,p) level

(Bond lengths are in angstroms and angles are in degree) (2)

(1) Initial Isomerization and Dissociation of C_3HF_7

Figure 2.36 presents the potential energy surface curves for the isomerization and dissociation of C_3HF_7 at the CCSD(T)/cc-pVTZ//M06-2X/6-311G(d,p) level. For ease of discussion, we set the energy of the reactant C_3HF_7 as the reference zero point. It can be observed that C_3HF_7 has three isomerization pathways and three cleavage pathways.

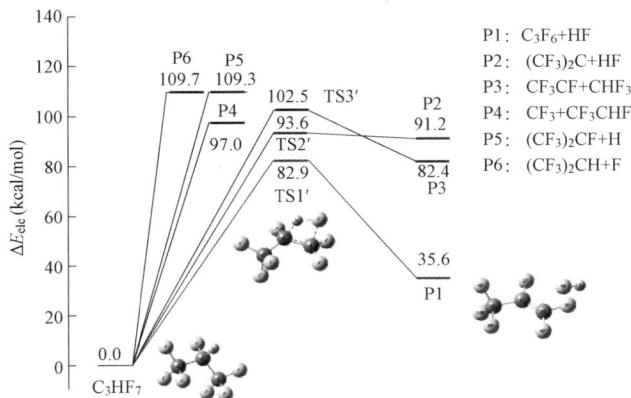

Figure 2.36   The potential energy surface of C_3HF_7 isomerization and dissociation of based on the relative energies at CCSD(T)/cc-pVTZ//M06-2X/6-311G (d,p) level

Isomerization reactions are as follows:

- C_3HF_7 undergoes the transition state TS1', overcoming an electronic barrier of 82.9 kcal/mol, leading to HF elimination and the formation of product P1 (C_3F_6+HF). In this process, the H atom gradually moves away from the central C atom, approaches the F atom of the CF_3 group, and begins to move away from C_3F_6 after combining with the F atom to form HF. Transition state TS1' is a first-order saddle point with C1 symmetry and a virtual frequency of 1754.77i.

- C_3HF_7 passes through the transition state TS2', overcoming an electronic barrier of 93.6 kcal/mol, again resulting in HF elimination and the generation of product P2 $[(CF_3)_2C+HF]$. The energy of the product is 91.2 kcal/mol higher than that of C_3HF_7. In this process, the H atom gradually approaches the central F atom, and the angle $\angle H—C—F$ between these two atoms and the central C atom gradually decreases. After combining with the F atom, the H atom begins to move away from $(CF_3)_2C$. The angle $\angle H(11)—C(1)—F(2)$ in the C_3HF_7 molecule is 110.11°, and the bond lengths of the C—H and C—F bonds to be broken are 1.092 Å and 1.362 Å, respectively. In the transition state TS2', the angle $\angle H(11)—C(1)—F(2)$ decreases to 34.23°, and the bond lengths of the C—H and C—F bonds to be broken increase to 1.539 Å and 1.763 Å, respectively. The $(CF_3)_2C$ carbene cannot exist stably and undergoes isomerization to C_3F_6, overcoming a smaller barrier of 14.1 kcal/mol.

- C_3HF_7 can also undergo the transition state TS3', overcoming an energy barrier of 102.5 kcal/mol, isomerizing to product P3 $(CF_3CF+CHF_3)$. In this process, the H atom moves to the C(4) atom on CF_3, and the energy of the product is 82.4 kcal/mol higher than that of C_3HF_7.

Additionally, C_3HF_7 has three channels for homolytic cleavage, generating products P4 (CF_3+CF_3CHF), P5 (CF_3CFCF_3+H), and P6 $[(CF_3CHCF_3+F)]$ through the cleavage of C—CF_3, C—H, and C—F bonds, respectively. The energies of products P4, P5, and P6 are 97.0 kcal/mol, 109.3 kcal/mol, and 109.7 kcal/mol higher than that of C_3HF_7.

Since the energy barrier for the pathway leading to the production of P1 (C_3F_6+HF) is the lowest among the isomerization reactions of C_3HF_7, the reaction pathway (pathway 1) leading to P1 (C_3F_6+HF) is the main reaction pathway for the isomerization of C_3HF_7, and the product P1 (C_3F_6+HF) is the main product of the decomposition of C_3HF_7.

From the C_4F_7N decomposition mechanism, it is known that the energy barrier for the isomerization of C_4F_7N to produce C_3F_6 is 98.0 kcal/mol, higher than the energy barrier for the isomerization of C_3HF_7 to produce C_3F_6 (82.9 kcal/mol). Kinetically, C_3HF_7 molecules are more prone to decompose to generate C_3F_6 than C_4F_7N. Therefore, the detected product C_3F_6 in the 200-500 ℃ overheating experiment is derived from the decomposition of C_3HF_7, and C_4F_7N molecules do not decompose within this temperature range. As the temperature increases, C_4F_7N begins to cleave into CF_3 and CF_3CFCN radicals. The CF_3CFCN radical isomerizes to CF_2CF_2CN, then absorbs heat and homolytically cleaves into CF_2CN, CF_2, CN, and C_2F_4. The CF_3 radical undergoes thermal cleavage to form CF_2 carbenes. The cleaved CF_2, CN, and CF_3 react through free radical coupling reactions to produce stable products C_2F_4,

CNCN, and CF₃CN. Therefore, at 550 ℃, C_2F_4, CNCN, and CF_3CN were detected. When the temperature rises to 650 ℃, the number of cleavage reaction channels for C_4F_7N increases, the variety of products becomes more abundant, and the product content increases as more C_4F_7N molecules participate in the cleavage reactions.

(2) Electron Structure Changes during the Isomerization Process of C_3HF_7

To comprehend the changes in the electron structure during the reaction process of C_3HF_7 isomerizing into C_3F_6 and HF, we conducted a detailed analysis using LOL (Localized Orbital Locator) contour plots and Mayer bond orders for the $C_3HF_7 \longrightarrow C_3F_6 + HF$ reaction. Figure 2.37 illustrates the LOL contour plot on the C(3)—C(1)—H(11) plane during the isomerization process of C_3HF_7. Higher numerical values indicate stronger localization in the respective areas, with white areas representing regions where the values exceed the upper limit of 0.750.

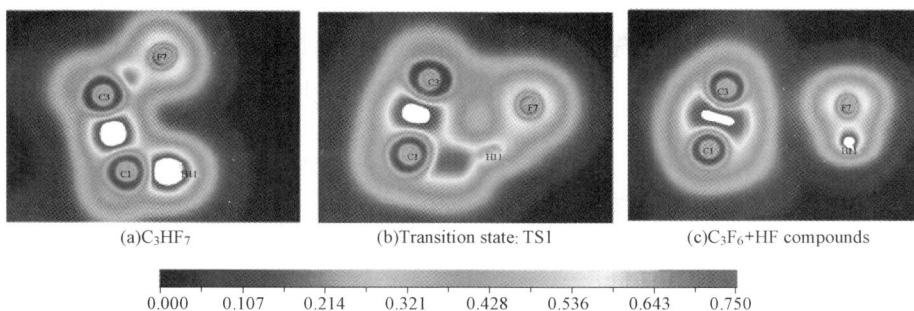

(a)C_3HF_7 (b)Transition state: TS1   (c)C_3F_6+HF compounds

0.000   0.107   0.214   0.321   0.428   0.536   0.643   0.750

Figure 2.37   The localized orbital locator (LOL) color-filled maps of the reaction
$C_3HF_7 \longrightarrow C_3F_6 + HF$ in the C(3)—C(1)—H(11) plane

From Figure 2.37(a), it can be observed that regions between C1—H11, C1—C3, and C3—F7 exhibit high localized values, indicating strong localized covalent bonds corresponding to C1—H11 bond, C1—C3 bond, and C3—F7 bond in the C_3HF_7 molecule. As H11 atom migrates, in the transition state TS1', r[C(3)—F(7)] is elongated from the original 1.332 Å to 1.837 Å, r[C(1)—H(11)] is elongated from the original 1.092 Å to 1.432 Å, and r[H(11)—F(7)] is shortened from the original 2.603 Å to 1.155 Å.

In Figure 2.37(b), the values between C3—F7 in the transition state decrease, indicating weak-ened localization, suggesting a reduction in covalent interaction between C(3) and F(7) atoms. Although the values between C1—H11 also decrease, the localization between them remains strong, indicating a robust covalent interaction in C1—H11 during the transition state TS1'. Simultaneously, the region between H11 and F7 starts to connect, and the values in the localized region increase, indicating the formation of a weak covalent interaction between H11 and F7. The geometric changes during the transition from C_3HF_7 to the transition state TS1' mainly involve the gradual separation of the F(7) atom from the C(3) atom, approaching the H(11) atom, while the distance between H(11) and C(1) atoms is elongated by only 0.34 Å.

In Figure 2.37(c), the region between H11 and F7 is fully connected, and the localization becomes stronger. The localized region between H11 and F7 is disconnected from C(1) and C(3), indicating that in the product C_3F_6+HF structure, H11 has moved completely away from the C(1) atom and F(7) has moved away from the C(3) atom. Strong covalent interaction has formed between H11 and F7, completing the isomerization process. At this point, the C1—C3 bond exhibits a flattened and concave localized domain, showing clear characteristics of multiple bonds.

As there are significant changes in Mayer bond orders for the C1—H11, C3—F7, C1—C3, and H11—F7 bonds during the C_3HF_7⟶C_3F_6+HF process, the Multiwfn 3.6 software package was used to calculate the Mayer bond orders for these four covalent bonds and plotted the bond orders as a function of IRC in Figure 2.38. It can be seen that in the C_3HF_7 molecule, the interaction between C(1) and H(11), C(3) atom forms a single bond, with Mayer bond orders around 1.0. The C(3) atom in the CF_3 group forms a single covalent bond with the F(7) atom, with a Mayer bond order of approximately 1.0. The distance between H(11) and F(7) atoms is 2.603 Å, indicating no covalent interaction, and the Mayer bond order is zero.

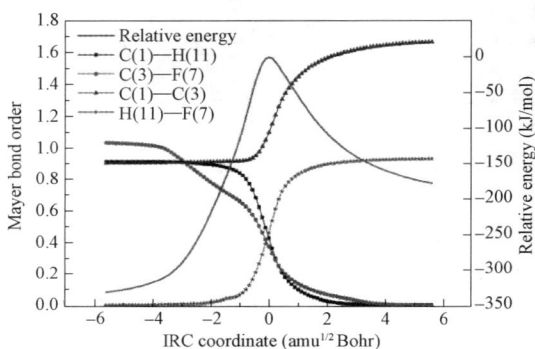

Figure 2.38   Mayer bond order variations of key bonds along IRC path in the isomerization reaction of C_3HF_7⟶C_3F_6+HF

Before $s=-1$(amu$^{1/2}$ Bohr), the bond orders of C1—H11, C1—C3, and H11—F7 remain nearly unchanged, while the bond order of C3—F7 has already started to decrease, decreasing from the original 1.0 to around 0.6. Additionally, from Figure 2.34, it is observed that in the optimized transition state TS1', r[C(3)—F(7)] is elongated from the original 1.332 Å to 1.837 Å. This indicates that the geometric structure change in the isomerization process begins with the CF_3 group, where the F(7) atom gradually moves away from the C(3) atom, weakening the covalent interaction between the F(7) and C(3) atoms.

After the reaction process reaches $s=-1$(amu$^{1/2}$ Bohr), the bond orders of C1—C3 and H11—F7 start to increase, the bond order of C1—H11 starts to decrease, and the bond order of C3—F7 continues to decrease. This suggests that as the reaction progresses, the H(11) atom in the C_3HF_7 molecule begins to approach the F(7) atom, the H(11) atom starts moving away from the C(1) atom, and the F(7) atom continues to move away from the C(3) atom. When

s=1(amu$^{1/2}$ Bohr), the bond order of H11—F7 increases from zero to around 1.0, while the bond orders of C1—H11 and C3—F7 decrease from 1.0 to around zero. This indicates that at this point, a covalent single bond has formed between H11 and F7, and the covalent single bonds of C1—H11 and C3—F7 have been completely broken. At the same time, the bond order of C1—C3 increases from 1.0 to around 1.7, indicating that the C—C single bond between C(1) and C(3) atoms has transformed into a C=C double bond. After this point, the bond orders remain almost unchanged, and the isomerization process concludes, yielding the products C_3F_6+HF.

Mayer bond orders show minimal changes at the beginning and end of the reaction, but they change significantly during the formation of new covalent bonds and the breaking of old covalent bonds. This indicates that there is little geometric structure change at the start of the reaction, the most substantial geometric changes occur around the formation of the transition state, and there is almost no further change in the geometric structure after the formation of the products. The changes in Mayer bond orders and LOL color maps clearly depict the electronic structure changes during the $C_3HF_7 \longrightarrow C_3F_6$+HF reaction, and through analysis, it is found that the two methods align well.

2.4.5 Fault Criteria of C$_4$F$_7$N/CO$_2$ Decomposition

2.4.5.1 Comparison of Decomposition Components

The qualitative comparison of decomposition components of C_4F_7N/CO_2 mixed gas under corona discharge, spark discharge, surface discharge, and overheating conditions is presented in Table 2.6.

Table 2.6   **Comparison of decomposition by-products of C$_4$F$_7$N/CO$_2$**

mixture under discharge and overheat conditions

Decomposition by-products	Partial discharge			Partial overheating		
	Corona discharge	Spark discharge	Surface discharge	200–500 ℃	550–600 ℃	>650 ℃
CO	●	●	●	●	●	●
CF_4	●	●	●	-	-	●
C_2F_6	●	●	●	-	-	●
C_3F_6	●	●	●	●	●	●
C_3F_8	●	●	●	-	-	●
$C_2O_3F_6$	trace	trace	trace	-	-	-
CHF_3	●	●	●	-	-	trace
C_2F_4	●	●	●	-	●	●
2-C_4F_6	trace	trace	trace	-	-	-
2-C_4F_8	trace	trace	trace	-	-	trace
i-C_4F_{10}	●	●	●	-	-	●

Continue

Decomposition	Partial discharge			Partial overheating		
by-products	Corona discharge	Spark discharge	Surface discharge	200–500 ℃	550–600 ℃	>650 ℃
CNCN	●	●	●	-	●	●
CF₃CN	●	●	●	-	●	●
C₂F₅CN	●	●	●	-	-	●

Note: "●" represents detected compounds, and "-" represents undetected compounds.

From the table, it can be observed that C_4F_7N/CO_2 mixed gas generates $C_2O_3F_6$ and 2-C_4F_6 under discharge conditions, while these two substances were not detected among the decomposition components under overheating conditions. $C_2O_3F_6$ is related to the residual O_2 in the chamber, where O_2 transforms into O_3 under discharge conditions, and then reacts with CF_3 radicals to generate $C_2O_3F_6$. However, under overheating conditions, O_2 is not sufficient to transform into O_3, preventing the formation of $C_2O_3F_6$. 2-C_4F_6 is generated by the deep decomposition of C_4F_7N, involving the composite formation of CCF_3 carbene. The generation of CCF_3 involves multiple secondary reactions, and the energy required for the reaction path leading to CCF_3 is very high. Under discharge conditions, the discharge zone temperature is generally around 2,000-3,000 K, providing sufficient energy for the generation of CCF_3. However, at 700 ℃ overheating conditions, C_4F_7N is insufficient to decompose and generate CCF_3 carbene. Therefore, the presence of $C_2O_3F_6$ and 2-C_4F_6 in the decomposition components can be used as indicators of discharge faults, and their absence indicates an overheating fault.

Furthermore, at temperatures ranging from 200 ℃ to 500 ℃, the overheating decomposition products are only CO and C_3F_6, while under discharge conditions, the variety of decomposition products is extensive. Therefore, the severity of discharge and overheating faults can also be assessed by analyzing CO and C_3F_6. If only CO and C_3F_6 are detected in the decomposed gas, it indicates a pure overheating fault, with an overheating temperature range of 200 ℃ $<T<$ 500 ℃. If not only CO and C_3F_6 but also nitrile compounds are detected, it suggests a higher temperature overheating fault (550 ℃ $<T<$ 600 ℃). If CF_4, C_2F_6, and other fully fluorinated hydrocarbons are detected, it indicates a discharge fault or a high-temperature overheating fault ($T>$ 650 ℃).

2.4.5.2 Content Ratios of Decomposition By-Products

Comparing the decomposition characteristics under discharge and overheating conditions reveals distinct differences. Overheating faults at 200 ℃ to 500 ℃ only produce CO, C_3F_6, and HF, while overheating decomposition components become more diverse at 650 ℃, similar to the components under discharge conditions. At temperatures above 650 ℃, the content of CO gas is not significantly different from the content of C_3F_6 gas, whereas, under discharge conditions, the CO content is much higher than that of C_3F_6. Therefore, the relative content ratio of decomposition components, the ratio of CO to C_3F_6 content ($c[CO]/c[C_3F_6]$) can be used to distinguish between discharge and overheating faults.

Based on the mechanism of decomposition product formation, C_2F_4 is mainly formed by the decomposition of CF_3 radicals, forming CF_2 carbenes, while CF_4 and C_2F_6 mainly originate from the coupling reactions of CF_3 radicals with F or CF_3 radicals. There is a "competition" relationship between CF_4, C_2F_6, and C_2F_4. Under discharge conditions, the energy required for the breaking of C—F covalent bonds in the CF_3 group comes from the kinetic energy gained by electrons in collision reactions. Therefore, this "competition" relationship differs in various discharge forms. From this perspective, the ratios C_2F_6/CF_4 and $C_2F_4/(CF_4+C_2F_6)$ can be used as characteristic quantities. The content ratio C_2F_6/CF_4 is denoted as $v[C_2F_6]/v[CF_4]$, and since C_2F_4 cannot be quantified, the ratio of peak areas obtained from characteristic fragment ions was used, denoted as $v[C_2F_4]/v[CF_4+C_2F_6]$, where CF_4, C_2F_6, and C_2F_4 have mass-to-charge ratios of $m/z=69$, 119, and 81, respectively.

Furthermore, C_2F_4 can also be formed by the isomerization of CF_3CF carbenes, and C_3F_6 can be generated by the combination of CF_3CF carbenes with CF_2, and there is a mutual generation between C_2F_4 and C_3F_6 through isomerization reactions. Therefore, there is also a "competition" relationship between C_2F_4 and C_3F_6, and the ratio of peak areas, denoted as $v[C_2F_4]/v[C_3F_6]$, can be used as another characteristic quantity.

Considering both discharge and overheating faults, four characteristic quantities are proposed: $c[CO]/c[C_3F_6]$, $c[C_2F_6]/c[CF_4]$, $v[C_2F_4]/v[CF_4+C_2F_6]$, and $v[C_2F_4]/v[C_3F_6]$. The analysis of their variations under overheating and discharge conditions is performed. The mass-to-charge ratios for C_2F_4 and C_3F_6 are $m/z=81$ and 169, respectively.

(1) Characteristic quantities under overheating

Figure 2.39 shows the changes in the ratios $c[CO]/c[C_3F_6]$, $c[C_2F_6]/c[CF_4]$, $v[C_2F_4]/v[CF_4+C_2F_6]$, and $v[C_2F_4]/v[C_3F_6]$ as a function of overheating temperature for a 10% C_4F_7N/90% CO_2 mixed gas over overheating times ranging from 1 to 6 hours.

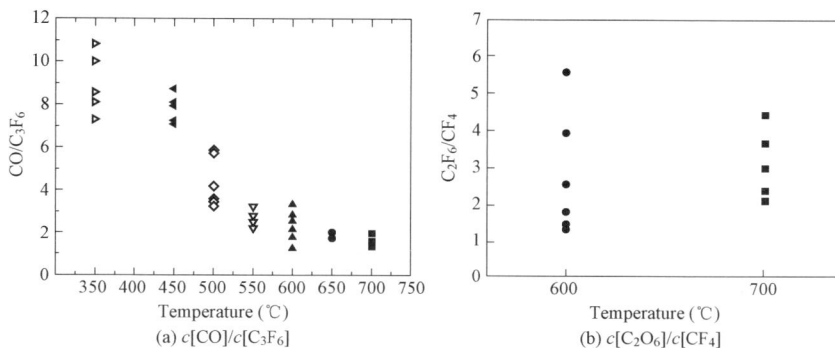

Figure 2.39　Relationship between the characteristic ratio and temperature under partial overheating (1)

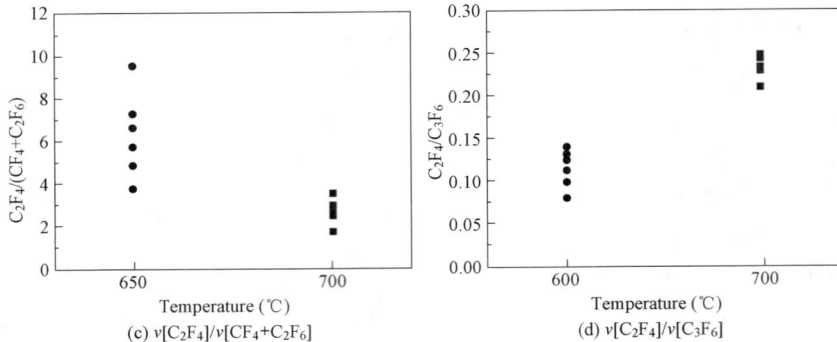

Figure 2.39 Relationship between the characteristic ratio and temperature under partial overheating (2)

As the temperature increases, the content ratio $c[CO]/c[C_3F_6]$ shows a decreasing trend, and the $c[CO]/c[C_3F_6]$ ratio is less than 12 in the temperature range of 350-700 ℃. CF_4 and C_2F_6 gases are only produced under high-temperature overheating faults ($T>650$ ℃), and the content of CF_4 is the lowest among the four quantifiable perfluoroalkanes, with a low content of C_2F_6. However, C_2F_4 gas is already produced at 550 ℃, but its content is much lower than that of C_3F_6. Therefore, the ratio $c[C_2F_6]/c[CF_4]$ and the ratio $v[C_2F_4]/v[CF_4+C_2F_6]$ are greater than 1, while the ratio $v[C_2F_4]/v[C_3F_6]$ is less than 1.

(2) Characteristic quantities under corona discharge

Figure 2.40 depicts the trends of the ratios $c[CO]/c[C_3F_6]$, $c[C_2F_6]/c[CF_4]$, $v[C_2F_4]/v[CF_4+C_2F_6]$, and $v[C_2F_4]/v[C_3F_6]$ under corona discharge as a function of applied voltage. Here, the concentration of C_4F_7N is 10%, and the needle electrode materials are pure aluminum and stainless steel.

As the applied voltage increases, the ratio $c[CO]/c[C_3F_6]$ shows a decreasing trend and gradually stabilizes. Under an aluminum needle electrode, the ratio $c[CO]/c[C_3F_6]$ eventually stabilizes around 70, while under a stainless steel needle electrode, the ratio $c[CO]/c[C_3F_6]$ stabilizes around 30. The experiments with a pure aluminum needle electrode result in higher values of $c[CO]/c[C_3F_6]$, $c[C_2F_6]/c[CF_4]$, and $v[C_2F_4]/v[C_3F_6]$ compared to those with a stainless steel electrode. As the voltage increases, C_4F_7N molecules gain energy, and the probability of overcoming the 98.0 kcal/mol barrier for isomerization to generate C_3F_6 increases. When the voltage is sufficiently high for C_4F_7N to overcome the barrier and generate C_3F_6, the rate of C_3F_6 production slows down. Therefore, as the voltage increases, the ratio $c[CO]/c[C_3F_6]$ decreases and stabilizes. Since the experiments with an aluminum needle electrode result in higher concentrations of CO, C_2F_4, CF_4, and C_2F_6 gases compared to those with a stainless steel needle electrode, while the concentration of C_3F_6 is very close, the ratios $c[CO]/c[C_3F_6]$, $c[C_2F_6]/c[CF_4]$, and $v[C_2F_4]/v[C_3F_6]$ obtained with the aluminum needle electrode are higher than those with the stainless steel needle electrode.

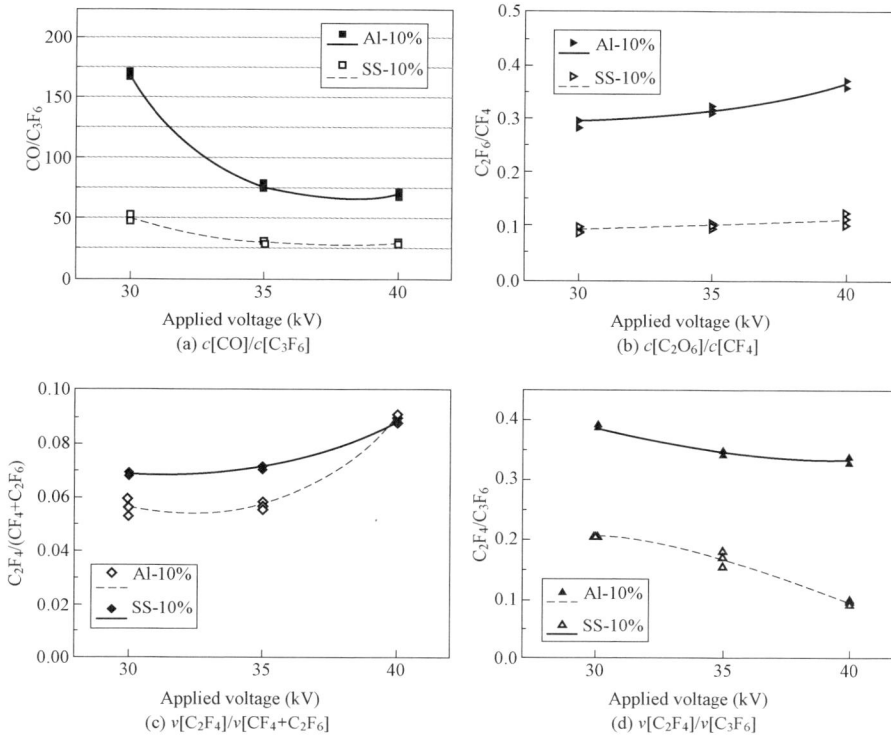

Figure 2.40　Relationship between the characteristic ratio and applied voltage under corona discharge

(3) Characteristic quantities under spark discharge

Figure 2.41 illustrates the trends of the ratios $c[CO]/c[C_3F_6]$, $c[C_2F_6]/c[CF_4]$, $v[C_2F_4]/v[CF_4+C_2F_6]$, and $v[C_2F_4]/v[C_3F_6]$ as a function of the mixture ratio under spark discharge conditions. It can be observed that with the increase in the volume fraction of C_4F_7N, the ratio $c[CO]/c[C_3F_6]$ gradually decreases and tends to stabilize. When $k(C_4F_7N) = 10\%$, the ratio $c[CO]/c[C_3F_6]$ is approximately 40, and the ratio $v[C_2F_4]/v[CF_4+C_2F_6]$ is in the range of approximately 0.15 to 0.2. Compared to corona discharge and surface discharge, the ratio $v[C_2F_4]/v[CF_4+C_2F_6]$ under spark discharge conditions is relatively large. In the breakdown form of discharge, electrons can gain higher kinetic energy at the moment of breakdown, leading to increased production of CF_2 carbenes from the cleavage of CF_3 groups. This, in turn, elevates the content of C_2F_4 generated through free radical coupling reactions, resulting in a relatively large value of the ratio $v[C_2F_4]/v[CF_4+C_2F_6]$.

(4) Characteristic quantities under surface discharge

Figure 2.42 shows the trends of the ratios $c[CO]/c[C_3F_6]$, $c[C_2F_6]/c[CF_4]$, $v[C_2F_4]/v[CF_4+C_2F_6]$, and $v[C_2F_4]/v[C_3F_6]$ as a function of applied voltage under surface discharge defect conditions, with $k(C_4F_7N)$ set to 10%. It can be observed that with the increase in applied voltage, the ratio $c[CO]/c[C_3F_6]$ shows a decreasing trend and eventually stabilizes between 140 and 150. Under surface discharge, the ratio $v[C_2F_4]/v[C_3F_6]$ is significantly higher than that under corona discharge, while the ratio $v[C_2F_4]/v[CF_4+C_2F_6]$ is comparable to that under corona discharge. This is mainly because under surface discharge, the content of C_3F_6 is the lowest among the four quantifiable perfluoroalkanes, whereas under corona discharge and spark discharge, the content of C_3F_6 is only surpassed by CF_4 and is higher than that of C_2F_6 and C_2F_8.

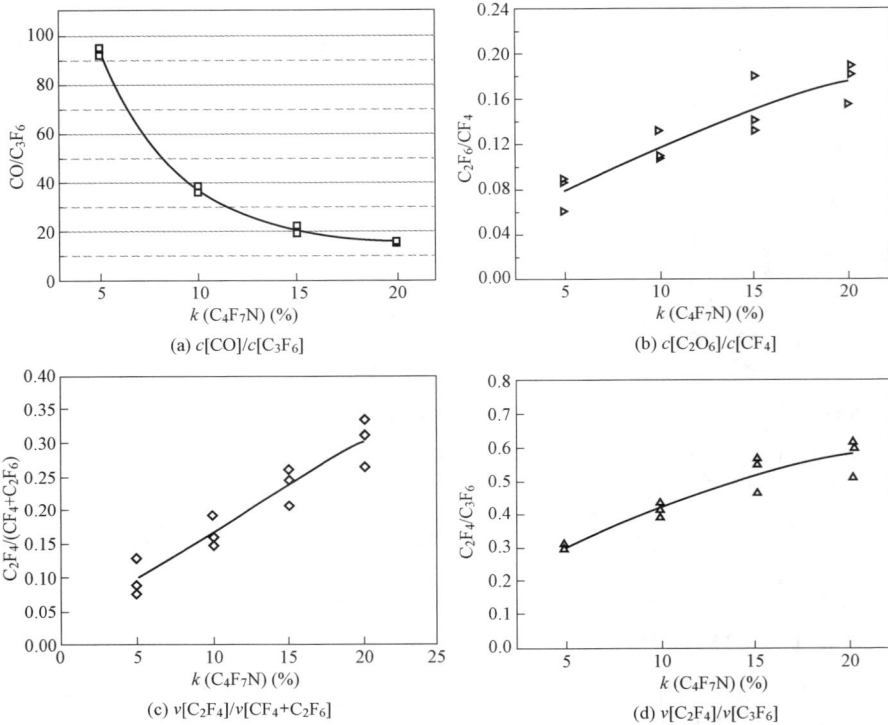

(a) $c[CO]/c[C_3F_6]$

(b) $c[C_2O_6]/c[CF_4]$

(c) $v[C_2F_4]/v[CF_4+C_2F_6]$

(d) $v[C_2F_4]/v[C_3F_6]$

Figure 2.41   Relationship between the characteristic ratio and mixing ratio under spark discharge

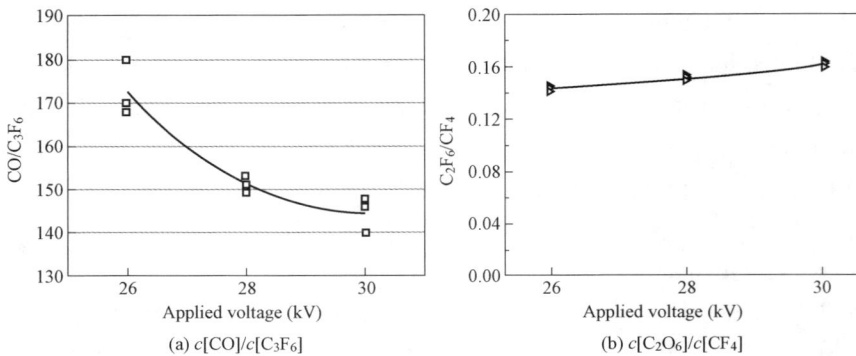

(a) $c[CO]/c[C_3F_6]$

(b) $c[C_2O_6]/c[CF_4]$

Figure 2.42   Relationship between the characteristic ratio and applied
voltage under surface discharge (1)

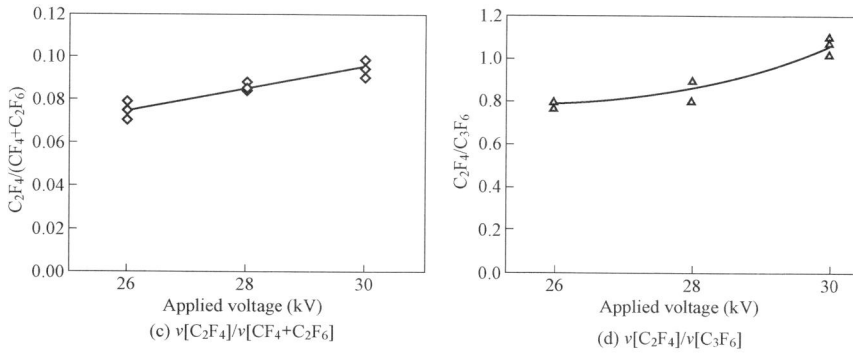

Figure 2.42　Relationship between the characteristic ratio and applied
voltage under surface discharge (2)

Comparing the decomposition component ratios $c[CO]/c[C_3F_6]$ under the three discharge defects and overheating defect conditions for the 10% C_4F_7N/90% CO_2 mixture, it can be noted that under overheating conditions, the ratio $c[CO]/c[C_3F_6]$ is consistently below 12, while under the three discharge conditions, the ratio $c[CO]/c[C_3F_6]$ is consistently above 30. Therefore, the numerical values of the content ratios can be used as a preliminary means to differentiate between discharge and overheating faults.

2.4.5.3　Set of Ratio-Based Characteristic Features

When $k(C_4F_7N)$ is in the range of 8% to 15%, there are significant differences in the variations of the content ratio $c[CO]/c[C_3F_6]$ and the peak area ratio $v[C_2F_4]/v[CF_4+C_2F_6]$ under various discharge conditions. Therefore, taking the content ratio $c[CO]/c[C_3F_6]$ and the peak area ratio $v[C_2F_4]/v[CF_4+C_2F_6]$ as a "feature pair", denoted as $v[C_2F_4]/v[CF_4+C_2F_6]$, $c[CO]/c[C_3F_6]$, the distribution of various discharge types in a two-dimensional space is shown in Figure 2.43. The three discharge types exhibit distinct regions in the two-dimensional space, with corona discharge and spark discharge showing a basically "L"-shaped distribution along the coordinate axes, and surface discharge located outside the "L" shape. This "feature pair" also has a certain ability to differentiate between different electrode materials in the corona discharge form. This reflects that the content ratio $c[CO]/c[C_3F_6]$ and the peak area ratio $v[C_2F_4]/v[CF_4+C_2F_6]$

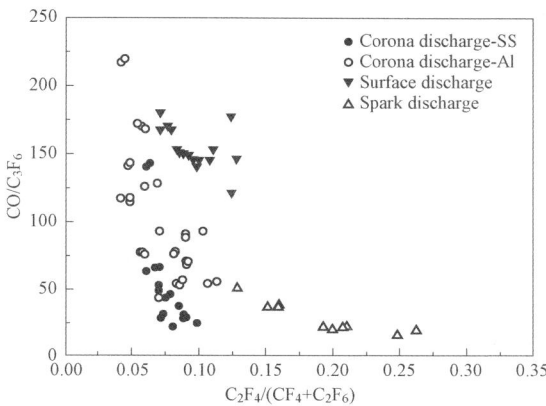

Figure 2.43　Relationship between the discharge type
and characteristic pair $v[C_2F_4]/v[CF_4+C_2F_6]$,
$c[CO]/c[C_3F_6]$

are feasible criteria for identifying discharge types.

Figure 2.44 shows the distribution of the three-ratio "feature set" composed of the ratio $c[C_2F_6]/c[CF_4]$, peak area ratio $v[C_2F_4]/v[C_3F_6]$, and $v[C_2F_4]/v[CF_4+C_2F_6]$ in a ternary coordinate system. It can be observed that high-temperature overheating faults and discharge faults have distinct distribution areas. Overheating faults are mainly distributed in the region where $v[C_2F_4]/v[C_3F_6]=0$ and $c[C_2F_6]/c[CF_4]=1$, while discharge faults are distributed in the region where $v[C_2F_4]/v[C_3F_6]>0.25$ and $c[C_2F_6]/c[CF_4]<0.75$. This is mainly because under high-temperature overheating faults ($T \leqslant 650$ ℃), although C_2F_4, CF_4, and C_2F_6 gases are produced, their concentrations are much lower than that of C_3F_6, making the ratio $v[C_2F_4]/v[C_3F_6]$ almost zero. According to the overheating decomposition experiment, the concentrations of CF_4 and C_2F_6 are comparable, making the ratio $c[C_2F_6]/c[CF_4]$ almost 1. In addition, the fault points of corona discharge, spark discharge, and surface discharge are also relatively concentrated. The three-ratio "feature set" in the ternary coordinate system clearly distinguishes different fault types and has a certain degree of recognition for electrode materials and overheating temperatures.

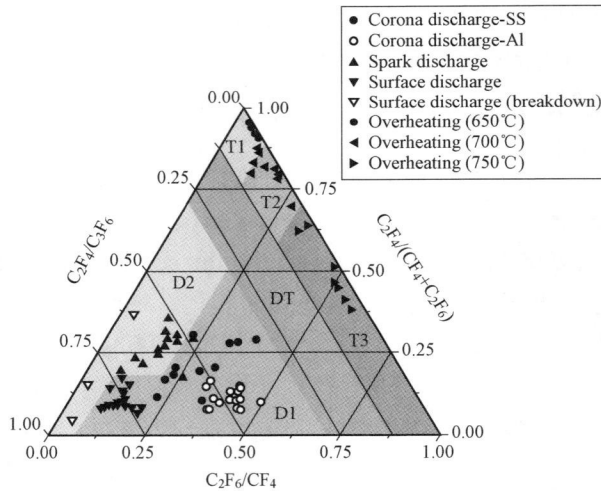

Figure 2.44   Distributions of discharge and overheating defect regions

Drawing inspiration from the David ternary diagram, the fault distribution areas displayed by the three-ratio "feature set" in the ternary coordinate system were preliminarily divided. The region where high-temperature overheating faults are located is designated as T, with subre-gions T1 representing overheating faults at 650 ℃, T2 at 700 ℃, and T3 above 700 ℃. Since the experimental data points in this study did not fall into the region between discharge and overheating faults, the region DT represents an uncertain area without discharge or overheating faults. Region D indicates the presence of discharge faults, with D1 representing discharge faults with relatively low discharge energy and D2 indicating the presence of breakdown-type discharge faults. Compared to spark discharge, corona discharge, and local surface discharge have relatively low discharge energy and stable characteristic ratio values. Therefore, corona discharge and local surface discharge are classified into region D1. However, compared to corona discharge, spark discharge has higher energy, less stable characteristic ratio values, and more scattered data points, with breakdown-type surface discharge data points clearly separated from the concentrated region of local surface discharge data points. Therefore, spark discharge and breakdown-type surface discharge are classified into region D2.

2. 5　Compatibility Characteristics of C_4F_7N and Solid Materials

The internal solid materials in gas-insulated electrical equipment mainly include metal materials, rubber sealing materials, solid insulation materials, and absorbents.

This section focuses on the compatibility characteristics of solid absorbents, rubber sealing materials, and epoxy resin under normal conditions (C_4F_7N/CO_2 gas atmosphere) and fault conditions (C_4F_7N/CO_2/decomposition by-products).

2. 5. 1　Solid Absorbent Materials

2. 5. 1. 1　Experimental Platform and Procedure

Due to the 90% volume fraction of the buffer gas CO_2, scanning the fragment with a massto-charge ratio $m/z=44$ would saturate the detector, making it impossible to detect. Therefore, the peak area of CO_2 gas at mass-to-charge ratio $m/z=16$ was selected for integration, and the pressure gauge readings were observed to assess the absorption performance of the absorbent for CO_2 gas.

(1) Absorbent Material and Quantity

The 3A, 4A, and 5A molecular sieves and activated alumina γ-Al_2O_3 spherical particles were used in the experiment. The particle size of the absorbent is 3-5 mm.

According to the Chinese standard GB/T 34320—2017, a 10% absorbent dosage ratio was used for the experiment, meaning the absorbent mass was 10% of the C_4F_7N/CO_2 mixed gas, equivalent to 2% of the SF_6 gas mass under the same conditions.

(2) Experimental Setup

Figure 2.45 illustrates the schematic diagram of the absorption experiment platform. The volume of the overheating device is approximately 24 L, and the absorption experiment tank is made of 304 stainless steel with a volume of about 1.35 L.

(3) Absorption Experiment Procedure

Overheating experiments were conducted with 10% C_4F_7N/90% CO_2 mixed gas at 0.6 MPa. Then, the gas after the experiment was injected into the experimental tank containing the absorbent. The specific experimental procedure is as follows.

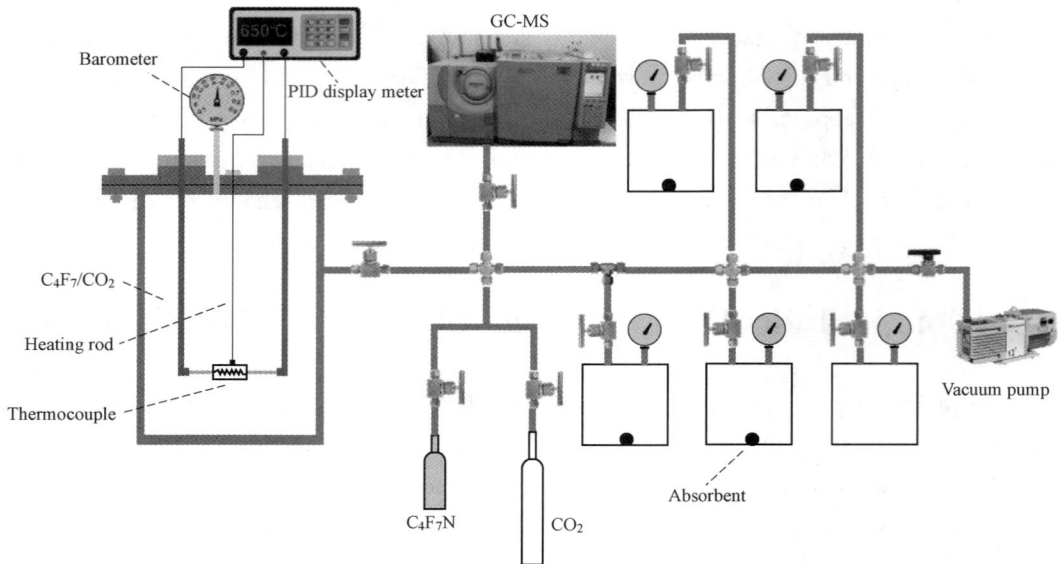

Figure 2.45   Schematic diagram of absorption experimental platform

- Use a halogen leak detector to check for leaks in all experimental tanks. If there is no leakage, proceed to the next step.
- Wipe the inner walls of the experimental tank clean with anhydrous ethanol and air-dry them. Weigh 1.6 g of absorbent, bundle it with a pure cotton cloth, and place the absorbent in four small experimental tanks.
- Set up a control group without absorbent. Insert the heating rod into the overheating device.
- After assembling all the experimental tanks, evacuate the overheating device and the small experimental tanks containing the absorbent for 1 hour and then close all valves.
- Open the CO_2 gas cylinder, introduce CO_2 gas into the overheating device. Stand for 20 minutes, evacuate, and repeat the process four times to thoroughly clean the overheating device, eliminating impurities such as moisture and oxygen that may interfere with the overheating experiment.
- Introduce 10% C_4F_7N into the overheating device, and finally fill it with CO_2 gas until the gauge pressure is 0.5 MPa.
- Let it sit for 12 hours before starting the local overheating experiment, which lasts for 3 hours.
- After the overheating experiment, let it sit for 8 hours to allow the decomposition products to diffuse evenly, and the gas temperature to drop to room temperature.
- Clean the small experimental tanks twice with high purity (99.999%) Helium gas, eva-

cuate for 1 hour each time to eliminate impurities such as moisture in the experimental tanks.

- After 8 hours, open the valves of the small experimental tanks and inject the gas from the overheating experiment into the small experimental tanks.
- Once the pressure in the overheating device and the small experimental tanks is balanced, close all valves. At this point, the pressure is 0.35 MPa (absolute pressure 0.45 MPa), and the absorption characteristic experiment begins.
- The five experimental tanks are placed at room temperature for 15 days, and the gas changes in the four small experimental tanks containing absorbent are periodically and quantitatively analyzed using GC/MS.

2.5.1.2 Theoretical Calculation of Molecular Size Parameters

For a molecule to be adsorbed, its dynamic diameter must be smaller than the pore diameter of the molecular sieve, allowing the molecule to enter the crystal cavity and be adsorbed. This property imparts high selectivity to molecular sieves. Molecular sieves themselves are highly polar materials with effective absorption of polar molecules, while their absorption capacity for non-polar molecules is weaker. Therefore, the absorption on molecular sieves is correlated with the dynamic diameter, size, shape, and polarity of molecules. To better explain the absorption principles of the absorbents, quantum chemistry theory was used. Using the Multiwfn 3.6 software package, its quantitative molecular surface analysis function is utilized. Both a spherical model and a rectangular box model describe the molecular shapes of C_4F_7N, CO_2, and various decomposition components, calculating the size parameters for each molecule.

Initially, the Gaussian16 software was used to perform geometry optimization for each molecule at the M06-2X/6-311G(d,p) level, obtaining the corresponding wavefunction files. Subsequently, the Multiwfn software was employed to load the wavefunction files, construct molecular surfaces, and analyze surface properties. Bader proposed using an isosurface with an electron density of 0.001 a.u. as the van der Waals surface in the gas phase, which has been widely accepted. Therefore, in the quantitative molecular surface analysis, the 0.001 a.u. isosurface was used as the van der Waals surface for the molecules.

For CF_4, which has a tetrahedral spatial configuration, the spherical model is suitable for describing the molecule. The distance between the two farthest points on the CF_4 molecule surface was taken as the molecular diameter. Other molecules are mostly planar or long-chain molecules, suitable for description using a rectangular box model, where the length, width, and height of the rectangular box represent the molecular shape.

Figure 2.46 displays the van der Waals surfaces and dimensions of C_4F_7N, CO_2, and various decomposition components.

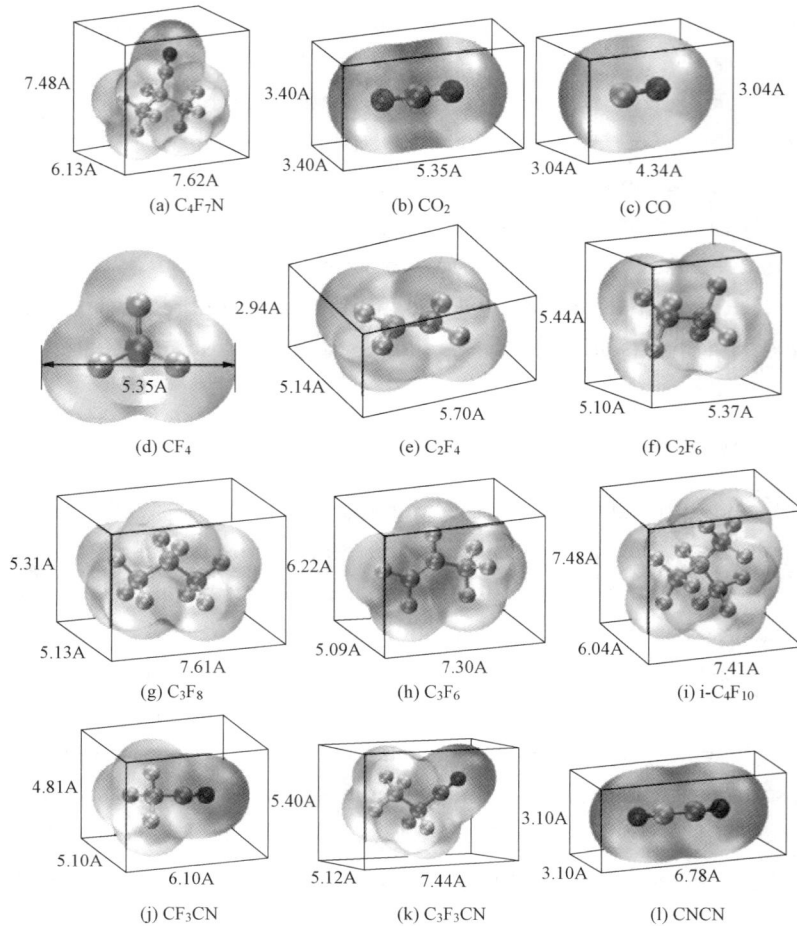

Figure 2.46   The van der Waals surface and dimensional parameters of molecules

2.5.1.3   Absorption Characteristics of CO_2 and C_4F_7N

(1) C_4F_7N

Figure 2.47 shows the variation of chromatographic peak areas of C_4F_7N with absorption time for four absorbents. At 0.45 MPa, γ-Al_2O_3 can adsorb C_4F_7N, while 3A, 4A, and 5A molecular sieves exhibit weaker absorption capabilities for C_4F_7N. This is primarily because molecular sieves have relatively uniform pore sizes, and when the dynamic diameter of a molecule is smaller than the pore size of the molecular sieve, it can enter the crystal cavity and be absorbed. The reported window pore sizes for 3A, 4A, and 5A molecular sieves are approximately 3 Å, 3.6-4 Å, and 4.2-4.4 Å, respectively. These values are smaller than the dynamic diameter of C_4F_7N, making it difficult for C_4F_7N molecules to enter the interiors of 3A, 4A, and 5A molecular sieves. However, γ-Al_2O_3 has an uneven pore size distribution, ranging from 1 to 40 Å, with a significant proportion in the range of 4 to 10 Å. As shown in Figure 2.47,

C_4F_7N molecules can enter the pores of γ-Al_2O_3, and the CN group in C_4F_7N can form weak interactions with the Al atoms in γ-Al_2O_3, allowing γ-Al_2O_3 to firmly adsorb C_4F_7N molecules [15].

(2) CO_2

Figure 2.48 shows the variation of chromatographic peak areas of CO_2 gas with absorption time for four absorbents. The changes in the peak area of CO_2 gas during the experiment are not significant, indicating weak absorption performance of the four absorbents for CO_2 gas. The dynamic diameter of CO_2 is greater than the pore size of the 3A molecular sieve, so the 3A molecular sieve does not absorb CO_2 gas. In the experiment, we observed that the performance of 4A and 5A molecular sieves for absorbing CO_2 gas is relatively weak. However, in industry, 4A and 5A molecular sieves are commonly used to absorb CO_2 gas in mixed gases under high temperature and pressure conditions to achieve gas purification. Temperature has a significant impact on molecular sieves, and the size of molecular sieve crystal cavities decreases as the temperature decreases. Therefore, for some molecules, molecular sieves only function at specific temperatures. In comparison to industrial absorption conditions, our experimental process involved lower temperatures and pressures, with a temperature of approximately 25 ℃ and a pressure of 0.45 MPa. Additionally, the amount of absorbent used in the experiment was much less than that used in industry, leading to significant differences between our experimental results and actual industrial conditions.

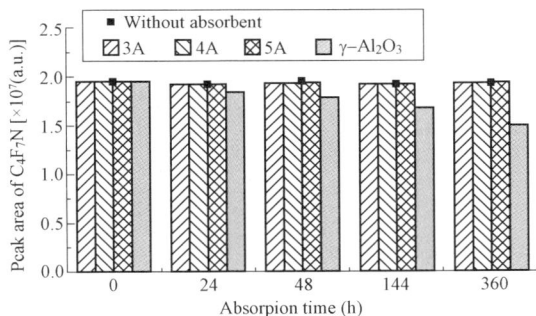

Figure 2.47 Absorption characteristics of activated alumina and molecular sieves for C_4F_7N

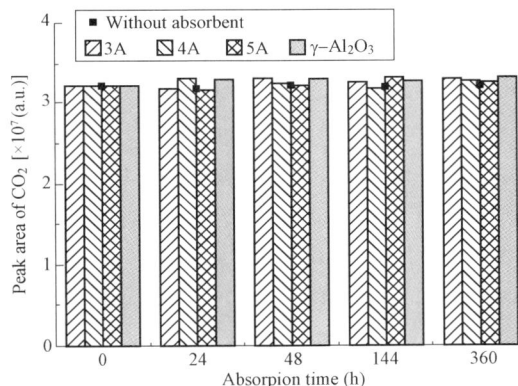

Figure 2.48 Absorption characteristics of activated alumina and molecular sieves for CO_2

2.5.1.4 Absorption Characteristics of decomposition by-products

Based on the local overheating decomposition characteristics, the decomposition components of C_4F_7N/CO_2 overheating decomposition can be categorized into three types.

- Small inorganic gas CO.
- Perfluorocarbons (PFCs) such as CF_4, C_2F_6, and C_3F_8.
- Nitrile gases CF_3CN, C_2F_5CN, and CNCN.

(1) CO

Figure 2.49 illustrates the variation of the chromatographic peak areas of CO gas over time under the absorption of 3A, 4A, and 5A molecular sieves, and γ-Al$_2$O$_3$. The peak area of CO gas remains nearly constant, indicating a weak absorption capacity of these four absorbents for CO gas. Due to the presence of feedback π bonds in CO molecules, the polarity of CO molecules is relatively weak. Additionally, the high absorption-free energy of the 4A molecular sieve for CO molecules results in weak electrostatic interactions between CO and the molecular sieve, relying only on weak van der Waals forces for absorption.

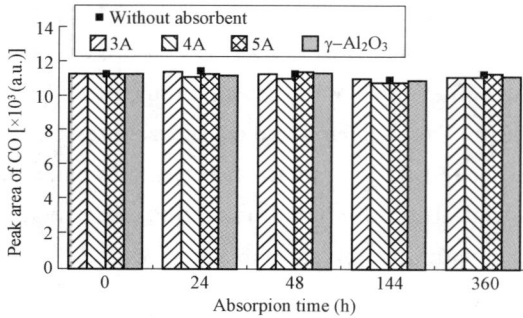

Figure 2.49   Absorption characteristics of activated alumina and molecular sieves for CO

As shown in Figure 2.46, the size of CO molecules is much smaller than the pore size of γ-Al$_2$O$_3$, and the linear structure of CO molecules allows for relatively easy ingress and egress through the pores of γ-Al$_2$O$_3$. Therefore, the absorption capacity of γ-Al$_2$O$_3$ for CO is comparatively weak.

(2) PFCs

We compared the chromatographic peak areas of perfluorocarbons (PFCs) after 360 hours, both without absorbents and under the influence of absorbents. The trend of peak area variation over time is depicted in Figure 2.50. In the absence of absorbents, the gas content of PFCs remains essentially the same, indicating a weak absorption capacity of the four absorbents for PFCs. This is primarily due to the large size of these molecules compared to the pore size of the molecular sieves. Additionally, CF$_4$, C$_2$F$_4$, and C$_2$F$_6$ are non-polar molecules, while C$_3$F$_6$, C$_3$F$_8$, and i-C$_4$F$_{10}$ have branched structures and larger sizes, making it challenging for them to enter the crystal pores of the molecular sieves. Furthermore, strong electrostatic repulsion exists between these PFC molecules and the framework of the molecular sieves, making absorption difficult. Consequently, the three types of molecular sieves exhibit almost no absorption of these PFC gases among the decomposition components.

As depicted in Figure 2.46, under the influence of γ-Al$_2$O$_3$, the content of these PFCs remains nearly unchanged. This is mainly attributed to the small molecular size of these PFCs, which allows for relatively easy

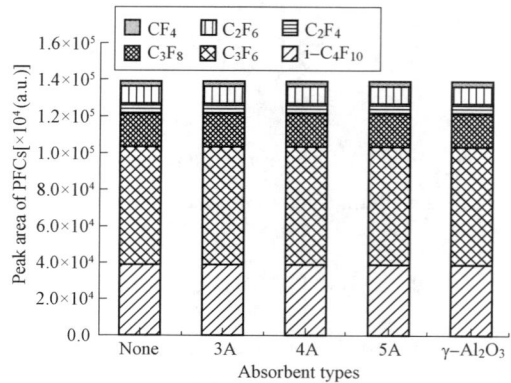

Figure 2.50   Absorption characteristics of activated alumina and molecular sieves for PFCs

ingress and egress through the pores of γ-Al$_2$O$_3$. Consequently, the absorption capacity of γ-Al$_2$O$_3$ for PFCs is relatively weak.

(3) Nitrile gases

Figure 2.51 depicts the variation in peak areas of nitrile products (CF$_3$CN, C$_2$F$_5$CN, and CNCN) over absorption time under the influence of different absorbents.

From Figure 2.51(a), it is evident that all four absorbents can adsorb CNCN, with the absorption capacity ranking as follows: 5A＞γ-Al$_2$O$_3$＞4A＞3A. After one day of absorption, CNCN is completely absorbed by a 5A molecular sieve, and at 48 hours, γ-Al$_2$O$_3$ almost entirely absorbs CNCN. In contrast, the absorption capacity of 3A and 4A molecular sieves for CNCN gas is relatively weak, and their absorption performance tends to saturate with increasing absorption time.

As shown in Figure 2.51(b), the 5A molecular sieve and γ-Al$_2$O$_3$ can absorb CF$_3$CN gas, while 3A and 4A molecular sieves exhibit minimal absorption of CF$_3$CN gas. Furthermore, γ-Al$_2$O$_3$ demonstrates a higher absorption efficiency for CF$_3$CN gas compared to 5A molecular sieve.

From Figure 2.51(c), it is observed that only γ-Al$_2$O$_3$ adsorbs C$_2$F$_5$CN gas, while the other three absorbents show no significant absorption of C$_2$F$_5$CN gas.

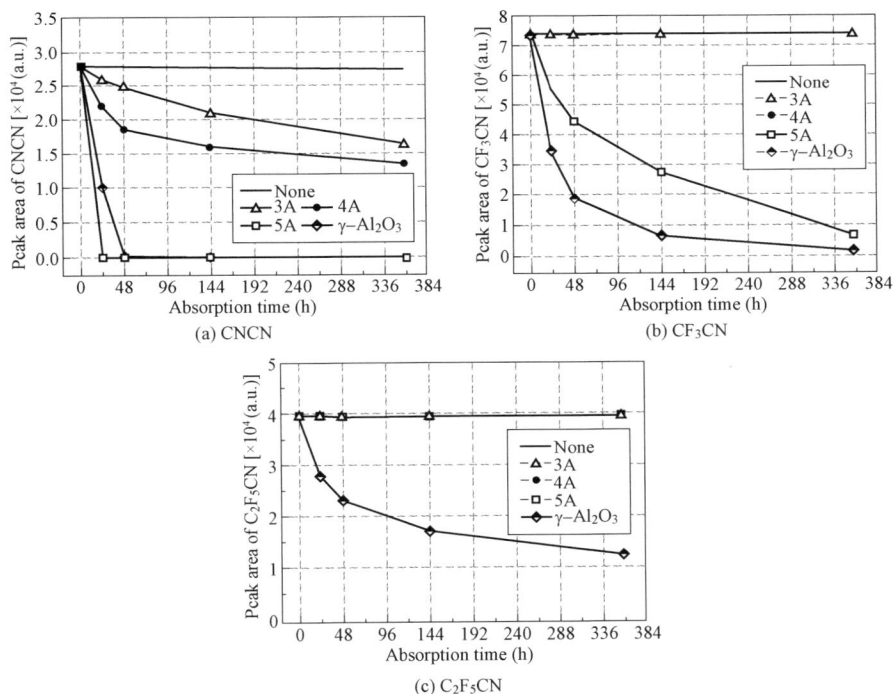

(a) CNCN

(b) CF$_3$CN

(c) C$_2$F$_5$CN

Figure 2.51　Absorption characteristics of activated alumina and molecular sieves for R-CN

It is observed that CNCN molecules have a linear structure with a molecular dynamic diameter of approximately 0.3 nm, smaller than the pore sizes of 5A and 4A molecular sieves but close to the pore size of the 3A molecular sieve. Consequently, CNCN molecules can easily enter the crystal pores of 4A and 5A molecular sieves but face difficulty entering those of the 3A molecular sieve. Therefore, in the absorption process of CNCN gas, the absorption efficiency of the 3A molecular sieve is lower than that of the 4A and 5A molecular sieves. Additionally, the CN group in CNCN molecules is electronegative, capable of forming weak interactions with Na^+ ions in the molecular sieve crystals, facilitating the absorption of CNCN by the molecular sieve. As the selected absorbents have the same particle size, another reason for the difference in CNCN absorption capacity among the three molecular sieves might be their distinct specific surface areas. The specific surface areas of 5A, 4A, and 3A molecular sieves are approximately 312.95 m^2/g, 168.58 m^2/g, and 33.00 m^2/g, respectively. Due to the higher surface area per unit volume of 5A molecular sieve, its ability to absorb CNCN gas surpasses that of 3A and 4A molecular sieves.

CF$_3$CN and C$_2$F$_5$CN molecules are larger and cannot enter the crystal pores of 4A and 3A molecular sieves, leading to minimal absorption of these two nitrile gases by 4A and 3A molecular sieves. However, the size of CF$_3$CN molecules is close to the pore size of the 5A molecular sieve, enabling them to enter the crystal pores. Additionally, the strong polarity inside the crystal pores of the 5A molecular sieve allows for strong electrostatic and van der Waals interactions with the polar CF$_3$CN molecules, firmly absorbing them within the crystal pores. As indicated in Figure 2.46, the size of C$_2$F$_5$CN molecules exceeds the free aperture of the 5A molecular sieve, preventing the absorption of C$_2$F$_5$CN by the 5A molecular sieve.

From Figure 2.46, it is evident that the molecular sizes of these three nitrile gases range from 3 to 8 Å, allowing them to enter the interior of γ-Al$_2$O$_3$. The CN group in these nitrile gases forms weak interactions with Al atoms in γ-Al$_2$O$_3$, facilitating their absorption by γ-Al$_2$O$_3$.

Furthermore, the variation in gas peak areas over time in the control group without absorbents indicates that the content of decomposition gases in the tank remains nearly unchanged within 15 days, suggesting a relatively weak performance of the C$_4$F$_7$N/CO$_2$ mixed gas.

2.5.2 Rubber Sealing Materials

In gas-insulated electrical equipment, the sealing rings come into contact with insulating gas on one side and are exposed to the atmosphere on the other side. In the event of an internal discharge or overheating faults, the decomposition of C$_4$F$_7$N/CO$_2$ will produce gases such as CF$_4$, C$_2$F$_6$, C$_3$F$_6$, C$_3$F$_8$, HF, CNCN, CF$_3$CN, etc. Despite their relatively low concentrations, these decomposition products, due to their weak composite ability, will persist and corrode rubber, accelerating the aging process of the rubber and posing a threat to the long-term safe and stable operation of the equipment. Commonly used rubber sealing materials in electrical equipment include ethylene propylene diene monomer rubber (EPDM), nitrile rubber (NBR), chloroprene rubber (CR), and fluoro rubber (FKM). Therefore, this section investigates the compatibility of these four rubber types (EPDM, NBR, CR, and FKM) under normal conditions (C$_4$F$_7$N/CO$_2$ gas atmosphere) and in the presence of defects (C$_4$F$_7$N/CO$_2$/decomposition by-product atmosphere). To eliminate the impact of temperature on the test results, a helium (He atmosphere) control group was set up. Additionally, to exclude the influence of the thermal decomposition of the C_4F_7N/CO_2 mixed gas itself on gas detection, another control group was established by filling the test tank with 10% C_4F_7N/90% CO_2 mixed gas and heating it at 70 ℃ for 28 days [16, 17].

The four types of rubber used in the experiment are ethylene propylene diene monomer rubber (EPDM, model HX807), nitrile rubber (NBR, model 5080-1), chloroprene rubber, and fluororubber (FKM, model F210).

Following the standards GB/T 528—2009 and GB/T 7759.1—2015, chemical composition changes and compression performance of rubber samples were determined using two forms of samples: dumbbell-shaped and cylindrical as shown in Figure 2.52. The dumbbell-shaped samples had a test length of 12 mm, a thickness of 2 mm, a narrow part width of 2 mm, a total length of 35 mm, and an end width of 6 mm. The cylindrical rubber samples had a diameter of 29 mm and a height of 12.5 mm. Dumbbell-shaped samples were used for observing surface morphology and detecting chemical composition, while cylindrical samples were employed for measuring compression permanent deformation.

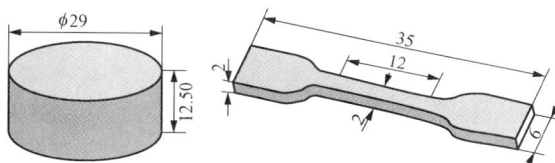

Figure 2.52   Size parameters of rubber samples (mm)

(1) Gas GC/MS Analysis

Gas samples from the stainless-steel gas tank before and after the experiment were analyzed using a gas chromatography-mass spectrometry (GC/MS) instrument. The detection methods for gas chromatography (GC) and mass spectrometry (MS) are the same as those described in Section 2.4.3, with the addition of quantifying COS gas by increasing the mass-to-charge ratio (m/z) to 60.

(2) Surface Morphology Analysis

* Macroscopic Morphology Observation:

  Macroscopic observations were conducted to check for surface phenomena such as cracking, wrinkling, shrinkage, or frosting on the rubber samples after the test.

* Microscopic Morphology Analysis:

  Scanning electron microscopy with energy-dispersive X-ray spectroscopy (SEM/EDS) was used to observe the microscopic morphology of rubber surfaces before and after the test. The rubber surface's chemical composition and element composition were determined. The samples were cut into small pieces with a thickness of 2 mm and a width of 3 mm and affixed to the sample stage using conductive adhesive. Since rubber is a non-conductive polymer material, surface charging can make imaging and focusing challenging. To improve SEM image quality, a gold coating was applied to the sample surface using a vacuum sputtering device before detection.

(3) Fourier Transform Infrared Spectroscopy (FTIR) Analysis

The Nicolet iS50 Fourier transform infrared spectrometer produced by Thermo Fisher Scientific was used to analyze the structural information of rubber samples before and after compatibility accelerated testing. The spectral range was 4,000 cm^{-1} to 400 cm^{-1}, optical resolution was 2 cm^{-1}, wavenumber accuracy was 0.241 cm^{-1}, and scanning was performed 16 times. Since rubber is an opaque material, attenuated total reflectance (ATR) sampling technology was employed for surface analysis.

(4) Mechanical Performance Analysis

The mechanical properties of rubber are crucial in its practical application. Compression permanent deformation, hardness, tensile strength, and other mechanical properties play important roles in different usage environments. In this study, compression permanent deformation, which is most relevant to static sealing, was chosen as the evaluation index. The changes in the mechanical properties of rubber before and after the test were analyzed to provide reference information for the selection and optimization of rubber seals for electrical equipment using C_4F_7N/CO_2 mixed gas.

According to the Chinese standard GB/T 7759.1—2015, rubber compression permanent deformation tests were conducted with a compression time of 28 days and a compression rate of 25%. Each type of rubber underwent three parallel tests (three samples compressed). After 28 days, the samples were removed from the compression device and left at room temperature for 3 hours, first, the gas changes after the test were measured using GC/MS. Subsequently, a desktop digital rubber thickness gauge was used to measure the heights at five positions in the center of the cylindrical rubber sample. The measurement range of the thickness gauge is 0-12.7 mm, with an accuracy of 0.01 mm, a pressing foot diameter of ϕ6 mm, and an applied pressure of 22 kPa$\pm$5 kPa. The compression permanent deformation C was calculated to analyze the changes in mechanical properties of the four types of rubber in three different atmospheres. The compression permanent deformation C is given by Equation (2.12).

$$C = \frac{h_0 - h_1}{h_0 - h_s} \times 100\% \tag{2.12}$$

Where h_0 —— the initial height of the sample, mm;

   h_1 —— the height of the sample after recovery, mm;

   h_s —— the height of the limiter, mm.

2.5.2.1   Gas Composition Changes

Before the compatibility test, the decomposition gases in the atmosphere of $C_4F_7N/CO_2/$ decomposition by-products mainly included CO (1,640 μL/L), CF$_4$ (513 μL/L), C$_2$F$_6$ (47 μL/L),

C_3F_6 (3 μL/L), C_3F_8 (6 μL/L), as well as C_2F_4, CF_3CN, CNCN, and C_2F_5CN.

After the experiment under three different atmospheres, the detected gases included CO (m/z=12), CO_2 (m/z=44), CH_4 (m/z=16), COS (m/z=60), CS_2 (m/z=76), C_3F_6 (m/z=69), and CHF_3(m/z=69). CO_2 gas was detected under the He atmosphere after the experiment, but its quantity was not analyzed as CO_2 served as a buffer gas.

From Figure 2.53, it can be observed that EPDM rubber samples produced CO, CO_2, and trace amounts of CS_2 gas in the helium atmosphere. In both C_4F_7N atmospheres, EPDM generated CO, CH_4, CS_2, COS, and C_3F_6 gases. Moreover, in the C_4F_7N atmospheres, the concentrations of CH_4, CS_2, and CO gases were higher than those in the helium atmosphere.

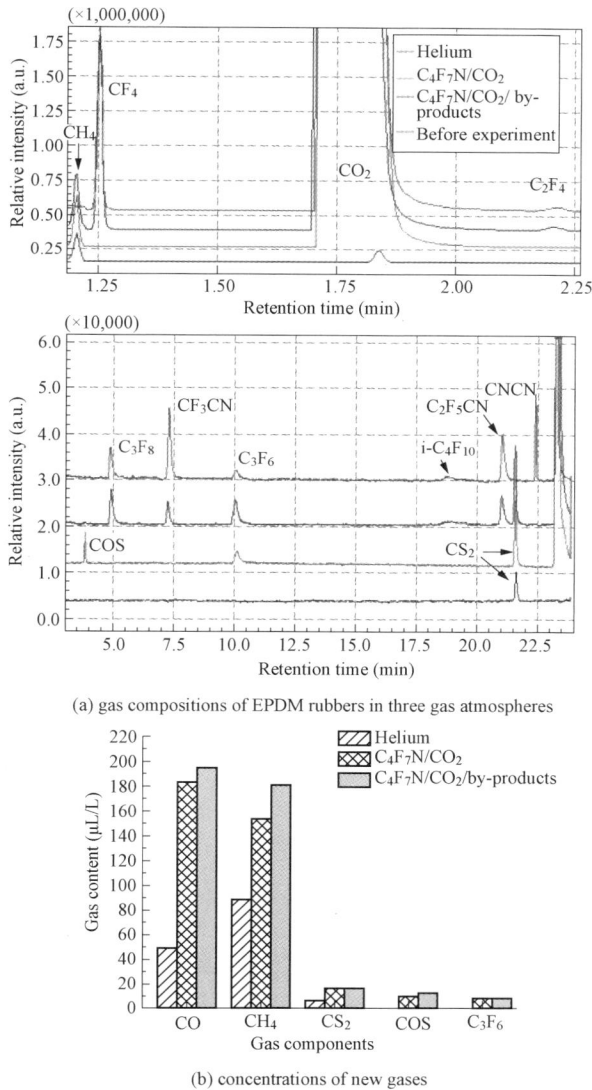

(a) gas compositions of EPDM rubbers in three gas atmospheres

(b) concentrations of new gases

Figure 2.53   Qualitative and quantitative results of EPDM rubbers after

compatibility test in three atmospheres

From Figure 2.54, it can be observed that NBR rubber samples underwent significant thermal decomposition in the helium atmosphere, generating large amounts of CO, CO_2, and CH_4 gases. The content of CO reached up to 333 μL/L, and the CH_4 was 178 μL/L. In the presence of C_4F_7N, NBR also produced C_3F_6 gas. In the atmosphere of C_4F_7N/CO_2/decomposition by-products, it generated a substantial amount of CO gas, approximately 10 times higher than in the helium atmosphere and six times higher than in the C_4F_7N/CO_2 gas atmosphere. Additionally, the types of gases produced by NBR in the presence of C_4F_7N were fewer than those generated by EPDM.

From Figure 2.55, it is evident that CR rubber samples underwent thermal decomposition in the helium atmosphere, producing CO, CO_2, and CH_4 gases. In the atmospheres of C_4F_7N/CO_2 and C_4F_7N/CO_2/decomposition by-products, CR also generated CS_2, COS, and C_3F_6 gases. The order of gas concentrations among these six gases in the three atmospheres, from highest to lowest, is C_4F_7N/CO_2/decomposition by-products atmosphere$>C_4F_7N/CO_2$ atmosphere$>$He atmosphere. This indicates that C_4F_7N and its decomposition products reacted with CR rubber samples.

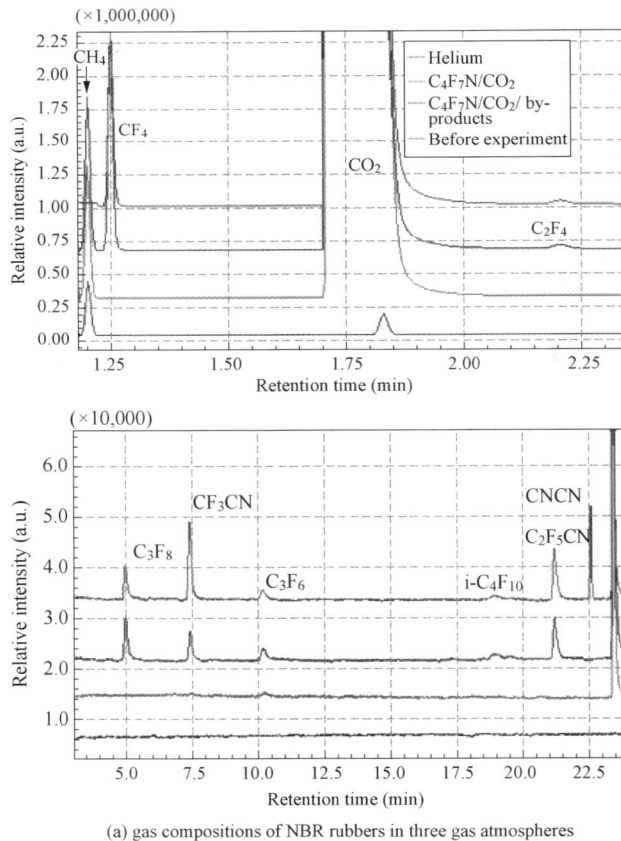

(a) gas compositions of NBR rubbers in three gas atmospheres

Figure 2.54   Qualitative and quantitative results of NBR rubbers after compatibility test in three gas atmospheres (1)

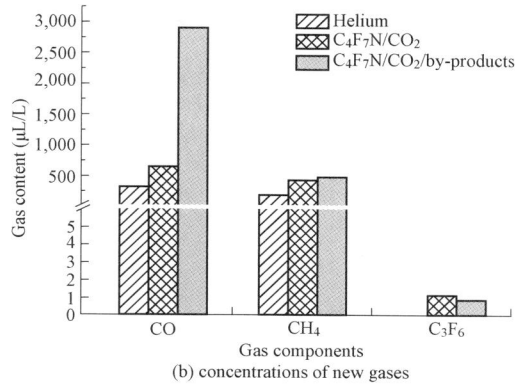

(b) concentrations of new gases

Figure 2.54   Qualitative and quantitative results of NBR rubbers after
compatibility test in three gas atmospheres (2)

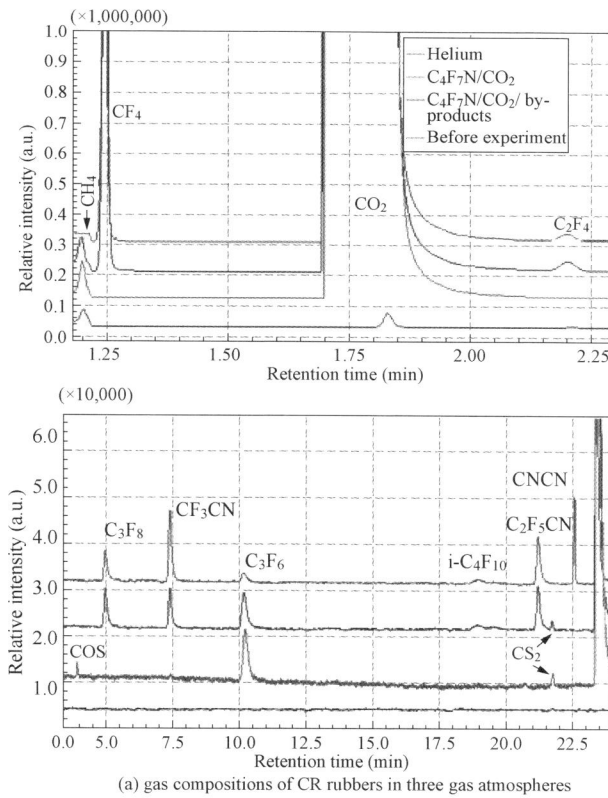

(a) gas compositions of CR rubbers in three gas atmospheres

Figure 2.55   Qualitative and quantitative results of CR rubbers after
compatibility test in three gas atmospheres (1)

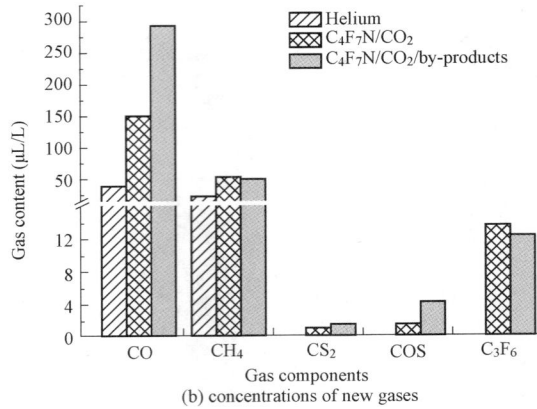

Figure 2.55   Qualitative and quantitative results of CR rubbers after

compatibility test in three gas atmospheres (2)

From Figure 2.56, it can be observed that FKM rubber samples underwent minimal thermal decomposition in the helium atmosphere, producing a small amount of CO and trace amounts of CHF$_3$ gas. In the presence of C$_4$F$_7$N, it also generated C$_3$F$_6$, indicating a direct

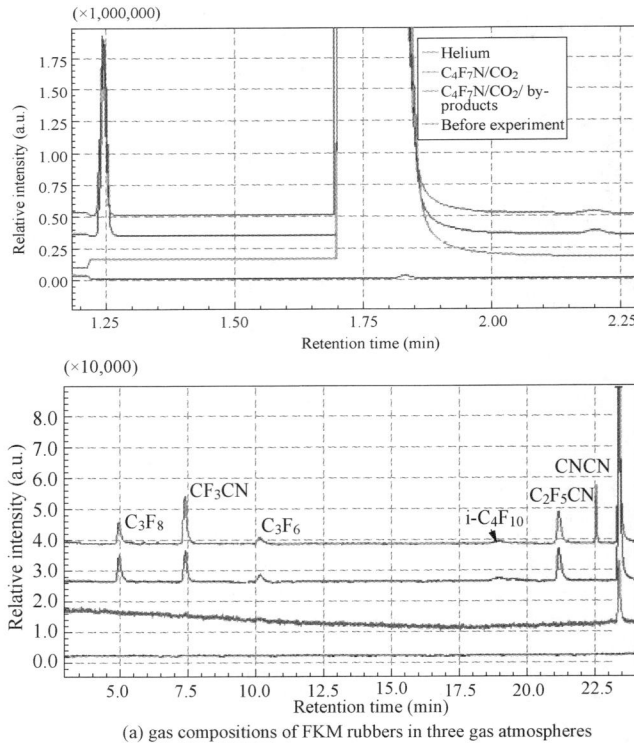

Figure 2.56   Qualitative and quantitative results of FKM rubbers after compatibility

test in three gas atmospheres (1)

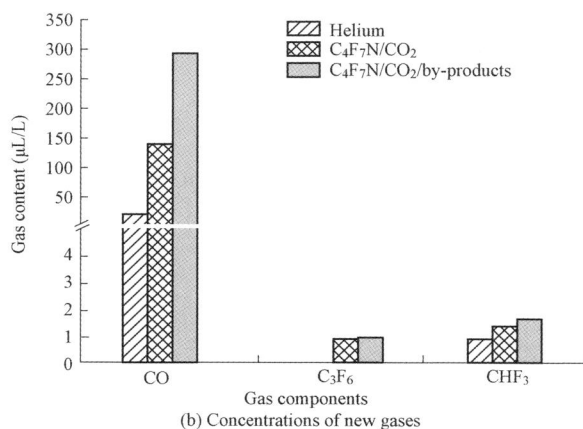

Figure 2.56   Qualitative and quantitative results of FKM rubbers after compatibility

test in three gas atmospheres (2)

association between the production of C_3F_6 and C_4F_7N and its products. In the monomer of FKM, F atoms partially replace H atoms, and the reactive H free radicals react with CF_3 groups generated by the thermal cleavage of C—C bonds, producing CHF_3. Comparing the types and concentrations of gases generated by the four rubbers in the three atmospheres, it can be observed that FKM, compared to the other three rubbers, produced the least amount of new gases and had the fewest types of newly generated gases after the compatibility test.

Rubber, during the formation of the elastic body, incorporates a certain amount of compounding agents. Common compounding agents include vulcanization accelerators (sulfur, accelerator M, accelerator NA-22, accelerator DM), vulcanization activators (magnesium oxide, zinc oxide), antioxidants (antioxidant D), fillers (carbon black), plasticizers (stearic acid, dibutyl sebacate), etc. Most vulcanization accelerators contain C and S elements, such as accelerator M (CAS 149-30-4, $C_7H_5NS_2$), accelerator DM (CAS 120-78-5, $C_{14}H_8N_2S_4$), and accelerator NA-22 (CAS 96-45-7, $C_3H_6N_2S$). These vulcanization accelerators mostly contain C-S-S-C disulfide bonds in their chemical structure, and these disulfide bonds are prone to breakage after heating, leading to the generation of CS_2. All four rubbers are sulfur-cured, commonly using sulfur as a vulcanization accelerator. Therefore, the types of cross-linking bonds are mostly polysulfide bonds, which are prone to breakage upon heating, further generating CS_2. Stearic acid, a rubber additive, contains the R—COOH group, which may decompose into reactive OH free radicals upon heating. COS can be generated by the reaction between CS_2 and reactive OH free radicals, as shown in Reaction (2.13).

$$CS_2 + OH\bullet \longrightarrow (S_2C—OH) \longrightarrow COS + SH\bullet \tag{2.13}$$

After the compatibility accelerated test, CH_4 gas was generated in EPDM, NBR, and CR under all three atmospheres. It is speculated that this might be due to the thermal decomposition of additives containing CH_3 groups in the rubber, resulting in the formation of CH_3

radicals. These CH_3 radicals further combine with H radicals to generate CH_4 gas. However, the specific impact of which additive is currently uncertain and requires further in-depth research.

C_3F_6 gas was detected in the C_4F_7N/CO_2 and C_4F_7N/CO_2/decomposition by-products atmospheres after the compatibility test, while it was not detected in the He atmosphere. This indicates a direct correlation between the generation of C_3F_6 gas and the insulation medium C_4F_7N. The C_4F_7N/CO_2 mixture without solid material, when heated at 70 ℃ for 28 days, did not produce new gases. From this, it can be inferred that the detected C_3F_6 gas after the test is produced by the reaction between rubber and C_4F_7N gas.

Figure 2.57 shows the total gas production from the four rubber types in C_4F_7N/CO_2 atmospheres. Under the C_4F_7N/CO_2 atmosphere, NBR rubber samples produce the highest total gas amount, reaching 1,075 μL/L, while FKM rubber samples produce the least amount, approximately 155 μL/L. The total gas production from all rubber types in the C_4F_7N/CO_2/decomposition by-product atmosphere is higher than that in the C_4F_7N/CO_2 atmosphere. In particular, the total gas production from NBR rubber samples under C_4F_7N/CO_2/decomposition by-product atmosphere is 3,413 μL/L, approximately three times the total gas production under C_4F_7N/CO_2 atmosphere. These results indicate that the reaction between C_4F_7N/CO_2 and NBR rubber is most significant, and the decomposition components promote the generation of more gases from the rubber, with the decomposition products having the greatest impact on NBR rubbers.

Figure 2.58 illustrates the chromatographic peak areas of nitrile compounds in the decomposition components before and after the compatibility test in C_4F_7N/CO_2/decomposition by-product atmosphere. After the compatibility test, the content of nitrile compounds in all four rubber types shows varying degrees of reduction. Among them, the content of CF_3CN and C_2F_5CN gases in the EPDM rubber is the lowest, while the content in the FKM rubber is the highest. This indicates that these two nitrile compounds react to a greater extent with the EPDM rubber and to a lesser extent with the FKM rubber. It is worth noting that CNCN gas was not detected in any of the four rubber types after the test, indicating a complete reaction of CNCN with all four rubbers.

Table 2.7 compares the types and quantities of gases produced by four rubber types in the atmospheres of C_4F_7N/CO_2 and C_4F_7N/CO_2/decomposition by-products. In the C_4F_7N/CO_2 atmosphere, EPDM rubber, and CR rubber produce five gases: CO, CH_4, CS_2, COS, and C_3F_6. NBR rubber produces three gases: CO, CH_4, and C_3F_6, while FKM rubber produces three gases: CO, C_3F_6, and CHF_3. NBR rubber produces the highest total amount of gases, and FKM produces the least. The order of total gas production from high to low is NBR rubber $>$ EPDM rubber $>$ CR rubber $>$ FKM rubber. Compared to the C_4F_7N/CO_2 atmosphere, the types of gases produced by the four rubber types remain unchanged in the C_4F_7N/CO_2/decomposition by-product atmosphere, but the total gas production significantly increases. The total gas production from NBR rubber is approximately 3.5 times that in the C_4F_7N/CO_2 atmosphere, with a substantial increase in CO content. Overall, FKM rubber exhibits the weakest reaction with C_4F_7N/CO_2, producing the least amount of gas. NBR rubber shows the strongest reaction with C_4F_7N/CO_2, and the influence of the fault decomposition product on NBR rubber is significant. EPDM rubber and CR rubber also produce toxic gases COS and CS_2 additionally. Therefore, from the perspective of gas production, FKM rubber has the least impact on the C_4F_7N/CO_2 mixture.

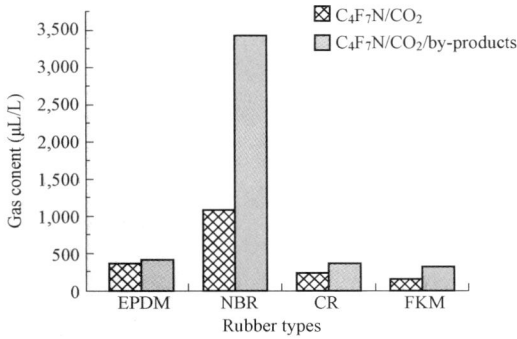

Figure 2.57 Total gas content of four rubbers in C_4F_7N atmospheres

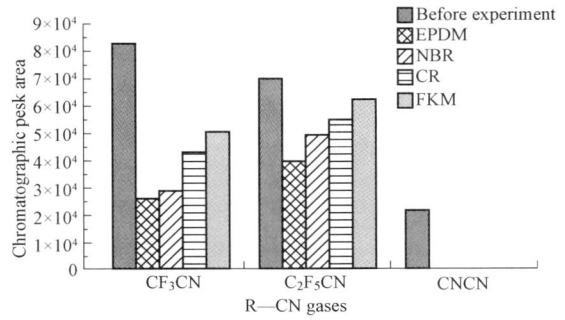

Figure 2.58 Peak areas of R—CN (R═CN, CF₃, C_2F_5) compound before and after compatibility test in C_4F_7N/CO_2/decomposition by-products

Table 2.7 Comparison of gases produced by rubber samples in C_4F_7N atmospheres after test

Gas atmosphere	Rubber type	Gas composition and concentration (μL/L)	Total concentration (μL/L)
C_4F_7N/CO_2	EPDM	CO(182), CH₄(152), CS₂(16) COS(10), C₃F₆(7)	368
	NBR	CO(653), CH₄(421), C₃F₆(1)	1,076
	CR	CO(151), CH₄(51), CS₂(1), COS(2), C₃F₆(14)	219
	FKM	CO(153), C₃F₆(1), CHF₃(2)	156
C_4F_7N/CO_2/ decomposition n by-products	EPDM	CO(194), CH₄(181), CS₂(16) COS(12), C₃F₆(7)	410
	NBR	CO(2922), CH₄(490), C₃F₆(1)	3,413
	CR	CO(292), CH₄(49), CS₂(2) COS(4), C₃F₆(13)	360
	FKM	CO(300), C₃F₆(1), CHF₃(2)	303

2.5.2.2 Appearance Changes

Figure 2.59 depicts the appearance of the four rubber types before and after the test in different atmospheres. NBR rubber exhibits a white surface and noticeable frosting in both C_4F_7N atmospheres but shows no significant changes in the He atmosphere. The surface morphology of the other three types of rubbers remains unchanged in all three atmospheres.

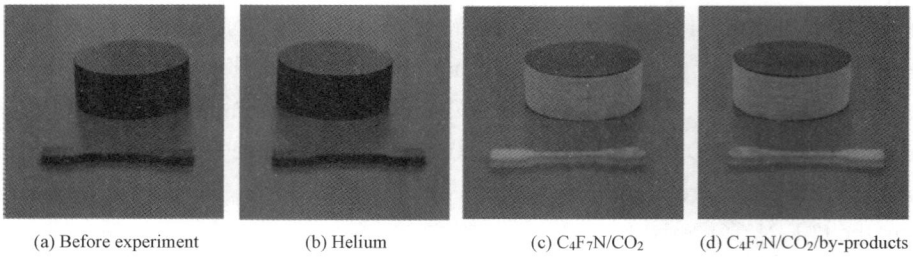

(a) Before experiment   (b) Helium   (c) C_4F_7N/CO_2 (d) C_4F_7N/CO_2/by-products

Figure 2.59   Appearance of NBR before and after compatibility test

2.5.2.3 SEM Analysis of Microscopic Morphology

Figure 2.60 shows the surface morphology of EPDM samples magnified by 350 times and 2,000 times before and after the compatibility test in three atmospheres. The original sample surface exhibits poor flatness with granular protrusions, possibly caused by additives in the rubber being encapsulated within, leading to uneven granular protrusions on the surface. The surface flatness of the rubber improves in the atmosphere of helium, with increased surface structural density. However, in both C_4F_7N atmospheres, the rubber surface develops cracks and becomes rough. At a magnification of 2,000 times, it is observed that many flaky substances precipitate on the rubber surface.

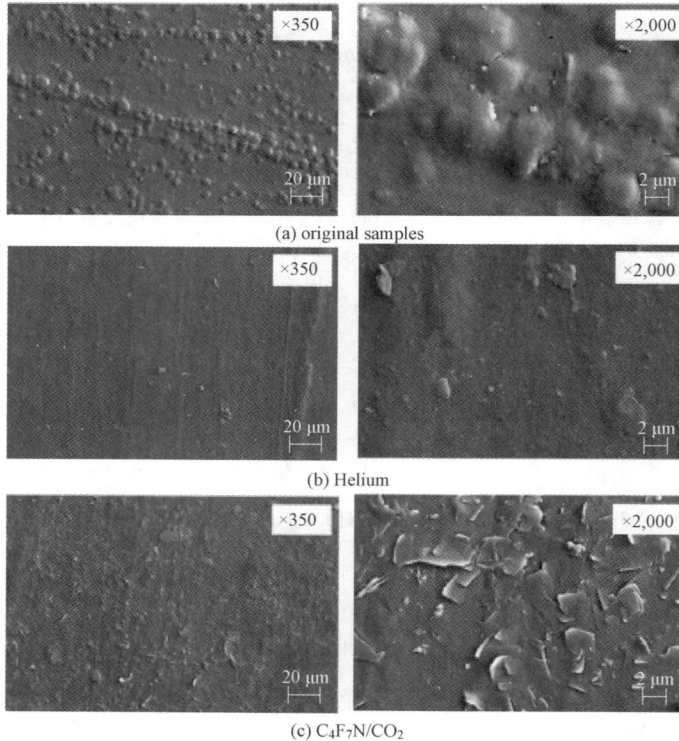

(a) original samples

(b) Helium

(c) C_4F_7N/CO_2

Figure 2.60   Surface morphology of EPDM rubber samples before and after compatibility test (1)

(d) C₄H₇N/CO₂/decomposition by-products

Figure 2.60   Surface morphology of EPDM rubber samples before and after compatibility test (2)

To determine the elemental composition of the flaky precipitates on the surface of EPDM rubber samples, energy-dispersive EDS measurements were conducted on the precipitates. Figure 2.61 presents the EDS elemental analysis results of EPDM rubber surfaces in C_4F_7N atmospheres. The rubber surface underwent gold sputter coating before detection, resulting in the detection of a significant amount of Au element in the EDS results. In addition to Au, the rubber surface detected C, O, and Zn elements. Since H-type EPDM rubber is polymerized from ethylene, propylene, and a third monomer, 1,4-hexadiene, its monomers have a carbon-hydrogen structure, explaining the abundant detection of C element in EDS. The ASTM-standard EPDM rubber formulation includes small amounts of inorganic and organic additives such as zinc oxide, carbon black, stearic acid, and sulfur accelerators, in addition to EPDM monomers. Hence, it is inferred that the O and Zn elements detected by EDS originate

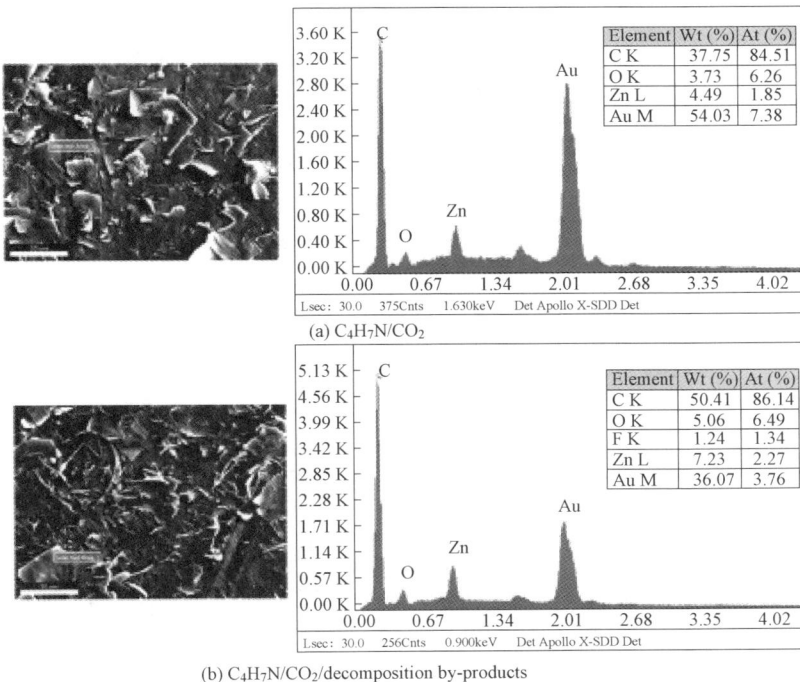

(a) C₄H₇N/CO₂

(b) C₄H₇N/CO₂/decomposition by-products

Figure 2.61   EDS results of EPDM rubber samples in C_4F_7N atmospheres after compatibility test

from additives or by-products of reactions in the rubber. The content (wt%) of C, O, and Zn elements in the solid precipitates on the surface of EPDM rubber exposed to $C_4F_7N/CO_2/$ decomposition by-product atmosphere is higher than that in C_4F_7N/CO_2 atmosphere, and trace amounts of F element are detected. This suggests that more solid material precipitates in the presence of fault decomposition products in the C_4F_7N/CO_2/decomposition by-product atmosphere.

Figure 2.62 displays SEM images of NBR rubber samples magnified by 350 times and 2,000 times before and after the compatibility test in three atmospheres. The original sample surface has good flatness, with no apparent change in the sample surface under He atmosphere. However, in the presence of C_4F_7N, a layer of white powdery substance forms on the surface of NBR rubber, and at 2,000 times magnification, a dense layer of flaky precipitates is observed adhering to the rubber surface. Since the NBR rubber's surface under He atmosphere does not exhibit "frosting" and the surface morphology remains unchanged, the influence of temperature on the rubber surface morphology can be ruled out. Therefore, it is speculated that C_4F_7N molecules may induce the precipitation of additives in the rubber or that the additives in the rubber react with C_4F_7N molecules to produce by-products that precipitate on the surface. This reduces the content of additives in the rubber, correspondingly decreasing the solubility of additives and ultimately leading to frosting on the rubber surface.

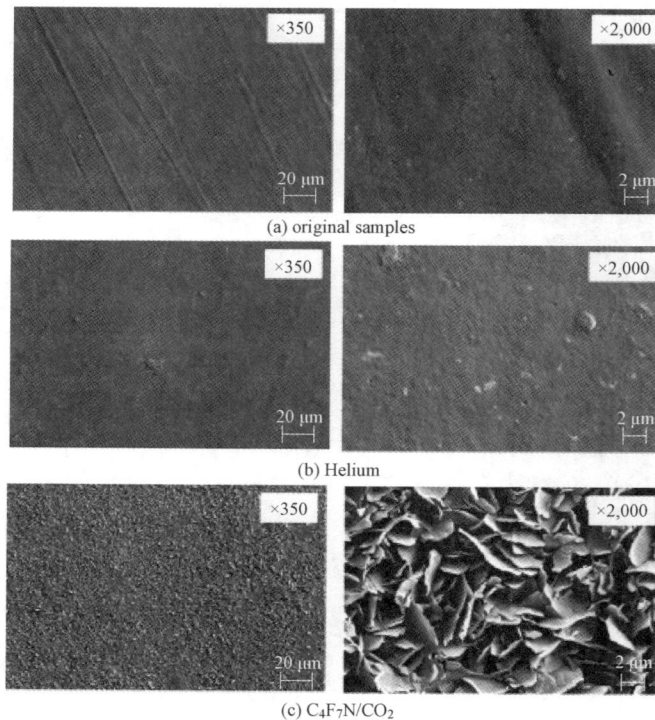

(a) original samples

(b) Helium

(c) C_4F_7N/CO_2

Figure 2.62   Surface morphology of NBR rubber samples before and after compatibility test (1)

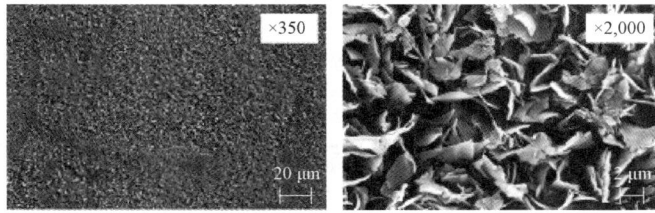

(d) C₄H₇N/CO₂/decomposition by-products

Figure 2.62   Surface morphology of NBR rubber samples before and after compatibility test (2)

Figure 2.63 displays the EDS elemental analysis of the NBR rubber sample surface. In addition to the Au element, the surface of the rubber also detected C, O, and Zn elements, indicating that the precipitated solid material contains C, O, and Zn elements. Similar to EPDM, the base formula of NBR rubber contains small amounts of inorganic and organic additives, such as zinc oxide, gas black, stearic acid, and sulfurization accelerators, in addition to butadiene and acrylonitrile monomers. Therefore, it is inferred that the C, O, and Zn elements in the white precipitate on the rubber surface after the test may originate from additives or reaction by-products in the rubber. Furthermore, the C, O, and Zn element contents (Wt%) on the NBR rubber's surface in the C_4F_7N/CO_2/decomposition by-product atmosphere are significantly higher than those in the C_4F_7N/CO_2 atmosphere. This suggests that more solid material has precipitated on the NBR rubber's surface in the presence of fault decomposition products.

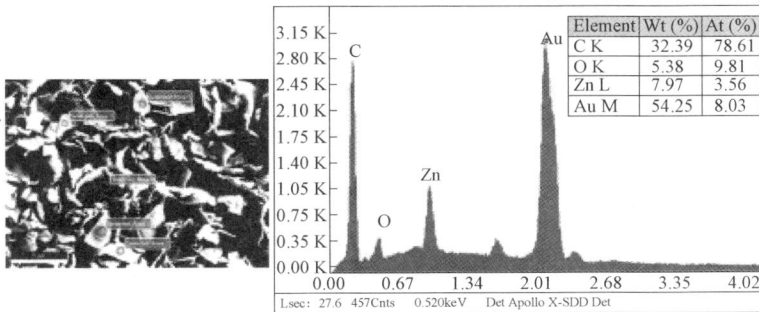

Element	Wt (%)	At (%)
C K	32.39	78.61
O K	5.38	9.81
Zn L	7.97	3.56
Au M	54.25	8.03

(a) C₄H₇N/CO₂

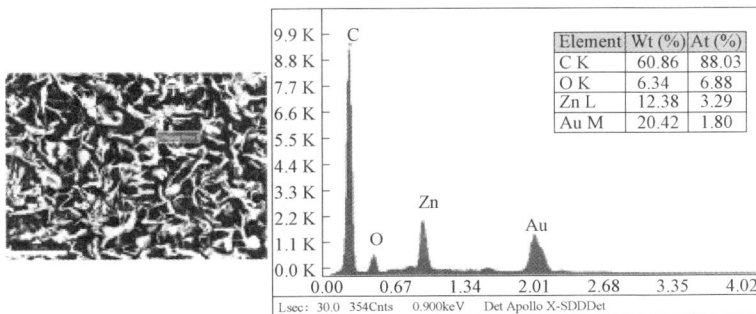

Element	Wt (%)	At (%)
C K	60.86	88.03
O K	6.34	6.88
Zn L	12.38	3.29
Au M	20.42	1.80

(b) C₄H₇N/CO₂/decomposition by-products

Figure 2.63   EDS results of NBR rubber samples in C₄F₇N atmospheres after compatibility test

Figure 2.64 presents SEM microscopy images of CR rubber samples magnified 350 times before and after the compatibility tests under three different atmospheres: He, C_4F_7N/CO_2, and C_4F_7N/CO_2/decomposition by-products. In the He gas atmosphere, the surface smoothness of the CR sample deteriorated, displaying protrusions. This phenomenon may be attributed to the gradual migration of inorganic additives (magnesium oxide and zinc oxide) from the interior of the rubber to the surface upon heating. In comparison, there was no significant difference in the surface morphology of the rubber in the C_4F_7N/CO_2 atmosphere, as shown in Figure 2.64(b), (c), and (d), compared to the He gas atmosphere. This suggests that the surface protrusions are primarily induced by temperature, and the components of C_4F_7N/CO_2 and its decomposition by-products do not significantly affect the surface morphology of the CR samples.

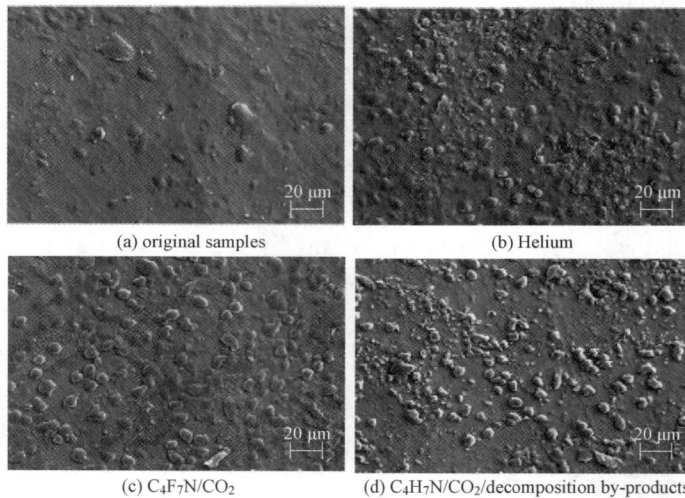

(a) original samples	(b) Helium
(c) C_4F_7N/CO_2	(d) C_4H_7N/CO_2/decomposition by-products

Figure 2.64   Surface morphology of CR rubber samples before and after the compatibility test (×350)

Figure 2.65 displays SEM microscopy images of FKM rubber samples magnified 350 times before and after the compatibility testing under three different atmospheres.

In the He atmosphere. The surface of the specimen became smooth and polished. Comparatively, the surface smoothness of the rubber under both C_4F_7N atmospheres was slightly reduced, although no other noticeable changes were observed when compared to the rubber in

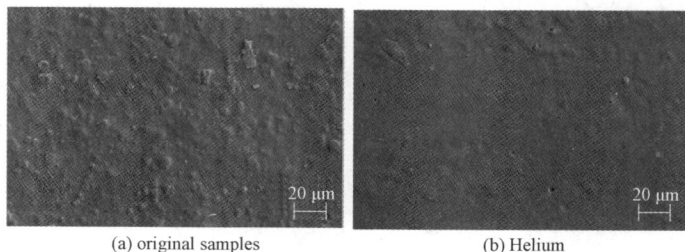

(a) original samples	(b) Helium

Figure 2.65   Surface morphology of FKM rubber samples before and after the compatibility test (×350) (1)

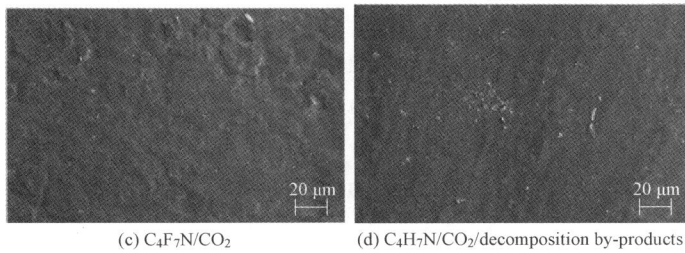

(c) C_4F_7N/CO_2 (d) C_4H_7N/CO_2/decomposition by-products

Figure 2.65   Surface morphology of FKM rubber samples before and after the compatibility test (×350) (2)

the He gas atmosphere. These results indicate that C_4F_7N/CO_2 and its decomposition products did not significantly affect the surface morphology of FKM samples.

Table 2.8 compares the macroscopic and microscopic morphological changes of four rubber types in the atmospheres of C_4F_7N/CO_2 and C_4F_7N/CO_2/decomposition by-products. CR rubber and FKM rubber samples showed no significant changes in appearance in both C_4F_7N atmospheres, with only a slightly rougher microscopic morphology compared to that in the He gas atmosphere. Although there was no apparent change in the appearance of EPDM rubber samples, SEM analysis revealed the presence of a small number of flake-like substances. NBR rubber samples exhibited severe "frosting" and the precipitation of a dense layer of flake-like substances. Therefore, from the perspective of surface morphology, FKM, and CR rubber samples performed better in C_4F_7N/CO_2 and C_4F_7N/CO_2/decomposition by-product atmospheres, while NBR rubber samples exhibited the most significant changes.

Table 2.8   Morphology comparison of rubber samples in C_4F_7N atmospheres after test

Gas atmosphere	Rubber type	Macroscopic Morphological Changes	Microscopic Morphological Changes
C_4F_7N/CO_2	EPDM	—	Becomes rough, with precipitation of flake-like substances
	NBR	Frosting	Precipitation of a dense layer of flake-like substances
	CR	—	Surface roughening and precipitation of flake-like substances
	FKM	—	Slight decrease in surface smoothness
C_4F_7N/CO_2/de composition by-products	EPDM	—	Becomes rough, with precipitation of flake-like substances
	NBR	Frosting	Precipitation of a dense layer of flake-like substances
	CR	—	Surface roughening and presence of protrusions
	FKM	—	Slight decrease in surface smoothness

Note: "—" indicates no significant change; Microscopic morphological changes are relative to rubber specimens in the He gas atmosphere.

2.5.2.4 Fourier Transform Infrared Spectroscopy (ATR-FTIR) Analysis

Figure 2.66 illustrates the ATR-FTIR analysis used to characterize the microstructural changes in four rubber samples before and after the 70 ℃ compatibility accelerated test. Under different atmospheres, the four rubber types exhibit multiple absorption peaks in the frequency range of 4,000 to 500 cm^{-1}.

Figure 2.66(a) shows the FTIR detection results of the surface chemical composition of the EPDM sample in different atmospheres. The original EPDM sample exhibits strong absorption peaks at 2,915 cm^{-1}, 2,847 cm^{-1}, and 1,733 cm^{-1}, corresponding to the asymmetric stretching vibration of —CH$_2$, symmetric stretching vibration of —CH$_2$, and stretching vibration of the carbonyl group (—C═O), respectively. The EPDM rubber sample contains zinc stearate (CAS 540-10-3), pentaerythritol tetraricinoleate (CAS 14450-05-6), and 3-methyl-4-isopropy-lphenol (CAS 3228-02-2). Therefore, the absorption peak at 1,733 cm^{-1} is attributed to the carbonyl group (—C═O) of the ester-type additives. Compared with the absorption spectra of the original sample and the EPDM sample in He atmosphere, it is observed that the absorption peak of the carbonyl group (—C═O) at 1,733 cm^{-1} disappears in the atmospheres of C$_4$F$_7$N/CO$_2$ and C$_4$F$_7$N/CO$_2$ decomposition by-products.

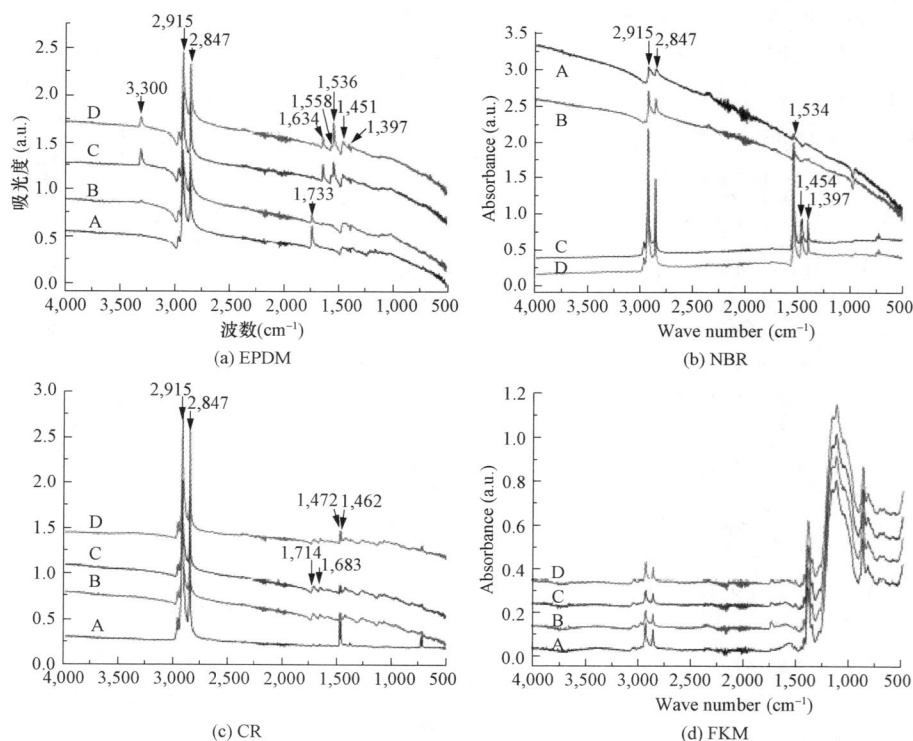

Figure 2.66　ATR-FTIR spectra of rubber samples before and after compatibility test

Simultaneously, characteristic peaks of asymmetric and symmetric stretching vibrations of carboxylate ions (R—COO—) appear at 1,536 cm^{-1} and 1,451 cm^{-1}, and a peak of symmetric deformation vibration of —CH$_3$ appears at 1,397 cm^{-1}. Therefore, it is inferred that carboxylates are present in the EPDM rubber surface deposits. Combining the EDS detection results from the previous section, it can be concluded that the deposits on the rubber surface in the C_4F_7N atmospheres are zinc stearate. Furthermore, in the presence of C_4F_7N, the FTIR spectrum of the EPDM sample shows additional peaks at 3,300 cm^{-1}, 1,634 cm^{-1}, and 1,558 cm^{-1}, which are characteristic peaks of the amide structural unit. Specifically, the peak at 3,300 cm^{-1} corresponds to the N—H stretching vibration of R_1—CONH—R_2, the peak at 1,634 cm^{-1} corresponds to the C=O double bond stretching vibration of R_1—CONH—R_2, and the peak at 1,558 cm^{-1} corresponds to the N—H bending vibration of amide II. After comparing with the Sadtler standard infrared spectrum, it is found that these three peaks are typical characteristic peaks of the rubber additive ethylenediamine bis (stearoylamide). Combined with the SEM images, it can be inferred that in the presence of C_4F_7N, the EPDM surface deposits ethylenediamine bis (stearoylamide) after the test. Additionally, comparing the FTIR detection results of the rubber samples in the atmospheres of C_4F_7N/CO_2 and C_4F_7N/CO_2/decomposition by-products, it is found that the absorption peaks at 3,300 cm^{-1}, 1,634 cm^{-1}, and 1,558 cm^{-1} in C_4F_7N/CO_2 atmosphere are stronger than those in C_4F_7N/CO_2/decomposition by-product atmosphere, while the peaks at 1,733 cm^{-1}, 1,536 cm^{-1}, and 1,451 cm^{-1} are weaker. This indicates that the discharge decomposition components promote the deposition of zinc stearate and inhibit the deposition of ethylenediamine bis (stearoylamide).

Figure 2.66(b) shows the FTIR detection results of the surface chemical composition of the NBR sample in different atmospheres. The absorption spectrum of the original NBR sample is almost the same as that in the He atmosphere, indicating that no new functional groups are formed on the surface of the NBR sample after the test in the He atmosphere. However, in the presence of C_4F_7N, the absorption peaks of the NBR sample are significantly different from those of the original sample and the sample in the He atmosphere. Sharp absorption peaks appear at 1,454 cm^{-1} and 1,397 cm^{-1}, and the peaks at 2,915 cm^{-1}, 2,847 cm^{-1}, and 1,534 cm^{-1} are greatly enhanced. According to the literature, 1,534 cm^{-1}, 1,454 cm^{-1}, and 1,397 cm^{-1} are the three characteristic peaks of zinc stearate, corresponding to the antisymmetric and symmetric stretching vibrations of the carboxylate ion COO in the carboxylic acid salt. Combined with the EDS detection results, it can be determined that the "frosting" phenomenon on the surface of the NBR sample after the test is caused by the migration of a large amount of zinc stearate from the rubber to the surface. Zinc oxide and stearic acid are common vulcanization accelerators in rubber, and zinc stearate is a reaction product of zinc oxide and stearic acid. C_4F_7N promotes the reaction between zinc oxide and stearic acid to some extent, producing zinc stearate.

Figure 2.66(c) shows the FTIR detection results of the surface chemical composition of the CR sample in different atmospheres. Compared with the original sample, the absorption peaks of the CR sample at 1,462 cm^{-1} (bending vibration of —CH$_2$) and 722 cm^{-1} (in plane swinging vibration of —(CH$_2$)$_n$—, $n>4$) are weakened in the three atmospheres, indicating a decrease in the number of methylene groups after the compatibility accelerated test in the three atmospheres. Combined with the GC/MS detection results, it is known that CH_4 gas is produced after the test, leading to the weakening of the absorption peaks of methylene. Comparing the FTIR detection results of the CR sample in the three atmospheres, it is found that the FTIR absorption peaks on the surface of the CR sample are not significantly different in the three atmospheres, indicating that the chemical composition of the surface of the CR sample after the compatibility accelerated test in both C_4F_7N atmospheres is basically the same, and C_4F_7N/CO_2 and its decomposition products do not have a significant impact on chloroprene rubber.

Figure 2.66(d) shows the FTIR detection results of the surface chemical composition of the FKM sample in different atmospheres. The FTIR absorption peaks of the FKM sample in the three atmospheres before and after the test show no significant changes, indicating that there is no significant change in the chemical structure of the FKM sample in the three atmospheres after the compatibility accelerated test.

Table 2.9 lists the compiled positions of the Fourier Transform Infrared (FTIR) spectroscopy absorption peaks, vibrational modes, and their corresponding chemical structures.

Table 2.9   **Absorption peaks of ATR-FTIR and their corresponding functional groups and chemical structures**

Wavenumber (cm^{-1})	Vibrational Mode	Functional Group
3,300	N—H stretching vibration	R_1—CONH—R_2, amide
2,915	C—H asymmetric stretching vibration	R_1—CH_2—R_2
2,847	C—H symmetric stretching vibration	R_1—CH_2—R_2
1,733	Carbonyl group (C=O) stretching vibration	R—CO—O, Ester
1,634	Carbonyl group (C=O) stretching vibration, amide I band	R_1—CONH—R_2, Amide
1,558	N—H bending vibration, amide II band	R_1—CONH—R_2, Amide
1,536	Asymmetric stretching vibration of carboxylate ions	R—COO—, Carboxylate ion
1,454	Symmetric stretching vibration of carboxylate ions	R—COO—, Carboxylate ion
1,397	C—H symmetric deformation	R—CH_3
722	C—H in-plane rocking	—$(CH_2)_n$, $n>4$

The ATR-FTIR detection results of the four rubbers in the atmospheres of C_4F_7N/CO_2 and C_4F_7N/CO_2/decomposition by-products are compared, as shown in Table 2.10. The surface chemical composition of EPDM and NBR samples in the two C_4F_7N atmospheres has changed, with EPDM rubber in both C_4F_7N atmospheres precipitating the additives, zinc stearate, and ethylenediamine bis (stearoylamide), while NBR rubber precipitates zinc stearate. The surface infrared absorption peaks of CR rubber and FKM rubber samples are basically the same as those in the He atmosphere, indicating good performance.

Table 2.10   **Comparison of FTIR results of rubber samples
in C_4F_7N atmosphere after test**

Gas atmosphere	Rubber type	FTIR	Generated Chemical Substances
C_4F_7N/CO_2	EPDM	New Absorption Peak	Zinc stearate and ethylenediamine bis (stearoylamide)
	NBR	New Absorption Peak	Zinc stearate
	CR	Same as He Atmosphere	—
	FKM	Same as He Atmosphere	—
C_4F_7N/CO_2/decomposition byproducts	EPDM	New Absorption Peak	Zinc stearate ethylenediamine (stearoylamide)
	NBR	New Absorption Peak	Zinc stearate
	CR	Same as He Atmosphere	—
	FKM	Same as He Atmosphere	—

2.5.2.5 Compressive Permanent Deformation Rates

Figure 2.67 shows the compressive permanent deformation rates of the four rubber samples after testing in three different atmospheres. The compressive permanent deformation rates of EPDM, CR, and FKM samples did not exhibit significant changes in the three atmospheres. However, for the NBR sample, the compressive permanent deformation rates in both C_4F_7N atmospheres were higher than those in the He atmosphere, with the highest deformation observed in C_4F_7N/CO_2/decomposition by-product atmosphere. Additionally, based on the FTIR and GC/MS test results, it was found that the NBR sample in both C_4F_7N atmospheres experienced severe frosting, accompanied by the highest gas generation. It is speculated that prolonged exposure to C_4F_7N atmospheres may lead to a sparse cross-linking structure and relaxed molecular bonds within the rubber matrix, resulting in deteriorated mechanical properties at the macroscopic level.

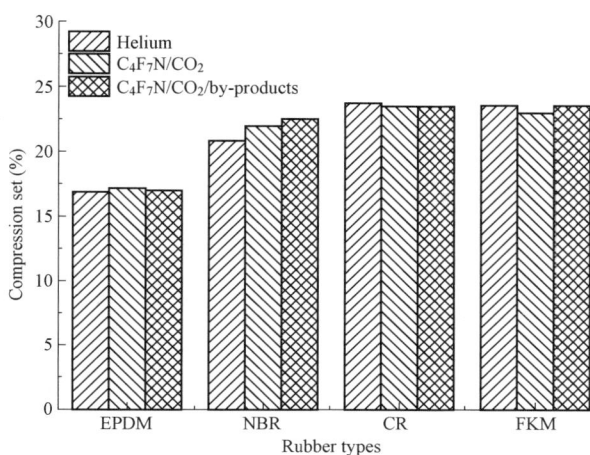

Figure 2.67   Compression set of four rubbers after compatibility test in three atmospheres

Although the surface morphology and chemical composition of the EPDM sample changed significantly after testing, its compressive performance did not deteriorate. Therefore, by adjusting the formulation of EPDM to reduce the likelihood of reaction with C_4F_7N, EPDM can be used as a sealing material for C_4F_7N/CO_2 mixture in electrical equipment. Comparing the compressive permanent deformation of the four rubbers in the He atmosphere, it can be observed that the compressive permanent deformation of the FKM and CR samples is larger than that of the EPDM sample at the same size, indicating that the compressive performance of FKM and CR rubbers is not as good as EPDM rubber. Therefore, when selecting FKM rubber as a sealing material in practical applications, it is necessary to change the dimensions to achieve optimal sealing performance.

2.5.3 Epoxy Resin Materials

2.5.3.1 Test Materials

As shown in Figure 2.68, the test samples consist of bisphenol A epoxy resin. Cylindershaped samples are used for surface flashover characteristic tests, and disc-shaped samples are used for surface resistivity measurements. The cylindrical samples have a height of 10 mm and a diameter of 15 mm, while the disc-shaped samples have a thickness of 1 mm and a diameter of 40 mm. The sample fixture is shown in Figure 2.69. Experimental preparation and vacuuming process:

- Before the experiment, gently wipe the surface of the epoxy resin samples with anhydrous ethanol to remove surface impurities.
- Clean the surface with deionized water to remove any residual anhydrous ethanol.
- Place the samples in a 70 ℃ aging oven for 12 hours to ensure thorough drying.
- After removal from the oven, let the samples stand at room temperature for an additional 12 hours before use.

Figure 2.68　Two types of epoxy resin samples

Figure 2.69　The sample fixture

Cleaning of stainless-steel fixtures and test vessel:

- Before the experiment, clean the stainless-steel fixtures thoroughly using an ultrasonic cleaning machine containing an anhydrous ethanol solution.
- Wipe the stainless-steel test vessel with anhydrous ethanol.
- Place the fixtures and test vessel in an 80 ℃ oven for 24 hours.
- After drying, remove the test vessel and fixtures, allowing them to stand at room temperature for approximately 12 hours.

Vacuuming and Gas Washing Procedure:

- Follow the vacuuming and gas washing procedure outlined in Section 4.2.1, which is the same as the process for rubber samples. Prepare six sets of parallel cylindrical epoxy samples and three sets of parallel disc-shaped samples.
- Place the epoxy samples in three different atmospheres:

C_4F_7N/CO_2 mixture.

C_4F_7N/CO_2/decomposition by-product mixture, and helium (He) gas.

2.5.3.2 Testing and Characterization Methods

(1) Gas GC-MS Analysis

Analyze the gas components according to the GC/MS detection method outlined in Section 2.5.2.

(2) Surface Morphology Analysis

1) Macroscopic Morphology Observation:

Macroscopically observe whether there are phenomena such as fractures, cross-sections, or discoloration on the epoxy resin samples after the test.

2) Microscopic Morphology Analysis:

Utilize the SEM method as described in Section 2.5.2 to determine microscopic morphological changes before and after the test.

(3) Fourier Transform Infrared Spectroscopy (FTIR) Analysis

Employ the ATR-FTIR method outlined in Section 2.5.2 to analyze the surface chemical composition of cylindrical epoxy samples in different atmospheres before and after the test.

(4) Electrical Performance Testing and Analysis

Figure 2.70 shows the surface flashover test circuit, comprising a YDTW-300 power frequency non-partial discharge test transformer with an output voltage of 300 kV and a capacity of 100 kVA, a 10 kΩ protective resistor, a 500 pF capacitive voltage divider with a division ratio of 24,340:1, and a stainless steel test vessel. The cylindrical epoxy sample is placed between two parallel plate electrodes, ensuring close contact between the electrodes and the upper and lower surfaces of the sample.

Before each test, the electrode surfaces are polished using 1,000-grit and 18,000-grit abrasives, followed by wiping with alcohol and cleaning in an ultrasonic cleaner containing deionized water. After removal, they are dried in a vacuum oven for later use. Once the electrodes and the experimental vessel are assembled, the chamber is evacuated to below 50 Pa, filled with CO_2 gas to 0.1 MPa, and evacuated again, and this process is repeated twice to thoroughly clean the experimental chamber. Then, a pre-prepared 9% C_4F_7N/91% CO_2 gas mixture is introduced to an absolute pressure of 0.4 MPa, and after a 10-minute waiting period, the test begins.

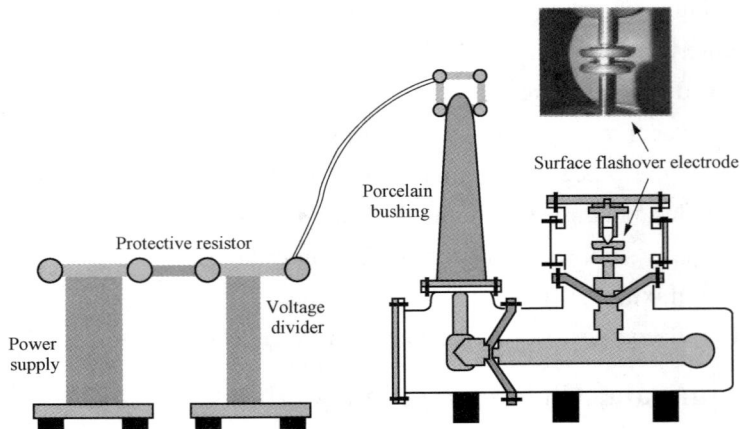

Figure 2.70　Surface flashover test platform

During the test, a direct voltage boost method is employed until the flashover occurs, and the voltage peak recorded by the oscilloscope is taken as the flashover voltage for that test. The procedure is repeated five times at each pressure, with a 5-minute interval between each test, and the average value is taken as the result for that pressure. Subsequently, the gas inside the experimental vessel is gradually released, and flashover tests are performed at pressures of 0.35 MPa, 0.3 MPa, and 0.25 MPa in sequence.

The surface resistivity of the circular epoxy samples before and after the test is measured using a HIOKI SM7110 high resistance meter. Figure 2.71 shows a physical image of the high resistance meter. The measurement electrode uses the SEM-8311 flat test electrode with a main electrode diameter of 19.6 mm, an inner protective electrode diameter of 24.1 mm, and an outer diameter of 28.8 mm. During the measurement, a voltage of 1 kV is applied with a frequency of 2 Hz. The laboratory temperature is 26 ℃, and the relative humidity is 20%. Before measurement, the high-resistance meter is preheated for 30 minutes. When measuring surface resistivity, the sample surface is not treated, only wiped with anhydrous ethanol, and then dried before placing it into the epoxy sample. Each type of atmosphere is tested once for three circular epoxy samples, and then the average value is taken as the measurement result of the surface resistivity for that atmosphere.

2.5.3.3　Gas Composition

The gas before and after the compatibility test was analyzed using GC/MS, and the chromatographic results are shown in Figure 2.72. No new gases were generated after the test in the three atmospheres, but in the C$_4$F$_7$N/CO$_2$/decomposition by-product atmosphere, the content of the decomposition component CNCN decreased after the test.

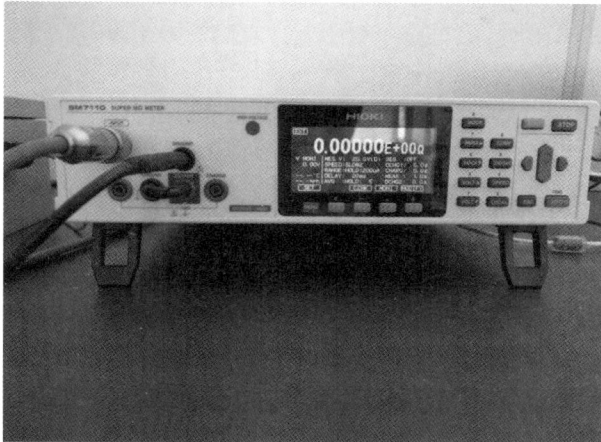

Figure 2.71   HIOKI SM7110 high resistance meter

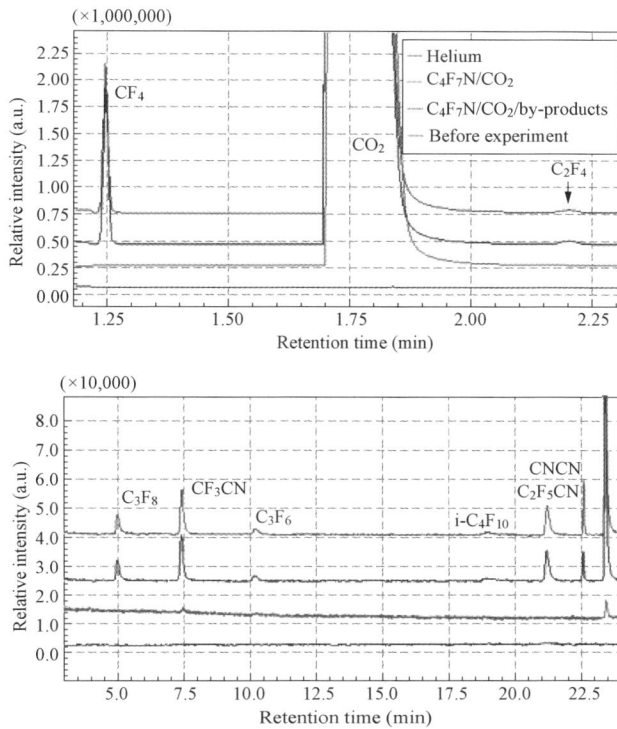

Figure 2.72   GC-MS results of epoxy resin samples in three atmospheres

Figure 2.73 shows the content of nitrile products CF$_3$CN, C$_2$F$_5$CN, and CNCN in C$_4$F$_7$N/ CO$_2$/decomposition by-product atmosphere before and after tests. After the compatibility test, the content of CF_3CN and C_2F_5CN gases remained essentially unchanged, but the CNCN gas content significantly decreased. It is speculated that some of the CNCN gas may be absorbed by the α-Al_2O_3 particles filled in the epoxy resin, leading to a reduction in CNCN content.

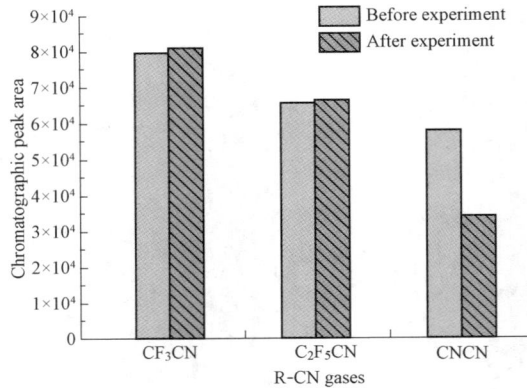

Figure 2.73　Peak areas of R—CN (R=CN, CF_3, C_2F_5) compounds before and after compatibility test in C_4F_7N/CO_2/decomposition by-products

2.5.3.4　SEM Microscopic Morphology Analysis

Figure 2.74 displays SEM microscope scanning photos of the surface morphology of epoxy resin samples magnified 4,000 times before and after the compatibility accelerated test. Compared to the samples before the test, the surface smoothness of the epoxy samples deteriorated under all three atmospheres after the test.

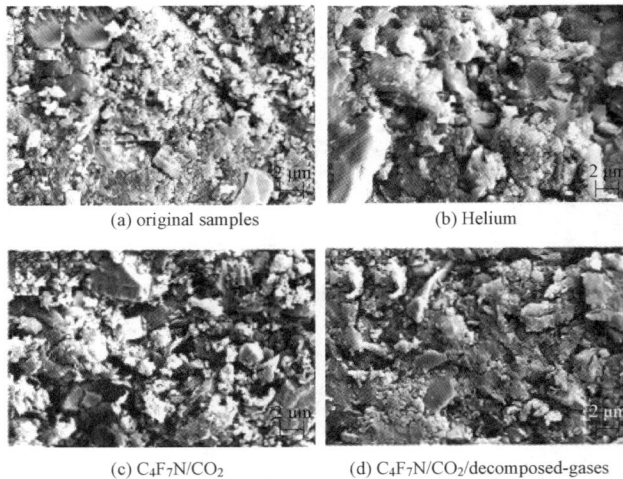

(a) original samples

(b) Helium

(c) C_4F_7N/CO_2

(d) C_4F_7N/CO_2/decomposed-gases

Figure 2.74　Surface morphology of epoxy resin samples before and after compatibility test ($\times$4,000)

2.5.3.5　Fourier Transform Infrared Spectroscopy (ATR-FTIR) Analysis

Figure 2.75 presents the ATR-FTIR detection results of the surface chemical composition of epoxy resin samples before and after the experiment.

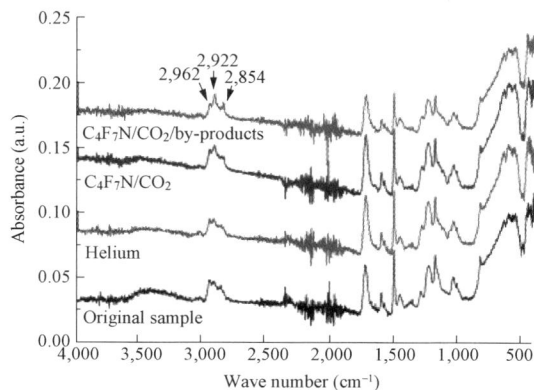

Figure 2.75　ATR-FTIR spectra of epoxy resin samples before and after compatibility test

The infrared spectra of the epoxy samples under the three atmospheres after the test are highly similar to the spectra before the test, with no new absorption peaks observed. This indicates that the surface chemical composition of the epoxy samples under the three atmospheres after the test is essentially the same, and no new functional groups have been generated. Compared to the FTIR detection results under the He gas atmosphere, the absorption peak intensity at 2,962 cm^{-1} (—CH$_3$ asymmetric stretching vibration) decreases, while the absorption peaks at 2,922 cm^{-1} (—CH$_2$ asymmetric stretching vibration) and 2,854 cm^{-1} (—CH$_2$ symmetric stretching vibration) both increase in intensity for the epoxy samples tested in C$_4$F$_7$N atmospheres. This suggests a decrease in the number of methyl groups and an increase in the number of methylene groups on the surface of the epoxy samples after the compatibility test in C$_4$F$_7$N atmospheres. Since no new substances were detected after the test under the three atmospheres, we speculate that the monomeric structure of the epoxy resin may have changed.

2.5.3.6　Electrical Property

The surface resistivity of epoxy samples before and after compatibility testing was measured, and the surface resistivity of epoxy samples in the three different atmospheres remained essentially the same. The surface resistivity of the original sample was $2.19 \times 10^{17} \Omega$, while in atmospheres of helium, C$_4$F$_7$N/CO$_2$, and C$_4$F$_7$N/CO$_2$/decomposition by-products, the surface resistivity of epoxy samples was $3.58 \times 10^{17} \Omega$, $2.92 \times 10^{17} \Omega$, and $2.78 \times 10^{17} \Omega$, respectively.

The variation of flashover voltage with pressure for epoxy samples before and after compatibility testing is shown in Figure 2.76. The flashover voltage of epoxy in C$_4$F$_7$N atmospheres is essentially the same as that in helium atmosphere, indicating that C$_4$F$_7$N and its decomposition components have minimal impact on the flashover characteristics of epoxy samples. Moreover, the flashover voltage of epoxy samples in the three atmospheres after testing is higher than that before testing. This could be attributed to the curing reaction of polymers in the epoxy resin at 70 ℃, resulting in increased crosslinking between molecular chains, densification of the internal matrix, and consequently, higher flashover voltage of epoxy samples after testing compared to before testing.

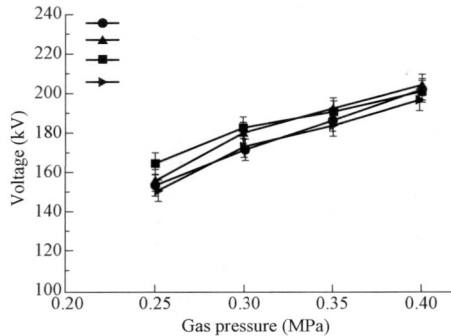

Figure 2.76   The Surface flashover voltage of epoxy resin samples before and after compatibility test

2.6 Review Questions

Q1: How many types of insulation media? Select one type of insulation medium and introduce its characteristics and application scenarios.

Q2: Please introduce the basic concepts of quantum chemistry theory and its applications in electrical engineering.

Q3: What are the requirements for replacing SF_6 gas as an insulation medium or arc extinguishing medium? Please introduce three or more SF_6 alternatives and their main advantages/disadvantages.

Q4: Please evaluate the insulation performance and arc extinguishing performance of C_4F_7N gas.

Q5: What are the main decomposition gases of C_4F_7N under partial discharge or overheating conditions? Please evaluate the compatibility of C_4F_7N with rubber and epoxy resin materials.

Bibliography

[1] James, Ron E and Su, Qi. Condition assessment of high voltage insulation in power system equipment, volume 53. IET, 2008.

[2] Samat, J and Lacaze, D. Micro-particles in transformer oil and dielectric withstand effects. Alstom

Review, 11: 47-57, 1988.

[3] Antonov, GI and Working Group Particles in Oil International Conference on Large High Voltage Electric Systems. Effect of particles on transformer dielectric strength. Cigré, 2000.

[4] Rabie, Mohamed and Franck, Christian M. Assessment of eco-friendly gases for electrical insulation to replace the most potent industrial greenhouse gas SF_6. Environmental science & technology, 52(2): 369-380, 2018.

[5] Chaohai, ZHANG and Dong, HAN and Kang, LI and others. SF_6 alternative techniques and their applications and prospective developments in gas insulated transmission lines. High Voltage Engineering, 43(3): 689-698, 2017.

[6] Nechmi, Houssem Eddine and Beroual, Abderrahmane and Girodet, Alain and Vinson, Paul. Fluoronitriles/CO_2 gas mixture as promising substitute to SF_6 for insulation in high voltage applications. IEEE Transactions on Dielectrics and Electrical Insulation, 23(5): 2587-2593, 2016.

[7] Li, Yi and Zhang, Xiaoxing and Chen, Qi and Zhang, Ji and Li, Yalong and Xiao, Song and Tang, Ju. Influence of oxygen on dielectric and decomposition properties of $C_4F_7N-N_2-O_2$ mixture. IEEE Transactions on Dielectrics and Electrical Insulation, 26(4): 1279-1286, 2019.

[8] Li, Zhichuang and Ding, WD and Gao, KL and Liu, YS and Liu, W and Guo, YJ. Surface flashover characteristics of epoxy insulator in C_4F_7N/CO_2 mixtures under lightening impulse voltage. High Voltage Engineering, 45(4): 1071-1077, 2019.

[9] Kieffel, Yannick and Biquez, François and Ponchon, Philippe and Irwin, Todd. SF_6 alternative development for high voltage switchgears. In 2015 IEEE Power & Energy Society General Meeting, pages 1-5. IEEE, 2015.

[10] Lutz, B and Juhre, K and Kuschel, M and Glaubitz, P. Behavior of gaseous dielectrics with low global warming potential considering partial discharges and electric arcing. In CIGRE Winnipeg Colloquium, Winnipeg, Canada, 2017.

[11] Zhou, Zhenrui and Han, Dong and Zhao, Mingyue and Zhang, Guoqiang. Review on decomposition characteristics of SF_6 alternative gases. Trans. China Electrotech. Soc., 35(23): 4998-5014, 2020.

[12] Zhao, Mingyue and Han, Dong and Zhou, Zhenrui and Zhang, Guoqiang. Experimental and theoretical analysis on decomposition and by-product formation process of $(CF_3)_2CFCN$ mixture. AIP Advances, 9(10), 2019.

[13] Zhao, MY and Han, D and Rong, WQ and Zhang, GQ and Huang, H and Liu, ZE. Decomposition characteristics of binary mixtures of $(CF_3)_2CFCN$ buffer gases under corona discharge. High Volt. Eng, 45: 1078-1085, 2019.

[14] Zhao, Mingyue and Han, Dong and Zhao, Weikang and Zhou, Zhenrui and Zhang, Guoqiang. Experimental and theoretical studies of C_3F_7CN/CO_2 mixture decomposition under overheating fault. CSEE Journal of Power and Energy Systems, 8(3): 941-951, 2020.

[15] Zhao, M and Han, D and Zhou, L and Zhang, G. Absorption characteristics of activated alumina and molecular sieves for C_3F_7CN/CO_2 and its decomposition by-products of overheating fault. Trans. China Electrotech. Soc., 35(9), 2020.

[16] Zheng, Zheyu and Li, Han and Zhou, Wenjun and Yuan, R and Liu, W and He, J. Compatibility of eco-friendly insulating medium C_3F_7CN and sealing material epdm. High Volt. Eng, 46(1): 335-341, 2020.

[17] Wang, Hao and Yan, Xianglian and Han, Dong and et al. Experiments for compatibility characteristics of C_4F_7N/CO_2 and its gas byproducts with commonly used rubber sealing materials. High Voltage Engineering, 48(7): 2625-2634, 2022.

3

Atomic Emission Spectrometry

3.1 Introduction

Atomic emission spectroscopy (AES) is a technique employed to analyze elements by observing the distinct spectral lines emitted when atoms in an excited state transition back to the ground state. This method is suitable for both in situ detection and positive material identification. AES encompasses various categories, such as X-ray Fluorescence (XRF), Optical Emission Spectroscopy (OES), Laser-Induced Breakdown Spectroscopy (LIBS), and others. Each of these technologies facilitates rapid, precise, and dependable analysis.

Laser-induced breakdown spectroscopy (LIBS) is a technique within atomic emission spectroscopy (AES) that employs a laser-generated plasma as the source for vaporization, atomization, and excitation[1]. Unlike conventional AES methods, which rely on physical devices (such as electrodes or coils) to create the vaporization/excitation source, LIBS utilizes focused optical radiation to generate plasma, offering numerous advantages. Notably, it enables in situ and remote interrogation of samples without requiring any prior preparation. In its basic operation, a LIBS measurement involves creating a laser plasma on or within the sample, followed by the collection and spectral analysis of the emitted light from the plasma. Qualitative and quantitative analyses are then performed by observing the positions and intensities of the emission lines.

Although the LIBS method has been known for 40 years, its early focus before 1980 primarily revolved around understanding the fundamental physics of plasma formation. However, over time, the analytical capabilities of LIBS have become increasingly apparent. Despite a few instruments based on LIBS being developed, they have not gained widespread use. Nevertheless, there has been a recent resurgence of interest in the method across a wide range of applications. This renewed interest can be attributed mainly to significant advancements in the technology of components such as lasers, spectrographs, and detectors used in LIBS instruments, along with emerging requirements to conduct measurements under conditions that are not feasible with conventional analytical techniques. LIBS has demonstrated a detection sensitivity for many elements that is comparable to or even surpasses that of other field-deployable methods.

3.2 Basic Principles

LIBS, an analytical method that emerged concurrently with the invention of the laser, has had a varied history. Initially, the ablation resulting from the laser pulse's interaction with the sample surface was utilized as a sampling technique, complementing electrode-generated sparks, as it allowed for the ablation of all materials and offered micro-sampling capabilities due to the finely focused laser pulse. Later, it became evident that the laser plasma generated during ablation could serve as its own excitation source.

The heightened interest in LIBS can be attributed to several factors. Firstly, there is a growing demand for a novel method capable of analyzing materials under conditions that were previously inaccessible using conventional analytical techniques. This demand is partly fueled by new regulations mandating stringent monitoring of materials and operations to safeguard the health and safety of workers and the public. Additionally, there is a need for enhanced industrial monitoring capabilities to boost efficiency and lower production costs. Secondly, significant advancements have been made in recent years in minimizing the size and weight of lasers, spectrographs, and array detectors, while simultaneously enhancing their capabilities. This progress enables the development of compact and robust instrumentation suitable for applications beyond laboratory settings.

3.2.1 Atomic Emission Spectroscopy

The primary objective of AES is to ascertain the elemental composition of a sample, regardless of whether it is in solid, liquid, or gaseous form. The analysis conducted through AES can vary from a straight-forward identification of the atomic constituents present in the sample to a more intricate determination involving the relative concentrations or absolute masses of these elements. The fundamental steps involved in AES are as follows:

- Atomization/vaporization of the sample to produce free atomic species (neutrals and ions), excitation of the atoms.
- Detection of the emitted light.
- Calibration of the intensity to concentration or mass relationship, determination of concentrations, masses, or other information.

The examination of emitted light forms the basis of analysis in AES, as each element possesses a distinct emission spectrum that serves as a unique "fingerprint" for the species. Extensive compilations of emission lines are available, aiding in the identification of elements. The position of the emission line(s) helps identify the element(s), while the intensity of the line(s), when properly calibrated, allows for quantification. The procedures and instrumentation employed in each step of AES are dictated by the characteristics of the sample and the type of analysis required, whether it be identification or quantification. It's important to note that since the initial step in AES involves atomization/vaporization, AES methods are generally unsuitable for determining the composition of compounds in a sample. However, in specific cases, information about molecular origins can be obtained.

The origins of AES can be traced back to the experiments conducted by Bunsen and

Kirchhoff around 1860, wherein atomization and excitation were achieved using a simple flame. Subsequently, more robust and controllable methods of excitation were developed, involving the use of electrical current to analyze the sample. Among the notable methods for vaporization and excitation are electrode arcs and sparks, the Inductively Coupled Plasma (ICP), Direct Coupled Plasma (DCP), Microwave-Induced Plasma (MIP), and hollow cathode lamps. These traditional sources typically necessitate substantial laboratory support facilities and some degree of sample preparation prior to analysis. In certain instances, innovative sampling methods have been devised for these sources to address specific applications. For instance, an air-operated ICP facilitates the direct analysis of particles present in the air, while particles collected on a filter can be introduced into the hollow electrode of a conventional spark discharge. However, these methods have seen limited usage for various reasons. LIBS represents an extension of the vaporization/excitation approach to optical frequencies.

3.2.2 Laser-induced breakdown spectroscopy (LIBS)

In LIBS, a vaporizing and exciting plasma is generated by a high-power focused laser pulse. A typical set-up for LIBS is illustrated in Figure 3.1. The laser pulses are directed onto the sample through a lens, and the resulting plasma light is collected either by a second lens or, as depicted in Figure 3.1, via a fiber optic cable. The light gathered by either component is then conveyed to a frequency-dispersive or selective device before being detected. Each firing of the laser yields a single LIBS measurement. However, it is common practice to combine or average the signals from multiple laser plasmas to enhance accuracy and precision and to mitigate non-uniformities in sample composition. Depending on the specific application, adjusting the time resolution of the spark may improve the signal-to-noise ratio or help to discriminate against interference from continuum, line, or molecular band spectra.

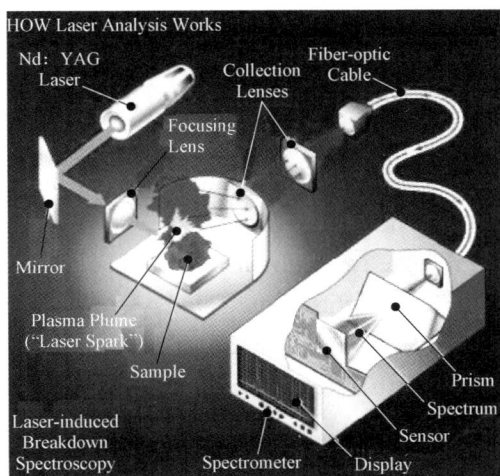

Figure 3.1　Diagram of a typical laboratory LIBS apparatus[1]

Because the laser-induced plasma constitutes a pulsed source, the resulting spectrum under-goes rapid temporal evolution. The temporal progression of a laser-induced plasma is depicted schematically in Figure 3.2(a). Initially, the plasma light is primarily characterized by a "white light" continuum, exhibiting minimal intensity variation across wavelengths. This continuum arises from bremsstrahlung and recombination radiation emitted as free electrons and ions recombine within the cooling plasma. Integrating the plasma light over the entire emission duration can lead to significant interference from this continuum light, hindering the detection of weaker emissions from minor and trace elements in the plasma. To address this issue, LIBS measurements typically employ time-resolved detection. This approach enables the exclusion of the strong white light present at early times by initiating the detector after this white light has substantially diminished in intensity, while atomic emissions are still observable. The key parameters for time-resolved detection include t_d, representing the interval between plasma formation and the commencement of plasma light observation, and t_b, indicating the duration over which the light is recorded [see Figure 3.2(a)].

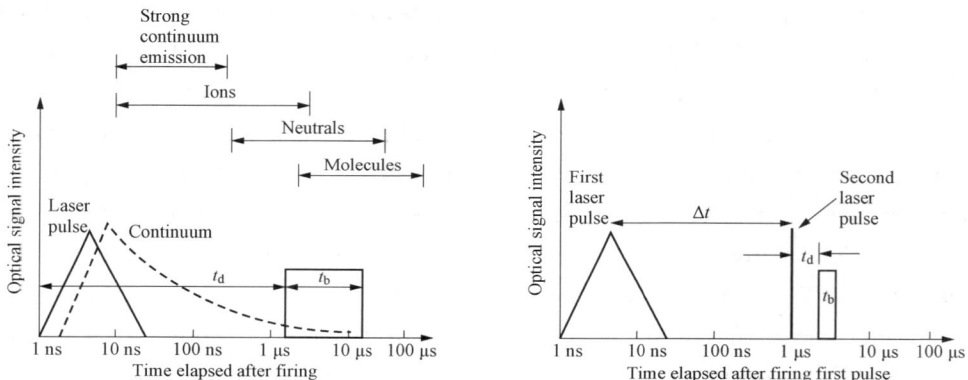

(a) The important time periods after plasma formation during which emissions from different species predominate. The box represents the time during which the plasma light is monitored using a gatable detector. Here t_d is the delay time and t_b is the gate pulse width. The timing here corresponds to an RSS experiment

(b) Important timing periods for a double-pulse RSP measurement. Here Δt is the time between the closely spaced double pulses [2]

Figure 3.2　Key parameters of LIBS [2]

The majority of LIBS measurements are conducted by using the RSS (repetitive single spark) in which a series of individual laser sparks are formed on the sample at the laser repetition rate (e.g. 10 Hz). In some cases, to enhance detection capabilities, the RSP (repetitive spark pair) is used. The RSP is a series of two closely spaced sparks (e.g. typically 1-10 μs separation) used to interrogate the target at the laser repetition rate. The timing arrangement, in this case, is shown in Figure 3.2(b). Note that t_d is measured from the second laser pulse in this case. The spark pair may be formed by two separate lasers or by a single laser.

3.2.3　The Physics and Chemistry of the Laser Plasma

The schematic described in Figure 3.3 illustrates the life cycle of a laser-induced plasma on a sample surface. This cycle involves two primary steps leading to breakdown due to optical excitation. Initial Electron Generation: The process begins with the presence or generation of a few free electrons. These electrons serve as initial receptors of energy through three-body collisions with photons and neutrals. Avalanche Ionization: In the focal region, there is avalanche ionization. Initially, free electrons are accelerated by the electric fields associated with the optical pulse between collisions. This acceleration acts to thermalize the electron energy distribution. As electron energies increase due to collisions, ionization occurs, producing more electrons, further energy absorption, and an avalanche effect. In the classical perspective, the acceleration of free electrons by the electric fields associated with the optical pulse leads to the thermalization of electron energy distribution. However, in the photon picture, absorption occurs due to inverse bremsstrahlung. The breakdown threshold is typically defined as the minimum irradiance needed to generate a visible plasma. This threshold represents the point at which the energy absorbed by the system is sufficient to initiate plasma formation and become observable.

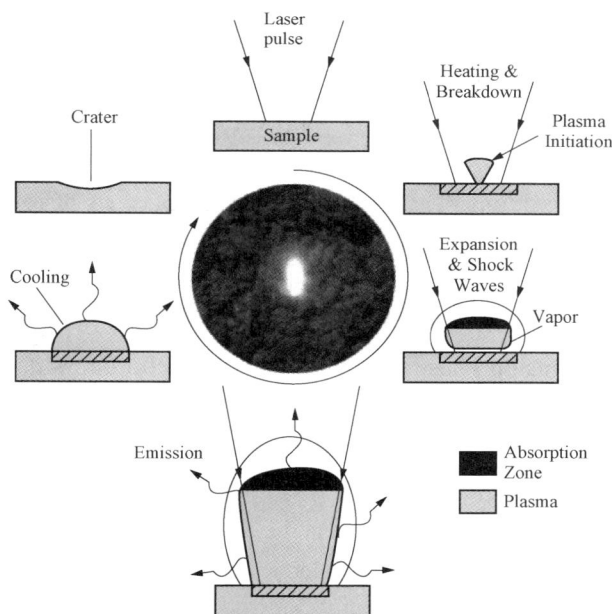

Figure 3.3   Life cycle diagram showing main events in the LIBS process [2]

Following breakdown, the plasma expands outward in all directions from the focal volume. However, the rate of expansion is highest towards the focusing lens due to the direction from which the optical energy enters the plasma. This nonisotropic expansion results in a pearor cigar-shaped appearance of the plasma. Initially, the rate of plasma expansion is rapid, typically on the order of 10^5 m/s. This expansion is driven by the sudden release of energy and the rapid increase in temperature of the plasma. The loud sound that one hears is caused by the shock wave emanating from the focal volume. As the plasma rapidly expands, it creates a pressure wave in the surrounding medium, resulting in the audible sound wave. This shock wave is a physical manifestation of the rapid expansion and energy release associated with the formation of the plasma.

The evolution of a plasma through its transient phases involves interactions with its surroundings and can be modeled using three main approaches: laser-supported combustion (LSC), laser-supported detonation (LSD), and laser-supported radiation (LSR) waves. Laser-Supported Combustion (LSC): In the LSC model, the plasma is characterized by relatively low temperature and density. This model predicts that the plasma and its boundary with the surrounding atmosphere are transmissive enough to allow incoming laser radiation to penetrate. The LSC model is suitable for low irradiances used in LIBS experiments. Laser-Supported Detonation (LSD): Similar to LSC, the LSD model also describes a plasma with relatively low temperature and density. It predicts that the plasma and its boundary are transmissive enough to permit laser radiation penetration, especially for wavelengths shorter than that of the CO_2 laser (10.6 μm). The LSD model is particularly relevant for LIBS experiments conducted at low irradiances. Laser-Supported Radiation (LSR) Waves: In contrast to LSC and LSD, the LSR model considers a more opaque plasma, where radiation is the dominant energy transfer mechanism. This model is typically applied to plasmas with higher temperatures and densities. In LIBS experiments utilizing low irradiances, the LSC and LSD models closely match experimental observations. These models describe plasmas at relatively low temperatures and densities, where the plasma and its boundary with the ambient atmosphere allow for sufficient transmittance of laser radiation, especially for wavelengths shorter than that of the CO_2 laser.

Throughout the expansion phase, the plasma emits useful emission signals. It cools and decays as its constituents give up their energies in a variety of ways. The ions and electrons recombine to form neutrals, and some of those recombine to form molecules. Energy escapes through radiation and conduction.

As is well known, plasma temperatures can be determined in a variety of ways, including spectroscopic and probe methods. There are electron, excitation, and ionization temperatures, to name just a few. Each is determined based on different diagnostics and measurements, and they may or may not remain the same. The situation is complicated by the transient nature of pulsed plasmas. Generally, pulsed plasmas do not start in equilibrium, but evolve to that state. Often electrons start at a much higher kinetic temperature and eventually equilibrate with heavier atoms and ions through collisions. Physics tells us that momentum transfer is small in collisions between bodies of very different masses, hence the time scale for equilibration between electrons and atoms can be quite long.

Throughout the expansion phase, the plasma emits emission signals that provide valuable information. As the plasma cools and decays, its constituents relinquish their energies through various processes. These include recombination of ions and electrons to form neutrals, and subsequent formation of molecules from some of these neutrals. Energy also escapes through radiation and conduction.

Determining plasma temperatures is crucial for understanding its behavior. Various methods, including spectroscopic and probe techniques, can be employed for this purpose. Different temperatures, such as electron, excitation, and ionization temperatures, can be measured based on distinct diagnostics. However, these temperatures may or may not remain the same throughout the plasma's evolution.

Pulsed plasmas present a particularly complex scenario due to their transient nature. They typically do not start in equilibrium but evolve towards it over time. Initially, electrons may have much higher kinetic temperatures compared to heavier atoms and ions, but they eventually equilibrate through collisions. However, because momentum transfer is small in collisions between particles of vastly different masses, the timescale for complete equilibration between electrons and atoms can be relatively long. This temporal evolution adds another layer of complexity to the characterization of pulsed plasmas.

3.3　Forming the LIBS Plasma in Gases, Liquids, and on Solids

Laser-induced breakdown spectroscopy (LIBS) initially gained attention for its ability to analyze solids like metals and geological samples. However, its versatility soon became apparent, leading to its application in analyzing a wide range of samples beyond solids. LIBS can now analyze gases, liquids, particles suspended in gases or liquids, as well as coatings on solids. This adaptability has made LIBS a valuable tool in various fields including environmental monitoring, industrial process control, and even in forensic science.

3.3.1　Gases

In gases, less energy is consumed during the atomization process, allowing more energy to be available for excitation of atoms. This can lead to a higher ratio of excited ions to neutral atoms. Generally, higher irradiance results in a greater initial ratio of ions to neutral atoms. The energy required for breakdown (the point at which a gas becomes ionized) is typically slightly higher in gases compared to on surfaces, unless there are particulate matters present in the gas. The volume of plasma formed depends on factors such as the energy per pulse and the wavelength of the laser. For instance, a nominal pulse energies of 200 mJ, the plasma length would be largest for high-energy CO_2 laser pulses (5-8 mm in wavelength), and smallest for 266 nm Nd:YAG pulses (1-5 mm in wavelength). Molecular Dissociation: Plasma can completely dissociate molecular gases, allowing for the determination of the composition of the original gases. Plasma Temperature: Plasma generated early in its lifetime can exhibit high temperatures, often around 20,000 K or even higher.

3.3.2 Liquids

By forming the laser plasma on the liquid surface or on liquid drops, researchers can effectively analyze liquids. If the liquid is transparent at the laser wavelength, plasma can even be formed within the bulk liquid. Compared to LIBS analysis in air, the plasma formed within the bulk liquid tends to decay more rapidly, resulting in broader emission lines and lower temperatures, typically starting at around 7,000-12,000 K. To enhance detection limits for certain elements in aqueous media, a double-pulse technique can be employed. In this method, two sequential laser pulses, separated by microseconds, are used to interrogate the same volume of the sample. The first pulse creates a vapor cavity, which is then probed by the second pulse, mimicking an analysis conducted in air. This technique is applicable to both bulk liquids and liquid drops. Additionally, the addition of an absorber to the liquid can enhance plasma formation, thereby improving the overall analysis. These techniques highlight the versatility of LIBS in analyzing various types of samples, including liquids.

3.3.3 Particles

Particles entrained in liquids (hydrosols) or gases (aerosols) are indeed significant for environmental monitoring, and there are two primary strategies for gathering information about them. The first method involves direct monitoring within the ambient medium, where the laser spark is employed to ablate or vaporize the aerosol. Sufficient energy is usually retained to excite the particles, yielding strong spectra and reasonable detection limits. However, incomplete vaporization of the particles is a concern, which can complicate quantification efforts. The second strategy entails capturing the particles on filters, followed by subsequent interrogation of the filter surface using the laser spark. This approach is essentially a special case of solid sampling, which allows for further analysis. Both strategies are valuable for determining particle compositions and loading, particularly in environmental applications where understanding the presence and concentration of specific particles is crucial for monitoring air and water quality, assessing pollution levels, and ensuring public health and safety.

3.3.4 Solids

Breakdown on surfaces and ablation indeed involve complex processes that depend on various factors, including pressure above the surface and incident laser wavelength. At low pressures above the surface, breakdown can be initiated by multiphoton ionization, while at high pressures, inverse bremsstrahlung becomes predominant, both leading to avalanche ionization. The breakdown thresholds for surface targets are significantly lower compared to gas targets, typically by two to four orders of magnitude. Studies have extensively examined the evolution and build-up of plasma over time, position, and incident laser wavelength. The choice of laser wavelength plays a crucial role in the interaction with the surface. Longer wavelengths, such as those from the 10.6 µm CO_2 laser, are absorbed to a greater extent in the plasma above the surface due to absorption varies the square of the wavelength (λ^2). This absorption effectively shields the surface from absorption of the trailing edge of the laser pulse. Conversely, shorter wavelengths, such as the 248 nm from a KrF laser, result in a higher percentage of laser energy impacting the surface directly.

Absolutely, understanding the intricate dynamics involved in laser-induced breakdown spectroscopy (LIBS) is indispensable for optimizing its techniques across a wide array of applications. Whether it's material analysis, environmental monitoring, or other fields, grasping the nuances of LIBS enables researchers to fine-tune parameters, enhance sensitivity, and expand the scope of its capabilities.

The process of laser-induced breakdown spectroscopy (LIBS) on surfaces relies heavily on the ablation of material into the plasma volume. The amount of material ablated per pulse serves as a critical factor influencing the resulting spectral signals and analytical performance. The range of ablated mass, from 5 ng to 80 ng per pulse for aluminum, highlights the variability in ablation efficiency depending on factors such as laser parameters, target material properties, and environmental conditions. This variation underscores the importance of understanding and controlling the ablation process to achieve reliable and accurate analysis.

The detailed fundamental studies regarding laser ablation are crucial for advancing our understanding of the process and optimizing its applications, particularly in laser-induced breakdown spectroscopy (LIBS). Here are some key insights derived from these studies: Crater Dimensions and Analytical Results: The dimensions of the crater formed during ablation have a significant impact on the analytical results obtained through LIBS. The ratio of crater depth-to-diameter, known as the aspect ratio, influences fractionation effects, especially when it exceeds a certain threshold, typically greater than 6. Understanding these effects is essential for accurate elemental analysis. Laser Energy Coupling and Aspect Ratio: During crater formation, the coupling of laser energy to the solid surface increases. Notably, when the aspect ratio exceeds 5, the amount of energy coupled into the surface can increase dramatically, up to 10 times that for a flat surface. This enhanced energy coupling affects the dynamics of plasma formation and emission, influencing the observed spectral signals. Plasma Temperature and Electron Density: Recent studies have focused on investigating the plasma temperature and electron density at different points within the forming crater. These parameters provide valuable insights into the dynamics of plasma formation and evolution, contributing to a deeper understanding of LIBS processes. Repetitive Sparks for Depth Profiling: Repetitive sparks at the same location have been utilized for depth profiling applications. This technique allows for the determination of coating thicknesses and the characterization of layered structures, such as paint layers on paintings. By analyzing the elemental composition at different depths within the sample, researchers can extract valuable information about its composition and history. Overall, these fundamental studies play a critical role in advancing the

capabilities of LIBS for various applications, from material analysis to cultural heritage conservation. By elucidating the complex interactions between laser radiation and solid surfaces,
researchers can optimize experimental protocols, improve analytical accuracy, and unlock new
avenues for scientific discovery and technological innovation.

When laser energy is focused on solids in a vacuum, higher stages of ionization are
achieved compared to the same intensity in ambient conditions. The absence of background
atoms or molecules alters the post-breakdown expansion dynamics of the plasma, leading to
unique spectroscopic signatures. In the realm of femtosecond pulse lengths, ablation studies
are advancing rapidly.

Femtosecond pulses can create high aspect ratio craters due to their ultra-short duration.
Comparisons between craters generated by nanosecond and femtosecond pulses provide
valuable insights into the effects of pulse duration on ablation dynamics and surface morphology. Interrogating solids underwater using LIBS presents specific challenges and opportunities. Significant emission is observed only when employing a double-pulse technique. In a
single-pulse scenario, most of the energy is utilized in forming a vapor cavity on the surface,
leaving insufficient energy for excitation. By using a second pulse to interrogate the vapor
cavity, a plasma with excitation properties similar to that of a single laser spark formed on
metal in air can be generated, enabling underwater LIBS analysis.

These observations underscore the versatility of LIBS across various environments and
pulse regimes. Understanding the intricacies of LIBS under different conditions is crucial for
its successful application in diverse fields, ranging from materials science to environmental
monitoring and beyond.

3.4　Components for a LIBS Apparatus

3.4.1　General

The main components of a laser-induced breakdown spectroscopy (LIBS) instrument is
shown in Figure 3.1,[1] a LIBS Instrument typically include: laser, optical system, spectral
selection, detector. The specifications and configurations of these components can vary depending on the specific application requirements. Factors to be considered when designing
or selecting a LIBS instrument include: elements to be monitored (number and type), sample
characteristics (compositional complexity, homogeneity, etc.), type of analysis (e.g. a qualitative versus quantitative measurement), state of the samples (e.g. gas, liquid, or solid).
By carefully considering these factors and tailoring the instrument design accordingly, researchers can develop LIBS systems optimized for their specific analytical needs and applications.

3.4.2 Laser Systems

The key parameters in the laser systems of LIBS include: pulse energy, pulse repetition rate, beam mode quality, size and weight of the laser system, cooling and electrical power requirements. Regarding the wavelength of the laser beam, while it may not be a primary consideration in all cases, it can impact factors such as eye safety, spectral range coverage, and material interaction efficiency. For instance, certain materials may absorb laser energy more efficiently at specific wavelengths, leading to enhanced LIBS signal generation. All these parameters collectively define the performance and practicality of a LIBS laser system, the choice of wavelength can still be important depending on the specific requirements and constraints of the application.

Nd: YAG Lasers are highlighted as the preferred choice for LIBS due to their reliability, convenience, and ability to generate powerful pulses necessary for creating laser-induced plasmas. Both flashlamp-pumped pulsed and Q-switched Nd:YAG lasers, with pulse widths typically falling in the range of 6-15 nm. Compact Nd:YAG lasers are also noted for their suitability in portable instrumentation. The fundamental wavelength of Nd:YAG lasers is 1,064 nm, but they can be converted to shorter wavelengths (532, 355, and 266 nm) using passive harmonic generation techniques with crystals. While shorter wavelengths may offer advantages in terms of increased energy coupling into specific samples, the 1,064 nm wavelength is often preferred due to its highest power density. Pulsed CO_2 lasers (emitting at 10.6 μm wavelength) and excimer lasers (with typical wavelengths of 193, 248, and 308 nm) are mentioned as alternatives to solid-state lasers for LIBS. However, they are less commonly used due to higher maintenance requirements and the need for special optical materials, as their wavelengths lie in the far infrared and ultraviolet spectral regions, respectively.

The introduction of femtosecond lasers into Laser-Induced Breakdown Spectroscopy (LIBS) represents a significant advancement in the field. Femtosecond lasers emit extremely short pulses, typically in the range of tens to hundreds of femtoseconds. This ultra-short duration allows for precise control over the laser-material interaction process. With femtosecond lasers, the surface irradiation is completed before the plasma expands significantly. This means that there is no shielding effect commonly observed with longer pulses, where the expanding plasma can absorb subsequent laser pulses. As a result, the energy deposition and subsequent ablation and crater formation processes differ from those observed with longer pulse durations. The separation of ablation from plasma formation, facilitated by femtosecond lasers, presents a new avenue for analyzing these two distinct parts of the LIBS technique. However, as noted, there is currently no conclusive evidence on how this separation affects the analytical results. The use of femtosecond lasers in LIBS has the potential to provide valuable insights into the laser-material interaction process and may lead to improvements in analytical capabilities. However, more research is necessary to fully understand and harness the advantages of femtosecond laser technology in LIBS applications.

3.4.3   Methods of Spectral Resolution

The basis of a LIBS measurement is the collection and analysis of an emission spectrum. The emission lines of the elements are tabulated in various sources. Important properties of a spectrometer are (1) the resolution, the minimum wavelength separation at which two adjacent spectral features can be observed as two separate lines, and (2) the width of the spectrum that can be observed. The specifications for these depend on the particular problem at hand. Typically a wider band of observable spectrum is needed when several elements are being monitored simultaneously.

In the case of the narrow-bandpass filter, only a single narrow wavelength band is passed through the wavelength-selective element. Ideally, the bandpass corresponds to the emission line width. The transmitted light is then detected using a photon detector. The advantage of the fixed-wavelength filter is its very small size and low weight and cost.

When monitoring multiple features at widely different wavelengths, the echelle spectrographs are being used in Laser-Induced Breakdown Spectroscopy (LIBS) measurements. Echelle Spectrograph consists of a coarse diffraction grating that spectrally disperses light in the usual manner. Additionally, a prism, with its dispersion perpendicular to that of the grating, is used to stack orders vertically over one another. This arrangement creates a two-dimensional display of wavelength vs order. The advantage of using an echelle spectrograph is its capability to monitor a large wavelength range with reasonable spectral resolution. When operating over a large wavelength interval, it's essential to calibrate the instrument response carefully. Factors such as the grating blaze (which determines the efficiency of the grating), transmission and reflection coefficients of the optics, and the wavelength response of the detector all influence the sensitivity of the instrument across different spectral regions. There is evidence suggesting that using an echelle spectrograph with an array detector can significantly enhance analytical figures of merit for LIBS analysis. The echelle spectrograph offers advantages in LIBS measurements by enabling the monitoring of a broad wavelength range with reasonable spectral resolution. However, proper calibration and consideration of factors affecting instrument sensitivity are crucial for accurate and reliable measurements. The adoption of echelle spectrographs, especially when combined with array detectors, represents a promising avenue for improving the analytical capabilities of LIBS systems.

The choice of spectral resolution method in Laser-Induced Breakdown Spectroscopy (LIBS) depends on various factors related to the analysis requirements. Complexity of the sample, the number of elements present in the sample and the abundance of their emission lines determine the necessary spectral resolution. Samples with many elements or with numerous emission lines for each element may require higher spectral resolution to distinguish between closely spaced lines. Number of elements to be monitored, the more elements that need to be monitored, the greater the demand for a spectral resolution method that can handle simultaneous or sequential detection of multiple elements. Simultaneous or sequential monitoring, depending on the experimental setup and analysis goals, elements can be monitored either simultaneously or sequentially. Simultaneous monitoring requires spectral resolution methods capable of handling multiple emission lines simultaneously, while sequential monitoring may allow for lower spectral resolution methods. Location of emission lines, the specific wavelengths of emission lines of interest influence the choice of spectral resolution. If emission lines of interest are closely spaced, higher spectral resolution may be necessary to resolve them accurately.

3.4.4　Detectors

The type of detector used in Laser-Induced Breakdown Spectroscopy (LIBS) measurements is closely related to the method of spectral selection employed in the setup. Photomultiplier Tubes (PMT) are highly sensitive and can detect low levels of light by amplifying the signal produced when light hits a photosensitive material. They are often used in conjunction with fixed filters or monochromators for spectral selection. PMTs are particularly useful when high sensitivity and low noise are crucial. Photodiodes generate a current proportional to the incident light and are simpler and more robust than PMTs. They are also used with fixed filters or monochromators, and they offer a cost-effective solution for many applications where extreme sensitivity is not required. Photodiode Arrays (PDA) consist of multiple photodiodes arranged in a linear or two dimensional array, allowing them to capture spatial information about the light pattern. PDAs are used with spectrographs to record continuous spectra across a range of wavelengths. Charge Coupled Devices (CCD) are highly sensitive and capable of detecting low levels of light with high resolution. They consist of an array of pixels that accumulate charge in response to incident light, which is then read out and converted into a digital signal. CCDs are widely used in LIBS due to their ability to provide detailed spectral information across a broad wavelength range. Charge Injection Devices (CID) are similar to CCDs but allow for non-destructive readout, meaning that the same pixel can be read multiple times. This feature can be advantageous in certain applications where dynamic range and exposure control are important. Array detectors, when used with spectrographs, offer the advantage of capturing the entire spectrum simultaneously, providing a comprehensive analysis of the emitted light from the plasma generated in LIBS. This is essential for identifying and quantifying multiple elements present in the sample. The choice of detector ultimately depends on the specific requirements of the LIBS application, including sensitivity, spectral resolution, and cost considerations.

The process of time-resolved detection of the plasma light is crucial to obtaining accurate and clear spectral data. This is primarily to avoid the interference caused by the intense spectrally broad white light emitted during the initial microseconds (0-1 μs) after the plasma formation. To achieve this, a microchannel plate (MCP) is used in conjunction with array detectors like PDA, CCD, and CID. The MCP acts as a light valve positioned in front of the array detector. It plays a critical role in controlling the timing of light detection. When the MCP is activated, the incident light is amplified and then strikes the photosensitive array (PDA, CCD, CID). This process significantly increases the detectable light signal, improving the detector's ability to capture weak spectral lines that would otherwise be lost in noise. By controlling the timing of the MCP activation, these intensified detectors can effectively isolate the spectral emissions of interest, avoiding the contamination from early-time broad-spectrum emissions and thereby enhancing the clarity and accuracy of the LIBS measurements.

3.5 Industrial Application of LIBS

3.5.1 LIBS Detection in Electric Power Grids

3.5.1.1 Introduction

Silicone rubber has significantly enhanced the field of outdoor high-voltage insulation, offering superior performance compared to traditional ceramic materials.[3] This advancement is largely attributed to silicone rubber's unique surface behavior in the presence of water. Even when contaminants deposit on its surface, Silicone rubber exhibits excellent hydrophobic properties, which means it repels water effectively. This characteristic is maintained even under adverse field conditions and after prolonged exposure to environmental contaminants. This property helps in preventing the formation of conductive water films on the insulator surface, thereby reducing leakage currents and the risk of flashovers. The performance of silicone rubber in high-voltage insulation applications is closely linked to its material formulation. Room Temperature Vulcanized (RTV) Silicone Rubber is often applied to improve the pollution performance of ceramic insulators.[4] By coating ceramic insulators with RTV silicone rubber, the overall hydrophobicity of the system is enhanced, which helps in reducing the accumulation of pollutants and the associated electrical discharge activities. This ease of application allows for quick and effective maintenance and enhancement of existing ceramic insulators, extending their service life and improving their performance under polluted conditions.

Monitoring the thickness of room-temperature vulcanized (RTV) silicone rubber coatings is crucial for electric utilities to ensure the reliability and performance of high-voltage insulators. Traditional methods of measuring RTV thickness are often challenging and time-consuming. However, Laser-Induced Breakdown Spectroscopy (LIBS) offers a promising solution for this task by providing a time-efficient and online analysis method. A laser is focused on the RTV coating surface, ablating small craters into the material. The depth of these craters is directly related to the number of laser pulses. The emitted light from the ablated material is captured and analyzed using a spectrometer. This provides the emission spectra of the RTV coating, which contains information about its composition and thickness. Studies have shown a linear relationship between the depth of the laser-ablated craters and the number of laser pulses. This linearity allows for precise calibration and measurement of the coating thickness based on the number of pulses applied. Scanning Electron Microscope (SEM) is used to observe and measure the crater's depths, confirming the linear relationship and ensuring accuracy in the LIBS measurements. LIBS provides a real-time and in situ method for monitoring the thickness of RTV coatings. This capability is particularly valuable for electric utilities that need to quickly assess the condition of their insulators without taking them out of service. By measuring the coating thickness, LIBS can help identify areas where the RTV coating has deteriorated due to powdering, surface peeling, aging, or other forms of damage. This information is critical for maintenance decisions. When LIBS detects that the RTV coating has significantly thinned or disappeared, it indicates the need for maintenance actions such as reapplying the RTV coating or replacing the silicone rubber insulators. This proactive approach helps prevent failures and extends the lifespan of high-voltage insulation systems. Figure 3.4 shows RTV coatings have been used widely on surface of insulators.

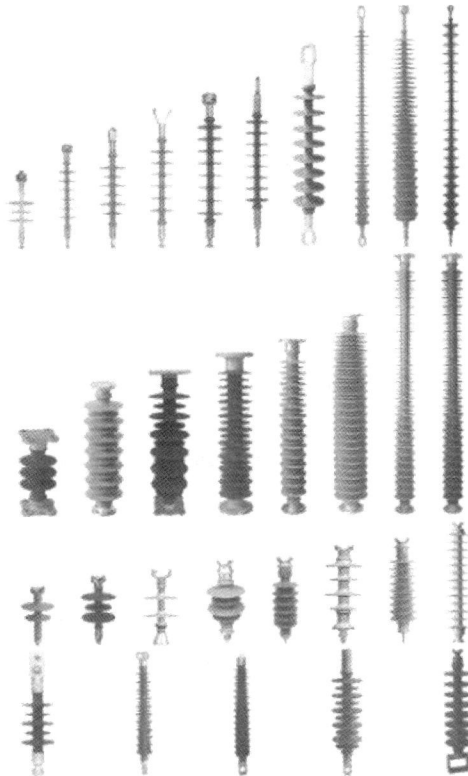

Figure 3.4   Room temperature vulcanized silicone rubber (RTV)
coatings have been widely used on the surface of insulators

In high-voltage transmission and distribution systems, the primary insulating component is the surrounding atmospheric air. This choice is made for several key reasons, which contribute to the overall efficiency and performance of the electrical transmission network. Utilizing increased voltage levels is essential to minimize power losses during transmission. Higher voltages reduce the current for a given power level, which in turn reduces resistive losses in the conductors. Higher transmission voltages contribute to the stability of the power system. Stability improvements include better voltage regulation and reduced risk of voltage collapse under high load conditions. Atmospheric air is an abundant and free resource, making it an economical choice for insulation in high-voltage systems. Unlike other insulating materials, air does not incur direct costs for procurement and installation. One significant disadvantage of using air as an insulating medium is its inability to provide mechanical support for high-voltage conductors. Gases and liquids generally lack the necessary mechanical strength to maintain the physical integrity of the system under stress conditions such as wind, ice, or seismic activity. To address the mechanical support limitation, solid insulators are used in conjunction with atmospheric air. These insulators provide the necessary mechanical properties to support and maintain the structure of high-voltage transmission systems. Solid insulators, often made from materials like porcelain, glass, or polymer composites, are capable of withstanding mechanical loads and environmental stresses. They provide the essential structural support to keep conductors in place and ensure the stability of the transmission system. An effective insulation system in high-voltage applications is typically composed of both gas and solid dielectrics. The atmospheric air acts as the primary insulating medium, while solid insulators provide mechanical support and additional electrical insulation as needed.

The performance of the gas-solid interface in high-voltage insulators is crucial for the efficiency and reliability of the entire high-voltage installation. The insulator must perform efficiently under various service conditions, including environmental stresses and electrical loads. A single insulator failure can result in prolonged outages, making reliability a paramount concern.

Over time, insulators can accumulate contaminants that may become conductive, especially in coastal areas. This contamination can lead to surface leakage currents and flashovers, compromising the insulator's performance. The pollution of high voltage insulators is a major cause of power outages in many transmission and distribution systems.

Silicone rubber insulators exhibit hydrophobic (water-repellent) properties, which prevent the formation of conductive water films on the insulator surface. This is particularly advantageous in preventing electrical conductivity due to surface contamination. Compared to traditional ceramic materials like porcelain and glass, which are hydrophilic (water-attracting) and thus more vulnerable to pollution, silicone rubber insulators offer significantly improved performance under polluted conditions.[5]

The introduction of silicone compounds has revolutionized outdoor transmission and distribution insulating systems. The key advantage is the material's ability to maintain hydrophobicity even after long-term exposure to pollutants. Silicone rubber insulators are less prone to contamination-related performance degradation, making them more reliable in harsh environmental conditions.

Room Temperature Vulcanized (RTV) Silicone Rubber Coatings can be applied to the surface of traditional ceramic insulators to imbue them with the beneficial properties of silicone rubber, such as hydrophobicity. RTV coatings help ceramic insulators resist wetting and subsequent contamination, thus improving their performance under polluted conditions.

The effectiveness of RTV coatings depends on their formulation and the fillers used. These components determine the coating's durability and hydrophobicity. Proper application procedures are crucial to ensuring that the RTV coating adheres well to the ceramic surface and performs effectively in service. The longevity and performance of RTV-coated insulators are influenced by the environmental conditions they are exposed to, such as UV radiation, temperature fluctuations, and mechanical stress.

3.5.1.2 Pollution of High Voltage Insulators

Pollution of high voltage insulators is a pervasive issue for outdoor high voltage installations worldwide and is often the primary cause of power outages. The process is typically considered through a six-stage mechanism:

The first stage is the deposition of contaminants. Wind, acid rain, dust, industrial pollution and biological materials (e.g., bird droppings) deposit contaminants on the insulator surface. These contaminants can either possess or develop electrical conductivity, which is critical for the subsequent stages.

The second is moisture deposition. In coastal areas, sea salts are the primary contaminants. These salts become conductive when diluted in water. Mechanisms such as fog, dew, condensation, and light rain deposit moisture on the insulator surface, wetting the contaminants and forming a conductive layer.

The third is formation of surface conductivity and leakage current. The wetted, conductive contaminants form a path for leakage current to flow across the insulator surface. The film of contaminants acts as a distributed resistance, with its value depending on the amount of contamination and the degree of wetting.

The forth stage is the formation of dry bands. The flow of leakage current generates joule heating, which causes localized drying of the insulator surface, especially in areas with small radii from the insulator axis of symmetry. These dried areas, known as dry bands, increase the resistance in those zones.

The fifth is dry band arcing. The initial insulating surface transitions into a series of electrolytic resistances. The voltage distribution along the insulator leakage path changes, leading to intense stress in certain areas. Dry band arcs form as the voltage stress causes discharges over the dry bands. These arcs do not cover the entire leakage path and do not immediately lead to flashover.

The sixth is flashover and complete breakdown. Under favorable conditions, such as an optimal combination of surface film conductivity and gas discharge properties, the dry band arcs can propagate, resulting in a complete flashover. This event bridges the entire leakage distance, causing a sustained arc and leading to a power outage.

To prevent the breakdown and outage, using hydrophobic materials like silicone rubber for preventing the formation of conductive films, thus reducing the stages leading to flashover. Regularly cleaning insulators to remove contaminants reduces the risk of pollution-related failures. Applying Room Temperature Vulcanized (RTV) silicone rubber coatings on ceramic insulators can provide a hydrophobic surface, improving performance in polluted environments. Optimizing insulator design to minimize contaminant accumulation and enhance self-cleaning properties can help mitigate pollution effects.

Figure 3.5   Flashover of a 150 kV
post-porcelain insulator[3]

Dry band arcing and, finally, a flashover on a 150 kV post porcelain insulator during an artificial pollution test is illustrated in Figure 3.5.

3.5.1.3   Silicone Coatings for High Voltage Ceramic Insulators

Given the vulnerability of outdoor high-voltage insulators to pollution, enhancing the surface properties of insulators is essential for improving system efficiency and reliability. Coatings can be employed to interfere with the pollution mechanism, either by reducing contaminant accumulation or by minimizing wetting.

Grease coatings encapsulate contaminants, preventing them from forming conductive paths on the insulator surface. These coatings are typically based on hydrocarbons or silicones. Silicones are preferred due to their higher thermal stability, making them suitable for various climatic conditions. The ability of grease to encapsulate contaminants is limited by the amount of grease applied. As contaminants accumulate, the grease becomes saturated. In environments with moderate to heavy pollution, grease coatings may become saturated in less than six months, necessitating frequent replacement. Once saturated, the performance of greased insulators is significantly inferior to that of ungreased ceramic insulators.

Silicone-based coatings impart hydrophobic (water-repellent) properties to the insulator surface, reducing the degree of wetting and formation of conductive films. By maintaining a hydrophobic surface, silicone coatings prevent the establishment of a continuous conductive path, thus delaying or preventing leakage current formation and subsequent stages of the pollution mechanism. Silicone coatings are stable over a wide range of temperatures, making them suitable for diverse climatic conditions.

The capacity of grease or other coatings to encapsulate contaminants is finite. Once saturation is reached, the coating must be replaced, which is both time-consuming and costly. Frequent replacement of saturated coatings can be a significant expense for utilities, especially in heavily polluted environments.

Coatings may degrade over time due to environmental factors such as UV radiation, temperature fluctuations, and mechanical wear. Regular monitoring and maintenance are required to ensure continued performance. Effectiveness Post-Saturation: After saturation, the insulating properties of the coated surface may degrade, necessitating reapplication or replacement of the coating to maintain reliability.

In addition to grease coatings, Room Temperature Vulcanized (RTV) silicone rubber coatings offer a distinct approach to managing pollution on high-voltage insulators. Unlike grease coatings, which encapsulate contaminants, RTV coatings leverage a different mechanism to enhance the surface properties of insulators.

RTV coatings incorporate a mechanism wherein hydrophobic molecules migrate to the surface. When contaminants accumulate on the RTV-coated surface, these hydrophobic molecules migrate into the contamination layer, imparting water-repellent properties. This molecular migration makes the surface of the contaminants hydrophobic, thereby reducing surface wetting and delaying the development of the pollution mechanism. RTV coatings are known for their durability and ability to maintain hydrophobic properties over an extended period. This helps in preventing the formation of conductive films on the insulator surface. RTV coatings are resistant to various environmental factors such as UV radiation, temperature fluctuations, and mechanical wear, which contributes to their longevity.

RTV coatings are applied to the surface of ceramic insulators to enhance their performance under polluted conditions. The process involves coating the insulator surface with a layer of RTV silicone rubber, which cures at room temperature to form a protective, hydrophobic layer. Figure 3.6 illustrates a porcelain post-insulator with an RTV coating. The bulk of the porcelain is white, but the RTV-coated surface appears brown, with a visible 5 mm white coat indicative of the coating's presence.

3.5.1.4 RTV Coatings Formation

Composite materials used in outdoor insulating systems are complex formulations composed of a base polymer and various additives known as fillers. The fillers are incorporated to tailor the material properties to meet specific application requir-

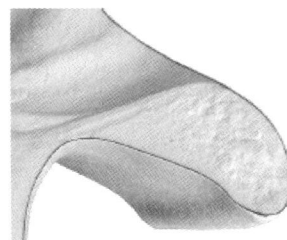

Figure 3.6 Section of a 150 kV post-porcelain coated with a 5 mm RTV coating [3]

ements, which start with the intrinsic properties of the base polymer.

The base polymer forms the primary matrix of the composite material, providing the fundamental characteristics such as flexibility, hydrophobicity, and thermal stability. Example: Polydimethylsiloxane (PDMS) is a commonly used base polymer in RTV silicone rubber coatings for high-voltage insulators due to its outstanding properties.

Fillers are added to enhance specific properties such as mechanical strength, thermal conductivity, and electrical insulation. Common fillers include silica, alumina, and other particulate materials, which can improve the mechanical and thermal performance of the composite.

Polydimethylsiloxane (PDMS) is the base polymer used in many silicone formulations for outdoor high-voltage systems, including RTV coatings and composite insulators.

The PDMS polymer chain consists of silicon-oxygen (Si-O) bonds, with two methyl groups attached to each silicon atom. This structure provides PDMS with its unique properties. The repeating unit of PDMS is depicted in Figure 3.7, illustrating the silicon-oxygen backbone and the side methyl groups.

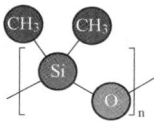

Figure 3.7 Representation of the polydimethy lsiloxane monomer[3]

PDMS has excellent hydrophobic properties, which help to repel water and prevent the formation of conductive films on the insulator surface. Thermal Stability: It can withstand a wide range of temperatures, making it suitable for various climatic conditions. The flexible nature of PDMS allows it to maintain performance under mechanical stress and environmental variations.

To achieve the desired performance in outdoor insulating systems, the formulation of composite materials must be optimized based on the anticipated service conditions. This involves determining the appropriate type and amount of both the base polymer (PDMS) and the fillers.

The environmental conditions, such as temperature extremes, UV exposure, and pollution levels, influence the formulation requirements. Performance Goals: The composite material should meet specific performance criteria, such as mechanical strength, electrical insulation properties, and resistance to environmental degradation.

3.5.1.5 Application of RTV Coatings in High-Voltage Substations

High-voltage substations are crucial for the operation of a high-voltage transmission system. An outage at a substation can significantly impact power system stability and operation, often more so than an outage on a transmission line, as substations serve as power system nodes. Pollution performance is a key concern for transmission system operators, particularly in coastal power systems where environmental contaminants can affect equipment.

Many maintenance methods have been implemented to ensure fault-free substation performance under pollution conditions. However, the application of RTV (Room Temperature Vulcanizing) coatings has been proven to be the most effective method for already installed equipment for several reasons: (1) Improved Pollution Performance, RTV coatings exhibit hydrophobic surface behavior, which resists the development of leakage current and dry band arcing even in the presence of contaminants. This significantly improves pollution performance by preventing conductive paths on the insulator surface; (2) Ease of Application, RTV coatings are available in the form of paint, making them easy to apply to insulators regardless of profile geometry or insulator use. This is particularly convenient in substations where insulators serve additional roles, such as in current and voltage transformers, bushings, and other components with varied housing sizes and shapes; (3) Extended Application Lifetime, compared to silicone grease, RTV coatings offer a longer application lifetime, potentially exceeding ten years depending on environmental conditions. Unlike silicone grease, which encapsulates contaminants, RTV coatings possess hydrophobicity transfer capabilities. This changes the initial hydrophilic behavior of the contamination film to hydrophobic, maintaining effective performance over a longer period; (4) Cost-Effectiveness and Preventive Maintenance, the longevity of RTV coatings allows for large-scale application at a reasonable cost. Additionally, their application can be scheduled at convenient times for the utility, making them a viable preventive maintenance measure. Figure 3.8 illustrates the RTV coatings provide greater hydrophobicity capability on the surface of a clean insulator.

(a)　　　　　　　(b)

Figure 3.8　Surface hydrophobicity in a clean (a) and in a contaminated (b) RTV coated surface[3]

3.5.1.6　Application of LIBS for RTV Coatings Detection

After long-term operation, RTV (Room Temperature Vulcanizing) coatings can deteriorate due to environmental factors such as flashover, rain, sunlight, and strong wind, leading to powdering, surface peeling, and aging. This degradation weakens the anti-pollution flashover capability of the coatings.[6] Consequently, it is essential to measure the thickness of RTV coatings accurately to ensure safe and effective operation. Typically, the thickness should be between 0.3 mm and 0.5 mm.[7, 8] However, environmental factors during the application can result in non-conformant thickness. Therefore, precise measurement methods are urgently needed. Figure 3.9 and Figure 3.10 show the deterioration of RTV coating on the surface of the retired or aged insulators.

An actual example of the detection of RTV coating thickness and its contamination

Figure 3.9   A retired composite insulator string from a 500 kV AC power transmission line[6]

will be described below: A Nd: YAG laser (Quantel) was chosen for emission pulsed laser with the wavelength $\lambda = 1,064$ nm, the pulse width of 7 ns, and the frequency of 2 Hz, and focused on the surface of the samples by a 50 mm focal lens. The diameter of the laser beam at the surface of the sample was 100 μm. The energy of a single pulse was adjustable from 40 to 90 mJ. Connected with the laser, the detection part captured the emission spectra of the plasma with every shot in the air. Owing to the unequal distribution of laser shot energy, the ablation crater edges were not smooth, even the crater was separated into two parts in samples when the laser pulses were 50 times. With the increasing number of shot pulses, the crater became larger and deeper, and the edges looked more smooth.[9]

Figure 3.11, Figure 3.12 and Figure 3.13 show the emission spectrum of the plasma excited by the laser on a sample. According to the database of the National Institute of Standards and Technology (NIST), USA, elements such as carbon (C), oxygen (O), aluminum (Al), silicon (Si), and zinc (Zn) were identified in the emission spectra. Strong spectral peaks of these specific elements were listed in Table 3.1.[8] For the same type of RTV coating, the emission spectrum remains consistent after different numbers of laser pulses, unless the RTV coating is punctured. If the RTV coating is punctured, there will be a significant change in the emission spectra, which marks the end of the measurement.[8, 10]

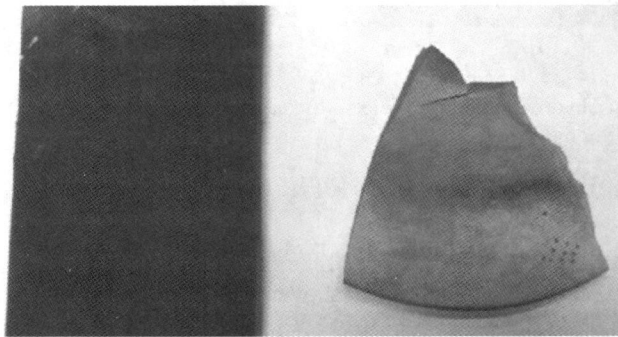

Figure 3.10   New (left) and aged (right) silicon rubber samples [6]

Table 3.1   **Peaks in Spectra**

Spectrum Line	Wavelength (nm)	Spectrum Line	Wavelength (nm)
C(I)	247.85; 283.67; 396.14	Al(I)	212.34; 221.01; 237.21; 394.40; 396.15
O(I)	777.41; 795.22; 844.64	Fe(I)	243.82; 2,434.03; 245.22
Si(I)	251.61; 252.85; 263.13; 288.16		

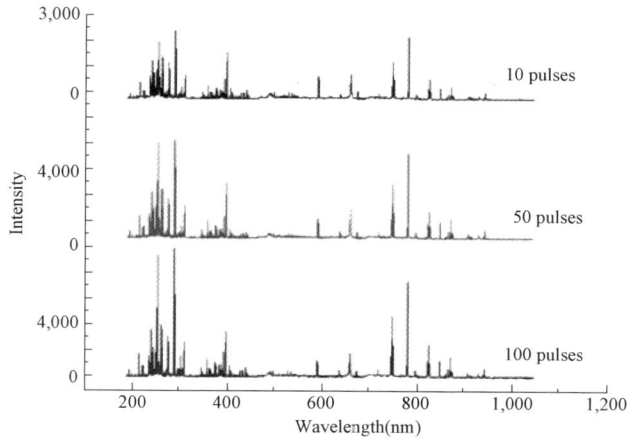

Figure 3.11   The emission spectrurm of the plasma excited by laser on RTV[8]

3.5.2 LIBS Application in Thermal Power Plants

3.5.2.1 Introduction

Laser-Induced Breakdown Spectroscopy (LIBS) is a powerful analytical technique that can be employed to monitor various aspects related to the use of coal in thermal power plants, the by-products of coal combustion, and alternative fuel sources. The main applications of LIBS include: (1) LIBS can determine the elemental composition of coal, ensuring its quality and suitability for combustion. (2) Detecting trace elements such as sulfur, mercury, and arsenic helps in assessing potential environmental impacts. (3) LIBS can analyze the chemical composition of both bottom ash and fly ash. This information is vital for understanding the properties of ash and its potential uses or disposal methods. By detecting harmful elements in ash, LIBS aids in environmental compliance and pollution control. (4) LIBS can analyze coal used as feedstock in gasification processes, ensuring its suitability and optimizing the gasification process. (5) LIBS can monitor the composition of synthetic gas produced, ensuring its quality and safety. (6) LIBS can evaluate the elemental composition of biomass to ensure it meets the required standards for use as a fuel. Similar to coal, the by-products of biomass combustion can be monitored for harmful elements. LIBS can analyze the ash to predict its slagging and fouling tendencies. This helps in taking preventive measures to avoid performance degradation in boilers. (7) LIBS can be used in situ to monitor the deposition of slag and fouling on heat transfer surfaces, facilitating timely maintenance and cleaning.[12]

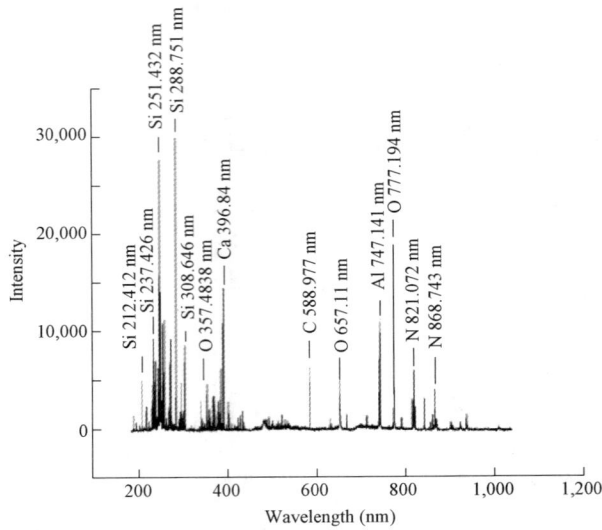

Figure 3.12　The emission spectrum after the 20th shot on a RTV sample [11]

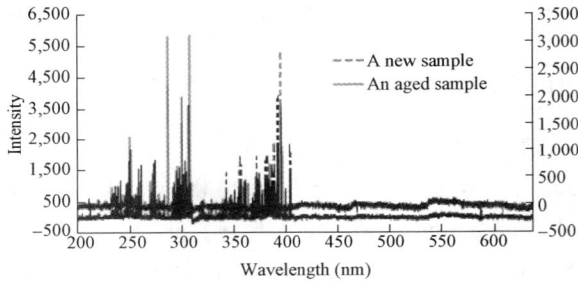

Figure 3.13　Comparison between spectra of fresh and aged RTV samples [6]

3.5.2.2　Mineral Matter in Coal and Its Impact on Combustion Processes by LIBS

Coal is a complex mixture of organic and inorganic materials, varying significantly in composition based on rank and geographic locality. The inorganic components can be present as separate phases or bound to organic components. The inorganic materials, or mineral matter, in coal are crucial for end-users, particularly in power generation. The presence of minerals such as pyrite (FeS_2) is problematic. High pyritic content in coal contributes to sulfur dioxide (SO_2) emissions during combustion, leading to environmental pollution. Furthermore, pyrite reduces the ash fusion temperature, causing molten ash to adhere to boiler surfaces, reducing heat transfer efficiency and increasing corrosion risk. Boiler manufacturers and utility operators utilize empirical indices to evaluate coal's impact on steam generators. These indices, derived from coal-ash analyses, inform various aspects of boiler design and operation. Regular laboratory analyses, following standardized coal sampling and techniques, are essential for accurate assessment.

Several techniques are employed to determine the chemical composition of coal, including: (1) Atomic Absorption/Emission Spectroscopy. (2) X-ray Fluorescence (XRF). (3) X-ray Diffraction (XRD). (4) Inductively Coupled Plasma (ICP)-Atomic Emission Spectrometry. (5) Scanning Electron Microscopy (SEM). (6) Electron Probe Microanalysis (EPMA).

Specific ASTM standards detail these methods: ASTM D-3682: Analysis of major and minor elements in coal ash via atomic absorption/emission spectroscopy. ASTM D-4326: X-ray fluorescence analysis for coal ash. ASTM D-6349: Inductively coupled plasma (ICP)-atomic emission spectrometry for major and minor elements in coal ash.

LIBS is emerging as a promising technology for coal mineral analysis. This technique relies on laser interaction with the coal matrix and offers an alternative to traditional methods, meeting the power generation industry's need for accurate and efficient coal analysis. Understanding and analyzing the mineral matter in coal is essential for optimizing combustion processes and minimizing environmental impacts. Regular use of standardized analytical techniques and the adoption of new technologies like LIBS can significantly enhance the efficiency and sustainability of coal utilization in power generation.

3.5.2.3 Experimental Setup for LIBS Analysis of Coal Mineral Matter Composition

An experimental setup utilizing LIBS was assembled to determine the mineral matter composition in coal under controlled laboratory conditions.[12] The LIBS system was designed to achieve several objectives: Sparking the coal sample using a system laser within a controlled atmosphere. Displacing the sample to obtain a collection of laser shots. Resolving the spectra using a spectrometer. Processing the acquired spectral data.

The LIBS system utilized a Q-switched Nd:YAG laser (Big Sky Laser-CFR-400) capable of emitting ultraviolet (UV), visible, and near-infrared (IR) pulses. UV pulses were directed into a beam dump using a 266 nm beam splitter. Visible and near-IR pulses, at wavelengths of 1,064 nm and 532 nm respectively, were directed into the sample chamber using corresponding laser mirrors. The pulse energies were set at 100 mJ for 1,064 nm and 180 mJ for 532 nm. An f/4 lens focused the laser light onto the coal sample to generate the LIBS spark. Light emitted from the sample was collected using f/2 mini-lenses. UV and visible-grade fused silica optical fibers guided the emitted light to the spectrometer.

Equipped with a sample cart mounted on a motorized XY stage, controlled by Measure-Solid software. Maintained a controllable non-oxygen atmosphere (helium) to preserve sample integrity from air elements.

The optical spectrometer contained an Echelle-type grating, allowing for high-resolution spectra collection over a broad wavelength range (200-780 nm). Timing for spectral data collection was managed using MeasureSolid software.

The analysis focused on the following elements due to their involvement in fouling processes in coal-fired boilers: Aluminum (Al), Calcium (Ca), Potassium (K), Magnesium (Mg), Sodium (Na), Iron (Fe), Silicon (Si), and Titanium (Ti).

Specific compounds were added to control the concentration of targeted elements: silica, alumina, titanium oxide, iron oxide, magnesium carbonate tetrahydrate, potassium bromide, sodium chloride, and calcium carbonate.

Spectral data were acquired from synthetic samples with single-element additions. Each sample underwent three tests, with 50 spectral traces collected per sample using the Echelle spectrometer. Carbon content was measured and used as a reference.

The LIBS experimental setup provided a controlled and precise method to analyze the mineral matter composition in coal. By focusing on key elements involved in fouling processes, the setup aimed to enhance the understanding of coal properties and improve the efficiency and sustainability of coal-fired power generation.

3.5.2.4 LIBS Off-Line Coal Analysis at a Power Plant

To perform an analysis of coal using Laser-Induced Breakdown Spectroscopy (LIBS) at a coal-fired thermal power plant, it is crucial to ensure accurate sampling and preparation. The variability in coal feedstock can significantly affect the performance and efficiency of the power plant, especially concerning high-temperature slagging, which necessitates retroactive remedial actions[12].

Obtain grab samples from the conveyor belt supplying coal to the feeder, generally, collecting more than 20 grab samples for each type of coal to ensure a representative analysis. Split the collected samples into two portions. Send one portion to a laboratory for standard ASTM analyses. Retain the second portion for LIBS analysis. Crush the retained samples to ensure uniformity. Sieve the crushed samples to a particle size smaller than 60 mesh. Dry the sieved samples to remove superficial moisture. Set up the LIBS system according to the manufacturer's specifications. Calibrate the LIBS system using standard reference materials to ensure accuracy. Analyze the prepared coal samples using the LIBS system. Record the spectral data generated for each sample. Conduct ASTM standard analyses on the split samples sent to the laboratory. This typically includes determining the elemental composition and mineral content of the coal. Compare the elemental composition results from the LIBS analysis with the ASTM results. Identify any discrepancies and analyze the potential causes. Evaluate the mineral composition data to assess the susceptibility of coal to high-temperature slagging. Use this information to predict and mitigate slagging issues proactively. Based on the slagging potential assessment, implement necessary retroactive measures to mitigate the impact of slagging fuels. This may include adjusting the combustion process, modifying fuel blends, or using additives.

Continuously monitor the coal feedstock using LIBS to detect any changes in composition. Adjust the operational parameters of the power plant as needed to optimize performance and reduce slagging risks. By following this structured approach, the variability in coal feedstock can be managed more effectively, and the impact of high-temperature slagging can be minimized, ensuring more efficient and reliable operation of the coal-fired thermal power plant.

3.5.2.5　LIBS On-Line Coal Analysis at a Power Plant

To address the challenges posed by fuel variability and delays in obtaining fuel analysis at a coal-fired thermal power plant, a Laser-Induced Breakdown Spectroscopy (LIBS) system can be implemented for a full-scale, over-the-belt installation. This installation can help provide real-time analysis and improve the efficiency and reliability of the plant operations.[12]

Laser and spectrometer should be installed at a distance from the measurement point. Designed to convey the laser and plasma light to the system, providing enough depth of field to compensate for changes in the height of the coal layer due to conveyor belt sagging. Belt Speed need to be configured for 500 ft/min. Measurement point is located after a transfer point where the mixed coal stream presents a coal top size of no more than 2 inches. This location ensures the surface measurement nature of LIBS is effectively utilized.

Preliminary Calibration need to be conducted in the laboratory. Additional calibration performed on-site. Conducted at a laser pulse frequency of 5 pulses per second. Conducted over 73 hours to evaluate the analyzer's performance. Focus on the accuracy of iron measurement due to its impact on ash fusibility temperatures. Iron content in ash ranged from 7.77.

Advantages of LIBS technology include: (1) Provides immediate feedback on coal composition, enabling proactive adjustments to the combustion process. (2) Helps manage increased coal prices and stringent environmental regulations by optimizing fuel feedstock. (3) Capable of handling a wide range of coals, including those with varying elemental compositions. (4) Avoids the certification and maintenance issues associated with nuclear-based instruments. (5) Suitable for installation at various points in the coal supply chain, from mines to power plants.

By implementing a LIBS system for over-the-belt analysis, coal-fired power plants can achieve better control over fuel quality, enhance operational efficiency, and comply with environmental regulations. This technology represents a promising solution for the dynamic challenges of modern coal-fired power generation.

The LIBS system could be assembled for a full-scale, over-the-belt installation at a coal-fired thermal power plant. This plant fires a range of bituminous coals that arrive via rail trains from a range of suppliers. An important problem with the plant feedstock is the fuel variability and the delay in obtaining fuel analysis from the on-site laboratory, especially, when up to 50% of the unloaded coal goes directly to the boilers. For this particular installation, the LIBS system was set up with the excitation laser and spectrometer at a distance from the measurement point.

3.5.3 LIBS Application in Nuclear Power Stations

The remote, non-contact capabilities of LIBS make it particularly attractive for situations where information is required on the elemental composition of materials within radioactive environments.[2] For example, there are many cases throughout the nuclear industry where compositional analysis of a plant component is required during routine inspection campaigns. Options for carrying out these inspections are often limited to the removal of a physical sample for laboratory analysis; however, this is sometimes not possible as the removal of a sample may compromise the mechanical integrity of the component being examined. Compositional analysis of highly radioactive process materials during spent fuel reprocessing poses major problems since the removal, transport, and chemical analysis of samples require highly specialized equipment and facilities. Analysis of components and process materials in situ can, therefore, offer significant savings in cost and time compared with physical sampling and laboratory analysis. Another benefit of in situ analysis is that, unlike the use of wet-chemical laboratory techniques, no secondary waste is produced.

The remote, non-contact capabilities of Laser-Induced Breakdown Spectroscopy (LIBS) offer significant advantages for the compositional analysis of materials in radioactive environments. This method is particularly beneficial in the nuclear industry, where routine inspections often require detailed elemental composition data without compromising the integrity of the plant components.

Reducing the need for personnel to enter hazardous radioactive areas, minimizing exposure risks. Allows for the analysis of components without physically removing samples, thereby maintaining the mechanical integrity of the components. Eliminates the need for sample removal, transport, and extensive laboratory analysis, offering significant savings in both cost and time. Unlike wet-chemical laboratory techniques, LIBS does not produce secondary waste, making it a more environmentally friendly option. During routine maintenance and inspection, LIBS can be employed to analyze the elemental composition of various components directly. Provides real-time data that can be used to assess the condition and performance of the components.

In highly radioactive environments, such as during the reprocessing of spent nuclear fuel, LIBS can be used to analyze process materials. This is particularly useful since the removal and transport of samples in such environments require highly specialized and costly equipment and facilities. The LIBS system must be robust and capable of functioning reliably in the presence of high radiation levels. Ancillary optics and laser delivery systems need to be designed to withstand harsh environmental conditions. Regular calibration is essential to maintain the accuracy of the LIBS system in such demanding applications. In situ calibration techniques may need to be developed to ensure consistent performance.

Examples from the nuclear industry show successful application of LIBS for analyzing the elemental composition of plant components during routine inspections.[2] LIBS has been employed to analyze highly radioactive process materials, providing valuable data without the challenges associated with physical sampling.

LIBS technology presents a compelling solution for the nuclear industry, offering a safer, cost-effective, and environmentally friendly method for compositional analysis in radioactive environments. By enabling in situ analysis, LIBS eliminates the need for physical sampling, reduces secondary waste, and provides real-time data critical for maintaining the safety and efficiency of nuclear facilities.

The integration of fiber optic cables into a LIBS instrument to facilitate remote analysis has garnered significant interest in the scientific community. A fiber optic probe LIBS system retains the core LIBS hardware components while incorporating a remote probe connected to the laser and optical detector via optical fibers. This configuration allows for remote elemental analysis of materials, offering flexibility and enhanced capabilities in various applications, particularly in hazardous or hard-to-reach environments. The incorporation of fiber optic cables within a LIBS system significantly enhances its capabilities for remote elemental analysis. By allowing for the safe, flexible, and real-time examination of materials in challenging environments, fiber optic probe LIBS systems are poised to become invaluable tools in industries ranging from nuclear power to environmental monitoring. The ongoing development of more sophisticated probe designs and calibration techniques will further expand the applications and reliability of this innovative technology.

3.5.4 LIBS Application in Metal Alloys Industries

3.5.4.1 In Situ Compositional Analysis of Steel Pipes at High Temperature

A transportable LIBS instrument has been developed as part of the European Community 5th Framework project called LIBSGRAIN to measure the composition of steel pipes while in service at high temperatures (around 700 ℃). [1, 2] This instrument is designed to provide real-time compositional analysis without disrupting normal plant operations. The instrument incorporates a compact Q-switched Nd:YAG laser operating at 1,064 nm and producing a pulsed output of 0-50 mJ in 7 ns with a pulse repetition rate of 10 Hz. A specially designed optical head is used to scan the laser beam over the surface of the steel pipe. An echelle spectrograph with a spectral range of 200-800 nm is used to record the plasma emissions and dedicated software is used to produce two-dimensional false-color compositional maps of the selected regions of the steel pipe. The maps have been used to display the spatial distribution of Fe, Mn, Cu, Al, Ti, and Ni, and may be used to help estimate the probability of failure of the steel component in advance so that remedial measures may be taken before an actual failure. The LIBS instrument was successfully deployed at a steel and iron plant. Scans of a welded section of a steel pipe revealed compositional changes caused by element migration.

Demonstrated potential for predicting operational life and ensuring the integrity of steel components. This transportable LIBS instrument represents a significant advancement in the field of in-service material analysis, offering a robust, non-invasive method for monitoring the health of critical infrastructure components such as steel pipes in high-temperature environments.

The metal-producing and processing industries are continuously striving for increased productivity and improved product quality. This quest has generated a demand for advanced measurement methods capable of analyzing the chemical composition of materials quickly and, ideally, in real-time. Laser-Induced Breakdown Spectroscopy (LIBS) emerges as an ideal solution for this requirement. LIBS enables the simplification or even elimination of sample taking by allowing direct measurements on specimens within the production line. The laser itself can handle sample preparation: it first ablates surface layers like oxides, scale, oil residues, and coatings, which do not represent the bulk material composition. Once these layers are removed, the laser acts as an excitation source for spectrochemical analysis. The current analytical capabilities of LIBS for multi-element analysis are on par with, or even superior to, traditional methods. One of the key advantages of LIBS is its ability to perform non-contact measurements over large distances. This opens new application fields, which are challenging for conventional methods such as X-ray fluorescence analysis or spark emission spectrometry. Examples include product identification testing and online analysis of liquid steel. By integrating automated online measurement for quality assurance directly within the process, manufactures can achieve faster and more efficient feedback, leading to significant cost savings and enhanced competitiveness.

In the oil and gas industry, pipelines and process piping face increasing exposure to corrosion as fields become more sour, and wells operate at greater depths with higher pressures and hotter products. To meet the demands for corrosion resistance and mechanical stability, a variety of steel grades are used in the production of pipe fittings. These materials range from high alloy steel grades to nickel-based alloys. Using the incorrect steel grade can result in corrosion, leading to severe damage and costly consequences. The rising quality standards, especially in the nuclear industry, along with environmental responsibilities, necessitate the precise identification of materials used in each produced pipe fitting. To address this need, an automatic inspection machine was developed, leveraging Laser-Induced Breakdown Spectroscopy (LIBS). The machine can inspect and identify over 35 different material grades, effectively preventing any mix-up of material grades. This automatic inspection process ensures that each fitting meets the required specifications, thereby enhancing quality assurance and minimizing the risk of corrosion-related failures.

The conventional inspection method for pipe fittings is sparking optical emission spectrometry (spark OES). The fittings had to be cleaned and a considerable percentage of the fittings had to be measured two or three times before they had been correctly identified. The duration of one inspection was 4 s and the electrode of the spark OES had to be cleaned every three measurements. The throughput was 60 fittings per hour. Moreover, the surface of the fittings had to be abraded after the spark discharge. In the third job step, the fittings had to be marked with an inkjet printer or by electrolytic etching. In view of a material inspection of each produced pipe fitting, the economic efficiency had to be improved.

The conventional inspection method for pipe fittings, sparking optical emission spectrometry (spark OES), has several limitations. Fittings require cleaning before inspection, and a significant percentage of fittings must be measured two or three times to ensure accurate identification. Each inspection takes approximately 4 seconds, and the electrode of the spark OES needs cleaning after every three measurements. This results in a throughput of only 60 fittings per hour. Additionally, after the spark discharge, the fitting surfaces must be abraded, and the fittings must be marked using an inkjet printer or electrolytic etching in a subsequent step. The need to inspect each produced pipe fitting economically efficiently led to the development of a more advanced solution.

The laser-based inspection system simplifies the process, combining inspection and marking into a single step. Cleaning and abrading the fittings are no longer necessary. The workflow begins with an operator placing the workpieces on a table-type circular conveyor, which rotates 90 degrees to position the pipe fitting in front of the inspection machine's measuring window.

A triangulation sensor, aligned with the pulsed laser beam used for LIBS, measures the distance to the specimen. If the surface is within a specified tolerance range, the pulsed laser is activated to generate plasma. The inspection time is 2 seconds, during which 100 spectra are generated and evaluated. If the material grade matches the expected one, the fitting is marked by an inkjet marker in the next cycle. The fitting is then transferred to a conveyor belt by a gripper and transported to the packaging station. If there is a material mix-up, the fitting is not marked and is sorted out by the conveyor belt.

The light emitted from the laser-induced plasma is collected by a fiber optic cable and sent to a spectrometer equipped with 12 photomultipliers for elements such as Fe, Ni, Cr, Mo, Ti, Cu, Nb, Al, and W. The photomultiplier signals are processed by a multi-channel electronic device with fast gateable integrators and analog-to-digital converters for time-resolved spectrometry.

This advanced inspection machine can identify more than 30 different steel grades, including alloy steels, stainless steels, Duplex, Super Duplex, 6MO grades, high nickel alloys, titanium, and clad steels. It can measure Fe and Ni concentrations up to 100.

3.5.4.2 Liquid Steel Analysis

In steel works as shown in Figure 3.14, the direct analysis of liquid steel aims to provide rapid elemental analysis of steel composition, particularly during secondary metallurgical processes. This approach facilitates achieving the target composition more accurately, reduces process times, and minimizes energy and material consumption, ultimately enhancing steel quality and productivity. A goal is the online analysis of melt composition.

Figure 3.14   Raw material is fed into a steel furnace in ordinary iron and steel works

Currently, conventional analysis methods are offline and discontinuous. The analysis process involves several steps: taking a sample from the liquid steel, solidifying it, transporting it to a container laboratory or central laboratory, preparing the sample, and analyzing it using methods like spark optical emission spectrometry (spark OES) or combustion techniques. Despite recent advancements in container laboratory installations and automation reducing analysis times to about 3-5 minutes, direct laser analysis using LIBS (Laser-Induced Breakdown Spectroscopy) can potentially cut these times by approximately 50%.

However, applying LIBS faces significant challenges, particularly in ensuring reliable access to the melt and effectively transmitting the analytical emission lines of key elements such as carbon (193.09 nm), sulfur (180.73 nm), and phosphorus (178.28 nm) in the short ultraviolet wavelength range. Three configurations are typically used for LIBS application: (1) Bottom-blow nozzles of a converter. (2) A bore in the side wall of a converter. (3) A lance immersed from the top of the melt.

Although LIBS has demonstrated detection limits of 250 μg/g for carbon in solid samples, this performance is sufficient for converter applications but not for secondary metallurgy. Recent improvements have shown the feasibility of achieving detection limits near and below 10 μg/g for critical elements.

The LIBS analyzer can determine up to 20 elements within 60 seconds, with the number of elements depending on the lines installed in the spectrometer. Calibration is performed by adding a defined amount of alloying material to the melt and then taking samples, which are analyzed conventionally by spark OES and combustion methods. This calibration ensures the accuracy and reliability of the LIBS measurements. Figure 3.15 shows a mobile LIBS instrument.

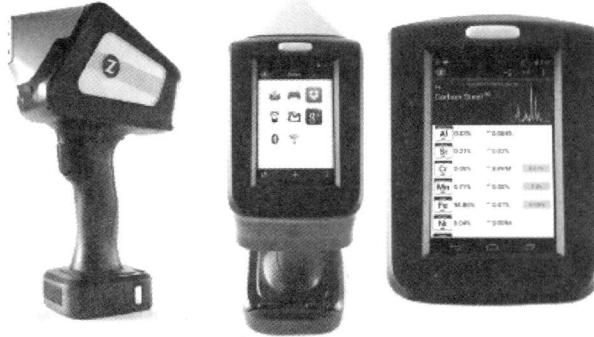

Figure 3.15   Mobile LIBS instrument for fast elemental analysis of metals and alloys

3.6　The Lowest Detectable Limit of LIBS

3.6.1　LIBS Lowest Detectable Limit

Although LIBS could be used to detect nearly all elements in the chemical periodic table, the lowest detectable limits are quite different from one another. Figure 3.16 shows detailed information about that. For all metal and alloy elements, LIBS could identify to nearly 1 μL/L.

3.6.2　LIBS Typical Suppliers

3.6.2.1　B&W TEK INC.

B&W Tek was founded in 1997 in the USA. B&W Tek is the global leader in innovative mobile spectroscopy solutions. We utilize our own key building blocks consisting of spectrometers, light sources, sampling accessories, and software capabilities to produce portable and handheld spectroscopy and laser instrumentation. B&W Tek provides Raman, LIBS, UV-Vis, and NIR solutions for the pharmaceutical, biomedical, physical, chemical, safety, security, and research communities.

(http://bwtek.com/)

3.6.2.2　Oxford Instruments

Oxford Instruments was founded in 1959 in the United Kingdom. Oxford Instruments is a leading provider of high-technology tools and systems for research and industry. We design and manufacture equipment that can fabricate, analyze, and manipulate matter at the atomic and molecular levels. Oxford Instruments' high-technology tools and systems are used across a wide range of applications in many markets. Wherever the key global issues of energy conservation, protection of the environment, national and personal security and health are being addressed, our products can be found.

(http://www.oxford-instruments.com/)

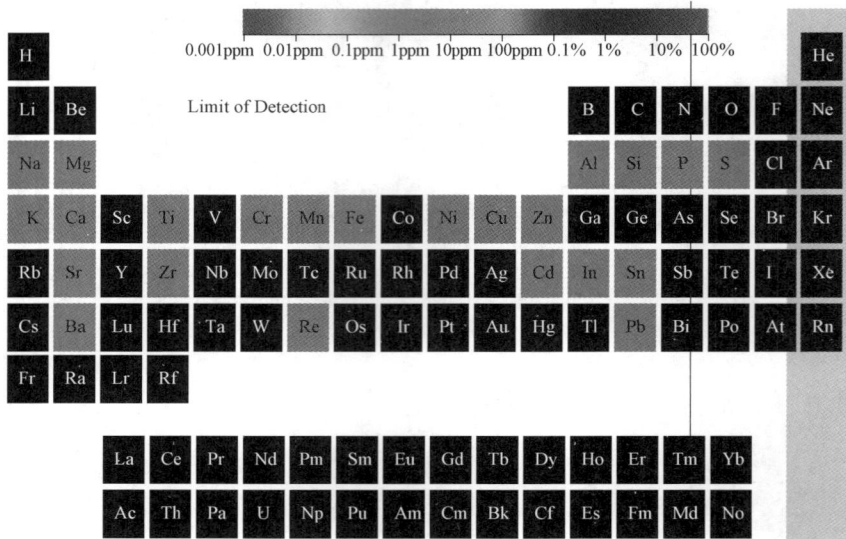

Figure 3.16   The lowest detectable limit by LIBS in the chemical periodic table

3.6.2.3 Ivea

IVEA was founded in 2005 in France. IVEA is a French company started in 2005 and located in the Paris area, that offers solutions for chemical elemental analysis by LIBS. The laser-induced breakdown spectroscopy technology is aimed at research centers and monitoring industrial processes. Those systems, developed under license from CEA and AREVA NC allow determining quickly the multi-elemental composition of solids, liquids, gases, and aerosols. The products developed by IVEA can cover a wide range of applications.

(http://www.ivea-solution.com)

3.6.2.4 Applied Photonics

Applied Photonics Ltd was founded in 1998 in the United Kingdom. Applied Photonics Ltd is a UK company specializing in technology based on Laser-Induced Breakdown Spectroscopy (LIBS). Since the company's formation in 1998, Applied Photonics has been supplying a range of LIBS products including bench-top, portable, fiber-optic, stand-off, and custom-designed systems. It also offers specialist services including contract R&D, feasibility studies, LIBS-based remote material characterization services to the nuclear industry, and full design and build projects.

(http://www.appliedphotonics.co.uk)

3.6.2.5 Progression

Progression was founded in 2002 in the USA. Progression is a leading manufacturer and provider of industrial Magnetic Resonance Testing (MR Testing) and Laser Induced Breakdown Spectroscopy (LIBS) analyzers. These techniques have been proven effective in several industries including petrochemicals, coal, biofuels, and minerals. In addition, we provide mass flow meter analysis and consulting, including custom sampling systems, two-phase mass flow monitors, and electrostatic measurement equipment based on the triboelectric effect. These are used for monitoring unique applications within the chemical process industry. The world's leading polymer manufacturers utilize Progression products every day to improve process efficiency and product consistency.

(http://www.progression-systems.com/)

3.6.2.6 AtomTrace

AtomTrace was founded in 2014 in the Czech Republic. AtomTrace was founded as a startup by the Central European Institute of Technology (Brno, CZ) to commercialize promising technologies that arise from research and development particularly in the field of material analysis by the LIBS technique.

(https://www.atomtrace.com/)

3.7 Review Questions

Q1: What are the basic steps of atomic emission spectroscopy? (see Sections 3.2.1).

Q2: What is the working principle of laser-induced breakdown spectroscopy? (see Sections 3.2.2).

Q3: Please give some examples of LIBS applications in the electrical power industry, such as in thermal or nuclear power plants, or in power transmission lines. (see Sections 3.5.1, 3.5.2 and 3.5.3).

Q4: Some scientists or engineers are trying to detect the aging of internal insulation parts in GIS, GIL, or GIT by LIBS or other spectrum analysis methods. Could you give some examples about that, and explain the advantages and disadvantages of these methods?

Bibliography

[1] Noll, Reinhard. Laser-induced breakdown spectroscopy. In Laser-Induced Breakdown Spectroscopy, pages 7-15. Springer, 2012.

[2] Andrzej. W. Mikziolek. Laser Induced Breakdown Spectroscopy. Cambridge University Press, 2006. doi:

10.1017/CBO9780511541261.

[3] Tiwari, Atul and Soucek, Mark D. Concise encyclopedia of high performance silicones. John Wiley & Sons, 2014.

[4] Zhiniu Xu. Study of RTV Coating Hydrophobicity and Its Influence on Insulator Electric Field and Pollution Flashover Characteristic. PhD thesis, North China Electric Power University, 2011.

[5] Junjie Wang. Research on hydrophobicity measurement and inleuence factors for room temperature vulcanizing silicone rubber (RTV). Master's thesis, North China Electric Power University, 2009.

[6] Ping Chen. Effect of surface roughness on signal intensity by laser induced breakdown spectral. Spectroscopy and Spectral Analysis, 39(6):96-104, Jun. 2019.

[7] Wang, Xilin and Wang, Han and Chen, Can and Jia, Zhidong. Ablation properties and elemental analysis of silicone rubber using laser induced breakdown spectroscopy. IEEE Transactions on Plasma Science, 44(11): 2766-2771, 2016.

[8] Wang, Xilin and Wang, Han and Jia, Zhidong and Wang, Liming and Zhang, Xinghai and Gan, Degang and Li, Yawei. A new method to measure RTV coatings thickness with laser induced breakdown spectroscopy (LIBS) technique. In 2016 IEEE Electrical Insulation Conference (EIC), pages 81-84. IEEE, 2016.

[9] Wang, Xilin and Hong, Xiao and Chen, Can and Wang, Han and Jia, Zhidong and Zou, Lin and Li, Ruihai. Elemental analysis of RTV and HTV silicone rubber with laser induced breakdown spectroscopy. In 2017 IEEE Electrical Insulation Conference (EIC), pages 9-12. IEEE, 2017.

[10] Wang, Xilin and Hong, Xiao and Wang, Han and Chen, Can and Zhao, Chenlong and Jia, Zhidong and Wang, Liming and Zou, Lin. Analysis of the silicone polymer surface aging profile with laser induced breakdown spectroscopy. Journal of Physics D: Applied Physics, 50(41): 415601, 2017.

[11] Wang, X and Wang, H and Zhao, C and Jia, Z and Zhou, J. Composition analysis of room temperature vulcanized material with laser induced breakdown spectroscopy technique. Trans. China Electrotech. Soc., 31: 96-104, 2016.

[12] Hark, R. Laser-Induced Breakdown Spectroscopy for Coal Analysis. Journal of Analytical Atomic Spectrometry, 2014.

4

Gas Chromatography & Photoacoustic Spectroscopy

4. 1 Introduction

The power transformer plays a crucial role in power conversion, facilitating the efficient transmission and distribution of electrical energy.[1, 2] For a step-up power transformer, electrical energy with specific voltage and current flows through the primary terminals, passes through the power transformer, and exits through the secondary terminals with adjusted voltage and current. By increasing the voltage and reducing the current, power transformers enable economical power transmission over long distances. Conversely, a step-down power transformer can lower the voltage and increase the current to deliver electricity to end users effectively.

During normal operation, the iron-core and coils of a power transformer (collectively known as the active part) generate losses that produce heat. The cooling system's role is to dissipate this heat at a steady, controlled rate to maintain an acceptable temperature rise between the iron-core and coil assembly and its surrounding medium. Additionally, the active part is subject to a wide range of electrical voltage stresses. The insulation system's function is to manage these stresses, keeping them below the permitted maximum levels under both normal and abnormal operating conditions. This is achieved through the careful selection and arrangement of insulation materials and insulation parts used in the transformer's active components.

In the late nineteenth century, it was discovered that immersing the iron-core and coil assembly of power transformers in mineral oil could enhance their cooling and electrical insulating performance. This innovation led to the development of liquid-filled transformers used today. Mineral oil, along with other types of insulating liquids, possesses excellent dielectric insulating properties. It provides a superior insulation medium when used to impregnate the Kraft paper typically wrapped around winding conductors and other elements of the windings.

The use of a liquid medium improves heat dissipation from the active part compared to using a gas like air. Moreover, the liquid-insulating medium is in direct contact with all the iron-core and coils assembly inside the transformer, enabling extensive analysis of the transformer's condition through properties of the insulating liquid.

Assessing the condition of power transformers primarily aims to identify abnormal behavior in a transformer and determine the type of failure mode causing it. For asset owners managing a fleet of power transformers, implementing a dissolved gas analysis (DGA) program provides significant value.[3, 4] It enables the detection of failure modes that, if undetected, could increase the operational risk to unacceptable levels. By identifying issues early and responding promptly, organizations can mitigate the financial, safety, environmental, and reputational consequences of major transformer failures.

4.2 Insulating Liquids

Power transformer manufacturers have consistently sought to gain a competitive edge through advancements in materials engineering and improved calculation accuracy for predicting transformer behavior. Insulating liquids, a critical component in transformers, have been a significant focus area for innovation. The insulating liquids in power transformers mainly include four categories: mineral oils, synthetic esters, natural esters, and nanofluids.

Mineral oil was the first liquid used for this purpose and remains the most popular option for filling liquid-immersed transformers due to its cost-effectiveness and good insulating properties. Manufacturers have developed refined versions with enhanced performance characteristics such as higher oxidation stability and better heat dissipation. Synthetic esters are engineered fluids that offer superior environmental benefits, including biodegradability, high fire points, and excellent dielectric properties. They are particularly suitable for use in eco-sensitive and high-temperature environments. Natural esters, derived from vegetable oils, are increasingly popular due to their environmental sustainability and excellent insulating properties. They also offer high moisture tolerance and good biodegradability. The incorporation of nanoparticles into insulating liquids has led to the development of nanofluids, which offer improved thermal conductivity and dielectric strength. These fluids can enhance the cooling efficiency of transformers, leading to more compact and efficient designs.

Chemical additives are used to enhance the properties of insulating liquids. Anti-oxidants, for example, can extend the life of the insulating oil by preventing oxidation. Other additives can improve viscosity, thermal stability, and electrical performance.

The development of more accurate and sensitive testing methods allows for better monitoring of the condition of insulating liquids. Techniques such as dissolved gas analysis (DGA) and furan analysis help in early detection of potential issues, enabling preventative maintenance and reducing the risk of transformer failure.

Advances in materials engineering have led to insulating liquids that are more compatible with other transformer components, such as solid insulation materials and metals. Additionally, the push for sustainability has driven the development of more eco-friendly insulating liquids that reduce environmental impact without compromising performance.

The continuous improvement in insulating liquids through materials engineering and precision in predictive calculations has been pivotal for transformer manufacturers. These advancements not only enhance the performance and reliability of transformers but also address environmental and sustainability concerns, offering manufacturers a competitive advantage in the market.

Crude oil undergoes various refining processes to produce transformer-insulating oil. Crude oils can be classified into two main groups, paraffinic and naphthenic, depending on the compounds that constitute them. Transformer oil is typically produced by the following processes: distillation, dewaxing, extraction, and hydrogenation.

While the actual composition of the oil molecules is relatively complicated, three main types of structures influence the properties of transformer oils and they are shown in Figure 4.1.[2] Transformer oil is considered naphthenic or paraffinic depending on which structure is more prevalent for a particular oil. In general, if the oil contains between 56% and 65% of carbon-bonded paraffin, the oil is considered paraffinic. If it contains between 42% and 50% of carbon-bonded paraffin, the oil is considered naphthenic. Oils in between these percentages are considered intermediate. The content of these compounds influences the various properties of the oil including, oxidation stability, viscosity, temperature stability, gas absorption, dielectric properties, etc.

(a) Paraffinic   (b) Naphthenic   (c) Aromatic

Figure 4.1   Schematic representation of the three main types of structures in transformer oil [2]

4.3 Decomposing and Relevant Chemical By-products

Inside the power transformer tank, there is an iron-core and coil assembly, and it was made of various materials, including iron, copper, stainless steel, paper, wood, cast resin, oil, painting, rubber, synthetic high polymer materials and so on. From the viewpoint of chemical reactions, the relationship between components and materials including: the copper from the windings, steel from the iron cores and clamping structures, hydrocarbons from the insulating oil, and cellulose from paper, insulating boards, and blocks (see Figure 4.2). Additionally, energy sources in the form of losses generated during normal operation, as well as energy contributed by abnormal conditions and faults, drive these compounds into decomposing and recombining to form new chemical by-products. This can be visualized as a chemical reactor where energy sources drive reactive materials to form measurable and quantifiable by-products. Identifying useful correlations between the energy injected into the system and the chemical behavior of the components enables the assessment of the transformer's condition.

Figure 4.2   Internal view of a typical power transformer

1—Iron Core; 2—Low Voltage Winding; 3—High Voltage Winding; 4—Tap Winding;

5—Leads; 6—Low Voltage Bushing; 7—High Voltage; 8—Yoke Clamping Part

For diagnosing the condition of the transformer, we focus on two main components of the insulation system: liquid insulation (oil) and solid insulation (paper and pressboard). Each of these elements contributes specifically to the formation of gases.[4]

The heat produced by losses and certain types of failures, as well as the energy from electrical discharges, contribute to the breakdown of oil molecules, producing various by-products. The energy required to break or form chemical bonds varies; For example, breaking a hydrogen-hydrogen bond requires about half the energy needed to form a carbon-carbon triple bond, as seen in acetylene. These by-products remain dissolved in the oil up to their solubility limits, beyond which they evolve as free gases (bubbles). Figure 4.3 shows the main gas produced as by-products of oil decomposition.

One of the most important components of the insulation system, aside from the oil, is the solid insulation system. This insulation, comprising paper, pressboard, and other solid materials, is made up of cellulose fibers. Cellulose is an organic compound made up of long chains of repeating building blocks. The black-colored atoms in Figure 4.4 are oxygen atoms. The same energy that causes the oil to break down similarly affects the paper. At a molecular level, all cellulosic materials are made up of long chains called polymers. Another source of oxygen is water.

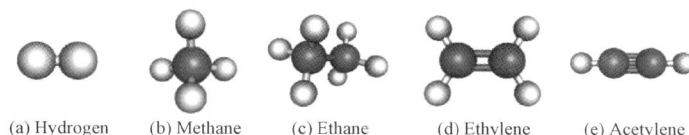

(a) Hydrogen (b) Methane (c) Ethane (d) Ethylene (e) Acetylene

Figure 4.3　Main hydrocarbon gases produced by the oil[2]

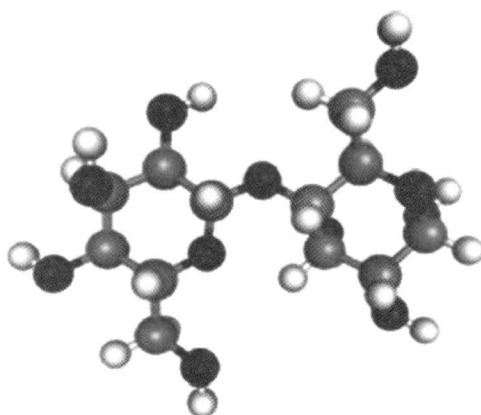

Figure 4.4　Cellulose molecule [2]

A brand-new transformer that has been properly manufactured and assembled will typically have less than 0.5% of residual moisture per dry weight of insulation throughout the system after commissioning. However, over the life of a transformer, additional water can enter the system via improperly maintained breathing systems or oil leaks.

In some cases, such as in free-breathing transformers, oxygen enters via the normal "breathing" process of the oil expanding and contracting due to temperature changes. Increases in temperature, either through normal losses or during abnormal events or faults, cause the C-O molecular bonds to break, releasing oxygen, carbon, and hydrogen into the oil. These elements recombine to form carbon monoxide (CO), carbon dioxide (CO_2), water (H_2O), and a range of compounds collectively known as furans. Neither water nor furans are part of gas analysis and are measured using other methodologies for detecting. Figure 4.5 shows that cellulose molecules can form relevant by-products due to overheating or electrical discharge.

These gas production mechanisms correlate with the chemical elements present inside the transformer and the levels of energy available to enable these chemical reactions. The transformer engineering community has developed several interpretation techniques that take advantage of this behavior.

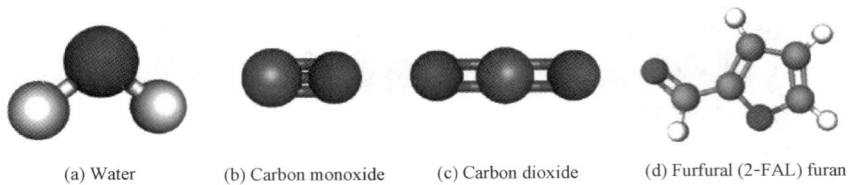

(a) Water   (b) Carbon monoxide   (c) Carbon dioxide   (d) Furfural (2-FAL) furan

Figure 4.5   By-products of cellulosic and hydrocarbon decomposition in the presence of oxygen[2]

The distinct energy levels at which these gases are produced have been crucial for inferring the type and intensity of the mechanisms causing them. Since the early 1970s, engineers have been studying the amount, proportion, and evolution rates of these gases dissolved in transformer oil to correlate them with known failure modes and their severity levels. Other materials present in the transformer also interact during these chemical reactions, contributing additional by-products beyond those from hydrocarbons in the oil. [5, 6]

4.4   Importance of Oil Sample Analysis

Ensuring the accuracy of transformer oil analysis is crucial for the correct diagnosis of the transformer's condition. The reliability of this analysis is underpinned by two key concepts: repeatability and reproducibility. Repeatability refers to the consistency of results when the same sample is analyzed multiple times by the same laboratory or analyst. Reproducibility, on the other hand, refers to the consistency of results when the same sample is analyzed by different laboratories or analysts.

While perfect repeatability and reproducibility are ideal, they are not practical due to inherent variations in testing processes and equipment. Understanding a laboratory's capability to consistently produce accurate results is vital for the end user, who relies on these results for interpreting the condition of the transformer.

In order to extract useful information from the gas production mechanisms explained in the previous section, the dissolved gases in the oil have to be extracted and quantified. As it might be expected, the results of the oil sample analysis are critical to the correct interpretation and diagnosis of the condition of a particular transformer. It is important that the results available for analysis are representative of the oil inside the transformer and as such, it is expected that they are repeatable and reproducible.

While each laboratory has to be evaluated by its potential clients, in general, laboratories ensure good quality repeatable, and reproducible results by:
- Adhering to well-known and reliable test methodologies.
- Regularly verifying the correct calibration of instruments.
- Obtaining and maintaining accreditation by a recognized assessment authority and.
- Participating in international round-robin tests.

The majority of modern laboratories around the world analyzing transformer oils use the gas chromatography technique. Two of the most well-known standards that prescribe how this technique should be applied are as follows:

- ASTM D3612-02(2009) - Standard test method for analysis of gases dissolved in electrical insulating oil by gas chromatography.
- IEC 60567 ED. 4.0 - Oil-filled electrical equipment - a sampling of gases and analysis of free and dissolved gases guidance.

Most modern laboratories employ gas chromatography (GC)[7] to analyze transformer oils. This technique involves injecting a gas sample of unknown composition into a gas chromatograph, which contains a capillary column with a stationary phase made of liquid or polymer. As shown in Figure 4.6,[8] an inert carrier gas, like helium or nitrogen, acts as the mobile phase. As the mobile phase traverses the stationary phase, the compounds within the sample separate and travel at different speeds based on their interaction with the stationary phase, resulting in distinct retention times. Detectors at the end of the column generate signals, with each compound's retention time corresponding to its identity, and the area under each signal's curve indicating the compound's quantity.

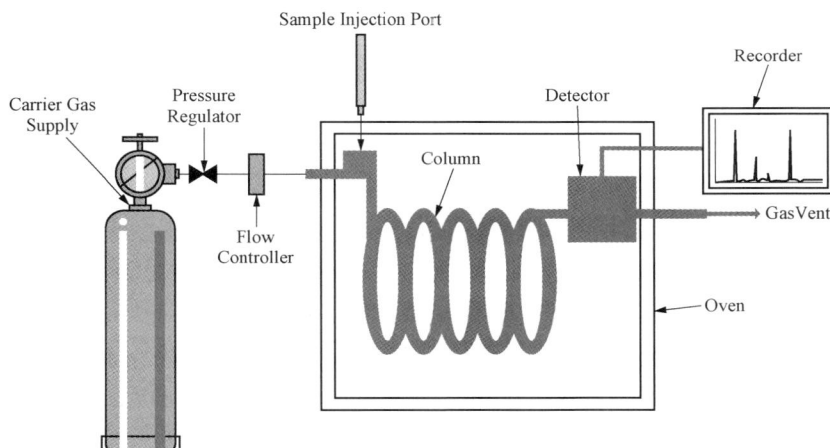

Figure 4.6   Simplified representation of how a gas chromatograph work[8]

Laboratories calibrate the GC using known gas mixtures, establishing baseline retention times and areas to identify compounds in unknown samples accurately. Before applying this methodology, dissolved gases must be separated from the oil sample. This is achieved through various techniques, often involving heating, mechanical agitation, and vacuum application.

ASTM-3612 outlines three methods (A, B, and C) for gas extraction:

Method A: Uses glassware and a vacuum to extract gases. Method B: Involves a stripping column with high surface area beads to strip gases from the oil, directing them to the analysis instrument. Method C (Headspace Sampling): Involves placing the sample in a sealed vial, allowing gases to migrate into the headspace, from which the proportion of gases is determined using mass equivalence equations. A typical gas chromatograph and auto-sampler setup, as shown in Figure 4.7. [8]

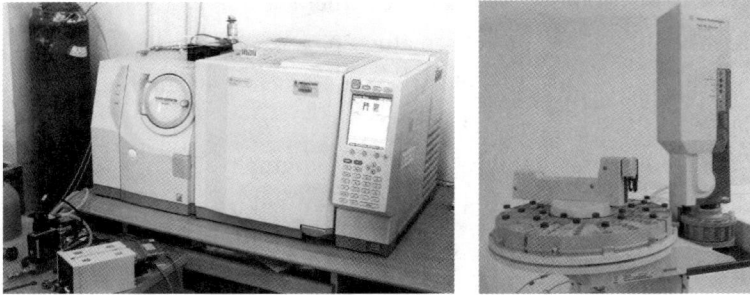

Figure 4.7   A typical gas chromatograph and auto-sampler

This method lends itself to automation of the gas extraction and analysis processes, which makes the laboratory more efficient and able to reduce lead times for analysis.

In Figure 4.7, a gas chromatograph and auto-sampler which is typically used to extract the gases. The results of gas chromatography are the concentration of each of the gases of interest and reported in $\mu L/L$.[9, 10]

4.5 Oil Sampling Guidelines and Industry Practices

The correct execution of the oil sampling process is an essential first step in a successful analysis and therefore interpretation of the condition of a particular transformer. Even the most careful laboratory analysis cannot fix the negative impact of a poor-quality sample. [3]

In regards to sampling, there are also useful guidelines and industry best practices published by the main technical bodies around the world. A few examples are as follows:

IEC 60475-2011: Method of sampling insulating liquids. This standard provides guidelines on how to properly sample insulating oils for accurate analysis.

IEEE Std 95-2002: Guide for Sampling and Testing of Oils Used in Transformers and Other Electrical Apparatus. This document offers best practices for oil sampling procedures and testing methods.

ASTM D923-15: Standard Practices for Sampling Electrical Insulating Liquids. This standard outlines methods for collecting representative samples of insulating oil for various types of analysis.

CIGRÉ Technical Brochure No. 227: Guidelines for Life Management Techniques for Power Transformers. Includes recommendations for oil sampling and diagnostics as part of transformer maintenance.

BS EN 60475:2011: Method of Sampling Insulating Liquids. This is the British adoption of the IEC standard, providing detailed instructions for sampling.

To ensure a representative sample is obtained every time, it is essential to select a sampling location that can be consistently accessed over the transformer's lifespan. This consistency provides reliable comparability of samples from the same transformer over time.

The most common location for routine oil sampling for condition monitoring purposes is the bottom drain or sampling valve. Another typical sampling location is the sampling valve of the Buchholz relay. When the transformer is equipped with such valves at ground level as shown in Figure 4.8, they are particularly useful for sampling gas in the event of a Buchholz gas accumulation alarm or trip. While other valves around the transformer, such as top filling valves, can also be used for oil sampling, they are not always easily accessible during routine inspections, especially when the transformer is energized.

Figure 4.8   A large capacity power transformer on site

4.6 Interpretation Techniques

The fundamental principle for inferring the condition of a transformer relies on the correlation between gas generation rates and proportions, and the energy that produces them.

Since the adoption of Dissolved Gas Analysis (DGA) as a diagnostic technique, numerous patterns and associations have been documented and used as diagnostic tools and guidelines.[3]

While there are a variety of interpretation techniques and guidelines, they generally rely on one or more of the following factors:

Gas Profile: The gas composition of a particular sample, typically the latest one. Gas Ratios: The relationship (ratios) between concentrations of different gases in a given sample. Rate of Change: The rate of change of these concentrations between samples taken at different times.

Transformers are complicated, and their construction varies from unit to unit. Factors such as design and manufacturing practices can impact the behavior of any particular unit. Therefore, while correlations found in a population of transformers provide generic insights and broad interpretation guidelines, in-depth knowledge of the specific characteristics of the transformer population being analyzed yields more accurate and confident condition assessments.

The interpretation of gases dissolved in transformer oil is often considered both an "art" and a "science." As individuals using DGA to monitor the condition of power transformers develop expertise and familiarity with a particular fleet or model of transformers, their ability to further analyze data, correlate with specific failure modes, and gain additional insights becomes more granular and specific.

Furthermore, DGA is only one tool necessary for a meaningful assessment of a transformer's condition. A comprehensive diagnosis requires multiple angles and all available data, including electrical tests, oil quality tests, external and internal inspections, paper condition, component conditions (e.g., on-load tap changer (OLTC), bushings), known design or manufacturing issues, loading, ambient conditions, undue electrical stresses (e.g., transients and through faults), and overloading. Analyzing all this information in a comprehensive manner represents the "art" component of dissolved gas interpretation that engineers in this industry often refer to.

4.6.1　Fault Types

Regardless of the interpretation method being utilized, all techniques aim to determine what, if any, failure mode is present in a transformer and its severity. Liquid-immersed transformers can experience a range of failure modes, which, based on their root origin, can be categorized into the following broad groups:

- Thermal Faults.
- Electrical Faults.
- Mechanical Faults.
- Electrical Faults.

The detailed explanation of above four items are listed below:

Thermal Faults: These involve overheating due to various reasons such as overloading, cooling system failures, or localized hotspots.

Electrical Faults: These include partial discharges, arcing, and other insulation breakdown phenomena.

Mechanical Faults: These involve physical damage or deformation of transformer components, often due to external factors like short circuits or transportation damage.

Chemical Faults: These include oil degradation, moisture ingress, or contamination, leading to a breakdown in insulation properties.

Depending on the actual location of the fault within the transformer, it might or might not involve solid insulation (i.e., paper, pressboard, etc.). In some cases, there might be multiple fault types in the same transformer, which should be taken into account when performing the diagnosis. Additionally, as faults evolve and become increasingly severe, they might transition from one category to another.

An important characteristic of using DGA to monitor the condition of transformers is that, in many cases, the technique is sensitive enough to provide an advanced warning (sometimes years in advance) of incipient failures developing in transformers. This allows the organization owning the transformer to investigate and correct the issue in a planned and efficient manner, rather than reacting to an after-the-fact fault, which typically causes major disruptions in the electrical network.

However, it is also worth noting that certain faults can develop in a matter of days or even hours. Although this represents a small percentage of all cases observed, in transformers that are sufficiently critical, this may prompt shorter sampling intervals, with online monitors providing the highest sampling and analysis frequencies, and therefore the earliest warnings currently available.

Beyond the information provided by a single analysis, the capacity of DGA to track the rate of gas growth over time is a valuable contribution to the condition monitoring and asse-ssment of power transformers. In addition to identifying the presence of a failure mode from a particular sample, the DGA interpreter can gain insights about the intensity and expected development of that failure mode by understanding the evolution of gases in that unit.

This makes DGA an ideal tool in the condition-based power transformer management toolbox of an organization. Faults often evolve in a predictable manner to the extent that they can be identified at very early stages of development, providing the asset owner with enough time to address the issue in a controlled way.

4.6.2　Techniques that Rely on the Gas Concentrations and Ratios

Throughout the history of DGA, researchers have found that the profile of each sample correlates to the type of condition that generated those gases. The most commonly known interpretation methods are as follows:
- Key Gas Method.
- Rogers Ratio Method.
- Dornenburg Ratio Method.
- Duval Triangles and Pentagons.

4.6.2.1　Key Gas Method

This technique focuses on the identification of specific key gases that are produced under certain fault conditions. Each gas is associated with a particular type of fault:

Hydrogen (H_2): Associated with partial discharges.

Methane (CH_4): Linked to low-temp- erature thermal faults.

Ethylene (C_2H_4): Indicative of higher temperature thermal faults.

Acetylene (C_2H_2): Associated with arcing or very high-temperature thermal faults.

Ethane (C_2H_6): Linked to general overheating.

Carbon Monoxide (CO) and Carbon Dioxide (CO_2): Often related to paper insulation degradation and thermal faults.

Total Combustible Gas (TCG) Analysis: This method sums the concentrations of all combustible gases (H_2, CH_4, C_2H_4, C_2H_2, C_2H_6, and CO) to provide an overall indication of the severity of the condition. Higher TCG levels typically indicate more severe faults.

Another group of techniques commonly used to analyze DGA results is those that rely on the relationship between two or more gases.

When ratios rather than absolute concentrations of gases are analyzed, these tend to be less sensitive to the volume of oil in the transformer. Ratios also have characteristic patterns that can be correlated with certain classes of faults which make them useful as interpretation techniques.

As mentioned by the IEEE interpretation guidelines: The use of gas ratios to indicate a single possible fault type is an empirical process based upon the experience of each individual investigator in correlating the gas analyses of many units with the fault type subsequently assigned as the cause for disturbance or failure when the unit was examined.

4.6.2.2 Rogers Ratios

As mentioned by the IEEE C57.104 guidelines, the methodology originally proposed by Dornenburg was subsequently confirmed by Rogers on European systems. The Rogers methodology comprises three ratios:

- Acetylene to ethylene.
- Methane to hydrogen, and.
- Ethylene to ethane.

In the case of Rogers, the fault diagnosis was stated as six different fault cases, numbered from 0 to 5. This number has sometimes been referred to as the "Rogers Code". The cases listed by the IEEE guidelines are as follows:

- 0——Unit normal.
- 1——Low-energy density arching - PD.
- 2——Arcing - high-energy discharge.
- 3——Low-temperature thermal.
- 4——Thermal $<700\ ℃$.
- 5——Thermal $>700\ ℃$.

Each ratio corresponds to different types of faults, such as partial discharges, thermal faults, and electrical faults.

4.6.2.3   Dornenburg Ratios

In a report published in 1970 as part of the CIGRE International Conference on Large High Tension Electric Systems, the member of the 15-01 Working Group included the results of investigations of decomposition gases produced under normal and abnormal operating conditions. In this report, three main fault types were analyzed:

- Oil decomposition due to PD.
- Oil decomposition due to arc discharges under oil.
- Oil decomposition due to thermal stress.

The report highlighted the empirical correlations between these fault types and certain gas ratios. The results were presented in a chart that included the ratios of methane to hydrogen on one axis and the ratios of acetylene to ethylene on the other. The IEEE guidelines also include the Dornenburg method and recommend the use of four gas ratios:

- Methane to hydrogen.
- Acetylene to ethylene.
- Acetylene to methane.
- Ethane to acetylene.

4.6.2.4   Duval Triangles and Pentagons

The current consensus is that the common types of faults detectable by DGA are as follows [11, 12]:

- Partial discharges (PD).
- Discharges of low energy (D1).
- Discharges of high energy (D2).
- Thermal faults (T1) below 300 ℃.
- Thermal faults (T2) between 300 and 700 ℃.
- Thermal faults (T3) above 700 ℃.
- Stray gassing (S) below 200 ℃.

One of the most widely recognized and utilized interpretation methods in the industry is the graphical representations created by Michel Duval. Figure 4.9 and Figure 4.10 present the Duval Triangle and Duval Pentagon methods respectively, which are used for diagnosing various faults types in oil-immersed power transformers based on dissolved gas analysis.

Michel Duval's graphical representations are am-

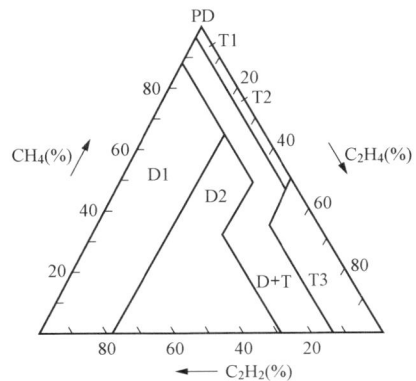

Figure 4.9   The classical Duval triangle for diagnosing various failure modes of transformers, bushings, and oil-filled cables

ong the most recognized and utilized methods for gas analysis in the electrical industry. Since 1970, Michel Duval has served as a senior scientist at Hydro Quebec's Institute of Research in Canada. In 1974, he developed a method to correlate the relationship between three gases using a triangular chart, known as the Duval Triangle. This innovative approach facilitated the identification of fault types in transformers and other electrical equipment by examining the ratios of these gases.

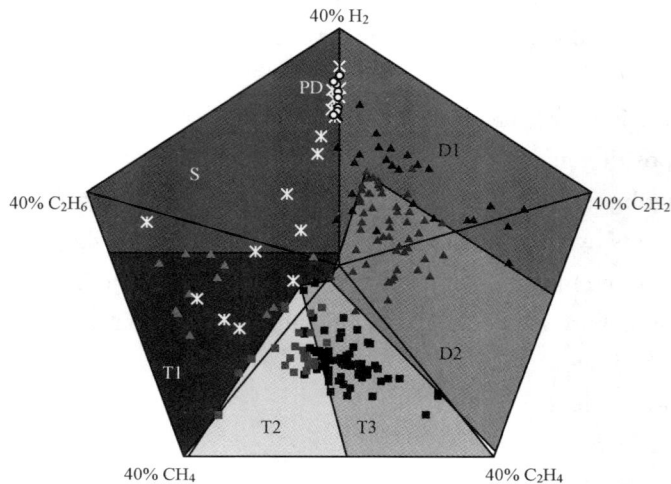

Figure 4.10   The Duval Pentagon for the six "basic" faults, PD, D1, D2, T3, T2, and T1, and stray gassing of mineral oil S (see text for meanings of the abbreviations)

The Duval Triangle allows for the correlation of three gases. This method identifies various fault types such as partial discharge, thermal faults, and electrical discharges based on the ratios of the three key gases. In 2008, the method was expanded to include six additional triangles. This extension enabled the identification of new fault types and included non-mineral fluids and online tap changers. Each triangle correlates specific gas combinations, improving diagnostic precision.

Duval introduced a pentagon-based representation to incorporate five gas ratios. These pentagons allow for a more comprehensive analysis of the gas content, providing a more detailed fault diagnosis.[11, 12] The method has been further refined over the years to enhance its applicability and accuracy. Figure 4.10 presents the new method using 5-gas ratios in a "pentagon" representation applicable to mineral oil-flled equipment.

For each Duval chart, whether triangular or pentagonal, the following steps are followed:

(1) Total Gas Calculation: The total amount of gases is calculated. (2) Percentage Calculation: The proportion of each gas relative to the total is calculated as a percentage. (3) Coordinate System Plotting: These percentages are used to plot a point in the triangular or pentagonal coordinate system.

Duval's methods provide a robust framework for fault detection and diagnosis in electrical equipment, ensuring reliable operation and maintenance of transformers and related systems. These graphical representations continue to be essential tools for engineers and technicians in the industry. For each chart, the total amount of gases is calculated, and then the proportion of each gas in relation to that total is calculated as a percentage. These percentages are then used to identify a point in the "triangular" or "pentagonal" coordinate system.

These triangles and pentagons have been implemented in various monitoring and diagnosis platforms. They have also been included in international guidelines; for example, Figure 4.9 is included in the IEC 60599 interpretation guidelines.

4.7 Online Dissolved Gas Analysis by Sensors and Chromatography

Power transformers are critical and costly assets in the electric power system, essential for ensuring reliable energy flow from the grid through transmission and down to the plant level. As a significant investment in both utility systems and industrial complexes, transformer condition assessment and management are of paramount importance. Each entity has unique investment levels and risk models for asset condition and assessment management, but the common element is the stratification of transformers based on their criticality. This variability lies in the prioritization lines and investment amounts allocated for condition monitoring at each level.[1, 2]

A simplified model below shows one approach to transformer condition management:
- Critical: Those transformers that, if failed, would have a large negative impact on grid stability, utility revenue, and service reliability of the critical facility. For example, generator step-up transformers (GSU) and transmission transformers that are part of critical power flow fall in this level, or the main transformers in a critical facility.
- Important: Those transformers that, if failed, would have a significant negative impact on revenue and service reliability of a utility system, or the production of the plant. Transmission substation transformers and major distribution substation transformers are generally at this level.
- Recoverable: Those transformers that, if failed, would have a low impact on revenue and reliability or the production of the plant. These are mainly smaller distribution substation transformers.

Transformer reliability has become increasingly important due to several factors: (1) Decreased Longevity: Modern power transformers do not last as long as those in the past. In the U.S., the average lifespan is about 40 years, with many transformers from the 1960s and 1970s nearing the end of their design life. In China, the average lifespan is about 20 years due to rapid economic development and frequent upgrades to power supply capacity. (2) Higher Loads: Increased electricity demand has placed higher loads on transformers, reducing their longevity. (3) Budget Constraints: The consolidation and deregulation of the electric industry have led to reduced budgets for maintenance and condition monitoring.

Effective transformer condition management involves selecting appropriate monitoring tools to avoid unplanned failures, reduce maintenance costs, and defer capital expenditures on replacements.

Here is a simplified model approach: (1) Critical Transformers: Advanced monitoring systems, real-time data analysis, and frequent inspections. (2) Important Transformers: Periodic condition assessments, moderate level monitoring systems. (3) Recoverable Transformers: Basic monitoring, less frequent inspections.

Early detection of potential issues to prevent sudden failures. Efficient allocation of maintenance resources based on transformer criticality. Extend the useful life of transformers, delaying the need for replacements.

By stratifying transformers based on their criticality and investing appropriately in condition monitoring tools, utility and plant managers can effectively manage transformer assets, ensuring reliability and optimizing investment returns.

4.7.1 Online Monitoring of Power Transformers

Online monitoring systems are essential for continuously assessing the condition of large, important, and critical power transformers. Several types of online monitoring systems are available on the market, each serving a specific purpose in transformer condition assessment, including: (1) Dissolved Gas Analysis (DGA). (2) Power Factor Monitoring of Bushings. (3) Leakage Current Monitoring of Lightning Arrestors. (4) Frequency Response Analysis (FRA) of Transformer Windings.

Dissolved Gas Analysis (DGA): Monitors the health of oil-filled power transformers by analyzing the gases dissolved in transformer oil. Traditionally, DGA tests were performed offline in laboratories at periodic intervals (quarterly, semi-annually, or annually). This practice began earnestly in the 1960s and has evolved to include online DGA tools. Online DGA allows for continuous monitoring without taking the transformer out of service. It helps utilities avoid unplanned failures, adopt condition-based maintenance, and extend the transformer's useful life, thereby deferring capital expenditures.

Power Factor Monitoring of Bushings: Assesses the condition of transformer bushings by monitoring their power factor. Bushings are external auxiliaries susceptible to environmental conditions. A failure in the bushing is considered a failure of the transformer.

Leakage Current Monitoring of Lightning Arrestors: Monitors the leakage current in lightning arrestors to ensure their proper functioning. Like bushings, lightning arrestors are external and vulnerable to environmental factors. Their failure impacts the overall reliability of the transformer.

Frequency Response Analysis (FRA) of Transformer Windings: Evaluates the mechanical integrity of transformer windings by analyzing their frequency response. This method helps detect winding deformations, short-circuits, and other mechanical issues that can lead to transformer failure.

Online testing is a critical management tool for the condition monitoring and assessment of the most critical and important transformers. The benefits of adopting online monitoring systems include: Provides real-time data and continuous monitoring, allowing for immediate detection of potential issues. Helps maintain or improve the reliability of transformers, even with decreased capital expenditures and aging infrastructure. Supports a shift from time-based to condition-based maintenance, reducing unnecessary maintenance activities and associated costs. Enables utilities to extend the useful life of transformers by addressing issues before they lead to failures. Helps in deferring the need for new investments by maximizing the lifespan of existing transformers.

While laboratory DGA remains a valuable tool, the integration of online DGA has become increasingly popular. The coexistence of these two approaches allows utilities to leverage the strengths of both: Laboratory DGA could provide detailed and precise analysis, often used for validating online DGA findings or for periodic comprehensive checks. Online DGA offers realtime monitoring and early warning of potential issues, reducing the risk of sudden failures and enabling proactive maintenance.

First-generation DGA products from the 1970s, as well as some current models, provided total combustible gas (TCG) or single gas (hydrogen) monitoring. While these products can indicate developing problems in transformers, they do not offer robust diagnostic capabilities. Over time, online DGA products have evolved to include multigas monitors capable of detecting and analyzing some or all of the eight fault gases identified in IEEE standards, thereby offering comprehensive diagnostic capabilities.

One kind of DGA product which could be used to monitor TCG or single gas (hydrogen) levels, it could indicate developing problems without providing detailed diagnostics. Another kind of DGA product, for modern multigas monitors, could be used to detect and analyze multiple gases as specified in IEEE standards, providing diagnostic insights into the condition of transformers.

The Hydran^{TM} 201Ti is a modern, continuous Dissolved Gas-in-oil Analysis (DGA) monitor known for its ease of setup and comprehensive monitoring capabilities, as shown in Figure 4.11. It provides essential data in accordance with IEEE Standard C57.104. Acts as an early warning system to detect failure conditions and minimize the risk of unplanned outages.

The Hydran^{TM} 201Ti exemplifies the advancements in online DGA technology, moving from basic gas monitoring to providing detailed diagnostics that are crucial for the effective management of power transformers. This evolution reflects the ongoing efforts to enhance the reliability and longevity of critical electrical assets. The 201Ti uses fuel cell technology (described as fixed instruments - method 3 in the standard) and is now available with a choice of either the worldrenowned "Hydran Composite Gas" sensor which responds 100% to Hydrogen and is also sensitive to Carbon Monoxide, Acetylene and Ethylene or the more basic "Hydrogen Only" sensor which focuses purely on Hydrogen gas generation. Because the monitoring unit mounts on a single valve and uses Dynamic Oil Sampling, there is no need for a pump or extra piping to connect to different valves. Due to its uncomplicated features and the easily understood information it provides, the 201Ti has been amongst the monitors of choice for many years, with one of the largest installed.

Figure 4.11   Outline of HYDRAN 201Ti for total combustible gas (TCG)
or single gas (hydrogen) monitoring

- Continually measures key fault gas to give you an insight into the transformer's condition.
- Choice of gas sensor: traditional "Composite gas" or more basic "Hydrogen only".
- Communicates gas concentration and gas rate of change values remotely to avoid site visits and enable remote supervision.

Newer online DGA products possess the unique capability to continuously trend multiple transformer gases and correlate them with key parameters such as transformer load, oil temperature, ambient temperature, and customer-specified sensor inputs. This advanced capability allows utilities to relate gas generation to external events, which is crucial for achieving reliability and financial goals in today's environment. In fact, some online DGA tools may offer better accuracy and repeatability than laboratory DGA, which can significantly enhance the timeliness and confidence of transformer asset managers when detecting incipient faults.

The advancements in online DGA monitoring technology provide utilities with unparalleled insights into transformer health. By continuously trending multiple gases and correlating them with key operational parameters, these tools enhance the ability to detect and address developing faults in real time. This not only improves reliability and financial performance but also provides a more accurate and timely understanding of transformer conditions. The integration of traditional and online DGA tests, along with emerging technologies like neural network diagnostics, ensures that transformer asset managers are well-equipped to manage and maintain their critical infrastructure.

Recently, the ability to automatically supplement traditional DGA diagnostic tests with online DGA tests has become available in the market. This new development offers users of online DGA monitors unprecedented insight into the nature and identification of developing faults. The tools are typically ratio-based, and the online data set enables the trending of fault gas ratios over time rather than relying on traditional static snapshots. Diagnostic outcomes can now be determined more quickly and with greater certainty than in the past.

One of the online tools that have recently become available for condition monitoring of transformers is the online DGA, such as Serveron® TM8 shown in Figure 4.12. The Serveron® TM8 is a self-contained fully automated closed-loop gas chromatograph designed to be mounted on or near the subject transformer. Through chromatography, the Serveron® TM8 generates individual measurements of each of the eight critical fault gases (hydrogen, nitrogen, carbon monoxide, carbon dioxide, methane, ethane, ethylene, and acetylene) found in transformer oil. The accuracy of the measurements is commensurate with what one would expect to receive from a traditional laboratory.

Based on an analytical platform built around a gas chromatogram (GC), the Serveron® TM8 takes the recognized laboratory technology onto an onsite transformer. Unprecedented accuracy and repeatability make the TM8 the perfect solution for monitoring mission-critical applications where compromise is not allowed. The TM8 will provide a laboratory-quality set of DGA results every 2 hours and up to once per hour under active fault conditions. With TM8 the engineers can simultaneously track DGA, moisture, temperature, and load on your transformer.

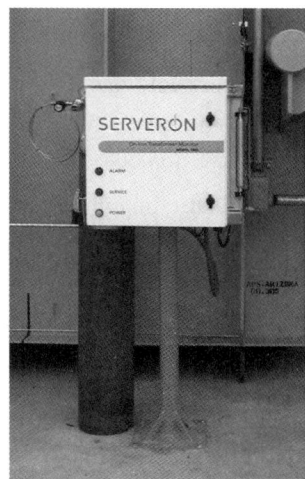

Figure 4.12   Outline of Serveron® TM8 online gas chromatography dissolved gas monitor

The features of Serveron®TM8 include: Correlates all 8 fault gases, moisture-in-oil, oil temperature, and ambient temperature to transformer load; TM8 software provides powerful tools to manage your fleet of DGA transformer monitors, focusing attention on alarming assets and simplifying the analysis of data which warrants further review; Automatic download of DGA data to a central database; Graphing and trending of DGA data; IEEE and IEC diagnostic tools such as Rogers Ratios, and the Duvals Triangles 1, 4&5 and Duvals Pentagon for advanced diagnostics.

The introduction of new Dissolved Gas Analysis (DGA) tools over the past few years has posed challenges for utility and plant managers in selecting the best approach to meet their needs. Transformer asset managers must make critical decisions based on DGA information, such as whether to take a transformer offline to avoid catastrophic failure. These decisions impact service reliabi- lity, revenue, and production.

Increased demand on older transformers accelerates fault development, increasing the risk of catastrophic failures. Even new transformers are not immune to failure, as indicated by the transformer reliability bathtub curve. Asset managers need to make reliability and revenu- eimpacting decisions more quickly and frequently than in the past. Managers must determine the type and amount of condition data required at each level of their condition management model to make informed decisions. In response to these needs, vendors have developed prod- ucts that support decision-making by providing timely, accurate, and reliable transformer DGA data and diagnostic tools. When selecting online DGA tools, transformer asset managers should consider the following: (1) Accuracy and Timeliness: Tools that provide continuous monitoring and real-time data can help detect faults early and improve decision timeliness. Tools that correlate gas data with transformer load, oil temperature, ambient temperature, and other sensor inputs offer a comprehensive view of transformer health. (2) Diagnostic Capa- bilities: Tools that detect and analyze multiple gases as specified in IEEE standards provide better diagnostic insights. Tools that use ratio-based diagnostics can trend fault gas ratios over time, providing dynamic and reliable fault detection. (3) Decision Support: Data Integration: The ability to automatically supplement traditional DGA tests with online data enhances the understanding of fault development. Emerging technologies like neural network diagnostics, although not yet standard, offer promising advancements in accuracy and reliability.

When selecting online DGA (Dissolved Gas Analysis) tools, transformer asset managers must prioritize fault coverage and diagnostic capabilities. While price is a factor, the relative value of the solution, as defined by its ability to provide comprehensive fault coverage and diagnostics, is of paramount importance. High-quality tools, though possibly more expensive, offer superior insights into transformer condition, thereby enhancing utility service reliability and revenue.

The investment strategy for online DGA tools should reflect the stratification model of transformer assets. For critical transformers, allocate the highest investment in online DGA tools with the most extensive fault coverage and advanced diagnostic capabilities. These trans- formers are vital for grid stability, revenue, and service reliability. Superior monitoring tools help prevent catastrophic failures and ensure continuous operation. For important transformers, invest in reliable online DGA tools that offer good fault coverage and diagnostics. These transformers, while not as critical as the top-tier, still play a significant role in maintaining utility service reliability and production efficiency. For recoverable transformers, Opt for basic

DGA tools that provide essential monitoring capabilities. These transformers have a lower impact on revenue and reliability, thus requiring fewer resources for condition monitoring. Given the current environment of increased loading on aging transformers, deferred capital expenditures, and heightened service reliability requirements, transformer asset managers should leverage the latest advancements in online DGA tools.

Transformer asset managers must focus on fault coverage and diagnostic capabilities when selecting online DGA tools, ensuring they get the best value for their investment. By strategically investing in high-quality tools for critical and important transformers, utilities can significantly enhance service reliability and revenue, while also optimizing maintenance costs and extending the useful life of their assets. Embracing the latest advancements in online DGA technology is essential for managing the aging infrastructure and meeting the increasing demand for electricity.

4.7.2　Online Monitoring of Bushings and Lightning Arrestors

The deterioration of oil and paper insulation in high voltage equipment is a significant concern. The normal aging process of such equipment typically spans 30-40 years and is influenced by thermal, electrical, and environmental factors. Transformer bushings, particularly oil-immersed condenser type bushings, are prone to failure due to the degradation of their internal capacitive layers. This process is gradual, with each layer progressively failing and burning through the kraftpaper insulation. The following points highlight the critical aspects of this issue: (1) Gradual failures occur slowly over time, often taking decades, and are detectable through regular monitoring. (2) Premature failures are relatively sudden and not easily detected by periodic offline tests. They require more sensitive monitoring techniques to identify early-stage issues.

Figure 4.13 shows three failed oil-impregnated condenser type bushings, the porcelain sleeves are all missing due to explosion and the condenser cores are burned also, only three copper tubes left.

Schering bridge technique utilizes voltage to detect millivolt-level changes indicative of insulation deterioration. This method is sensitive enough to identify early-stage failures that offline tests might miss. To manage the risk of premature bushing failures and ensure timely maintenance, continuous online insulation condition monitoring is essential. The continuous concern over the deterioration of oil and paper insulation in high voltage equipment necessitates advanced monitoring techniques. The Schering bridge technique and continuous online condition monitoring provide essential tools for early detection and proactive maintenance, thereby enhancing the reliability and longevity of high voltage equipment. By addressing insulation issues promptly, utilities can manage risks effectively and optimize their maintenance strategies. [1]

Figure 4.13   Three failed oil-impregnated condenser type bushings

The Dissipation Factor (DF), also known as tan δ, is a crucial parameter in assessing the condition of transformer bushings and lightning arrestors while the transformer remains energized. An online DF monitoring system has been developed to facilitate this continuous assessment.

The DF online monitoring system utilizes principles similar to the conventional Schering bridge used in laboratory settings. Data is collected under software control from transducers connected to the bushing DF/capacitive tap of the transformer. The acquired data is compared with data from another electrical phase to produce a DF value. Damaged or deteriorated dielectric materials lead to increased dielectric losses, which can cause additional heating and potentially lead to thermal runaway. This phenomenon involves the rapid breakdown of insulation due to an avalanche of failing dielectric layers. DF measurement captures dielectric losses, partial discharges, and treeing. High levels of partial discharge are often observed just after lightning or switching impulses and immediately before and during insulation failure. High DF levels indicate poor dielectric condition and the presence of partial discharges, which are critical to detect for timely maintenance. DF measurement is based on determining the phase shift between two voltage signals. The method is analogous to standard bridge methods, but utilizes software for precise angle difference determination. The measurement of DF is a well-established laboratory procedure used to evaluate insulation quality before commissioning new or refurbished highvoltage equipment. Extending this principle to online monitoring allows continuous assessment of insulation condition without taking the transformer offline. Continuous monitoring of DF provides real-time data on the condition of transformer bushings and lightning arrestors, enabling prompt detection of insulation issues.

By identifying high DF levels and the presence of partial discharges early, utilities can initiate maintenance procedures before catastrophic failures occur, thereby enhancing relia-

bility and reducing unplanned outages. Early detection of increased dielectric losses and potential thermal runaway conditions allows for timely intervention, preventing rapid insulation
breakdown. DF as a parameter is by nature a relatively slow-changing value and is an integral
characteristic depending on:

- Design, materials, and production technology.
- Operating voltages and temperatures.
- Aging of insulation related to design and operating conditions such as overvoltages,
  loading conditions, etc.
- Climatic/weather-related phenomenon.

The Dissipation Factor (DF) online monitoring system is designed to provide continuous
and accurate assessment of transformer bushings and lightning arrestors while the transformer
remains energized. The system calculates the DF of a unit as a relative value compared with a
reference voltage from another unit in service, thereby eliminating the need for a standard
capacitor.

The system uses a reference voltage from another unit in service, which does not need to
be associated with the same phase. The DF online monitoring system automatically makes the
necessary phase angle adjustments. By using relative measurements, the system reduces the
effects of ambient temperature, operating voltages, loading conditions, different aging characteristics, different designs, and varying operating conditions. The principle of cross-refere-
ncing units in a closed loop is used to confirm all measurements and isolate defective units
with increased confidence. The system requires a minimum of three units to be monitored to
ensure accurate relative measurements. All measurements are tested for integrity against three
parameters: RMS value, mean of the signal, and the calculated DF value. Only measurements
that pass the integrity tests are stored in the database, ensuring the reliability of the data. The
default monitoring period is set to once every 5 minutes. However, this setting can be adjusted
from as frequently as once per minute to as infrequently as once per day, depending on the
user's requirements. Sensors are configured through a user-friendly graphical interface during
installation. Each monitored device is entered into the database with an acquisition channel
number and descriptive text for easy identification. The condition of each monitored device is
displayed on the monitor screen, providing real-time status updates.

By using relative measurements, the system enhances the accuracy of the DF readings,
mitigating external influences that could skew the data. The closed loop system and rigorous
integrity testing ensure that faults are detected early, allowing for timely maintenance and
reducing the risk of catastrophic failures. The ability to adjust the monitoring frequency allows
users to tailor the system to their specific needs and operational conditions. The GUI simplifies the process of sensor configuration and system setup, making it accessible to a wide
range of users. The real-time display of device conditions enables continuous monitoring and
immediate response to any issues that arise.

4.8 Online Dissolved Gas Analysis by Photoacoustic Spectroscopy

4.8.1 Introduction

Photoacoustic spectroscopy (PAS) has emerged as an attractive and powerful technique wellsuited for various sensing applications. This technique leverages advancements in high-power radiation sources, sophisticated electronics, sensitive microphones, and digital lock-in amplifiers to achieve significant progress in detection capabilities. Recent research highlights that PAS can achieve trace gas detection at parts per trillion (pL/L) levels. Additionally, the photoacoustic (PA) detection of infrared absorption spectra using modern tunable lasers offers several advantages, such as simultaneous detection and discrimination of numerous molecules of interest.

PAS has demonstrated exceptional sensitivity, capable of detecting gas concentrations at ppt levels.[13, 14] This makes it a highly effective tool for monitoring trace gases in various environments. Utilizing modern tunable lasers, PAS can detect and discriminate between multiple molecules simultaneously. This ability is particularly advantageous for complicated mixtures where multiple analytes need to be monitored concurrently. The successful application of PAS in both gases and condensed matter has garnered attention from scientists and engineers across multiple disciplines. A substantial body of literature now exists on PAS, documenting its use and effectiveness in various contexts.

PAS has been extensively used for detecting various gases, including environmental pollutants, industrial emissions, and hazardous substances. PAS is also employed in liquid analysis, such as monitoring contaminants in water or analyzing biochemical samples. In solids, PAS can be used to study material properties, detect defects, and analyze compositions.

4.8.2 Fundamentals of Photoacoustics

4.8.2.1 Photo-thermal Phenomena

Photo-thermal spectroscopy encompasses a group of highly sensitive methods used to detect trace levels of optical absorption and subsequent thermal perturbations of samples in gas, liquid, or solid phases. The underlying principle of these spectroscopic methods is the measurement of physical changes—such as temperature, density, or pressure—as a result of a photo-induced change in the thermal state of the sample. The sample is subjected to optical excitation, typically using laser radiation or other forms of radiation. This radiation is absorbed by the sample, increasing its internal energy and placing it in an excited state. Some portion of the absorbed energy decays from the excited state in a non-radiative fashion, resulting in an increase in local energy. The increase in local energy leads to a temperature change in the sample or the coupling fluid (e.g., air). This temperature change can cause a density change. If the temperature change occurs faster than the sample or coupling fluid can expand or contract, it results in a pressure change. The physical changes induced by the photo-thermal effect can be measured using different techniques, which can be classified based on the type of physical change they measure: (1) Index of refraction change, photo-thermal interferometry (PTI) could be used to measure changes in the index of refraction. (2) Photo-thermal lensing (PTL) could be used to detect the deflection of a probe beam due to the refractive index gradient created by a temperature gradient. (3) Photothermal deflection (PTD) is a method similar to PTL, it measures the deflection of a probe beam passing through a region with a refractive index gradient. (4) Photoacoustic spectroscopy (PAS) is a method to measure the pressure changes induced by thermal expansion of the sample or coupling fluid. This technique is highly sensitive and can detect trace levels of gases and other analytes. PAS is one of the prominent photo-thermal techniques and has gained significant attention due to its capability to detect trace gas concentrations at parts per trillion (pL/L) levels.

4.8.2.2 Photoacoustic Spectroscopy

Photoacoustic spectroscopy (PAS) relies on periodic heating and cooling of the sample to produce pressure fluctuations, thereby generating acoustic waves. This is typically achieved using modulated or pulsed excitation sources. The acoustic waves detected in PAS are directly generated by the absorbed fraction of the modulated or pulsed excitation beam, making the signal from a PA experiment directly proportional to the absorbed incident power. The relationship between the generated acoustic signal and the absorbed power will vary based on the type of excitation source.

The application of PAS dates back to Alexander Graham Bell's invention of the photophone around 1880. In the late 1930s, Vengerov's study on the absorption in gases marked the first formal example of PAS. The technique gained popularity with laser applications, notably through the work of Kerr and Atwood, and was significantly advanced by Kreuzer, who demonstrated the detection of methane and ammonia at parts per billion (nL/L) and sub- nL/L levels, respectively, using laser excitation. While early work by Kreuzer highlighted the potential of PAS for gas analysis, the technique can also be applied to liquids and solids using both direct and indirect coupling methods. The acoustic wave generated in the sample is detected by a transducer in direct contact with the solid or liquid sample. This method is straightforward as the acoustic wave does not cross a high-impedance interface, allowing easy detection by the transducer. Involves a more complex process where the original acoustic wave is not the primary signal. This method was initially demonstrated by Bell and later explored by Parker, who observed increased signal contributions from cell windows in gaseous PAS experiments.

The gas-coupling method, described as the gas-piston model by Rosencwaig, explains this process: Periodic heating of the sample surface within a diffusion length generates a thermal wave. This thermal wave heats the gas layer directly above the sample surface, leading to periodic expansion in the gas layer. The expansion creates an acoustic wave detectable by standard microphones.

4.8.2.3　Experimental Arrangements for PA Detection

In comparison to other photothermal techniques that measure changes in refractive index or temperature using combinations of probe sources and detectors, PAS measures the pressure wave produced by sample heating. Although PAS experiments can take many forms, several key elements are constants. The main components of a PAS apparatus including: Excitation source, Sample cell, Pressure transducer, Signal Processing module, Data acquisition and processing module, and so on. Typically the excitation source is a laser or a filtered lamp. The source is modulated or pulsed to provide periodic heating of the sample. The modulated or pulsed radiation is directed at a sample cell. The sample cell can range from a simple sealed tube to a complicated resonant chamber or multipass cell. The design of the sample cell can affect the sensitivity and specificity of the PAS measurement. The pressure transducer Usually is a microphone with an appropriate frequency response. Detects the pressure wave created by the sample heating. The signal generated by the microphone is proportional to the amplitude of the pressure wave. Information about the phase and delay of the wave is also captured, providing additional data. The Signal Processing module is a lock-in amplifier, it Captures the amplitude and phase information of the pressure wave. Gated Accumulation (Boxcar Amplifier) is an alternative method to capture the signal information, particularly useful for pulsed sources. The data acquisition and processing module means a personal computer reads and records the voltage outputs from the lock-in amplifier. The data can be analyzed to extract meaningful information about the sample. Figure 4.14 shows the internal structure of photoacoustic spectroscopy by a broadband infrared emitter.

There are two main categories of light sources used for PAS: broadband sources, such as lamps, and narrow-band laser sources. The general category of lamps includes the following: arc lamps, filament lamps, and glow bars. These sources were some of the first sources used to study the phenomenon of PAS and have several advantages and disadvantages. The broadband output of these sources can be significant (e.g., from ultraviolet to infrared) and in most cases can potentially cover all regions of interest for optical excitation and subsequent PAS examinations. These sources are generally inexpensive and, depending on the overall wattage required, can be somewhat compact in size. Unfortunately, lamp sources also have low spectral brightness, require spectral selection through the use of filters or monochromators, and are usually restricted to low source modulation frequency and optical efficiencies. A typical example of broadband IR sources is shown in Figure. 4.15.

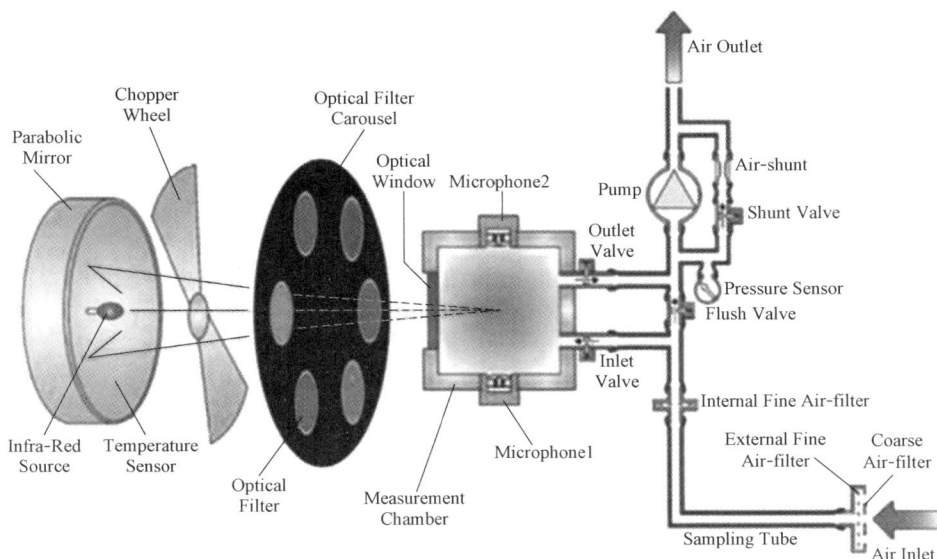

Figure 4.14   Internal structure of photoacoustic spectroscopy by a broadband infrared emitter [8]

Narrow-band laser sources include gas lasers, diode lasers, and quantum cascade lasers (QCLs). Gas lasers, such as CO_2 lasers, produce a very narrow and intense beam of light at specific wavelengths, making them highly efficient for PAS due to their high power and ability to be modulated. Diode lasers, which are semiconductor lasers, are compact and efficient, offering tunable wavelengths and making them versatile for various PAS applications. Quantum cascade lasers are another type of semiconductor laser that can produce mid-infrared light. They are powerful, tunable, and particularly useful for

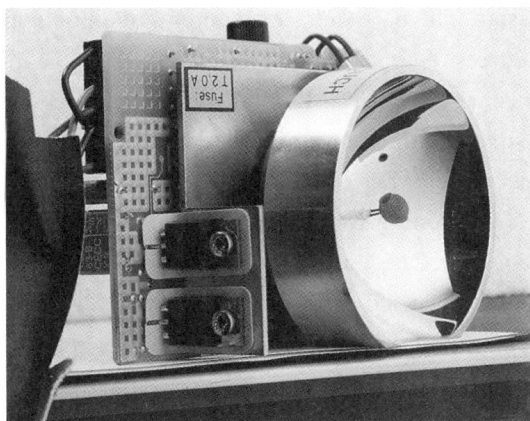

Figure 4.15   A typical example of broadband IR sources [8]

PAS because they can be modulated at high frequencies.

Laser sources are advantageous due to their high spectral brightness, coherence, and the ability to be modulated at high frequencies. This makes them ideal for PAS as they can provide high signal-to-noise ratios and sensitivity. However, they can be more expensive and complicated to operate compared to broadband sources.

The choice of light source in PAS experiments depends on the specific requirements of the application, including the desired wavelength range, modulation frequency, and sensitivity. Both broadband sources and narrow-band laser sources have their unique advantages and limitations, making them suitable for different types of PAS measurements.

Although Photoacoustic Spectroscopy with broadband IR sources is still common, laser sources have been mainly employed by modern PAS research[15]. As with lamps, lasers have numerous advantages but also present some disadvantages. Some of the key advantages of these sources are their large spectral brightness, collimated output, ease of modulation, and narrow spectral linewidth. Disadvantages include expense and limited tunability; however, these are not generic qualities of every laser architecture. As mentioned previously, PAS can be performed as a pulsed or modulated measurement with respect to light excitation. That allowance is seen vividly in the sources used for PAS experimentation. Early work in the detection of gases used a variety of these sources, including pulsed and continuous wave (CW) dye laser sources and CW laser sources, such as grating tunable CO and CO_2 and helium-neon (He-Ne) lasers. These sources usually provided reasonable or even high power levels, some limited tunability, and were based in the near-infrared or infrared wavelength regions. All of these features allowed for some level of spectroscopic studies to be performed.

Several PAS studies involving liquid and solid samples, which require more laser irradiation, used solid-state sources such as neodymium-doped yttrium aluminum garnet (Nd:YAG) lasers. The harmonics or pumping of an optical parametric oscillator enable the tuning of these sources. Semiconductor lasers based on a direct bandgap transition with various feedback mechanisms (e.g., distributed Bragg reflectors, distributed feedback reflectors [DFB]) were used for numerous studies, especially in regard to the study of atmospheric or small molecular gas targets that could be identified with tuning ranges from fractions to single-integer. This type of tuning was easily accomplished by the current and/or temperature tuning of the laser diode. Occasionally, efforts used other laser sources such as lead-salt diode lasers, which are centered in the infrared and theoretically can be tuned through mode hopping over a larger spectral band; however, these sources were plagued by cryogenic cooling requirements and low output powers (e.g., 0.1 mW typical).

In 1994, the quantum cascade laser (QCL) was developed and employed in laser PAS. In that year, Bell Laboratories first demonstrated the QCL as a new infrared laser source. Since that time, continuous and aggressive evolution has occurred. PAS sensing capability employing QCLs was identified early on, and demonstrations by Paldus et al. using these sources can be seen as early as 1999. Although QCLs took years to evolve into their current state, work continued on PAS studies using these sources throughout this development cycle. Furthermore, these sources, operating in low-duty cycles, have demonstrated that PAS based on lock-in amplification can still be performed and indeed shows great promise.

The advent of QCLs brought several advantages to PAS and infrared spectroscopy. QCLs offer a wide tunability range in the mid-infrared region, where many molecules have their characteristic absorption features. This tunability, combined with their high power output and narrow linewidth, makes QCLs ideal for detecting trace gases and other substances with high sensitivity and specificity. Additionally, the development of room-temperature QCLs has eliminated the need for cryogenic cooling, which was a significant limitation of earlier infrared laser sources.

QCLs have been successfully integrated into various PAS setups, enhancing the capability to detect low-concentration analytes in different environments. The use of QCLs in PAS has expanded the technique's applicability, enabling detailed spectroscopic studies and environmental monitoring with unprecedented precision and accuracy. As the technology continues to advance, it is expected that QCLs will further solidify their role as a crucial component in PAS and infrared spectroscopy, driving new discoveries and applications across multiple fields.

4.8.2.4   Optical Filters and Selective Absorption of Light Energy

When broadband IR sources are applied in the PAS instrument, the concentration of each targeted sample gas among all gas mixtures must be detected one by one. Each gas has its unique light energy absorption coefficient, which depends on the wave number of the incident light, as shown in Figure 4.16. Therefore, several optical filters are installed on the optical filter carousel, as shown in Figure 4.17. By changing the optical filters, a definite wavelength band of light energy can be selected to be absorbed by the targeted gas. When the input optical energy is modulated periodically, it results in periodic pressure changes in the PA cell as well. This setup allows for the selective detection of different gases based on their absorption characteristics. The use of optical filters ensures that only the specific wavelength band corresponding to the gas of interest is passed through, improving the specificity and accuracy of the measurements. By systematically rotating the optical filter carousel and modulating the input light, the PAS instrument can effectively isolate and measure the concentrations of multiple gases in a mixture. The periodic modulation of the light source leads to periodic heating and cooling of the gas in the PA cell, generating pressure waves that are detected by a microphone or another pressure transducer. The amplitude and phase of these pressure waves are analyzed to determine the concentration of the target gas. This method leverages the distinct absorption features of different gases, making PAS a powerful technique for gas detection and analysis.[13, 16]

4.8.2.5   Acoustic Resonators

A typical acoustic resonator, as shown in Figure 4.18. It is an essential element of a PA sensor, serving both as a container for the sample and as the detector. Therefore, the optimum design of a PA cell is necessary to facilitate signal generation and detection. To date, a variety of cell configurations have been reported for solid, liquid, and gas samples. These include cells operated at acoustically resonant and nonresonant modes, single and multipass cells, and cells for intracavity operation. Resonant cells are general designed to amplify the acoustic signal by matching the acoustic frequency of the cell to the modulation frequency of the light source. Resonant cells typically provide higher sensitivity due to the amplification of the acoustic wave within the cell. Non-resonant cells do not rely on resonance and are often simpler in design. They may be used when the modulation frequency cannot be matched to the cell's resonant frequency or when a broader frequency response is desired.[17]

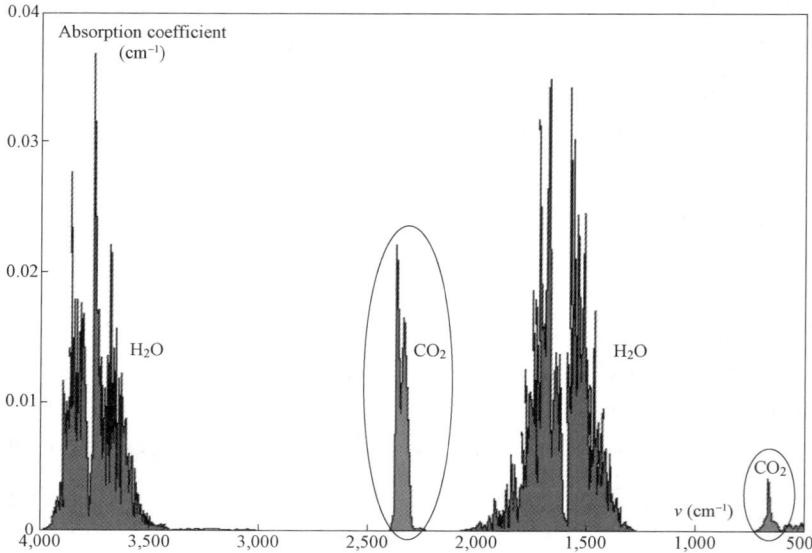

Figure 4.16   Optical energy absorption coefficient is closely related to the wave number of incident light

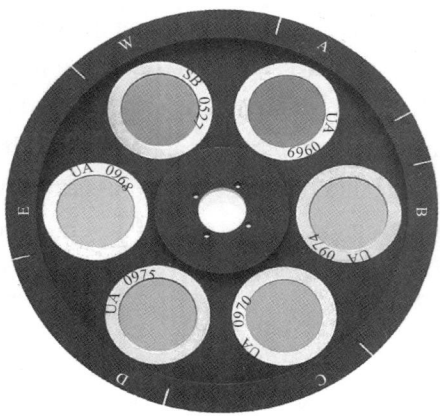

Figure 4.17   Six optical filters are installed on the optical filter carousel

Figure 4.18   A typical acoustic resonator (PA cell) operated at acoustically non-resonant modes

The detection sensitivity in photoacoustic spectroscopy (PAS) is often limited by the signalto-noise ratio (SNR). Achieving high sensitivity in non-resonant gas cells can be challenging due to various noise sources, such as amplifier noise and external acoustic noise, resulting in a small SNR. Additionally, light absorption at the cell windows and walls can create a background signal that is difficult to distinguish from the PAS signal generated by the gas sample.

Utilizing Brewster windows can minimize reflections, thereby reducing background noise. This help in attenuating unwanted acoustic noise within the cell. Minimizing the heating noise from windows by employing windowless or open cell designs. Using higher modulation frequencies can enhance the SNR by reducing the influence of low-frequency noise. By matching the modulation or pulse frequency to the acoustic resonance frequency of the PAS cell, resonant eigenmodes can be excited, amplifying the signal. Common resonators include cylindrical cells operating on longitudinal, azimuthal, or radial resonances, and Helmholtz resonators. Helmholtz resonators are widely used for their ability to amplify signals at specific resonance frequencies. Cylindrical resonant cells are used in conjunction with acoustic transmission line theory to optimize performance for trace gas detection. Research on miniaturized PA cells, including MEMS scale and macro resonant cells, has shown that miniaturization can be achieved without significant loss in signal. [18]

Miklos et al. design a differential PA cell with a fully symmetrical design reduces flow noise and electromagnetic disturbances. Gas flows through both tubes to produce similar flow noise, but the laser passes through only one tube, generating the PA signal in just one resonator. Differential amplification then suppresses coherent noise components, enhancing the SNR.

The condition and quality of the PA cell surface significantly impact the background signal due to scattering and molecule absorption. Investigating different cell materials and surface treatments or coatings can help minimize these effects, improving the overall sensitivity and accuracy of the PAS measurements.

By implementing these design and optimization strategies, PAS instruments can achieve higher sensitivity and more accurate detection of trace gases, expanding the applicability and effectiveness of the technique in various fields.

4.8.2.6 Detectors

In a PA instrument, the acoustic waves generated in a PA cell as a result of the absorption of radiation by a sample are detected by a pressure sensor. The appropriate choice of acoustic pressure sensor depends on the application, considering factors such as sensitivity requirements, ease of operation, and ruggedness.

The typical examples of miniature electret microphones include products from Knowles, Sennheiser, and Intricon Tibbetts. The best suppliers of condenser microphones include products from Bruel and Kjaer. These devices are user-friendly and straightforward to integrate into PAS setups. They are sufficiently sensitive for PAS studies involving solids, liquids, and gases. Their responsivity is only weakly dependent on frequency, making them versatile for various applications.

A lock-in amplifier is typically used to detect the small voltage produced by the microphone due to sample absorption of radiation. The detection threshold of a PAS system is generally influenced more by other noise sources (e.g., external noise, window heating, and absorption of desorbing molecules from the cell walls) rather than the microphone's responsivity or electrical noise.

Kauppinen et al introduced a cantilever-type pressure sensor made out of silicon, replacing the traditional capacitive microphoner, as shown in Figure 4.19. The sensor in the cantilever microphone is a flexible bar with typical dimensions of a few millimeters in width and length and a thickness of 5-10 μm. The cantilever is separated on three sides from a thicker frame with a narrow gap (3-5 μm) and moves like a flexible door due to pressure variations in the surrounding gas. A laser interferometer measures the displacement of the cantilever as it bends due to pressure changes without stretching. Cantilevers exhibit superior sensitivity compared to conventional microphones. However, PA cells containing cantilever sensors and the associated interferometric detection systems are significantly more expensive and fragile than cells equipped with conventional capacitive microphones.

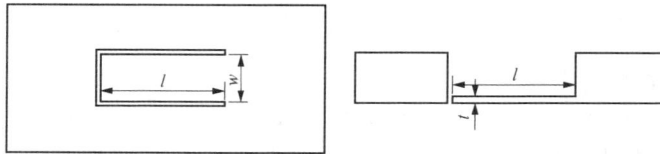

Figure 4.19　Dimensions of a cantilever-type pressure sensor

The choice between traditional microphones and advanced cantilever-type sensors depends on the specific requirements of the PAS application. While commercial microphones offer a balance of sensitivity, ease of use, and cost-effectiveness, cantilever sensors provide higher sensitivity but at the expense of increased complexity and cost. The ongoing development and optimization of these sensors continue to enhance the capabilities and applications of PAS in various fields.

Quartz-enhanced PAS (QEPAS) is another innovative PAS detection approach in which the traditional microphone is replaced with a quartz crystal tuning fork (TF). Kosterev et al. proposed inverting the typical resonant PAS approach, where the absorbed energy is usually accumulated in the resonant acoustic transducer. This approach eliminates the limitations imposed on the PAS cell by acoustic resonance conditions since the resonant frequency is determined by the TF. Consequently, the cell is optional in QEPAS and is utilized primarily to isolate the gas sample from the environment and control its pressure. This method allows for the examination of gas samples as small as 1 mm^3 in volume.

Quartz was chosen as a suitable material for TFs due to its status as a common, low-loss piezoelectric material that is both mass-produced and inexpensive. Additionally, quartz TFs are common in atomic force and optical near-field microscopy and are well characterized. Only the symmetric vibration of a TF (where the two prongs bend in opposite directions) is piezoelectrically active. Efficient excitation of this vibration is achieved when the excitation beam passes through the gap between the TF prongs. The pressure wave generated when optical radiation interacts with a gas excites a resonant vibration of the TF, which is then converted into an electric signal due to the piezoelectric effect. This electric signal, which is proportional to the gas concentration, is measured by a trans-impedance amplifier. The initial feasibility experiments performed by Kosterev et al. utilized a quartz-watch TF.

For PAS applications involving liquid and solid samples, the use of conventional microphones has been reported to be inefficient by Hordvik and Schlossberg and Farrow et al. Both groups identified issues with acoustic impedance mismatching at the solid-gas or liquid-gas interface, resulting in most of the acoustic energy being reflected or absorbed back into the sample rather than transferred across the boundary. Improvements in sensitivity were demonstrated through the use of piezoelectric transducers in contact with solid and liquid samples. These piezoelectric elements offer the advantage of good impedance matching. Similar to conventional microphone detection schemes, a lock-in amplifier is used to detect the voltage change produced by the piezoelectric sensor. This direct coupling method is simple, and numerous studies have employed piezoelectric elements in contact with liquids or solids for PAS detection. While more recent reports describing PAS detection of solid samples utilize conventional microphones and the indirect coupling method described by Rosencwaig, piezoelectric transducers remain widely used for liquid studies.

4.8.3 Photoacoustic Sensing Applications in Smart Grids

In smart grids, the photoacoustic spectroscopy is mainly used on conventional PAS sensing, including detection in the dissolved gas analysis of transformer oil, decomposition by-products of SF_6, and so on. Because PAS sensing has evolved tremendously since its initial application in power transmission and distribution industry. Figure 4.20, Figure 4.21 present on-line dissolved gas analysis (DGA) instruments based on photoacoustic spectroscopy technology, as offered by domestic and international supplies.

Dissolved Gas Analysis (DGA) is recognized as a powerful monitoring technique for detecting developing faults within transformer main tanks and associated oil-filled equipment.[19] Virtually all large transformers, old or new, have cores and windings immersed in oil, along with input and output electrical connections. The transformer windings are typically electrically insulated by thick layers of paper insulation wrapped around each part of each winding. The oil serves both as a heat dissipation medium and as an insulating medium. When the oil or paper insulation is stressed—such as under elevated temperature conditions associated with high load and/or fault conditions or even under normal operating conditions—it breaks down to form a range of byproducts and simple gases. These gases dissolve into the oil immediately following their creation and will remain there indefinitely (if they cannot escape from the electrical equipment via a breather or a leak).

Figure 4.20   An online dissolved gas analysis instrument based on photoacoustic spectroscopy is installed on-site

(a)

(b)

(c)

Figure 4.21   Enlarged photo of the online DGA by PAS instrument of Kelman (General Electric)

Traceable gas mixtures were provided directly to a calibrated photoacoustic gas analyzer and measured. This process was repeated ten times, and analysis for accuracy and repeatability was performed. This work was conducted in a laboratory environment, so it may not account for possible variability associated with environmental conditions. However, it does illustrate the ability of PAS detectors to accurately measure each of the target gases at medium concentrations.

PAS-based DGA instruments have been developed to address the shortcomings of online gas chromatography (GC)-based instruments. They provide a real alternative to GC by matching their performance and operating successfully in the field. Utilizing a technology historically designed for online application, PAS instruments are very stable and repeatable monitoring instruments suited for the tough environmental and operational demands associated with remote transformer monitoring. PAS is the new high-end standard for monitoring critical transformers. [14, 20]

- Standard gas-in-oil samples were prepared and tested on a portable DGA instrument that employs a PAS detector in conjunction with an automated headspace gas extraction system. Six samples were analyzed and a comparison was made between the standard sample quantity and the measured gas quantity. This provided an assessment of the capability of the instrument to measure gas-in-oil where the levels of gas are known.

- Oil samples from field transformers were collected and analyzed on the same portable instrument, with duplicate samples analyzed on a laboratory GC. This allows for the detection of any variability in measurement capability when dealing with real situations.

- Oil samples were collected from various transformers with online PAS-based systems. These samples were analyzed in a laboratory GC system and results were compared.

4.9 Typical Instrument Suppliers and the Measurable Scope of PAS

4.9.1 PAS Typical Suppliers

4.9.1.1 General Electric

GE introduces a PAS instrument as Kelman^{TM} DGA 900 on June 2017, it is a multi-gas online Dissolved Gas Analysis (DGA) and moisture analyzer, which once installed on a transformer will help detect and diagnose incipient faults and trend asset health via the monitoring of 9 critical gases plus moisture. This can be applied to any mineral oil, natural or synthetic ester fluid-filled power transformer used in the generation, transmission, distribution, and industrial applications. As the global average age of transformers continues to rise and grid architectures become more dynamic, the possibility of rapid aging, unplanned outages, and even catastrophic failure between off-line tests also increases, leading many asset owners to adopt online DGA monitoring strategies to increase network reliability and operational efficiencies. The Kelman DGA 900 is designed to cater to this need and utilizes an evolved implementation of GE's proven Photo-Acoustic Spectroscopy (PAS) measurement technology, providing laboratory challenging levels of precision and repeatability with no consumables and no need for frequent re-calibration.

(https://www.gegridsolutions.com/press/gepress/kelman_dga_900_announcement.htm)

(https://www.gegridsolutions.com/events/Whitepaper.ashx?id=1)

(https://www.gegridsolutions.com/events/Whitepaper.ashx?id=2)

4.9.1.2　Camlin Power

Camlin operates with the vision of bringing revolutionary products to life for a wide range of industries, including power and rail, and also has interests in a number of R&D projects in a variety of scientific sectors. At Camlin we believe in high-quality engineering and design, allowing us to develop market-leading products and services. In short, we love creating value for our customers by solving difficult problems. As of today, the Camlin operation spans over 20 countries across the globe. CAMLIN design and manufacture a range of value-adding power solutions to help our utilities prove a better service to their customers, and ultimately, keep the lights on around the world. We provide products and solutions across the electrical network - from high-voltage asset monitoring equipment for helping utilities manage their transformers, generators, and circuit breakers, to devices that can be retrospectively installed on low-voltage networks for fault location and network automation. Our products are sold through our companies, KELVATEK and CAMLIN POWER (United Kingdom & Ireland and International respectively).

(https://www.camlingroup.com/power/)

4.9.1.3　Finland Gasera Co., ltd

Finland Gasera co., ltd is a company oriented by the University of Turku in Finland by Prof. Kauppinen. Gasera One trace level photoacoustic spectroscopy multi-gases analyzer is a product for the power industry. Trace impurity gases such as hydrocarbons, H_2O, CO, CO_2, NO_x, N_2, H_2S, etc., are harmful to the production of high purity helium, high purity hydrogen, high purity nitrogen, other high purity gases. It can damage production equipment and affect product quality, and even pose a fatal threat to equipment that USES these gases. The DKF10-MH trace-level multi-component gas analyzer was developed by Duketech and Finland Gasera co., ltd. specializes in the monitoring and analysis of $\mu L/L$ micro-level and nL/L ultra-micro level impurity gases.

(https://www.gasera.fi/)

4.9.1.4 LumaSense Technologies

LumaSense Technologies A/S manufactures and markets instruments for gas monitoring, offering portable and installed gas and vapor monitors based on the infrared absorption and photoacoustic detection principle (PAS). The gas monitors are used for research studies as well as for continuous field measurements of multiple compounds from nL/L and μL/L levels to percent level concentrations.

(https://innova.lumasenseinc.com/manuals/1412i/)

4.9.1.5 m-u-t GmbH

m-u-t GmbH is part of the m-u-t group. the m-u-t group is a turn-key provider of efficient solutions in photonics. Contact-free optical measurement technology can optimize a large variety of strongly growing future markets in a resource-efficient and environmentally friendly manner. m-u-t GmbH develops and produces systems (as a module or stand-alone) for use in industrial environments, train systems, agriculture, and environmental technology as well as in medical and life science applications. Systems are developed customer-specific - from the idea stage all the way to the series production. m-u-t GmbH offers more than 20 years of expertise in optical measurement technology in a large range: from UV and VIS via NIR and MIR to LWIR. m-u-t GmbH develops, produces, and services series products for OEM customers.

(https://www.mut-group.com/en/)

4.9.1.6 SIGAS

SIGAS Measurement Engineering Corp. is mainly engaged in trace gas analysis equipment and provides complete gas monitoring systems, which are used in environmental monitoring and industrial process analysis. SIGAS dedicate to high-tech technology and try to establish the most outstanding team which will succeed based on practical knowledge, continuous innovation, and serious service. SIGAS offers a full range of technical services to customers, including project conception, development process, or final product molding.

(http://www.sigas-group.com/en/)

4.9.2 The Measurable Scope of PAS

One of the most important applications of PAS in the electric power industry is to detect dissolved gas in transformer oil. The lowest detectable limits are as shown in Table 4.1.

Table 4.1   The lowest detection level (LDL) of six gases by GE (Kelman)

Gas	Lower Detection Level in Gas (μL/L)
Carbon Dioxide (CO_2)	<1

Continue

Gas	Lower Detection Level in Gas (μL/L)
Carbon Monoxide (CO)	<1
Methane (CH_4)	<0.5
Ethane (C_2H_2)	<0.5
Ethylene (C_2H_4)	<0.5
Acetylene (C_2H_2)	<0.2

The PAS could also be used for environmental monitoring and industrial process analysis, and it can also be widely used in petroleum, chemical, metallurgy, mining, pharmaceutical, semiconductor processing, packaging industry, as well as the flue gas online monitoring, TOC value measurement of water, including agricultural greenhouse gas detection, gas safety monitoring, terrorist attacks, storage and transportation of dangerous goods. The measurable scope of various gases is shown in Table. 4.2, and detailed information could be found also at website of SIGAS Measurement Engineering Corp.: https://www.sigas-group.com/?Product_13/93.html

Table 4.2   The measurable scope of typical gases in various industries by PAS

Gas		Measurement rage	Linearity error	Accuracy	Background gas
NH_3	Ammonia	0-100 μL/L	0.5 μL/L / 0.1% FS	0.1 μL/L / 0.1% FS	Air or N_2
NH_3	Ammonia	0-1 g/m^3	0.1% FS	0.1% FS	Air or N_2
C_5H_{10}	Cyclopentane	0-100 μL/L	0.5 μL/L / 0.1% FS	0.1 μL/L / 0.1% FS	Air or N_2
C_3H_6O	Acetone	0-2,000 μL/L	0.1%	0.1% FS	Air or N_2
C_2H_6O	Ethyl Alcohol	0-2,000 μL/L	0.1%	0.1% FS	Air or N_2
C_3H_8O	Isopropyl Alcohol	0-1,000 μL/L	0.1%	0.1% FS	Air or N_2
CH_4O	Methyl Alcohol	0-500 μL/L	0.1%	0.1% FS	Air or N_2
NF_3	Cyclopentane	0-20 μL/L	0.5 μL/L / 0.1% FS	0.1 μL/L / 0.1% FS	Air or N_2
N_2O	Nitrous Oxide	0-500 μL/L	0.1%	0.1% FS	Air or N_2
C_2Cl_4	PERC	0-800 μL/L	0.1%	0.1% FS	Air or N_2
SO_2F_2	ProFume	0-1 g/m^3 0-1,000 μL/L 1-150 g/m^3	0.1%	0.1% FS	Air or N_2
CCl_3F	R11	0-500 μL/L	0.1%	0.1% FS	Air or N_2
$C_2Cl_3F_3$	R113	0-50 μL/L	0.5 μL/L / 0.1% FS	0.1 μL/L / 0.1% FS	Air or N_2
C_2F_6	R116	0-100 μL/L	0.5 μL/L / 0.1% FS	0.1 μL/L / 0.1% FS	Air or N_2
CCl_2F_2	R12	0-100 μL/L	0.5 μL/L / 0.1% FS	0.1 μL/L / 0.1% FS	Air or N_2
$C_2HCl_3F_2$	R122	0-100 μL/L	0.5 μL/L / 0.1% FS	0.1 μL/L / 0.1% FS	Air or N_2
C_2HClF_3	R123	0-100 μL/L	0.5 μL/L / 0.1% FS	0.1 μL/L / 0.1% FS	Air or N_2

Continue

Gas		Measurement rage	Linearity error	Accuracy	Background gas
$CClF_3$	R13	0-100 μL/L	0.5 μL/L / 0.1% FS	0.1 μL/L / 0.1% FS	Air or N_2
CH_4	R14	0-100 μL/L	0.5 μL/L / 0.1% FS	0.1 μL/L / 0.1% FS	Air or N_2
$CHCl_2F_2$	R22	0-100 μL/L	0.5 μL/L / 0.1% FS	0.1 μL/L / 0.1% FS	Air or N_2
CHF_3	R23	0-150 μL/L	0.5 μL/L / 0.1% FS	0.1 μL/L / 0.1% FS	Air or N_2
C_8H_8	Styrene	0-50 μL/L	0.5 μL/L / 0.1% FS	0.1 μL/L / 0.1% FS	Air or N_2
SO_2	Sulfur Dioxide	0-100 μL/L	0.5 μL/L / 0.1% FS	0.1 μL/L / 0.1% FS	Air or N_2
SO_2	Sulfur Dioxide	0-1,000 μL/L	0.1%	0.1% FS	SF_6
$C_8H_2OO_4Si$	TEOS	0-100 μL/L	0.5 μL/L / 0.1% FS	0.1 μL/L / 0.1% FS	Air or N_2
$C_2HCl_3F_2$	Toluene	0-100 μL/L	0.5 μL/L / 0.1% FS	0.1 μL/L / 0.1% FS	Air or N_2
C_2HCl_3	TRI	0-1,000 μL/L	0.1%	0.1%	Air or N_2
CH_3Br	Methyl Bromide	0-2 g/m^3	0.1%	0.1%	N_2 (only)

4.10 Review Questions

Q1: Which kind of properties should the liquid dielectric materials have in electrical apparatus? (see Section 4.2)

Q2: If an electrical apparatus is considered an enclosed chemical reactor chamber, which kind of chemical reactions will be taken place internally? please explain the reason. (see Section 4.3)

Q3: Please explain the working principle of chromatography. (see Figure 4.7)

Q4: Please explain what the dissolved gas analysis method is, and why it is the most effective way to predict the potential faults in the electrical apparatus. (see Section 4.7)

Q5: Please explain the working principle of photoacoustic spectroscopy. (see Section 4.8)

Bibliography

[1] Gill, Paul. Electrical power equipment maintenance and testing. CRC press, 2008.

[2] Liu, Jian. Electrical, Control Engineering and Computer Science: Proceedings of the 2015 International Conference on Electrical, Control Engineering and Computer Science (ECECS 2015), Hong Kong, 30-31 May 2015. CRC Press LLC, 2016.

[3] Abu-Siada, Ahmed. Power transformer condition monitoring and diagnosis. Institution of Engineering & Technology, 2018.

[4] Kulkarni, Shrikrishna V and Khaparde, SA. Transformer engineering: design and practice, volume 25. CRC press, 2004.

[5] Chudnovsky, Bella H. Transmission, distribution, and renewable energy generation power equipment:

Aging and life extension techniques. CRC Press, 2017.

[6] Hirschler, Marcelo M and others. Electrical insulating materials: International issues. ASTM, 2000.

[7] Piet de Coning and John Swinley. A Practical Guide to Gas Analysis by Gas Chromatography. Elsevier, 2019.

[8] Norazhar Abu Bakar, A. Abu-Siada and S. Islam. A review of dissolved gas analysis measurement and interpretation techniques. IEEE Electrical Insulation Magazine, 30(3):39-49, 2014.

[9] Ferrer, Imma and Thurman, Michael E. Advanced techniques in gas chromatography-mass spectrometry (GC-MS-MS and GC-TOF-MS) for environmental chemistry. Newnes, 2013.

[10] Hübschmann, Hans-Joachim. Handbook of GC/MS. Wiley Online Library, 2000.

[11] Duval, Michel and Lamarre, Laurent. The duval pentagon—a new complementary tool for the interpretation of dissolved gas analysis in transformers. IEEE Electrical Insulation Magazine, 30(6):9-12, 2014.

[12] Duval, Michel and Lamarre, Laurent. The new duval pentagons available for dga diagnosis in transformers filled with mineral and ester oils. 2017 Electrical Insulation Conference (EIC), Baltimore, MD, USA, 2017.

[13] Michaelian, Kirk H. Photoacoustic infrared spectroscopy. Wiley-Interscience,, 2003.

[14] Dumitraş, DC and Puiu, A and Cernat, R and Giubileo, G and Lai, A. Laser photoacoustic spectroscopy: a powerful tool for measurement of trace gases of biological interest at sub-ppb level. Molecular Crystals and Liquid Crystals, 418(1): 217-227, 2004.

[15] Demtröder, Wolfgang. Laser spectroscopy: vol. 2: experimental techniques, volume 2. Springer Science & Business Media, 2008.

[16] Michaelian, Kirk H. Photoacoustic IR spectroscopy: instrumentation, applications and data analysis. John Wiley & Sons, 2010.

[17] Suart, B. Infrared spectroscopy: Fundamental and applications. Google Scholar, 2004.

[18] Einfeld, Wayne. Photoacoustic infrared monitor [microform] : Innova AirTech Instruments type 1312 multi-gas monitor / by Wayne Einfeld. National Exposure Research Laboratory, Office of Research and Development, U.S. Environmental Protection Agency Las Vegas, Nev, 1998. URL http://purl.access.gpo.gov/GPO/LPS3058.

[19] Wu, Zhiying and Yu, Qingxu and others. Photoacoustic spectroscopy detection and extraction of discharge feature gases in transformer oil based on 1.5 μ tunable fiber laser. Infrared Physics & Technology, 58: 86-90, 2013.

[20] Yip, Bernard Cheuk-Yuen. Trace detection in gases using photoacoustic spectroscopy and Fabry-perot interferometry. PhD thesis, Iowa State University, 1984.

5

Tunable Diode Laser Absorption Spectroscopy

5. 1 Introduction

Tunable Diode Laser Absorption Spectroscopy (TDLAS) is a precise and sensitive technique for measuring the concentration of various gaseous species such as methane, water vapor, and more. This technique leverages the properties of tunable diode lasers (TDLs) and laser absorption spectrometry to achieve extremely low detection limits, often down to parts per billion (nL/L). Besides concentration measurement, TDLAS can also provide data on temperature, pressure, velocity, and mass flux of the observed gas, making it a versatile tool in gas analysis.[1, 2]

A typical TDLAS setup comprises the following components: tunable Diode Laser Light Source, transmitting Optics, absorbing Medium, receiving Optics, and detectors. The tunable Diode Laser Light Source can emit light at specific wavelengths that correspond to the absorption lines of the target gas species. Examples of such lasers include Vertical Cavity Surface Emitting Lasers (VCSELs) and Distributed Feedback Lasers (DFBs). The transmitting optics shape and direct the laser beam into the absorbing medium. This may include lenses and mirrors to focus and steer the laser beam accurately. The absorbing Medium is a gas sample under investigation, contained in a cell that allows the laser beam to pass through it. The receiving Optics collect the laser light that has passed through the absorbing medium and direct it to the detectors. This setup can include additional lenses and mirrors to optimize light collection. The detectors Typically are photo-diodes, these measure the intensity of the transmitted laser light. The detected signal intensity decreases when the laser light is absorbed by the gas, and this change is used for analysis.

5. 2 Basic Principles

TDLAS can be used to measure the concentration of certain gases. When the inlet light which has a particular wavelength passing through a gas mass, the energy of light was absorbed by the target gas, the light intensity variation is proportional to the amount of gas in the light path. In this way, the target gas concentration was measured. For example, the amount of water vapor in an air sample can be measured as shown in Figure 5.1.[1] A beam of light passes through the sample and the amount of light loss (absorbed by the target gas) is measured by the detector. The target gas does not absorb all light, but rather a certain amount of light of a specific wavelength. If the wavelength of the light is adjusted, the amount of light (power) observed on the output will change accordingly.[3]

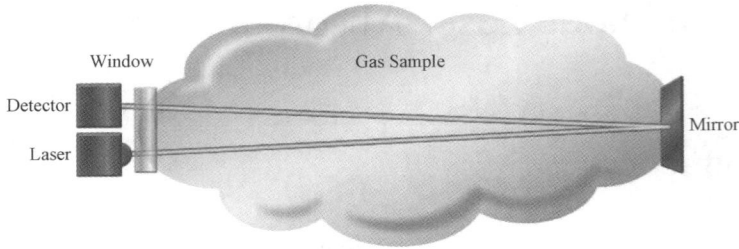

Figure 5.1   Working principle of tunable diode laser absorption spectroscopy

Figure 5.2 is an absorption spectrum diagram for water vapor with the wavelength on the X axis, and the wavelength range is between 1.3 and 2.8μm.[1] The transmittance is displayed on the Y axis. Transmittance is similar to absorbance except it shows how much light gets through the gas (the amount of light detected) rather than the amount absorbed. A value of 1.0 on the Y axis means all of the light is transmitted through and with zero absorption at that wavelength. There are three regions on the water spectrum graph where light is absorbed. A traditional absorption spectroscopy technique would be used to measure the absorption of one of these regions. This is a good technique as long as there are no other gases in the sample that absorb at the same wavelength. Figure 5.3 shows methane transmittance in the same wavelength range for measuring moisture in methane, [1] traditional techniques do not work because methane has a much stronger absorbance in the same regions on the graph and completely drowns out the measurement, this is so called interference between target gas the background gas.

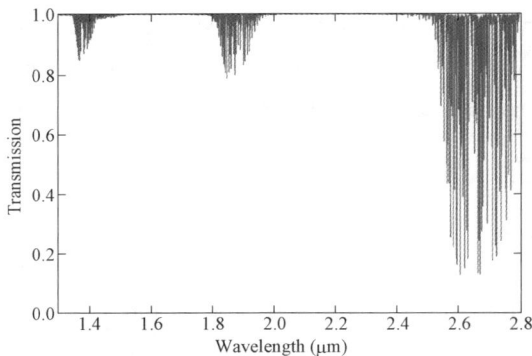

Figure 5.2   A diagram for water vapor (H_2O) transmission between wavelength 1.3 and 2.8 μm

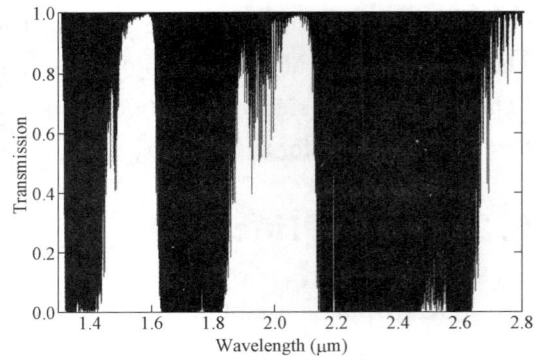

Figure 5.3   A diagram for methane (CH_4) gas transmission between wavelength 1.3 μm and 2.8 μm

Figure 5.4 shows the "zoom in" spectrum around the 1.9 micron wavelength.[1] At this location along the X axis, the peaks do not interfere very much (or not at all) between pure methane and water vapor. The individual peaks are about 0.3 nanometers wide and the moisture peak can be measured and discriminated from the methane. As long as these isolated or semi-isolated peaks can be found and the spectrometer used to measure the peaks has an adequate resolution (low signal-to-noise), these features can be analyzed to measure target gas concentrations down to part-per million (μL/L) or part-per-billion (nL/L) levels.

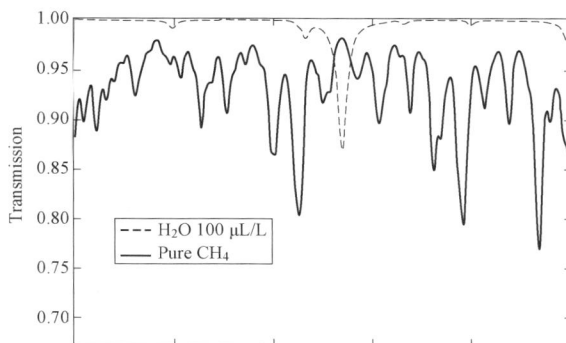

Figure 5.4   An enlarged diagram for identify the methane (CH₄) gas and water vapor around 1.9 μm wavelength

5.2.1 Laser Spectroscopy Theory

Using a prism, white light from the sun or a light bulb can be split into all the visible colors as well as electromagnetic radiation that is invisible to the eye, see Figure 5.5.[4] It is easy to see that Figure 5.2 and Figure 5.3 show only a tiny, narrow wavelength span within the range presented in Figure 5.5.

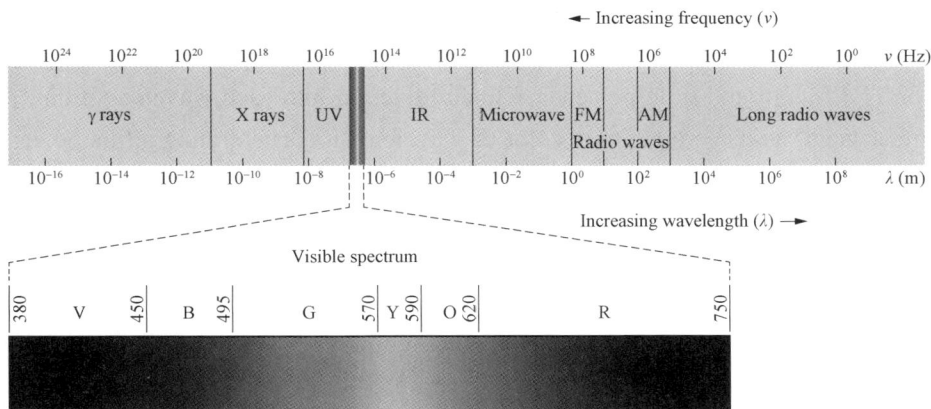

Figure 5.5   The electromagnetic spectrum where wavelength is given in nm and frequency is given in Hz

In this section, we first introduce the principles of molecular absorption spectroscopy. When a beam of light is transmitted through a gas sample, the light energy decreases if the wavelength of light matches the absorption wavelength of a gas sample. This energy decrease occurs because the energy is used to increase the internal vibrational and rotational energy levels of the molecules. For example, molecules such as oxygen (O_2) or water vapor (H_2O) absorb light only at specific wavelengths in the infrared or ultraviolet regions. These absorbing wavelengths are characteristic of the molecule and are referred to as its spectrum. When the energy of an incoming photon matches one of the internal energy levels of a molecule, the photon is absorbed, and the molecule transitions to a higher internal energy level. In the infrared region, the spectrum results from the excitation of vibrational and rotational energy levels in the molecule. In the visible and ultraviolet regions, the spectrum occurs when electrons orbiting the molecule change state, as illustrated in Figure 5.6.[2, 5]

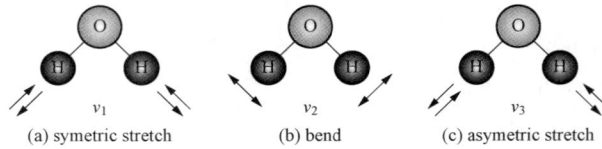

(a) symetric stretch   (b) bend   (c) asymetric stretch

Figure 5.6   Example of vibrational modes in water

Due to various unique properties, such as mass distribution and atomic bond strength, the absorption spectra in the infrared spectral region have a very distinct shape characteristic of each specific molecule. This region is commonly called the fingerprint spectral region. This distinctness can be used to identify the molecules in an unknown gas or to quantify the concentration of molecules in a gas mixture.

Generally speaking, the light source must match the absorption features of the gas to acquire any meaningful information about it. While a light bulb followed by a wavelength-selective filter, prism, or grating can theoretically be used to measure gas concentrations, it usually does not work very well. Splitting apart the different wavelengths with high enough resolution is difficult, and little power is actually diverted into each wavelength bin. A laser, on the other hand, emits a single pure color or wavelength, concentrating all its power at this single wavelength.

Laser spectroscopy is an extremely effective tool for the detection and quantification of molecular trace gases. The demonstrated detection sensitivity ranges from parts per million by volume (μL/L) to parts per billion by volume (nL/L) and even sub-nL/L levels, depending on the specific gas species and the detection method employed. The laser's narrow line width allows it to tune over a single line in the gas spectrum. Therefore, it is necessary to understand molecular lines, their shapes, and how they are influenced by pressure and temperature.

Tunable diode laser spectroscopy (TDLS) is a highly selective, sensitive, and versatile technique for measuring trace species. The diode laser source is ideal for optical spectroscopy due to its narrow line width, tunability, stability, compactness, and ability to operate at room temperature. In addition to measuring concentration, TDLS can determine temperature, pressure, velocity, and mass flux of the gas under observation by analyzing the detailed properties of absorption lines. TDLS has rapidly become the most commonly used laser-based technique for quantitative measurements of species in the gas phase. Today, it finds wide application in environmental monitoring, remote sensing, and process gas analytics.

The basic setup for tunable diode laser spectroscopy (TDLS) is straightforward, as depicted in Figure 5.7 [2]. Essentially, you require a tunable laser diode emitting at the appropriate wavelength, a gas cell, and a detector. The entire output power of the laser is focused into a very narrow region. The laser's wavelength can be finely adjusted by varying its temperature and/or current, allowing it to scan across the absorption line of interest. During measurement, the laser light passes through the gas sample, and the transmitted power through the sample is recorded as a function of the laser wavelength. A sharp absorption signal is observed when the laser emission wavelength aligns with a resonant absorption of the molecule.

5.2.2 Molecular Absorption Spectroscopy

The spectrum of a molecule comprises numerous spectral lines, often organized into bands spread across various wavelengths, where each line represents a specific proportion of absorbed light. For diatomic molecules, the energy states can be depicted in terms of potential energy relative to the distance between nuclei, as illustrated in Figure 5.8 . [2]

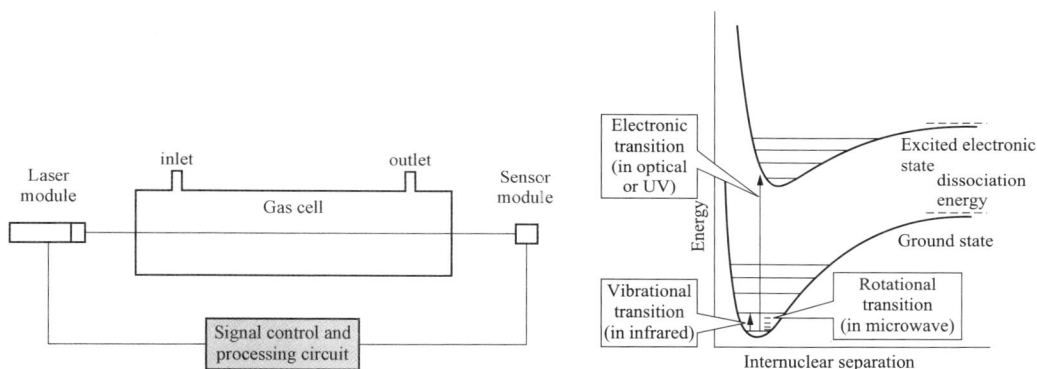

Figure 5.7   Basic setup for laser spectroscopy

Figure 5.8   Energy levels in a diatomic molecule

Electron transitions occur rapidly, with minimal change in internuclear distance. Infrared spectroscopy involves transitions between different vibrational energy levels within the same electronic state. Microwave radiation, on the other hand, induces rotational transitions between different rotational levels within the same vibrational state. Infrared spectra typically exhibit combined vibration-rotation transitions. For a molecule to absorb photons, it must possess a dipole moment, resulting from an uneven distribution of positive and negative charges among its atoms. This property enables the molecule to function as an antenna. This explains why nitrogen, lacking such uneven charge distribution, does not absorb infrared radiation, whereas molecules like hydrogen fluoride exhibit strong infrared absorption.

Generally, the vibrational transitions define the wavelength range of the spectrum, while rotational transitions contribute to its fine structure. Under high pressure or in large molecules, rotational lines may become less distinct. The diversity of states observed depends on the gas temperature, influencing the intensity of certain absorption lines.

5.2.3 Beer-Lamberts Law

The amount of light absorbed by a gas depends on the product of the number of molecules present, the absorption cross-section of the molecules at the specific wavelength, and the optical path length through the gas. This formula encapsulates how absorption spectroscopy quantifies the absorption of light by gases, where each factor contributes to the overall absorption intensity observed. Therefore, the gas concentration can be expressed as: [2]

$$\text{Concentration} = \frac{\text{How much light is absorbed by the gas molecules}}{[\text{Absorption coefficient for the gas}] \times [\text{Path Length}]}$$

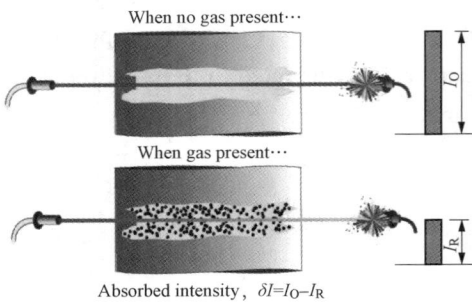

Figure 5.9   The power will be absorbed by a characteristic wavelength in the traveling path of the laser

It is necessary to distinguish the absorption of light by molecules from the absorption of light from other factors such as dust and dirt in the measurement path. To determine the gas concentration, we measure the amount of power absorbed at a characteristic wavelength, see Figure 5.9, and divide it by the cross-section and by the path length.

The absorption is given by Beer-Lambert's law, see Figure 5.10, which states that the fraction of photons absorbed over a path δL is constant. This constant is called the absorption coefficient.[6]

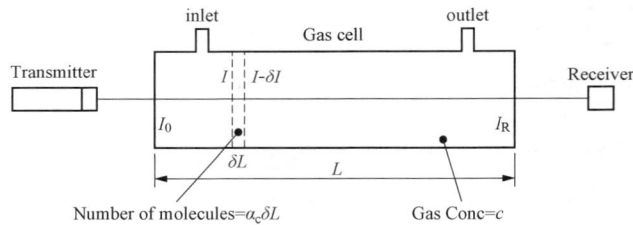

Figure 5.10   Beer-Lambert's law

$$\frac{\delta I}{I} = -\alpha(v)c\delta L \tag{5.1}$$

The absorption coefficient $\alpha(v)$ varies rapidly with the optical frequency v for gaseous species as the frequency of the light passes a resonance (an absorption line) in the molecule.

$$I_R = I_0 e^{-\alpha(v)cl} \tag{5.2}$$

For small values of the absorption $\alpha(v)L$ this is approximated with:

$$I_R = BG - I_0 Tc\alpha(v)L \tag{5.3}$$

Here, we distinguish between a background signal (BG) and a small absorption signal.

The BG represents the direct absorption signal and typically constitutes more than 99.9% of the total signal. Laser light absorption can originate from two main sources. Dust, dirt, and other contaminants exhibit broad spectral absorption characteristics, contributing a constant absorption profile across the laser's wavelength scan.

5.2.4 Spectroscopic Databases

The most widely used spectroscopic database is HITRAN http://www.cfa.harvard.edu/hitran/.[2] This extensive database, known as the HIgh-resolution TRANsmission molecular absorption database, was initially developed by the US Air Force in 1961 and later made public. The first edition, known as the McClatchey tape, was released on magnetic tape in 1973 and freely distributed. Today, the database is available for download from the HITRAN website, with the current edition being HITRAN 2008. Additionally, there is a free software package called HAWKS (HITRAN Atmospheric Workstation) that can manipulate, filter, and plot the line-by-line data and absorption cross-sections on Windows, UNIX, and MAC operating systems.

Another valuable resource is the vapor phase spectral library from Pacific Northwest National Laboratory (PNNL), which includes many organic compounds. The data can be accessed through a website hosted by NIST in the USA: http://webbook.nist.gov/chemistry/.

5.2.5 Unique Advantages of TDLAS

Trace-gas detection and monitoring could be performed by several techniques other than laser spectroscopy These techniques include mass spectrometry (MS), gas chromatography (GC) and Chemiluminescence detection.[1, 7, 8]

Mass spectrometry (MS) can provide very high sensitivity, its selectivity is often limited due to mass overlap between certain molecules, necessitating the pre-preparation of the molecular species to be detected. However, it should be noted that mass spectrometry is advantageous over laser spectroscopy for the detection of very large molecules.

Gas chromatography (GC) is a technique in which the molecules under test are transported through a column with a wall coated with a liquid or polymeric substance. The interaction between the types of gas molecules and the column wall causes different types in the sample gas mixture to leave the column at different times. This separation occurs due to the different absorption coefficients of different molecules. Thus, the molecular species can be qualitatively identified by the time they pass through the column. Quantitative detection is then performed, for example, with a flame photometric detector or mass spectrometer. Because of the principle of gas chromatography, only different kinds of continuous measurements can be made. In addition, the time required for measurement may be relatively high. Depending on the length of the column, the measurement time is usually 5 to 10 minutes or more. For near-site process measurements, this duration is often too long to detect rapid fluctuations in chemical processes.

Chemiluminescence depends on the chemical reaction of the molecule species of interest with the surface of the polymerized dye. The reaction excites the dye molecules, and when the excited molecules return to their high-energy ground state, a portion of this excitation energy is emitted in the form of photons (luminescence). Chemiluminescence has been used for in situ measurements of ozone and nitrogen oxides, and it offers the advantage of relatively fast measurements (about 10 Hz). The technique is highly selective, but it is only used to detect a small number of gas types. The chemical reactions that occur during the measurement are irreversible and require regular dye changes. Chemiluminescence is not an absolute measurement technique and requires regular calibration.

Tunable diode laser analyzers are among the most powerful process analyzers for fast, accurate measurements of gases directly in harsh environments, typically under high temperatures or pressures and varying transmission conditions. They generally exhibit minimal or no crossinterference from other gases, and their response time is very short. TDLAS analyzers continue to evolve, maintaining their technological edge in response to increasingly demanding process and emission control standards.

TDLAS offers a highly sensitive, highly selective, and fast time response trace gas detection technique for in-site trace-gas detection applications. With the use of the semiconductor laser diode laser and by precisely tuning the laser output wavelength to a set of single isolated absorption lines of the gas, the TDLAS technique offers gas concentration measurement with very high sensitivity and selectivity. The TDLAS analyzer requires very little maintenance and generally has very long calibration intervals. The detailed unique advantages of TDLAS could be explained as follows: (1) High Sensitivity: TDLAS is capable of detecting trace amounts of gases down to parts per million by volume (μL/L), parts per billion by volume (nL/L), and even sub-nL/L levels, depending on the specific gas species and detection method employed. (2) High Selectivity: TDLAS can distinguish between different molecular species with high precision due to its narrow linewidth, making it highly selective for specific gas molecules. (3) Fast Response Time: The technique allows for real-time monitoring and rapid measurements, making it suitable for dynamic processes and transient events. (4) Non-Intrusive: TDLAS measurements do not require direct contact with the gas sample, allowing for non-intrusive and remote sensing applications. (5) Quantitative Analysis: It provides accurate quantitative measurements of gas concentrations, as well as additional properties such as temperature, pressure, velocity, and mass flux of the gas under observation. (6) Minimal Interference: Due to its narrow spectral output, TDLAS minimizes interference from other gases and background noise, enhancing measurement accuracy. (7) Compact and Portable: The diode laser sources used in TDLAS are compact, lightweight, and can operate at room temperature, making the system portable and easy to deploy in various settings. (8) Versatile Applications: TDLAS can be used in a wide range of applications, including environmental monitoring, industrial process control, medical diagnostics, and scientific research. (9) Robustness: The technique is stable and reliable over long periods, with low maintenance requirements, making it suitable for continuous monitoring. (10) Compatibility with Harsh Environments: TDLAS systems can operate effectively in harsh environmental conditions, such as high temperatures, high pressures, and chemically reactive environments.

Above unique advantages make TDLAS a powerful and versatile tool for the detection and quantification of molecular trace gases in various applications. The improved measurement quality and low cost-of-ownership obtained by going to an in situ technology have led to a trend where TDLAS is replacing other technologies, including GCs, in an increasing number of applications.

5.3 Main Components of TDLAS and Relevant Techniques

5.3.1 Laser Sources of TDLAS

The TDLAS (Tunable Diode Laser Absorption Spectroscopy) sensor utilizes a laser diode as its light source, enabling the instrument to target very specific wavelengths with high reliability. As shown in Figure 5.11(a), the X-axis represents the wavelength and the Y-axis represents the detected power. The laser diode emits a spectrum that is approximately 0.03 nanometers wide, an order of magnitude smaller than any spectral feature being measured.[1, 9, 10]

Laser diodes can be "tuned," meaning the central wavelength of the spectral band can be adjusted by changing the current of the laser, as depicted in Figure 5.11(b). The laser in TDLAS is tuned across a wavelength range while it passes through the sample gas, measuring absorption within that range (the scanning wavelength range is about 1.5 microns or 15,00 nanometers). Since this scan range is much wider than the spectral feature, it is assumed that the center of the peak can be found within the scan. The stability of the diode laser ensures the system can automatically compensate for slight drifts in the X-axis over long periods, even years.

Important properties of Diode lasers include wavelength, power, coherence, cost, and operating temperature. For gas sensing applications, the most crucial property is spectral coherence; the laser must operate at a single emission wavelength. However, new types of single-mode lasers are under development. Efforts to improve mid-infrared diode lasers are of special interest to the molecular gas sensing co-mmunity because of the high sensitivity achievable in this region. These efforts focus on achieving room temperature operation and single-frequency output through novel device structures.

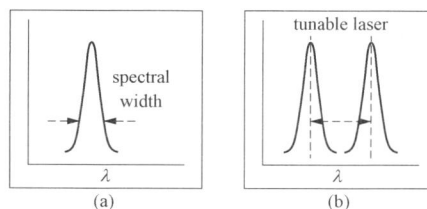

Figure 5.11   The center wavelength of the spectra band can be adjusted by changing the current to the laser

Key properties of diode lasers include wavelength, power, coherence, cost, and operating temperature. For gas sensing applications, the most crucial property is spectral coherence; the laser must operate at a single emission wavelength. However, new types of single-mode lasers are under development. Efforts to improve mid-infrared diode lasers are particularly important to the molecular gas sensing community due to the high sensitivity achievable in this region. These efforts focus on achieving room temperature operation and single-frequency output through novel device structures. The mid-infrared region is especially significant for molecular gas sensing because many molecular gases have strong absorption features in this wavelength range, allowing for highly sensitive detection and accurate quantification of trace gases. Achieving stable, highperformance mid-infrared diode lasers that can operate efficiently at room temperature without the need for complex cooling systems is crucial for the advancement of practical and reliable gas sensing technologies.

A distinct feature of the gas lasers listed in Table 5.1 [3] is their ability to yield high-pulse energies and, in some cases, very high average powers.

Table 5.1 **Spectral Emission Characteristics of Discretely Tunable High-Power Pulsed Lasers[3]**

Laser	Wavelength (nm)	Laser	Wavelength (nm)
ArF	193	Sr	430.5
KrF	248	Cd	533.7
XeCl	308	Cu	510.5
	308.2		578.2
XeF	351	Au	627.8
	353	Nd: YAG	1,064
	466~514	CO$_2$	10,532.09
N$_2$	337.1		10,551.40
HgBr	502		10,571.05
	504		10,591.04
Ca	373.7		

5.3.2 Laser Systems Signal Processing Techniques

In a typical diode laser spectrometer, the emission wavelength of the tunable diode laser is adjusted to cover the characteristic absorption lines of a species in the gas along the path of the laser beam. This tuning causes a reduction in the measured signal intensity, which is detected by a photodiode and used to determine the gas concentration and other properties, such as gas temperature. The main disadvantage of absorption spectrometry (AS) and laser absorption spectrometry (LAS) is that they rely on detecting a small change in signal superimposed on a large background. Any noise introduced by the light source or the optical system can significantly deteriorate the technique's detectability. As a result, the sensitivity of direct absorption techniques is therefore often limited to an absorbance of 10^{-3} - 10^{-4}, which is far from the theoretical limit that would be in the 10^{-7} range. This level of sensitivity is insufficient for many applications, which is why direct absorption spectroscopy (DAS) is typically not used in its simplest mode of operation.[2]

There are two technical methods to address the aforementioned problems. The first is to reduce the noise in the signal, and the second is to increase the absorption. The former can be achieved by using more advanced signal processing techniques, whereas the latter can be obtained by placing the gas inside a cavity where the light passes through the sample multiple times, thereby increasing the interaction length. It is also possible to enhance the signal by performing detection at wavelengths where the transitions have larger line strengths. This is achieved by using fundamental vibrational bands or electronic transitions instead of overtones.

If you were to graph the data with the current on the X-axis and the detected energy on the Y-axis, you would observe a positively sloped straight line [see Figure 5.12(a)]. By design, this straight line contains a small dip where the target gas absorbs light at the wavelength corresponding to that laser current. The center of the dip represents the wavelength of interest.[1] This dip indicates the specific wavelength at which the gas absorbs the laser light, and its depth can be used to determine the concentration of the gas. The position of the dip on the current axis allows for precise identification of the absorption feature, which is crucial for accurate gas sensing. This characteristic dip is key to the functionality of the TDLAS sensor, enabling it to detect and quantify trace gases with high sensitivity.

Since the scan range is much wider than the spectral feature, it is assured that the center of that peak will be found in the scan. This, combined with the inherent stability of diode lasers, explains how the system can automatically compensate for slight drifts in the X-axis over long periods (years). In TDLAS, the diode laser is tuned over a range of wavelengths, including the characteristic absorption lines of the target gas. As the laser wavelength sweeps through the absorption line, the target gas absorbs some of the laser light, leading to a reduc-tion in the detected signal intensity at specific wavelengths. This reduc-tion appears as a dip in the detected energy vs. laser current graph.

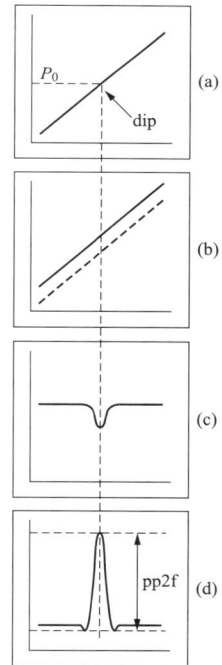

Figure 5.12 Working principle of the second harmonic phase sensitive detection method

To perform the absorption measurement, the system follows these steps: (1) Wavelength Tuning: The laser diode's current is varied to tune the emission wavelength across a range that includes the absorption feature of the target gas. (2) Signal Detection: A photodiode measures the intensity of the transmitted laser light.

The signal is recorded as the laser scans through the wavelength range. (3) Data Analysis: The recorded signal is analyzed to identify the dip in intensity, which corresponds to the absorption feature of the target gas. The depth and width of the dip provide information about the gas concentration and, potentially, other properties such as temperature. (4) Drift Compensation: Over long periods, slight drifts in the laser's emission wavelength may occur due to environmental changes or aging components. However, because the scan range is much wider than the absorption feature, the system can still detect the center of the absorption peak, ensuring accurate measurements despite these drifts. Above method allows for highly sensitive and stable detection of trace gases, making TDLAS a powerful tool for various gas sensing applications.

Compensating for possible contamination of the reflecting mirror must be considered. The amplitude of the dip (or peak) in the detected signal is a direct result of the absorbance but is not used alone to determine absorbance. Consider the effect on the curve in Figure 5.12(b) if the mirror becomes partially obstructed over time, assuming all other factors remain constant. The amplitude of the dip would change slightly, and the entire curve would shift downward in the same ratio [see Figure 5.12(b) - blue line]. Thus, changes in the target gas concentration alter the amplitude of the dip, and overall reductions in light (attenuation) change both the ampli- tude of the dip and shift the entire curve up or down.

Using a ratio of the absorbed light to the total light detected compensates for and elimi- nates the effect of a partially blocked mirror. In other words, if only 10% or 50% of the light passes through, the ratio remains unchanged, and the measurement is unaffected. However, if the mirror becomes very dirty, insufficient light will reach the detector, preventing power measurement.

Since the dip in the line is small relative to the overall curve, a direct measurement of the dip would result in a high signal-to-noise ratio. To achieve a more sensitive measurement and filter out noise, a method called "second harmonic phase-sensitive detection" is used. This electronic technique, effectively filters noise and conditions the curve. Starting with the dip in Figure 5.12(a), and correcting for the slope while greatly increasing the scale, the result is similar to the peak in Figure 5.12 (c). The second derivative of this curve results in a curve like that in Figure 5.12 (d). Using this "second harmonic" curve (the "2f" curve), the peak-to-peak (pp2f) value is measured. If you divide pp2f by the power (pp2f/P), this value is propor- tional to $(I_0 - I_R)/I_0$ which is proportional to the concentration.[1]

5.4 Cavity Enhancement Techniques

The lowest detectable limit often reflects the technique level of the instrument. The sensi- tivity of a laser spectrometer is closely related to the signal-to-noise ratio (SNR). The signal is determined by the gas concentration and the absorption path length. While increasing the gas concentration to achieve a higher signal level is not always feasible, especially when measuring very low concentrations are main tasks. An alternative way, the absorption path length can be extended using multipass absorption cells. These cells enhance the signal, as the absorption signal is directly proportional to the interaction length between the laser radiation and the trace gas. Multipass absorption cells increase the effective path length by reflecting the laser beam multiple times through the gas sample. This increased path length results in a stronger absorption signal, improving the spectrometer's sensitivity. There are different types of external cavities used for this purpose: Herriott and White cells, of non-resonant type (off-axis alignment), or of the resonant type, most often working as a Fabry-Pérot (FP) etalon. By employing these methods, particularly the use of multipass absorption cells, which typically can provide an enhanced interaction length of up to 2 orders of magnitude, the sensitivity of laser spectrometers can be significantly improved, allowing for the detection of very low gas concentrations with high accuracy. [2]

5.4.1 Multipass Spectroscopic Absorption Cells

Two conventional multipass cells are called the Herriott cell and White cell.

The Herriott cell first appeared in 1965, when Herriott invented a design that comprises only two concave spherical mirrors with an equal radius of curvature. The mirrors are separated by a distance close to their radius of curvature. The front mirror contains a coupling hole through which the laser beam is coupled into the cell, as depicted in Figure 5.13 (https://www. thorlabs. com/newgrouppage9.cfm?objectgroup_id=8169). If the laser beam is directed onto a circle on the rear mirror, the spots on the mirrors form a circular spot pattern. After a number of passes through the cell, the laser beam exits the cell through the same coupling hole but at an inverted angle. This reentrant condition depends on the mirror radius of curvature and the mirror's separation and acts to stabilize the optical path. Absorption path lengths of up to 50 meters are feasible for a cell with a base length of 0.5 meters and mirrors with a 10 cm diameter. The design was later improved using astigmatic mirrors, which enabled path lengths up to 250 meters for a cell with a base length of 1 meter.

The key elements of this design are: (1) Concave Spherical Mirrors: These mirrors have an equal radius of curvature and are crucial for the multiple reflections of the laser beam within the cell. (2) Coupling Hole: Located in the front mirror, it allows the laser beam to enter and exit the cell. (3) Circular Spot Pattern: The directed laser beam creates this pattern on the rear mirror, ensuring multiple passes through the gas sample. Reentrant Condition: Stabilizes the optical path, ensuring that the beam exits through the same hole it entered after multiple reflections.

These features significantly increase the effective absorption path length, enhancing the sensitivity of the gas detection system by allowing for more interactions between the laser light and the gas molecules. The improved design with astigmatic mirrors further extends this path length, making it suitable for detecting trace gases at very low concentrations.

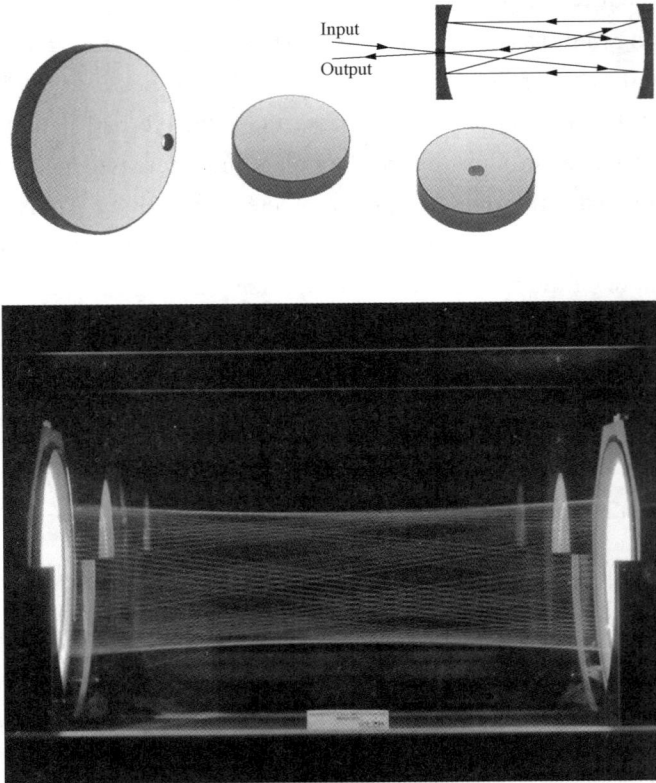

Figure 5.13　Schematic view of a multipass spectroscopic absorption cell and

a typical prototype demonstration

In 1943, John U. White first described his significant improvement over previous long path spectroscopic measurement techniques in his paper "Long Optical Paths of Large Aperture". A White cell is made up of three spherical concave mirrors with the same radius of curvature. The distance between the mirrors is equal to their radius of curvature.

In a White cell, the laser beam is reflected or passed between the mirrors several times, [9] as shown in Figure 5.14.[11] The animation on the website (https://en.wikipedia.org/wiki/Mult-ipass_spectroscopic_absorption_cells) shows a White cell in which the beam is reflected eight times. The number of passes can be easily adjusted by making a slight rotational change to mirrors M2 or M3. However, the total number of traversals must always be a multiple of four. When adding or removing passes, the entering and exiting beams do not change position, allowing the total number of passes to increase many times without changing the volume of the cell. This design allows for a larger total optical path length compared to the volume of the sample under test. The spots of different channels can overlap on mirrors M2 and M3, but must be distinct on mirror M1. If the input beam is focused on the M1 plane, then each round trip will also be focused on this plane. The tighter the focus, the more points on

M1 that do not overlap, and the greater the maximum path length.

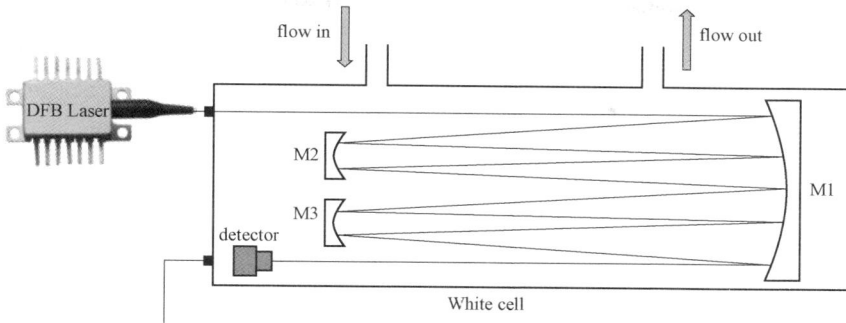

Figure 5.14   TDLAS experimental system by using white cell

At present, the White cell is still the most commonly used multipass cell, and it has many advantages: (1) Easily Controlled Number of Passes: The number of reflected channels can be simply adjusted, allowing flexibility in path length. (2) High Numerical Aperture: The design supports high numerical aperture, enhancing the amount of light collected. (3) Reasonable Stability: While not as stable as Herriott cells, White cells provide good stability in most applications. (4) Variable Path Length: The path length of White cells ranges from less than one meter to hundreds of meters, making them suitable for a wide range of applications.

Above features make the White cell a versatile and effective tool for improving the detection of trace gases by increasing the effective absorption path length and thus the sensitivity of the laser spectrometer.

5.4.2 Cavity Ring-down Spectroscopy (CRDS)

Cavity ring-down spectroscopy (CRDS)[12] is also called as cavity ring-down laser absorption spectroscopy (CRLAS). It is based on a measurement of the difference in the rate at which light intensity leaks out of a stable optical cavity with and without the absorbing gas present. Its application to molecular absorption measurements was first described by O'Keefe and Deacon at Los Gatos research in 1988. In most applications, the cavity is formed from a pair of high-quality plano-concave mirrors, see Figure 5.15, [2] often having reflectivities greater

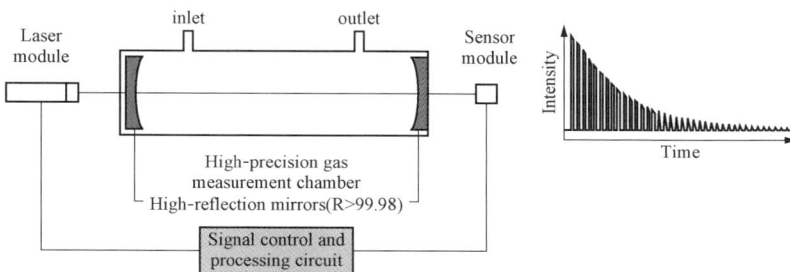

Figure 5.15   Cavity ring-down spectroscopy (CRDS)

than 0.999,9. It has been widely used to study gaseous samples that absorb light at specific wavelengths, allowing determination of mole fractions down to the parts-per-trillion level.

A typical CRDS setup consists of a laser used to illuminate a high-finesse optical cavity, and two highly reflective mirrors. A pulse of laser light is directed onto the back face of one of the mirrors, and the small amount of light coupled through the mirror into the cavity is repeatedly reflected backward and forwards between the two mirrors. The intensity of the trapped pulse will decrease by a fixed percentage during each round trip within the cell due to both absorption and scattering by the medium within the cell and reflectivity losses. The result is that the amount of light within the cavity decays exponentially with time.

When the laser is in resonance with a cavity mode, intensity builds up in the cavity due to constructive interference. The laser is then turned off to allow measurement of the exponentially decaying light intensity leaking from the cavity. During this decay, light is reflected back and forth thousands of times between the mirrors, giving an effective path length for the extinction on the order of a few kilometers. The exponential decay is recorded by positioning a sensitive photo detector behind the second mirror to detect the tiny amount of light leaking out of the cavity through the mirror on each pass.

If a light-absorbing material is placed in the cavity, the mean lifetime of the light decreases as fewer bounces through the medium are required before the light is fully absorbed or absorbed to some fraction of its initial intensity. A CRDS setup measures the time it takes for the light to decay to 1/e of its initial intensity, known as the "ringdown time." This ringdown time can be used to calculate the concentration of the absorbing substance in the gas mixture within the cavity.

Because the measured decay constant is independent of the absolute intensity of light injected into the cavity, CRDS has a distinct advantage over single-pass techniques that it is largely insensitive to noise caused by shot-to-shot fluctuations in the intensity of the pulsed laser source. The great utility of the CRDS method lies as much in the extremely high sensitivity as in the simplicity of the technique. Absolute concentrations are easily inferred from the absorption data that CRDS provides. CRDS concentration detection limits for many species have been demonstrated to be in the part-per-billion to part-per-trillion range.

5.5 Electric Power Industries Applications of TDLAS

5.5.1 Introduction

Some of the most common applications for TDLAS in industrial process analytics are presented in this subsection. For example, it is widely used for emission monitoring and combustion control in thermal power plants; In the iron and steel industry, TDLAS has proven valuable for various monitoring and control tasks; More recently, TDLAS analyzers have been introduced for different applications in the chemical industry. Nowadays, TDLAS has also found significant applications in the electric power industry for dissolved gas analysis of power transformers, SF_6 decomposition by-products analysis, and leak detection of gas-insulated switchgears (GIS). All above applications highlight the versatility and effectiveness of TDLAS in enhancing operational efficiency, safety, and environmental compliance across multiple industries.

5.5.2 TDLAS Applications in Dissolved Gas Analysis

The TDLAS technique can measure gas concentration with high resolution in a target spectral region by using a tunable sweeping diode laser source to scan across the specific absorption lines of the target gas. It is a direct absorption spectroscopy method based on the absorption characteristics at those wavelengths. Since the absorption lines of different gases are located at specific spectral bands, the characteristic absorption fingerprint of a particular species of interest can be identified. According to the Lambert-Beer absorption law (see section 5.2.3 of this chapter), the variation in the incident/emitted intensity is related to the path length of the laser beam and the concentration of the species at a specific wavelength, neglecting the scattering and reflection processes. The high resolution and specificity of TDLAS make it a powerful tool for detecting and quantifying trace gases in various industrial applications.

To measure the concentration directly is not feasible because the light intensity signal received by the photodetector (PD) is mixed with a large amount of background noise. The laser beam intensity is proportional to the drive current. To achieve harmonic modulation of the transmitted signal, the injection current of the laser is modulated using a cosine function.[13] By applying this modulation technique, the signal can be enhanced, and the noise can be filtered out, making it possible to accurately measure the concentration of the target gas. This approach allows for improved sensitivity and specificity in detecting and quantifying trace gases using TDLAS.

To establish the relationship between the gas concentration and the harmonic signal, the amplitude of the higher-order harmonic components, the second harmonic signal (2f) is preferred for detection and calculation. The signal processing is illustrated in Figure 5.16.[14] When there is no gas of interest absorbed in the light path, the photodetector (PD) output is proportional to the modulated intensity of the tunable laser source, as shown in Figure 5.16(a) and (b). Once the gas of interest is present, the absorption region appears during the PD detection cycle, and the absorption signal can be obtained through a lock-in amplifier, as shown in Figure 5.16(c) and (d). Since the output intensity of the laser varies with the injected modulation current, the 2f harmonic signal in Figure 5.16(e) is actually asymmetric around the central position.

For realization of higher sensitivity than by using Direct Absorption (DA) spectroscopy, the modulation technique can be included. It provides two main advantages: firstly, it measures a difference signal which is directly proportional to the species concentration and, secondly, it allows to shift of a measured signal to a higher frequency region, thereby offering a larger signal-to-noise ratio and thus higher sensitivity, see Figure 5.17. This figure describes the similar meaning of Wavelength Modulation Spectroscopy (WMS), and it was cited from the website of Combustion Energy Frontier Research Center of Princeton University (https:// cefrc.princeton.edu/sites/g/files/toruqf1071/files/Files/2013%20Lecture%20Notes/Hanson/ pLecture9.pdf).

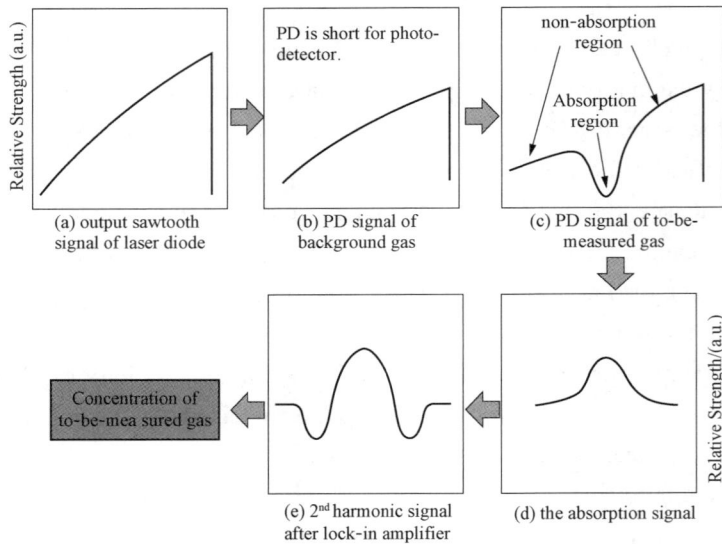

Figure 5.16　Illustration of signal processing through TDLAS, the horizontal axis of (a)~(d) is the wavelength (nm) and the horizontal axis of (e) is time (ms)

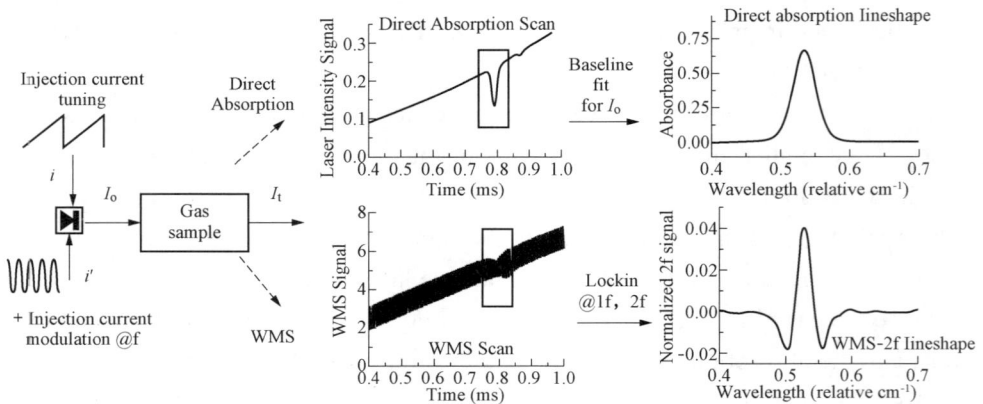

Figure 5.17　Direct Absorption (DA) & Wavelength Modulation Spectroscopy (WMS)

In practice, this technique is divided into two approaches. Wavelength Modulation Spect-roscopy, [15] is devoted to the case where the frequency of modulation is much lower than the absorption line width with detection at the modulation frequency ($1f$) or the next harmonics ($2f$, etc.). This typically corresponds to modulation frequencies from a few kHz to a few MHz. It has been applied since the early 1970s, with tunable diode lasers (TDL). On another side, Frequency Modulation Spectroscopy (FMS) applies to modulation frequencies that are comparable to or greater than the spectral feature of interest.

When the laser is modulated around its center frequency ωL at a frequency ωm with modulation amplitude $d\omega$, the instantaneous frequency is $\omega = \omega L + d\omega\cos(\omega mt)$. The intensity of the radiation transmitted through the absorption cell can then be expressed as a Fourier series expansion.

Due to the ease of the current modulation, WMS mostly was used with diode lasers. In the near-IR, it can be also performed by using commercial external modulators. However, due to the increasing need for trace gas detection, it is advantageous to work in the mid-IR region, where a lot of molecules have strong absorption bands.

Generally, a typical TDLAS hardware system is mainly composed of the light source and its drive unit, gas cell (absorption light path), photodetector (PD), control and data acquisition (DAQ) device, as shown in Figure 5.18.[14] The TDLAS system is relatively mature for detecting industrial gases in various applications. However, several challenges need to be considered for practical applications, including multiple gas measurements, cross-interference, high sensitivity, and vibration. Due to these complexities, it is not feasible to directly apply a similar device for online dissolved gas detection in a power transformer.

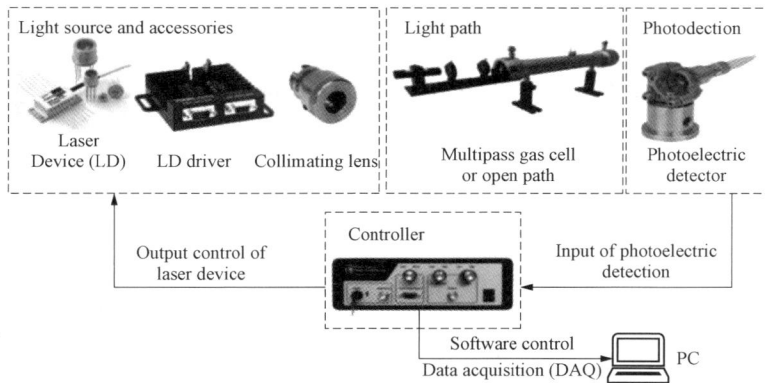

Figure 5.18   Main hardware components in a typical TDLAS system

In the TDLAS system, the laser serves as both a light source and a means of ensuring spectral subdivision. This requires the spectral width of the laser to be much narrower than the width of the absorption peak, as illustrated in Figure 5.19.[14] Rapid tuning, typically faster than 1 kHz, can be adapted to capture information across the entire absorption peak.

Transformer oil dissolved gas absorption bands are generally located in the mid-infrared (MIR) region. High detection sensitivity can be obtained using mid-infrared lasers. Near-infrared (NIR) lasers, while operating at room temperature and covering various absorption bands, generally have weaker gas absorption intensity. To increase detection sensitivity in the NIR region, long-path absorption cells and wavelength-modulation spectroscopy techniques can be utilized. Therefore, the near-infrared region, defined by the ASTM Working Group on NIR as 780 nm to 2,526 nm, is preferred for such applications. The transmission loss of the fiber is relatively low in NIR, especially optical telecom bands (O band: 1,260-1,360 nm; E band: 1,360-1,460 nm; S-band: 1,460-1,530 nm; C band: 1,530-1,565 nm; L band: 1,565-1,625 nm; U band: 1,625-1,675 nm), as shown in Figure 5.20.[14] The following central wavelengths: 1,653.72 nm for methane, 1,530.37 nm for ethyne, 1,620.04 nm for ethene, and 1,679.06 nm for ethane, are selected. A long optical path of 10.13 m by Herriott cell was achieved within 0.34 m mechanical length. The lowest detectable limits for four gases in this prototype are 0.5 μL/L (CH_4), 1.3 μL/L (C_2H_6), 1.6 μL/L (C_2H_4), and 0.06μL/L (C_2H_2) respectively. Figure 5.21 shows the online DGA prototype by TDLAS in the field of a 220 kV power transformer. [14]

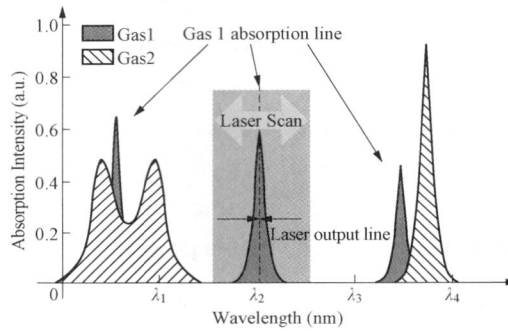

Figure 5.19   Schematic diagram of absorption line selection

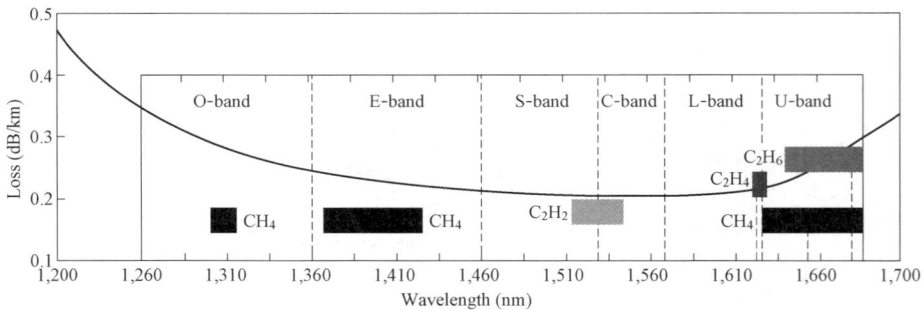

Figure 5.20   Absorption wavelength distribution of four hydrocarbon gases

5.5.3   CRDS Applications in GIS Faults Detection

The insulation defects in SF_6 gas-insulated switchgears, SF_6 gas-insulated transmission lines, and SF_6 gas-insulated power transformers can cause partial discharge or local overheat failures, leading to SF_6 decomposition. When the content of impurities such as trace oxygen or trace moisture is high, the iconic gases such as SO_2, HF, H_2S, CO, CF_4, SOF_2, SO_2F_2, SOF_4, COS, CS_4 and so on are generated. By detecting the content and variation trend of SF_6 decomposition byproducts, the cause of the local discharge and discharge quantity are judged, which is one of the research hot topics among domestic and overseas universities and institutes.[15, 16]

Figure 5.21   Developed online DGA equipment in the field (served for a 220 kV power transformer)

SF_6/N_2 mixture gas is not easy to liquefy at low temperatures and has higher breakdown strength under a homogeneous electric field. As a substitute gas for SF_6, long-distance gasinsulated transmission pipelines are widely used. However, the concentration of decomposition byproducts of SF_6 in the mixture gas is quite low, and the detection sensitivity of available instruments does not meet actual requirements. Currently, there have been numerous studies on the detection methods of SF_6 decomposition byproducts. However, these methods have limitations: (1) Semiconductor gas sensors: These have poor selectivity and are prone to poisoning. Additionally, their probes need to be replaced regularly due to a lifespan of one or two years. (2) Verification tubes: These require regular calibration and offer limited capabilities. Gas chromatography: This technique takes a long time from sampling to obtaining analysis results, making it unsuitable for online detection. Furthermore, the appropriate. (3) chromatographic column and carrier gas flow rate must be selected during measurement, and repeated measurements are necessary to ensure accuracy. (4) Gas chromatography-mass spectrometry (GC-MS): While GC-MS can detect a variety of gases with less sample consumption and higher sensitivity, the sampling level directly affects the detection results. Above limitations highlight the need for improved detection methods for SF_6 decomposition byproducts in SF_6/N_2 mixtures, especially for applications requiring high sensitivity and online detection capabilities.

Laser Cavity Ring Down Spectroscopy (CRDS) can be used for online monitoring of gas component detection with good selectivity, high precision (1μL /L), and fast response time (<5 s).

H₂S is a typical decomposition byproduct of SF_6. In the near-infrared band, H_2S has a rich absorption spectrum line, and its moderate absorption coefficient is in the order of 10^{-21}. This makes H_2S suitable for the detection of low-concentration gases. Additionally, in the medium infrared band, H_2S exhibits strong absorption in multiple bands, enabling the detection of even lower concentrations in this wavelength region.[17]

To set up a CRDS system using a Quantum Cascade Laser (QCL) for detecting gases like sulfur dioxide (SO_2), the system configuration includes several key components as depicted in Figure 5.22. [18]

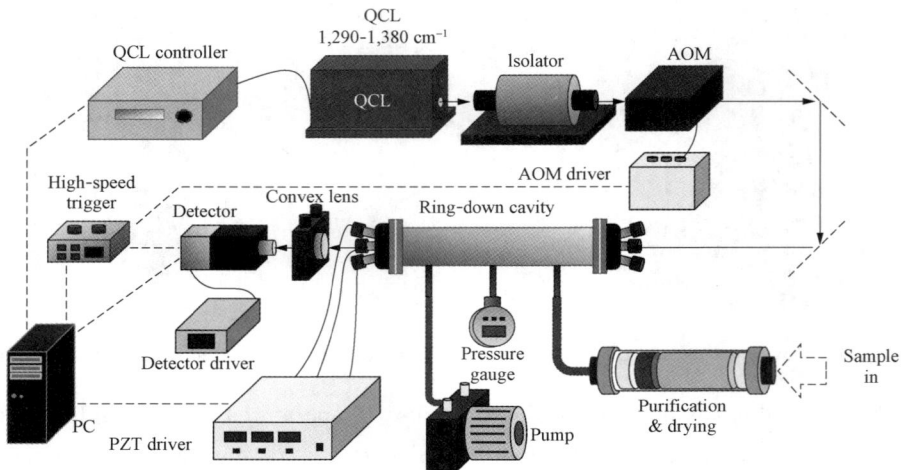

Figure 5.22　Schematic diagram of the CRDS experimental setup

The main components include: (1) Quantum Cascade Laser (QCL): This serves as the light source. The QCL is ideal for mid-infrared spectroscopy due to its ability to emit in this spectral region where many gases have strong absorption lines. (2) QCL Controller: This device manages the operation of the QCL, including its modulation and tuning to specific wavelengths. (3) Fast Detector: This is crucial for monitoring the intensity build-up in the cavity and recording the cavity ring-down (CRD) time. The possible detectors include: amplified photodiode, avalanche Photodiode, or photomultiplier, above detectors must be capable of fast response times, typically with bandwidths larger than 10 MHz, equating to response times shorter than 100 ns. This speed is necessary to accurately record the rapid decay of light intensity within the optical cavity.

The detection process could be described as follows: (1) Laser Emission: The QCL emits a beam of light at a specific wavelength that corresponds to an absorption line of the target gas (SO_2) in this case. (2) Optical Cavity: The emitted light is directed into an optical cavity formed by highly reflective mirrors. The light bounces back and forth within this cavity, building up in intensity. (3) Trigger Pulse and Beam Deflection: A trigger pulse is sent to switch off the QCL, quickly extinguishing the first-order diffracted beam. The fast detector then measures the exponentially decaying light intensity leaking from the cavity, known as the ring-down time. (4) Data Acquisition: The CRD time is recorded and used to determine the concentration of the absorbing gas. The absorption lines of SO_2 in the mid-IR region make it suitable for detection using this method. The detector must be fast enough to record the rapid decay in light intensity, with response times shorter than the ring-down time to ensure accuracy. This setup allows for highly sensitive detection of SO_2, leveraging the advantages of QCLs and the CRDS technique to provide accurate and reliable measurements.

The cavity ring-down cells can be divided into three options: straight cavity (two-mirror cavity), folded cavity, and ring cavity (multi-mirror cavity). The structure of the straight cavity is simple and the prototype cavity is shown in Figure 5.23. [17]

Figure 5.23   Prototype of the passive cavity cell

Only when the laser forms a stable resonance in the cavity can the continuous wave cavity attenuation spectrum be measured, which requires the cavity to be extremely stable degrees. Therefore, the material of the titanium alloy seat cavity is selected. The physical picture of the passive cavity is shown in Figure 5.23. [17] The cavity is 360 mm long and the inner hole is 1 mm in diameter. Both ends of the cavity are polished and the cavity mirror is fixed on both ends with light glue. By calculation, the laser has no diffraction loss due to the restriction of the inner diameter of the cavity.

The final experimental data are as follows: the relative error of the measurement of H_2S concentration in the range of 1μL/L to 10μL/L is less than 1μL/L, and the relative error of the measurement in the range of 10 μL/L to 50 μL/L is less than 10%, which meets the requirements of DL/T 1205—2013 "Test Method for Decomposition Products of Sulfur hexafluoride Electrical Equipment". The total assembly of the instrument is as shown in Figure 5.24. [17, 19]

Figure 5.24   Online laser cavity ring-down spectroscopy equipment in the field (served for a GIS with SF_6/N_2 mixture gas)

5.5.4 TDLAS Applications in SF₆ Gas Leak Detection

Sulfur hexafluoride (SF_6) is extensively used in high-voltage electrical equipment as an insulating and arc-quenching medium due to its excellent dielectric properties. However, SF_6 is also the most potent greenhouse gas, with a global warming potential (GWP) that is 23,900 times greater than that of carbon dioxide (CO_2) and an atmospheric lifetime of approximately 3,200 years. Consequently, minimizing SF_6 emissions is crucial for environmental protection.

Potential sources of SF_6 emissions occur from (1) Poor gas handling practices during the installation, maintenance, and decommissioning of equipment. Operation and maintenance of SF_6 gas carts used to remove, store, clean, and refill SF_6 gas. (2) Deterioration of SF_6-containing equipment fittings and materials over time due to chemical reactions, hardening, and corrosion. Studies have noted that approximately 10% of circuit breaker populations may experience leaks. Among these, 15% are minor leaks that can be repaired immediately, while 85% are significant leaks requiring scheduled repairs. Typical Leak Locations: (1) 73% of leaks occur at gas mechanisms. (2) 21% from worn or broken bushings. (3) 6% from gas tanks. Equipment repair or replacement is often necessary to mitigate leaks. As equipment ages, replacement may be more cost-effective and environmentally beneficial compared to repair. New equipment is designed with minimal to zero leak rates, following industry standards that recommend low leakage limits.

As electrical equipment ages and reaches the end of its operational service life, replacement rather than equipment repair may provide the more attractive SF_6 mitigation strategy. Many equipment manufacturers now guarantee minimal to zero leak rates for new equipment. Additionally, industry standards recommend that new equipment be built to low leakage limits. Figure 5.25 and Figure 5.26 shows a TDLAS SF_6 leak scanning instrument for GIS.

Figure 5.25   A TDLAS SF_6 leak scanning instrument for GIS

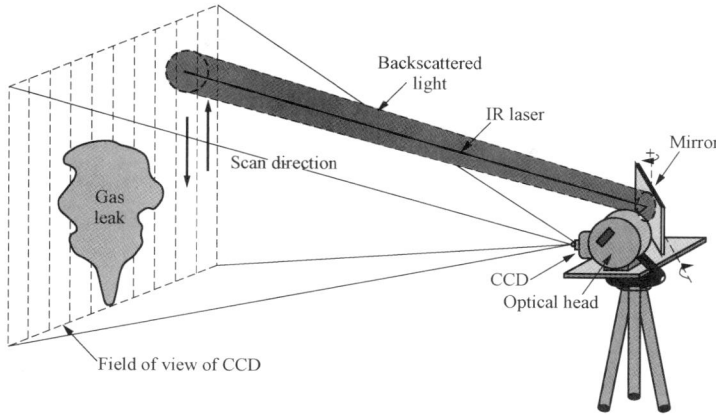

Figure 5.26   Working principle of a TDLAS SF$_6$ leak scanning instrument

5.5.5   TDLAS Applications in Coal-fired Power Plants

Oil and coal combined account for over 60% of the world's energy supply, with coal alone contributing about 27%. Coal-fired power plants, which convert the stored energy in coal to electricity, have an average efficiency of about 30%. There is significant potential for efficiency improvements through combustion optimization. Additionally, minimizing environmental impact through flue gas treatment is crucial, similar to methods used in Municipal Waste Incineration (MWI) plants. Figure 5.27 schematically depicts a coal-fired power plant, highlighting key measurement points for process analytics. Table 5.2 identifies important measuring points for a Tunable Diode Laser (TDL) analyzer. [20, 2]

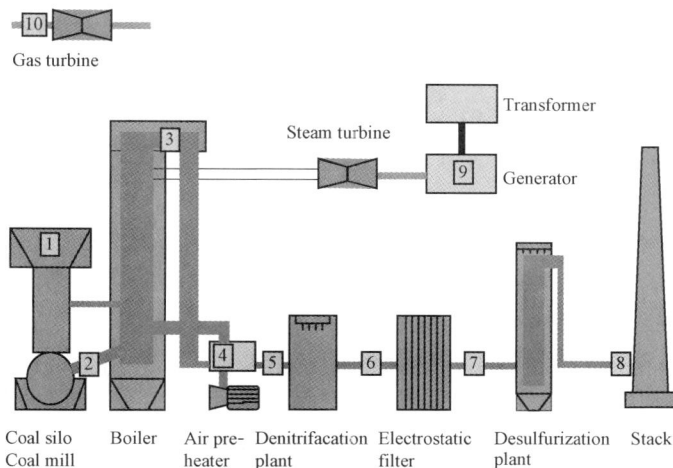

Figure 5.27   Schematic view of a coal-fired power plant

Table 5.2   Measurement locations in a coal-fired power plant

MP	Wavelength (nm)	Component	Measurement range
1	Coal silo	CO	0-5%
2	After coal mill	O_2	0-10%
3	Combustion chamber	O_2	0-10%
4	Air pre-heater	O_2	0-10%
5	de NO_x stage	CO, O_2	0-500 µL/L, 0-10%
6	Before electro filter	CO, O_2	0-500 µL/L, 0-10%
7	After electro filter	CO	0-500 µL/L
8	Stack	CO, O_2	0-500 µL/L, 0-10%

The arguments for oxygen monitoring for combustion optimization are very similar to those for an MWI plant: Combustion converts the primary chemical energy in coal into heat through oxidation at high temperatures. Oxygen, supplied as part of the combustion air, is essential for the combustion process. Perfect combustion requires just enough oxygen to burn all combustibles completely. In reality, an excess volume of oxygen is supplied due to insufficient mixing of fuel and oxygen. Excess oxygen increases NO_x content and energy losses due to dilution with cool air. Insufficient oxygen increases CO formation. The excess air value is a critical parameter for optimizing the combustion process and ensuring economic plant operation. The benefits of oxygen monitoring will result in higher efficiency by heating less excess air, leading to cost savings on electric power.

Real-time monitoring of oxygen concentration and gas temperature is possible with a TDL analyzer. TDL analyzers measure oxygen concentration and gas temperature simultaneously in the same gas volume by analyzing the relative strengths of the absorption lines. Monitoring CO levels in coal silos is crucial for preventing self-ignition. TDL analyzers measure CO concentration inside the coal silo, providing a rapid response.

High dust loads during coal filling necessitate short absorption paths to avoid measurement errors. Electrostatic filters trap dust particles through electrostatic attraction but can cause spontaneous discharges. It is vital to prevent flue gases with high CO concentrations from entering the filter to avoid ignition by electric sparks. Continuous, fast monitoring of CO content upstream of the filter ensures safe operation. By employing TDL analyzers for real-time monitoring of oxygen and CO levels, coal-fired power plants can achieve combustion optimization and enhance safety, contributing to both economic efficiency and environmental protection.

As shown in Figure 5.27, the coal combustion plant includes a denitrification stage very similar to the one used in a waste incinerator. However, for the larger coal combustion plants, the de NO_x stage uses selective catalytic reduction and the TDL application is to optimize this process by measuring ammonia and humidity in the same manner as in the MWI plant. The

NH_3 levels here are in the range of 0 - 10 μL/L.

5.6 Review Questions

Q1: Please describe the Beer-Lamberts Law by oral narration, and would you give one or two application examples using this law ? (Please refer to section 5.2.3)

Q2: Would you list the components for a TDLAS apparatus and the process of gas analytics by using it? (Please refer to section 5.3)

Q3: Please describe the cavity enhancement techniques? What are the advantages comparing other competive techniques? (please refer to 5.4.3 and 5.5.3)

Q4: Please draw a schematic map of Herriott cell, and describe the advantages by using multi-path cells? (Please refer to 5.4.2)

Bibliography

[1] Tunable laser diode spectroscopy. [Online]. Available: http://southeastern-automation. com/PDF/Spectra/ App%20Notes/Tech%20Note%20-%20TDLAS.pdf. Accessed: Tech Note Spectra Sensors.[TM]

[2] Tunable Diode Laser Spectroscopy - Theory and Background. Mettler-Toledo AG, 2012.

[3] Duarte, Frank J. Tunable laser applications, volume 150. CRC press, 2008.

[4] N. A. Ferlic. Forward scattering meter for visibility measurements. Master's thesis, University of Maryland, College Park, 2019.

[5] Kim, Young and Platt, Ulrich. Advanced environmental monitoring. Springer Science & Business Media, 2007.

[6] Baudelet, Matthieu. Laser Spectroscopy for Sensing: Fundamentals, Techniques and Applications. Elsevier, 2014.

[7] Gauglitz, Günter and Moore, David S. Handbook of spectroscopy. Wiley-VCH Weinheim, Germany, 2014.

[8] Garcia-Campana, Ana M. Chemiluminescence in analytical chemistry. CRC Press, 2001.

[9] Bösenberg, Jens and Brassington, David J and Simon, Paul C. Instrument development for atmospheric research and monitoring: lidar profiling, DOAS and tunable diode laser spectroscopy, volume 8. Springer Science & Business Media, 1997.

[10] Tasumi, Mitsuo. Introduction to experimental infrared spectroscopy: Fundamentals and practical methods. John Wiley & Sons, 2014.

[11] Wang, Shoulin and Gong, Weihua. Interference fringe suppression in tunable diode laser absorption spectroscopy based on ceemdan-wtd. Frontier in Physica, 2022.

[12] Berden, Giel and Engeln, Richard. Cavity ring-down spectroscopy: techniques and applications. John Wiley & Sons, 2009.

[13] Dyroff, Christoph. Tunable diode-laser absorption spectroscopy for trace-gas measurements with high

sensitivity and low drift. KIT Scientific Publishing, 2009.

[14] Jiang, Jun and Wang, Zhuowei and Han, Xiao and Zhang, Chaohai and Ma, Guoming and Li, Chengrong and Luo, Yingting. Multi-gas detection in power transformer oil based on tunable diode laser absorption spectrum. IEEE Transactions on Dielectrics and Electrical Insulation, 26(1):153-161, 2019.

[15] Ciaffoni, L and Cummings, BL and Denzer, W and Peverall, R and Procter, SR and Ritchie, GAD. Line strength and collisional broadening studies of hydrogen sulphide in the 1.58 μm region using diode laser spectroscopy. Applied Physics B, 92(4):627, 2008.

[16] Siciliani de Cumis, M and Viciani, S and Galli, I and Mazzotti, D and Sorci, F and Severi, M and D'Amato, F. Note: An analyzer for field detection of H_2S by using cavity ring-down at 1.57 μm. Review of Scientific Instruments, 86(5):056108, 2015.

[17] Xinwen Feng and Wenxia Li and Shikuan Chen and Yuandong Wang and Jie Zhang and Fei Li. Research on SF_6 and its decomposed product detection device based on crds. Electrical Measurement & Instrumentation, 55(9): 142-146, 2018.

[18] Tang, Jing and Li, Bincheng and Wang Jing. High-precision measurements of nitrous oxide and methane in air with cavity ring-down spectroscopy at 7.6 μm. Atmospheic Measurement Techniques, 2019.

[19] Xiaobo Liu and Zhezhe Pan. Study of "zero-emission" SF_6 monitoring technology by optical methods. Industrial Safety and Environmental Protection, 43(12):91-95, 2016.

[20] Jinyi Li and Fushuang Sun and Chenge Zhang and Zuowei Fu and Baoquan Yan and Honglian Li. Application and prospect of tunable laser absorption spectroscopy in coal-fired power plants. Laser Journal, 41(4): 8-17, 2020.

6

Infrared Thermal Imaging and Gases Leaking Detection

6.1 Introduction

Any object will emit electromagnetic radiation when its temperature is above absolute zero (−273 ℃). At absolute zero, electrons, atoms, and molecules occupy the lowest quantum energy states, making it impossible to transition between energy states and emit electromagnetic radiation. When temperatures exceed absolute zero, electromagnetic radiation is emitted, with the amount and distribution across the wavelength spectrum primarily depending on the object's temperature and the characteristics of its surface, known as emissivity. Emissivity is usually a function of wavelength and may also depend, to a lesser extent, on the object's temperature.

Figure 6.1 shows the distribution of emitted energy over the target electromagnetic spectrum at different temperatures.[1] The sun, at 6,000 K, appears almost white-hot because the energy it emits is concentrated in the visible spectrum, peaking at 0.5 μm. Other targets, such as a 3,000 K tungsten wire, an 800 K hot surface, and the 300 K environmental Earth (about 30 ℃), are also illustrated. It is evident that as the surface temperature decreases, not only does the emitted energy reduce, but the wavelength distribution also shifts to longer infrared wavelengths.

The visible wavelength range for human eyes is approximately 0.4 to 0.76 μm. However, not all objects emit electromagnetic radiation in this visible range at relevant temperatures. For example, four objects at various temperatures, a lamp filament at temperatures of 2,727 ℃ (i.e., 3,000 K), an ice block at the freezing point of water 0 ℃ (273 K), a cup of water at normal amb-

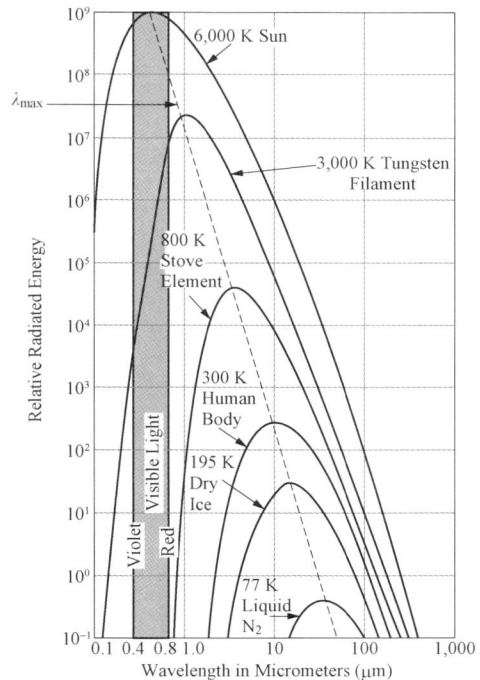

Figure 6.1   Blackbody curves at various temperatures

ient temperature 20 ℃ (293 K), and the boiling water at temperatures of 100 ℃ (373 K). From the curves in Figure 6.2,[2] we see that a lamp emits radiation to which the eye is sensitive. However, a cup of water at ambient temperature, or even at the temperature of boiling point, emits virtually no radiation in the band of wavelengths to which the eye is sensitive. Therefore, we can only see them directly if they are illuminated externally by sources such as the sun or an incandescent lamp. In other words, what we see is only the light reflected by the external light source, not the light emitted by the object's electromagnetic radiation.

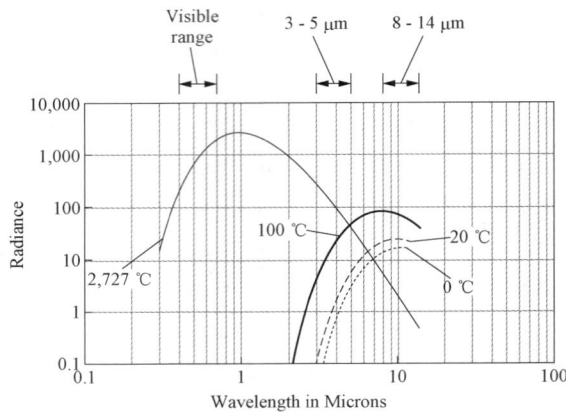

Figure 6.2   The radiance of a surface as a function of wavelength, for four different temperatures. Note that the radiance for T = 2,727 ℃ has been plotted on a scale reduced by a factor of 1,000 [2]

The thermal imaging is a technology converting images produced by the longer-wavelength "thermal" radiation emitted by these cooler bodies to visual wavelength images that we can see. Thermal imaging technology convert thermal radiation, which is invisible to the human eye, into visible images through the following steps: (1) Capture Infrared Radiation: Thermal imaging devices use infrared sensors or detectors to detect infrared radiation emitted by the surface of an object, typically in the wavelength range of 8 to 14 microns. These sensors convert the detected infrared radiation into electrical signals. (2) Convert Electrical Signals: The electrical signal captured by the sensors is directly proportional to the temperature of the object. Inside the thermal imaging device, electronic circuitry processes these electrical signals and converts them into digital signals. (3) Image Generation: The processed digital signals are sent to an image processing unit, where they are mapped onto a color or grayscale palette. Areas of higher temperature appear brighter or in warm colors (e.g., red, yellow), while cooler areas appear darker or in cool colors (e.g., blue, purple). (4) Display Image: The thermal image, representing the temperature distribution on the object's surface, is displayed on the device's screen. Users can directly observe and analyze the thermal characteristics of the object based on the displayed image.

Thermography has become a well-established technique for detecting leaks of various gases, including volatile organic compounds (VOCs), sulfur hexafluoride (SF_6), and carbon monoxide (CO). One of the significant applications of infrared (IR) thermal imaging is qualitative gas detection and identification. We shall delve deeper into how gaseous species influence IR imaging, particularly in the thermal infrared spectral range.

6.2 Key Factors and Characteristics of Thermal Imaging

Atmospheric transmission plays an important role in thermal imaging. In particular, many molecules present in the atmosphere absorb radiation in the wavelength range of interest. Of particular importance is the absorption of water vapor, carbon dioxide, oxygen, and carbon monoxide. Figure 6.3 shows a typical atmospheric transmission curve for thermal imaging.[2, 3]

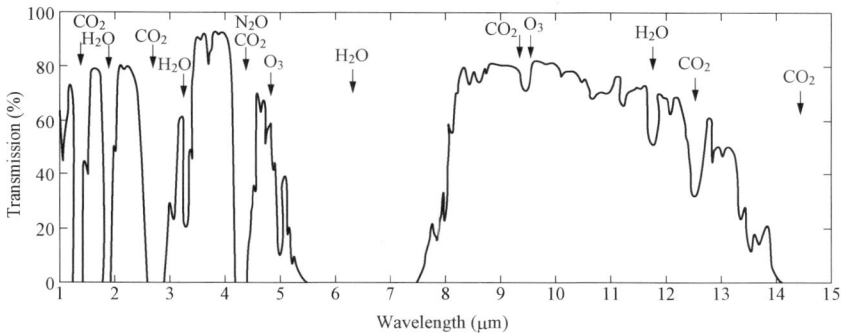

Figure 6.3   Example of the transmission through the atmosphere for a 1 km path

According to this curve, along with other considerations (such as the performance of radiation detectors), thermal imaging systems are generally operate in three useful wavelength bands. The first range extends from 0.8 to 1.7 μm, the second range extends from 3 to 5 μm, and the third range is between 8 and 14 μm. There are many reasons to deviate from these precise ranges. One important factor is the barrier to water vapor transmission from 5 to 8 μm, and another important factor is the performance of the detector. For example, the use of detector arrayscan greatly improve performance. However, for longer wavelength bands, viable photon detector arrays are currently limited to having a response extending no farther than about 11 μm. Of course, in applications where the atmosphere is not a limitation, different wavelength ranges can be used. Examples include measurements taken from space, situations where the path length is very short, or where the composition of the atmosphere can be controlled (for example, by using a pure dry nitrogen atmosphere).

There are three wavelength bands to be used in Thermal imagers, including SW (shortwave infrared) 0.8-1.7μm (please note that the SW is also specified as 0.9-1.7 in some thermal imaging literature), MW (midwave infrared) 3-5 μm, and LW (longwave infrared) 8-14 μm. The different infrared camera systems (some with additional accessories) used in most of the experiments described in this chapter are shown in Figure 6.4. They range from SW to MW to LW, as shown in Figure 6.5.

SW   FLIR SC2600

InGaAs
0.9-1.7 μm
640×512 pixels

MW   FLIR SC6000   Agema THV 550

InSb
1.5-5.5 μm
640×512 pixels

PtSi
3-5.5 μm
320×240 pixels

LW   FLIR T650sc   FLIR A35sc   FLIR SC2000

Bolometer   Bolometer
7.5-14 μm   7.5-14 μm   Bolometer
640×480 pixels   320×240 pixels   7.5-14 μm
                                               320×240 pixels

Figure 6.4   FLIR IR Cameras categories by wavelengths [4]

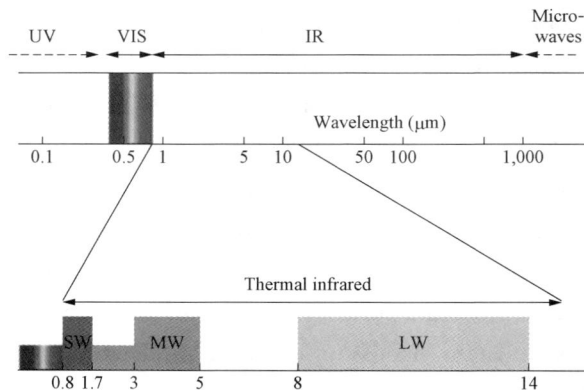

Figure 6.5   The wavelength range of SW, MW to LW in IR Cameras [4]

The choice of which of the three wavelength bands to use in thermal imaging depends on several factors, these include surface radiance, solar radiation, atmospheric transmission, and technology and commercial considerations. Here, we briefly describe the relevant factors[2]:

- Surface radiance: At ambient temperatures, the radiant flux emitted by a surface is greater in the 8-14 μm band than in the 3-5 μm band. However, the change in radiant flux for a given small change in the temperature of a surface is greater in the 3-5 μm band. Therefore, the relative importance of these two factors on the performance of the imager will depend on the design and mode of operation of the imager.

- Solar radiation: A small portion of solar radiation falls on the shorter end of the 3-5 μm band. In some applications, this can be an advantage, helping to improve image contrast and possibly providing additional information about the object. In other applications, such as when measuring the temperature of an object, it is simply unnecessary additional radiation. The sun's radiation can usually be eliminated by introducing a filter that reduces the wavelength range to which the imager is sensitive.

- Atmospheric transmission: When the imager is used for imaging over longer distances, the transmission characteristics of the atmosphere become very important. The degree of atmospheric attenuation of radiation varies greatly under different climatic conditions, with some conditions favoring the 3-5 μm band and others favoring the 8-14 μm band. Due to the longer wavelength, transmission through fog and smoke is better in the 8-14 μm band, which is particularly beneficial for surveillance applications.

- Technology and Commercial Considerations: In recent years, progress for the spectral region (0.8-1.7 μm) was due to intensive technology development of the III-V compound semiconductors based on the binary semiconductor compounds GaAs and InAs, the study of furnaces or molten glass and metals in various processes, and the thermal cameras whose spectral region 0.8-1.7 μm are beginning to replace other spot measurement devices such as pyrometers[4]. Therefore, technical developments influence the relative performance capabilities of the three bands, which can change over time. They also affect the complexity and cost of imagers, making commercial considerations important in choosing an imager and its operating band.

6.3 Main Components of Infrared Thermographers

Although infrared radiation (IR) is not detectable by the human eyes, an IR camera can convert it to a visual image that depicts thermal variations across an object or scene. IR covers a portion of the electromagnetic spectrum from approximately 760 to 14,000 nanometers (0.76-14 μm). At temperatures above absolute zero, all objects emit infrared light, and the amount of radiation increases with temperature. Thermal imaging is an imaging technique that is calibrated by an infrared camera to display the temperature value of an object or scene. Therefore, thermal imaging technology allows one to make a non-contact measurement of the temperature of an object. The main components of an infrared camera include optical elements, detectors, cooling or temperature stabilization devices for detectors, electronic components for signal and image processing, and a user interface with output ports, control ports, and image display (see Figure 6.7).[4]

Photon detectors of MW cameras lead to superior performance even at room temperature conditions. Therefore, if the highest temperature resolution is required, MW cameras should be used. For most regular analyses (e.g., in building or maintenance studies under well-defined conditions), the temperature resolution of bolometer cameras is sufficient, and these less expensive systems may be used for such thermal investigations. In contrast, SW camera systems suffer from missing thermal object radiation and are not suitable for room temperature range measurements. Therefore, MW cameras are almost always better suited for a given problem than LW cameras. However, above around 1,000 K, SW cameras become competitive. The preferred detectors with their wavelength ranges are shown in Figure 6.6.[4]

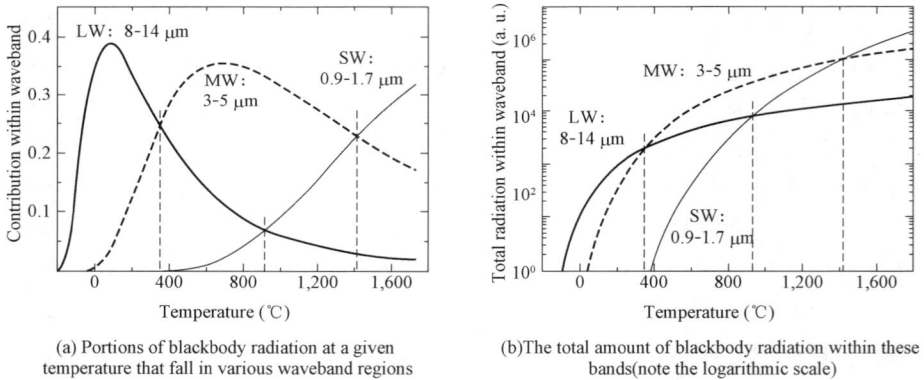

(a) Portions of blackbody radiation at a given temperature that fall in various waveband regions

(b)The total amount of blackbody radiation within these bands(note the logarithmic scale)

Figure 6.6　The three wavelength bands in thermal imaging

Two basic camera concepts for thermal imaging are scanning and staring systems (see Figure 6.8). In scanning systems, the image is generated as a function of time, row by row, similar to how TV screens work [see Figure 6.9(a)]. In staring systems, the image is projected simultaneously onto all pixels of the detector array [see Figure 6.9(b)].

IR line scanners are currently used for the thermal imaging of moving objects or scenes such as band processes (see Figure 6.9). For imaging, only the line scan mode is necessary. The second dimension of the thermal image is built up by the object or scene movement. Line scanners allow very high scan rates, up to 2,500 – 3,000 lines per second (for lower pixel numbers), variable pixel numbers in a measured line up to about 5,000 (for lower scan rates), and a large FOV up to 140°. Such line scanners often use one or two (at different temperatures) internal IR emitters with known emissivities and temperature for recalibration during each mirror revolution. For this purpose, the emitters are positioned outside the object FOV and the recalibration process is made between line scans (see Figure 6.9). [5, 6, 1]

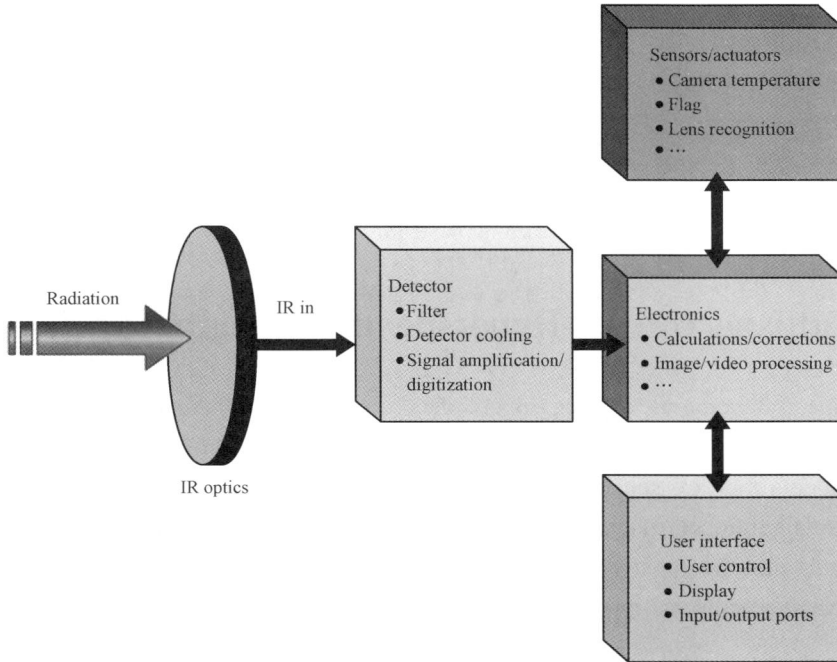

Figure 6.7　Block diagram with main IR camera components

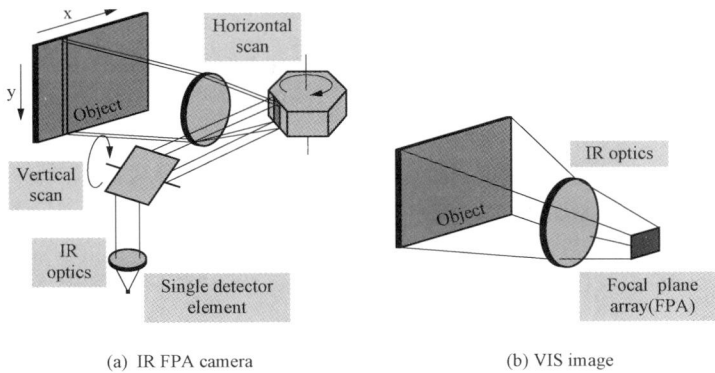

(a) IR FPA camera

(b) VIS image

Figure 6.8　Basic principles of image formation for scanning[4]

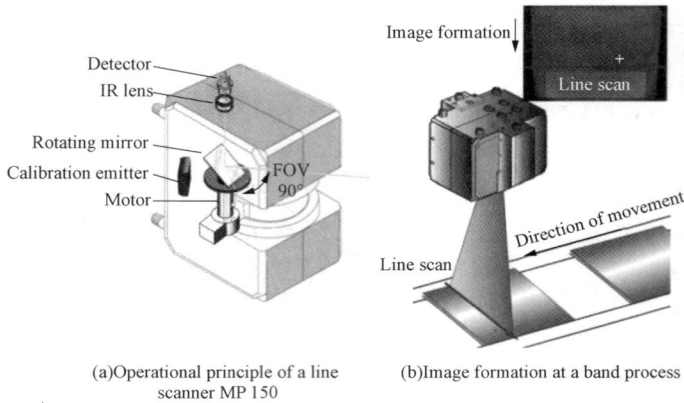

(a)Operational principle of a line
scanner MP 150

(b)Image formation at a band process

Figure 6.9   IR Line scanners are used for the thermal imaging of moving objects or scenes[4]

6.4   Multispectral vs Hyperspectral Imaging: Differences

Multispectral cameras capture image data at a variety of specific wavelength bands across the electromagnetic spectrum. The wavelengths can be separated with filters or detected with instruments that are sensitive to specific wavelengths. Multispectral imaging measures light in a small number (usually 3 to 15) of spectral bands. This technique uses discrete spectral bands that are mostly separated, for example, by using spectral filters or detectors with different spectral sensitivity distributions (see Figure 6.10).[7]

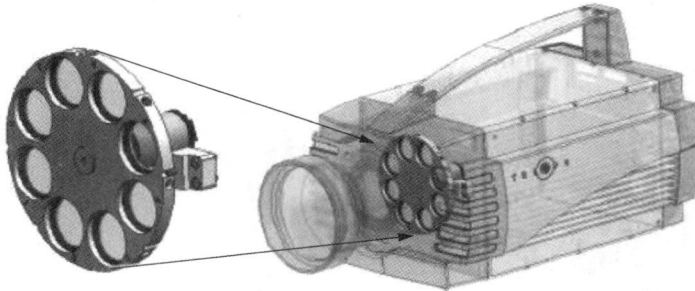

Figure 6.10   Typical outline of a multispectral infrared camera

Hyperspectral imaging is a special case of spectral imaging where there are usually hundreds of continuous spectral bands available. Hyperspectral cameras capture the light of a scene, dividing it into individual wavelengths or spectral bands. This process provides a two-dimensional image of the scene while recording spectral information for each pixel in the image. The result is a hyperspectral image in which each pixel represents a unique spectrum. This unique spectrum can be compared to a fingerprint. Because each material and compound reacts differently to light, their spectral characteristics are also different. Just as a fingerprint can be used to identify a person, a spectrum can identify and quantify substances in a scene. Hyperspectral imaging uses dispersion or interferometer instruments that, in principle, allow the detection of a continuous spectrum. The existing infrared spectroscopy technology provides excellent spectral characteristics for the system, making it possible to sample a complete continuum of predefined wavelength regions at excellent spectral resolution. The data is stored in a so-called 3D data cube, as shown in Figure 6.11. A cube refers to a given time and consists of a series of independent infrared images as a function of wavelength. Two dimensions are given by the spatial resolution of the infrared camera (two-dimensional focal plane array (FPA) sensor), and the third dimension is given by the wavelength or wavenumber scale generated by the spectrometer.

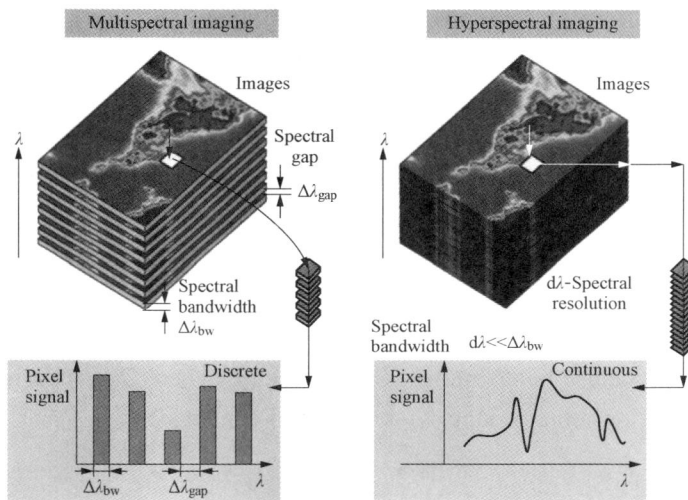

Figure 6.11   Schematic comparison of multispectral and hyperspectral imaging [4]

A distinction is made between multispectral and hyperspectral imaging depending on the number of bands used: up to 100 for multispectral and many more than 100, usually a continuous spectrum for hyperspectral imaging.

Each separate image represents the spatial radiance distribution within a narrow spectral band whose width is determined by the spectral resolution of the instrument. Therefore, every single pixel contains an IR spectrum, that is, it contains the same information as a point measurement with a non-imaging IR spectrometer. Such systems are becoming popular in gas imaging.

There are two types of spectrometer arrangements are used: slit-based and interferometer based systems. Both systems use FPA detectors but in different ways. In slit-based systems, single lines of the array are used (such detector linear arrays are used in line scanners), that is, one direction, for example, the horizontal, provides spectral information along a line of an object. Adjacent lines of the array receive signals of the same object line, but from a different spectral region. This is accomplished by using prismatic dispersion elements or grating. The

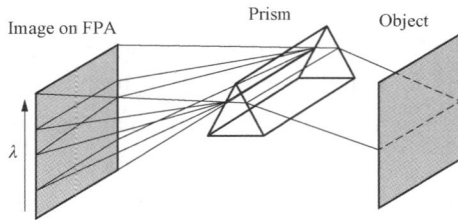

Figure 6.12 Principal setup of a slit-based system with a prism [4]

principal setup with a prism is shown in Figure 6.12.

IR hyperspectral imagers can capture both spatial and spectral information of an object scene simultaneously. Imaging spectrometers have recently become commercially available for general scientific use. Currently, this technology is applied to military, airborne (combustion processes), research, and environmental (chemical agent detection and investigation) applications. As a unique investigation tool for the identification of chemical substances or constituents of gas mixtures and solids, IR hyperspectral imaging will likely find many additional application fields.

Figure 6.13 presents an example in which a cloud consisting of an SF_6/NH_3 gas mixture is detected and identified with an LW (8-11 μm) hyperspectral camera.[4] The temperature distribution of the object scene is shown in grayscale. The SF_6 and NH_3 are identified from their spectral absorption signatures and depicted in different colors. Different spectral channels for the absorption features of NH_3 (yellow) and SF_6 (violet) easily allow one to distinguish several gas species simultaneously. Additionally, the temperature of the nongaseous opaque object is indicated by the grayscale (Image courtesy Telops Inc., Quebec, Canada. The website can be seen at https://www.telops.com/products/hyperspectral-cameras/).

Figure 6.13 Detection and identification of constituents in a gas mixture using hyperspectral imaging

6.5 Smart Grid Applications of Infrared Thermal Imagers

A small defect in electrical apparatus may quickly turn into a system-wide shutdown, halting power supplies until electrical engineers find and replace the faulty equipment. Infrared spectrum inspections can ensure that electrical apparatus operate normally and never reach that point. By identifying damaged components early, infrared energy audits allow you to plan for repairs and prevent downtime. They are essential for maintaining a productive business and an efficient electrical system.

Infrared thermal imagers provide unbeatable, highly accurate electrical inspection solutions. These systems quickly detect high temperatures, which are an early sign of failing equipment or fire hazards. They allow you to inspect the entire electric power system and see what the eye would normally miss. From facility monitoring to new home inspections, infrared thermal imaging helps keep things safe and moving forward.

The advantages of an infrared thermal imager include several key benefits: it allows for safe distance testing, ensuring personal safety and not affecting equipment operation; it enables contactless temperature measurement, preserving equipment integrity; it facilitates rapid scanning of large areas, saving time; it offers a wide temperature measurement range with high precision; and it accurately detects equipment defects. Therefore, an infrared thermal imager is crucial for implementing condition-based maintenance.

In general, electrical equipment failures can stem from various causes: unreliable internal electrical connections or poor contacts, increased dielectric losses, aging, cracking, or peeling of insulation, uneven voltage distribution or excessive leakage current, increased eddy current losses (or increased iron losses), insufficient oil levels or other heat-related issues, and faults such as overload or over-voltage during special operation modes. Design flaws in cooling systems can also contribute to equipment failures.

Factors affecting the accuracy of infrared measurements in power apparatus include: atmospheric absorption effects, influence of atmospheric dust and aerosols, wind speed, emissivity of surfaces, measurement angles, thermal radiation from nearby objects emitting their own heat, and impact from solar radiation.

To ensure accurate measurements, infrared temperature measuring must adhere to specific environmental requirements and restrictions:

- Avoid obstructions in the sight line, such as doors and cover plates.
- Do not conduct IR imaging examinations if the ambient temperature is below 5 ℃ or if relative humidity exceeds 85%.
- Prefer night-time imaging under overcast or cloudy conditions for optimal image quality. In thunderstorms, rain, fog, or snow, IR thermal imaging should be halted.
- Ensure wind speed is generally less than 5 m/s during inspections.
- During outdoor inspections, prevent direct sunlight or reflected light from entering the instrument, and avoid testing under lamplight. For indoor or nighttime inspections, turn off indoor lighting.
- When measuring the temperature of power equipment, prefer testing during peak load conditions and at least 30% of the rated load. Consider fully the influence of load conditions on testing results.

To achieve accurate measurements, the environmental conditions necessary for precise temperature measurement are:

- The wind speed should generally be less than 0.5 m/s.
- Power equipment should operate continuously for at least 6 hours; longer durations, preferably exceeding 24 hours, are recommended.
- Testing should occur on cloudy days or at least 2 hours after sunset.
- Ensure that the equipment being tested is balanced against background radiation and avoid nearby heat sources that could interfere, including human body heat.
- Avoid strong electromagnetic fields to prevent interference with the normal operation of the infrared thermal imager.

Figure 6.14 shows the effect of measuring readings by wind speed. When wind speed decreases from 15 mph to 0 mph, the highest temperature increases from 85 °F to 117 °F[8].

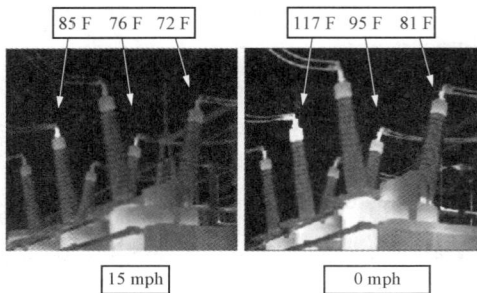

Figure 6.14   The wind speed has an obvious influence on readings of the thermal imager [8]

The IR thermal imager settings require adjusting the span/level values to ensure clear display of all objects in the same imaging, while avoiding overly bright and reddish images by adjusting brightness and contrast.

Figure 6.15 illustrates the importance of capturing all three phases in the same picture or maintaining consistent settings for two-phase detection when all three phases cannot be included in a single image.[8, 9]

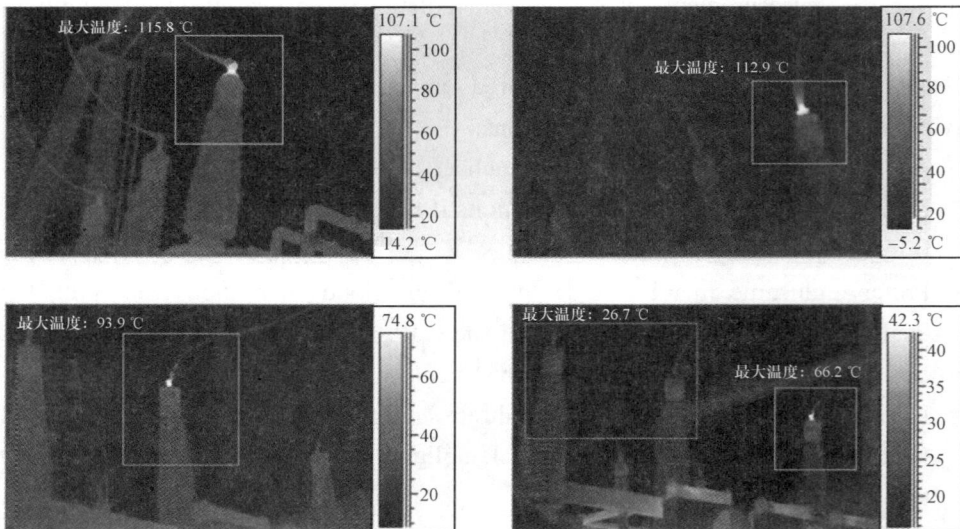

Figure 6.15   If two phases could not be captured in the same picture the setting must remain identical[8]

The temperature gradient among the three phases can be used as a criterion for predicting potential faults in various electric apparatuses. Detailed information is listed in Table 6.1.[8]

Table 6.1   The criteria for predicting power apparatus defects by
temperature difference among three phases

Type of electric apparatus	Relative temperature difference (%)		
	Ordinary defects	Major defects	Emergency defects
SF$_6$ circuit breaker	≥20	≥80	≥95
Vacuumed circuit breaker	≥20	≥80	≥95
Oil immersed bushings	≥20	≥80	≥95
High voltage switchgears	≥35	≥80	≥95
Air insulated circuit breaker	≥50	≥80	≥95
Disconnected switch	≥35	≥80	≥95
Other carrying current apparatus	≥35	≥80	≥95

The defects or potential faults are classified into three grades to determine different countermeasures and subsequent processing methods, including:

- Ordinary Defects: If a local overheat has a certain temperature difference but does not cause an accident, this kind of defect generally requires further observation of temperature curves. It should be addressed during a scheduled maintenance period to eliminate defects in a planned manner.
- Major Defects: Local overheating is more severe, and the temperature distribution gradient is larger. This kind of defect should be repaired as soon as possible. Necessary measures include strengthening inspection and reducing load current. For all voltage-type thermal equipment, monitoring should be increased, and other detection methods should be arranged. After confirming the defect's cause, actions to eliminate the defect should be taken immediately.
- Emergency Defects: The highest temperature of the equipment already exceeds the allowable temperature according to GB/T 11022 standards. This kind of defect should be addressed immediately. For all current hot-type equipment, the load current should be reduced immediately. For all voltage-type thermal equipment, if the defect is obvious, the equipment should be switched off and stopped immediately. If necessary, arrange other detection methods to further determine the nature of the defect.

The power transformer is one of the most critical apparatuses in the electric power grid, responsible for power transmission, distribution, and voltage conversion. The main components of power transformers include the core and coil assembly, cooling systems, voltage regulating devices (on-load tap changers or off-load tap changers), protection devices (gas relays, oil conservators, temperature measuring devices, etc.), and bushings.

According to the Chinese Electric Industry Standard DL/T 572—2010 (Operation specification for power transformers), the top oil temperature of oil-immersed power transformers should not exceed specified limits. For natural circulation cooling transformers, the top oil temperature should not exceed 85 ℃. When the temperature of the cooling medium is low, the top oil temperature will also decrease accordingly.

Infrared temperature measurement is one of the most effective means for online monitoring of transformer operating conditions while the transformer is electrified. Through infrared thermal imaging technology, a wide variety of defects can be detected in the transformer main bodies, oil conservators, bushings, coolers, and control units. There are many more defects suitable for infrared temperature detection; here we only list a few main examples, as follows:

- Side walls or adapters of the transformer tank experience local overheating due to strong leakage magnetic flux and eddy current losses.
- The cooling system is not working properly because oil circulation is obstructed by a closed valve.
- Improper sealing of the high voltage bushing results in internal overheating, and the surface temperature above the oil leakage point is close to the ambient temperature.
- The external or internal connection between two copper conductors is ineffective.
- The oil pump has a turn-to-turn short circuit issue.

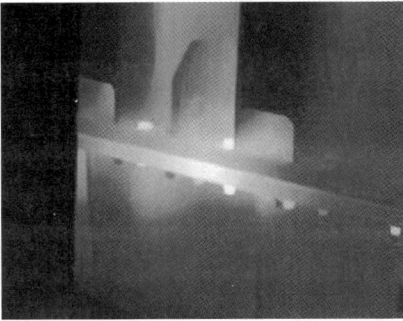

Figure 6.16 Temperature of bolts and nuts exceeds 100 ℃ due to eddy current losses[8]

Figure 6.16 and Figure 6.17 show the bolts and nuts overheating due to eddy current losses. In Figure 6.16, the highest temperature exceeds 100 ℃. Bolts and nuts overheating will accelerate the aging of the sealed gasket between the upper and lower tank of the oil-immersed power transformer, and furthermore, result in the generation of dissolved gas in transformer oils. The reasons for this are due to leakage magnetic flux of power transformers and improper design and installation of magnetic shielding plates. Ultimately, the above fastening pieces were replaced by stainless steel materials.

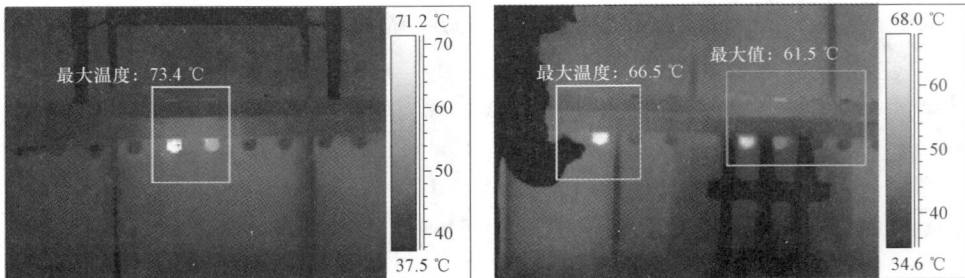

Figure 6.17 Bolts and nuts overheating due to eddy current losses[8]

Figure 6.18 show the abnormal temperature distribution in cooling systems. The cooling pipe could not run through the oil fluid smoothly due to the valve being switched off.

High voltage bushing of power transformers generally adopts an oil-paper capacitive structure, which is composed of copper conduct, oil-paper capacitive core, porcelain or composite insulator, and accessories. A copper conductor is used to lead out the high-voltage from the transformer tank, which plays the role of fixing and conducting the electric current. It is not only the skeleton of the capacitor

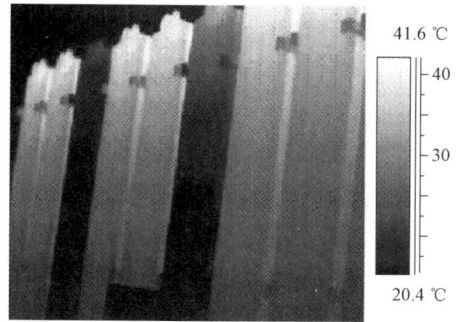

Figure 6.18   Temperature of two pieces of radiators much lower than that nearby due to the oil could not running through them [8]

core but also the first shielding of all capacitive shieldings. The capacitive core is the main insulation part of the bushing, it consists of aluminum foils with a semiconductor layer on the edge and an insulating medium between the shieldings wound on the copper conduct, forming a multi-layer concentric cylindrical capacitor. The hollow insulator is a container for the capacitor core and transformer oil, which acts as external insulation.

The common faults of the oil-immersed paper capacitive bushing are oil deficiency, moisture in oil, breakdown of the capacitance layer, degradation of insulation performance, and breakage of the measuring terminal.

There are three main hazards when oil deficiency of bushings occurs:

- When moisture content in the oil is high due to oil deficiency, air enters the bushing, causing the transformer oil and capacitor core to become damp. This reduces insulation levels, increases leakage current and dielectric loss, and leads to internal insulation heating. The deteriorated heat dissipation worsens the insulation aging, eventually leading to breakdown and potential explosion.

- Surface discharge becomes more likely. Severe oil shortage exposes the inner porcelain wall and high sections of the capacitor core to air, promoting surface discharge. This can escalate to thermal dissociation and slip flashover, generating hot gas that may cause the bushing casing or container to explode upon pole passage or surface flashover.

- Partial discharge occurs. Insufficient oil allows air to enter the bushing casing, where it dissolves into the oil and infiltrates the insulating paper. Under alternating electric fields, partial discharge ensues, potentially decomposing the insulating oil or paper. The characteristic thermal image of an oil-deficient bushing shows the highest temperature at the oil surface with a distinct horizontal boundary.

Figure 6.19 shows the oil shortage could be found easily by IR thermography. On top of bushing B, the brightness is much darker than the surrounding phases due to oil deficiency in it.

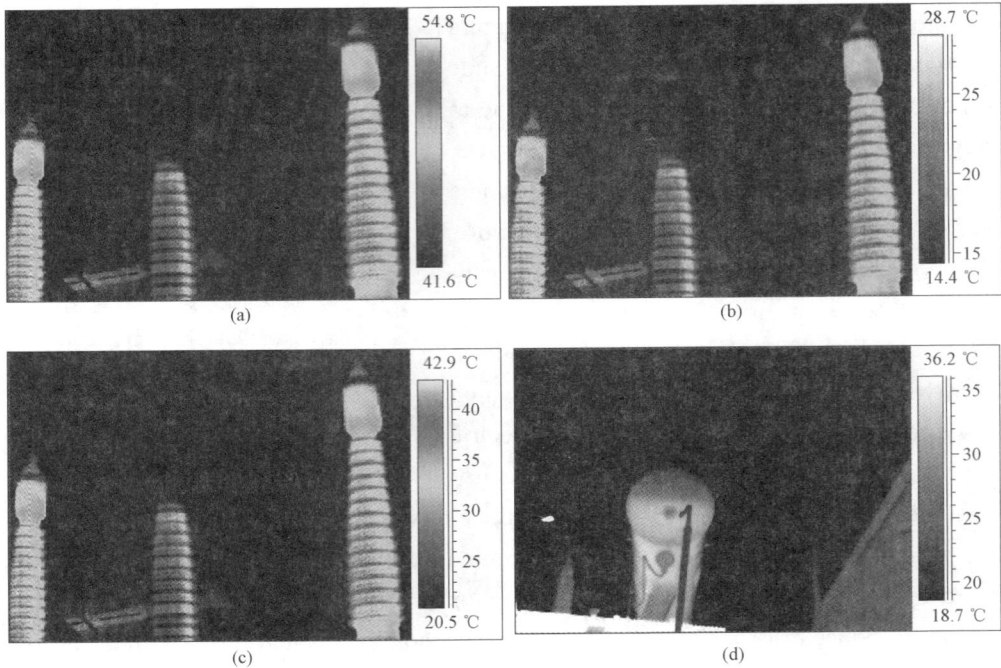

Figure 6.19   The brightness of bushing B top is much darker than
surrounding phases due to oil deficiency[8]

The grounding terminal in the shielding layers of a transformer bushing includes a lead for connecting it to ground, with the entire capacitive core immersed in insulating oil. This terminal, also known as a measuring terminal, allows for the measurement of capacitance and dielectric loss of the capacitor core to assess its insulation condition and overall internal insulation performance. Through this terminal, defects in the insulation of the conductor side and grounding shielding, such as dampness, degradation of insulation oil, or open or short circuits between capacitor shields, can be effectively identified. However, if the grounding terminal is open during normal operation of the power transformer, the grounding shielding layer can accumulate high voltage, potentially leading to equipment damage. Figure 6.20 illustrates the thermal imaging result showing a hot spot in the high-voltage bushing of the power transformer due to unreliable grounding of the capacitive shielding layer.

Figure 6.21 shows the highest temperature on the flange of the bushing is 75.3 ℃, about 50 K higher than the surrounding two phases. According to this figure and temperature distribution, predicting the internal connection between lead and rod has a problem. The high temperature was transmitted to the flange.

Figure 6.22 shows an example of oil pump has internal short circuit trouble. The temperature differences among oil pumps are larger than 10 K by thermal imaging due to turn-turn short circuit fault occurring inside an oil pump of a 330 kV main transformer.

Figure 6.20　The grounding terminal of the shielding layer is not reliably grounded causing a hot spot in the HV bushing of the power transformer[8]

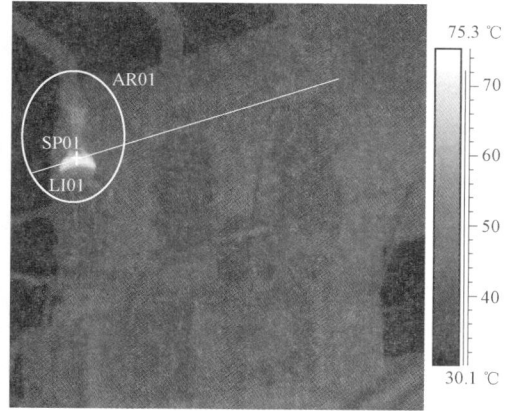

Figure 6.21　Internal lead and the external rod were unreliably connected on top of a transformer bushing[8]

Figure 6.22　The temperature differences among oil pumps are larger than 10 K by thermal imaging due to an inter-turn short circuit fault occurring inside an oil pump of a 330 kV main transformer[8]

6.6　Remote Gas Leak Detection by IR Imager

6.6.1　Infrared SF_6 Gas Leak Imager in Electric Power Grids

The leakage of SF_6 (sulfur hexafluoride) from gas-insulated electrical apparatus can significantly compromise the safety and efficiency of electrical equipment. The implications of SF_6 gas leakage including: (1) Reduction in Gas Pressure and Density: SF_6 is used as an insulating gas in high-voltage electrical equipment. Leakage leads to a decrease in gas pressure and density, which can affect the dielectric strength of the gas. (2) Loss of Insulation Ability: SF_6 primary role is to provide insulation. A reduction in its concentration can compromise the equipment's insulation properties, increasing the risk of electrical discharge and equipment failure. (3) Operational Safety: Ensuring the integrity of SF_6 insulation is crucial for the safe operation of electrical apparatus. Leaks can lead to dangerous situations, including short circuits, fires, and explosions.

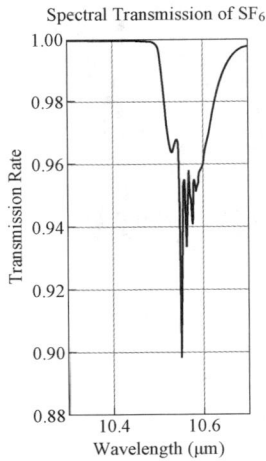

Figure 6.23 Spectral transmission of SF_6

There are many methods to monitor and detect SF_6 gas leakage currently. Here we mainly focus on remote gas leak detection by IR imager. To visually represent the spectral transmission of SF_6, a typical transmission spectrum would show significant absorption at specific wavelengths corresponding to the vibrational modes of the SF_6 molecule. This can be illustrated in Figure 6.23. [10, 11] The abscissa axis presents the wavelength of IR, in μm.

The spectral transmission of sulfur hexafluoride (SF_6) is an important aspect to consider, especially in applications involving gas detection and optical imaging. SF_6 is known for its strong absorption features in the infrared region, particularly in the 8-12 micrometer (μm) range, which is why it is often used as a reference gas in infrared spectroscopy and gas detection technologies.

The transmission spectrum of SF_6 can be characterized by sharp absorption bands, making it relatively easy to detect and measure using infrared spectrometers. These absorption bands are the result of the vibrational and rotational transitions of the SF_6 molecules.

Here are some key points about the spectral transmission of SF_6: (1) SF_6 exhibits strong absorption features in the mid-infrared region, particularly around 10.55 μm. This makes it suitable for detection using infrared spectroscopy. (2) High-resolution spectroscopy can be used to resolve individual absorption lines of SF_6, providing detailed information about the gas concentration and distribution. (3) SF_6's spectral characteristics are exploited in various applications, including leak detection in high-voltage electrical equipment, environmental monitoring, and as a tracer gas in atmospheric studies.

Working principle of SF_6 leak detection by infrared imaging:

- The IR imager is equipped with a spectral filter that transmits infrared radiation in the wavelength range of 10.3 to 10.7 μm. This range encompasses the primary absorption wavelength of SF_6, which is strongest at 10.55 μm.
- SF_6, like many gases, undergoes vibrational and rotational energy transitions when exposed to infrared radiation. These transitions are coupled to the field strength through changes in the molecule's dipole moment, allowing the imager to detect the presence of SF_6.
- Although the imager is finely tuned for SF_6, it can also detect other gases and vapors that exhibit similar infrared absorption characteristics, albeit with very high sensitivity

specifically tailored for SF_6.

The advantages of SF_6 IR leak imager:

- High sensitivity and accuracy, the lowest detectable limit is 0.001 ml/s, accurate locating leak spot.
- IR imager's thermal sensitivity is enhanced by adaptive temporal filter, achieving a sensitivity of less than 30 millikelvin (mK). This allows the imager to detect minute temperature differences caused by gas leaks.
- SF_6 The IR imager enables continuous monitoring of SF_6-insulated equipment without interrupting normal operation. This real-time detection is crucial for identifying leaks as soon as they occur, minimizing the risk of prolonged exposure and potential damage.
- Long-distance detecting and non-intrusive, safer to detector personnel. The technology allows for remote inspection and leak detection, eliminating the need to physically access or disassemble equipment. This non-intrusive approach ensures that maintenance can be performed without significant downtime.
- Through the detector's own screen directly observed leakage images, more intuitive.
- Applicability is wide and testing is not affected by environment and background.
- With temperature function at the same time, the instrument is combined with a high thermal imager.

Figure 6.24, Figure 6.25, and Figure 6.26 illustrate the effectiveness and versatility of the SF_6 IR leak imager in different settings and operating modes. The imager's capability to

Figure 6.24 SF_6 infrared (IR) leak imager is a valuable tool for visualizing gas leaks, especially against the backdrop of an open sky

detect leaks without the need for mirrors and its high sensitivity mode enhance its practicality and reliability, making it an essential tool for maintaining the safety and efficiency of SF_6 gas-insulated electrical apparatus and other applications. Detailed information could be found at a website (https://www.flir. asia/browse/industrial/gas-detection-cameras/).

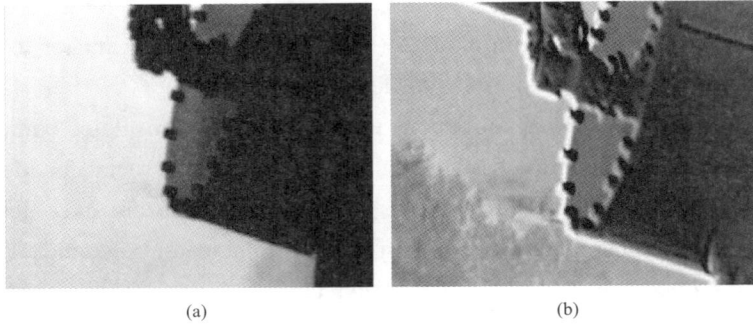

(a)   (b)

Figure 6.25   The setting of high sensitivity mode of SF_6 infrared light leakage imager
is very important for tiny concentration leakage

Figure 6.26   An application example of SF_6 IR leak imager indoors

6.6.2 Arbitrary Gas Leak Detection by Commercial IR Cameras

In this subsection, we present a selection of infrared (IR) images of different gases detected via absorption or emission processes of the gas between the camera and a background object under study. Figure 6.27 compares two camera types (GasFind MW and typical LW) for the detection of methane (CH_4) and propane(C_3H_8), flowing out at atmospheric pressure from a nozzle in front of a black body source at around 55 ℃. The two images of each gas were recorded nearly simultaneously, with the tube with the gas flow sometimes moved for a better view in live images. Three observations are noteworthy:

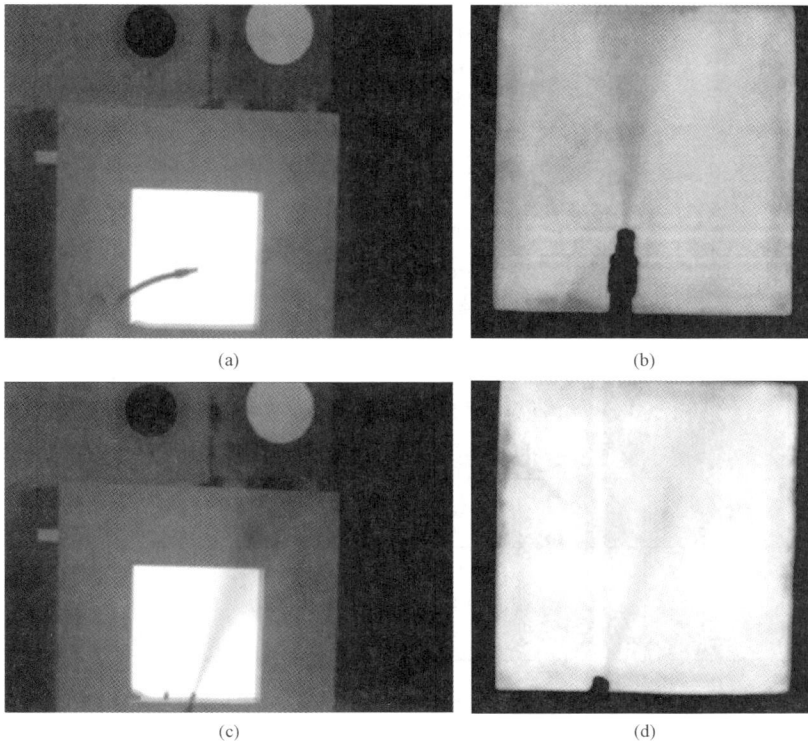

Figure 6.27   IR images of pure methane (a,b) and propane (c,d) flowing out of a nozzle in front

of a blackbody source. (a,c) MW FLIR GasFind camera images and (b, d) broadband

LW FLIR P640 camera images[4]

- Detection Wavelengths: It does not really matter which wavelength region is used for the detection of gases, the only prerequisite being that there must be an absorption band.
- Narrowband Filter Improvement: These images indicate the improvement which is due to the narrowband filter in the GasFind camera. Its images (a,c) have a much better signal contrast. In real-time observations, this is even more obvious. Please note, however, that here the two cameras have different spectral ranges.
- Spectral Differences: The differences between the spectra show up in the images. For CH_4, the MW detection improved considerably due to the much higher absorption constants.

Detecting gas leaks using commercial infrared (IR) cameras involves leveraging the unique properties of gases that absorb IR radiation at specific wavelengths. Here are some key points on how this works and its practical applications:

- How IR Cameras Detect Gas Leaks: Different gases absorb IR radiation at specific wavelengths. For example, methane CH_4 absorbs IR radiation strongly at around 3.3 μm.
- Camera Technology: Commercial IR cameras, designed for gas detection, are tuned to these specific wavelengths. They often use cooled detectors for higher sensitivity and accuracy.

- Imaging Process: When a gas leak occurs, the escaping gas absorbs IR radiation in its characteristic wavelength range. The IR camera detects this absorption as a change in the IR signal, creating a visual representation of the gas plume.

The gases that can be detected by IR cameras include: (1) Methane (CH_4), (2) Propane (C_3H_8), (3) Butane (C_4H_{10}), (4) Sulfur Hexafluoride (SF_6), and (5) Volatile Organic Compounds (VOCs), and many more. Detailed information could be found in Table 6.2 , it shows measurable gases and their suitable wavelength.[4] The website can be seen at https://www.flir.ca/products/flir-g-series/?vertical=optical%20gas&segment=solutions//.

Table 6.2   Characterization of commercially available optical gas imaging cameras using narrowband filters (e.g., FLIR, EyecGas); advertised list of detectable gases for leak detection and repair

Spectral range	MW	MW	MW	MW	LW	LW
Filter wavelengths in μm	3.2-3.4	3.8-4.05	4.2-4.4	4.52-4.67	8-8.6	10.3-10.7
Type of gas	Hydrocarbons		CO_2	CO	refrigerants	
Gases	Examples: • Benzene • Butane • Ethane • Ethylbenzene • Ethylene • Heptane • Hexane • Isoprene • Methyl ethyl ketone (MEK) • Methane • Methanol • MIBK • Octane • Pentane • 1-Pentane • Propane • Propylene • Toluene • Xylene • ...	Used in spectral region where there is no absorption by CO_2 or H_2O	CO_2	CO Also: • Acetonitrile • Acetyl cyanide • Arsine • Bromine isocyanate • Butyl isocyanide • Chlorine isocyanate • Chlorodimethylsilane • Cyanogen bromide • Dichloromethylsilane • Ethenone • Ethyl thiocyanate • Germane • Hexylisocyanide • Ketene • Methyl thiocyanate • Nitrous oxide • Silane	Freon MEK ClO_2 ... • R125 • R134A • R143A • R245fa • R404A • R407C • R410A • R417A • R422A • R507A	SF_6 NH_3 Also • ClO_2 • Ethylene • a large number of hydrocarbon chlorides, bromides, fluorides
Application areas	Natural gas Petrochemical Oil refining Offshore platforms	Industrial furnace surface inspection Looking through flames	CO_2 industries Enhanced oil recovery CCS	Petrochemical Chemical manufacturing steel industry	Food production Pharmaceutical Automotive Air conditioning	Electrical utilities Ammonia plants Fertilizer production

Arbitrary gas leak detection IR camera have many practical applications. In industrial plants: Regular inspection of pipelines, storage tanks, and other equipment in chemical plants, refineries, and natural gas processing facilities; In environmental monitoring: Detecting leaks in natural gas pipelines to prevent environmental contamination and reduce greenhouse gas emissions; For safety: Enhancing workplace safety by quickly identifying and addressing gas leaks, thereby reducing the risk of explosions and health hazards.

The advantages of arbitrary gas leak detection IR camera include: Allows for remote monitoring, which is safer and more convenient; Quick identification of leaks without the need for manual sampling; Provides a visual image of the gas plume, making it easier to locate and address the source of the leak.

High-quality IR cameras with gas detection capabilities can be expensive. IR cameras need to be specifically tuned for the gas of interest, limiting the range of detectable gases with a single device. Factors like humidity, temperature, and background IR radiation can affect detection accuracy.

The commercial VOC GasFind cameras shall detect a variety of gaseous species. Since the respective spectra sometimes shift around the fixed filter region, some gases may be detected slightly better than others. This may give rise to additional changes when detecting different gases.

6.7 Hyperspectral Imaging Applications in Electric Industry

Figure 6.28 shows an industrial example of SO_2 detection from a distance of about 1.5 km. [4] The spectra of the recorded pixels are compared to internally stored sets of spectral data of many gases. Whenever a measured spectrum coincides with a certain gas, this is marked in the image. Another example of a detailed study refers to a coal-burning power plant smokestack. The plume was strongly emissive across 1,800 to 3,000 cm^{-1}, and the spectra revealed simultaneous emissions of at least six gases: CO_2, CO, H_2O, NO, SO_2, and HCl.

In general, if the camera detects a gas-induced signal, it can be asserted only that one of many gases must be present. More detailed knowledge would be possible with a hyperspectral imaging system. Alternatively, once a leak is found, ground-based chemical or spectroscopic inspections can be conducted to definitely identify the relevant gas species. [12, 13]

Figure 6.28　Example of detection of industrial SO_2 emission from a distance of 1.5 km using a hyperspectral imaging system

In addition to playing an important role in gas composition detection, hyperspectral imaging is also unique in the study of solid insulating surface characteristics. Hyperspectral analysis is a novel assessment technique for dielectric surface aging, especially of materials used for highvoltage insulation.

Hyperspectral imaging (HSI) is a new technique available for diagnostics, allowing the spectrum to be obtained with fine wavelength resolution for each pixel in the image in a predefined range in a quasi-continuous way. The application field of this technique is very broad and spans various research areas. Hyperspectral images method was employed for on-line detection of the insulator pollution degree of transmission lines. Firstly, hyperspectral images of the samples with different pollution degrees were obtained by hyper-spectrometer. And then, after original hyperspectral images were corrected by black-and-white correction and multiplicative scatter cor-rection, hyperspectral curves from the region of interest (ROI) of corrected images were obtained. Finally, a multiclassification model of extreme learning machine (ELM) was built to realize the pollution degree classification of test samples. Consequently, the results of this study prove that hyperspectral technique has considerable potential for the non-contact detection of insulator pollution degree.[14, 15]

The diagnostics of electrical equipment based on the infrared part of the spectrum is typically applied, as hyperspectral imaging is a new measurement approach. Unlike the human eye, which focuses on visible light from red to blue wavelength bands with three color receptors (red, green, blue - RGB), spectral imaging divides the spectrum into many more subbands beyond the visible part. Multispectral imaging is a subarea focused on spaced spectral bands, as opposed to continuous spectral bands in hyperspectral systems. For each pixel in an image, a hyperspectral camera acquires the light intensity (reflectance) for a number of continuous spectral bands. Hyperspectral imaging is usually applied in the range of near-infrared (NIR), attributed to fundamental vibrational frequencies of molecules, up to the ultraviolet (UV) characteristic of electronic transitions.[16]

A reflectance spectrum - usually expressed as a percentage - reveals, for each wavelength in the predefined range, the ratio between the intensity of the reflected light and the incident light, measured with respect to a standard white reference. Such a spectrum can provide information on the specimen surface, for example, aged or subjected to partial discharges, useful for the identification of decomposition and erosion components, since the light that is not reflected is absorbed or transmitted depending on the chemical composition of the substrate material.

Hyperspectral imaging creates a novel dimension in the diagnostics of power equipment. It can identify spots on the insulating material surface that have deteriorated due to discharges.[17] HSI may be applied to the effects of different categories of discharges, such as corona, surface discharges, and plasma channels. Dielectric materials reveal different surface degradation severity in various spectral bands, depending on both electric and thermal stress and material properties. This implies the potential of hyperspectral-based diagnostics of power equipment, such as insulators, bushings, cable joints, and motors or transformer windings.

6.8 Review Questions

Q1: Who is Sir William Herschel? and please describe the experiments by him which lead to the discovery of infrared radiation ?

Q2: Please tell us the wavelength range of infrared ray?

Q3: Please describe the wavelength range of SW, MW to LW in IR cameras? If the background hot spot temperature is beyond 1,000 K, which one should be selected for high accurate measurement?

Q4: What is Hyperspectral infrared imaging technology? please give several actual application examples in industry?

Q5: Please give us several application examples of infrared imagers in smart grids?

Q6: Please give us several application examples of infrared imagers for remote gas leak detection?

Q7: Arbitrary gas leak could be detected by suitable IR cameras, is that right, why?

Bibliography

[1] Kaplan, Herbert. Practical applications of infrared thermal sensing and imaging equipment, volume 75. SPIE press, 2007.

[2] Williams, Thomas. Thermal imaging cameras: characteristics and performance. CRC Press, 2009.

[3] Salzer, Reiner and Siesler, Heinz W. Infrared and Raman spectroscopic imaging. John Wiley & Sons, 2014.

[4] Vollmer, Michael and Möllmann, Klaus-Peter. Infrared thermal imaging: fundamentals, research and applications. John Wiley & Sons, 2017.

[5] Lloyd, J Michael. Thermal imaging systems. Springer Science & Business Media, 2013.

[6] Gaussorgues, Gilbert and Chomet, Seweryn. Infrared thermography, volume 5. Springer Science & Business Media, 1993.

[7] Telops application notes, thermal infrared multispectral imaging of minerals. [Online]. Available: https://applied-infrared.com.au/images/pdf/Thermal%20Infrared%20Multispectral%20Imaging%20of%20Minerals%20(2015).pdf. Accessed: Nov. 2020.

[8] Cui Yong. Infrared thermometry techology. State Grid Technology Institute, Shandong Electric Power Company Inspection and Maintenance Subsidiary Corporation, 2014.

[9] Kamble, Pankaj S. and Chaudhari, Bhoopesh N. Application of infrared camera for monitoring of transformer bushing, 2017.

[10] Widger, Phillip and Haddad, Abderrahmane Manu. Evaluation of SF_6 leakage from gas insulated equip-

ment on electricity networks in great britain. Energies, 11(8): 2037, 2018.

[11] Bogue, Robert W. Remote gas detection using ambient thermal infrared. Sensor Review, 2003.

[12] Manolakis, Dimitris G and Lockwood, Ronald B and Cooley, Thomas W. Hyperspectral imaging remote sensing: physics, sensors, and algorithms. Cambridge University Press, 2016.

[13] He, Yuhong and Weng, Qihao. High Spatial Resolution Remote Sensing: Data, Analysis, and Applications. CRC press, 2018.

[14] Qiu, Yan and Wu, Guangning and Zhang, Xiao and et. al. An extreme-learning-machinebased hyperspectral detection method of insulator pollution degree. IEEE Access, 2019.

[15] Yin, Chengfeng and Wu, Guangning and et. al. Method for detecting the pollution degree of naturally contaminated insulator based on hyperspectral characteristics. High Voltage, 2020.

[16] Bleszynski, Monika and Mann, Shaun and Kumosa, Maciej. Visualizing polymer damage using hyperspectral imaging. Polymers, 12(9):2071, 2020.

[17] Florkowski, Marek. Hyperspectral imaging of high voltage insulating materials subjected to partial discharges. Measurement, page 108070, 2020.

7

Ultraviolet Spectroscopy & UV Camera

7.1 Introduction

Evaluating the insulation condition of electrical apparatuses is crucial for ensuring their reliability and longevity. One of the primary indicators used in this evaluation is partial discharge (PD) intensity and duration. PD is a localized electrical discharge that occurs in the insulation system of high-voltage equipment, and while it doesn't immediately cause failure, it can lead to insulation degradation over time, eventually resulting in complete breakdown. The intensity of a partial discharge is a measure of the energy released during the discharge event. It is often quantified by the magnitude of the pulse current generated by the discharge. Higher PD intensity typically indicates a more severe defect or a more advanced stage of insulation degradation. The duration of partial discharge refers to the time period over which the discharges occur. Persistent or long-duration PD can lead to significant insulation damage. Monitoring the duration helps in understanding the severity and progression of the insulation defect. By measuring PD intensity and duration, it's possible to detect and diagnose insulation defects at an early stage, allowing for preventative maintenance before catastrophic failure occurs. Continuous monitoring of PD allows for trend analysis, where increasing intensity or prolonged duration can signal worsening insulation condition, prompting timely intervention. Analyzing PD characteristics helps in estimating the remaining life of the insulation, thereby aiding in maintenance planning and avoiding unexpected downtimes. For newly manufactured equipment, PD testing is essential for ensuring that the insulation meets the required standards before it is put into operation.

The methods to measure PD intensity and duration could be catagries as follows: (1) Pulse Current Method. (2) Ultraviolet (UV) Imaging Method. (3) Ultrahigh-Frequency (UHF) Detection, and (4) Ultrasonic Method, and so on.

Pulse Current Method: Detects the pulse current generated by PD events. It's a traditional and widely used method but can be affected by external interference.

Ultraviolet (UV) Imaging Method: Detects UV light emitted by PD. It's less susceptible to electrical interference but relies on the intensity of UV emission, which can vary.

Ultrahigh-Frequency (UHF) Detection: Captures the high-frequency electromagnetic signals generated by PD. It's effective in noisy environments where other methods might struggle.

Ultrasonic Method: Uses sound waves produced by PD. This method is particularly useful in detecting PD in inaccessible areas or under high levels of electromagnetic noise.

In recent years, there has been growing interest in using ultraviolet (UV) signals as a characteristic quantity for detecting partial discharge (PD) in electrical insulation systems. This approach is based on the principle that partial discharges emit UV radiation, which can be detected and analyzed to assess the condition of the insulation.[1]

7.2 Basic Principles

During the discharge process, electrons in the high-voltage field gain energy and subsequently release this energy as they return to a lower energy state. This energy release occurs in various forms, including the emission of UV light. The intensity of the UV emission correlates with the severity of the discharge. UV imaging systems detect the UV light emitted during a discharge. These systems typically include a UV-sensitive camera or sensor capable of capturing the faint UV emissions that are otherwise invisible to the naked eye. The UV signals captured by the imaging device are processed and superimposed onto visible light images. This creates a composite image that clearly shows both the physical structure of the equipment and the location of any discharges. The intensity of the UV emissions in the image can also indicate the severity of the discharge. This combined UV-visible image allows operators to accurately identify the location and intensity of corona discharges, making it easier to assess the condition of high-voltage equipment. UV imaging provides a non-invasive and reliable method for monitoring the health of electrical apparatus, helping to prevent failures by enabling early detection of insulation issues.

7.2.1 The Origin of UV Band Structure

Johann Wilhelm Ritter was a German physicist who discovered the ultraviolet region of the spectrum and thus helped broaden humanity's view beyond the narrow region of visible light to encompass the entire electromagnetic spectrum from the shortest gamma rays to the longest radio waves. Detailed information could be found at website (https://www.britannica.com/biography/Johann-Wilhelm-Ritter/) and (https://en.wikipedia.org/wiki/Johann_Wilhelm_Ritter).

Atoms have discrete energy levels. When an atom absorbs UV light, it does so at specific wavelengths corresponding to the energy difference between these quantized levels. This results in an absorption spectrum characterized by very sharp lines, as each absorption event corresponds to a transition between two distinct energy levels.

Unlike atoms, molecules have complicated energy structures. In addition to electronic energy levels, molecules have vibrational and rotational energy levels. At room temperature, molecules can occupy many different vibrational and rotational states due to their thermal energy. Even at very low temperatures, molecular vibrations cannot be entirely "frozen out," meaning that molecules will still exist in various excited vibrational and rotational states. The energy differences between these vibrational and rotational levels are much smaller than those between electronic levels, and they are superimposed on the electronic levels.[2]

When a molecule absorbs UV light, it can undergo both electronic and vibrational-rotational transitions simultaneously, as shown in Figure 7.1. This means that a single electronic transition can be accompanied by many slightly different vibrational and rotational transitions. These transitions result in a multitude of closely spaced energy levels.

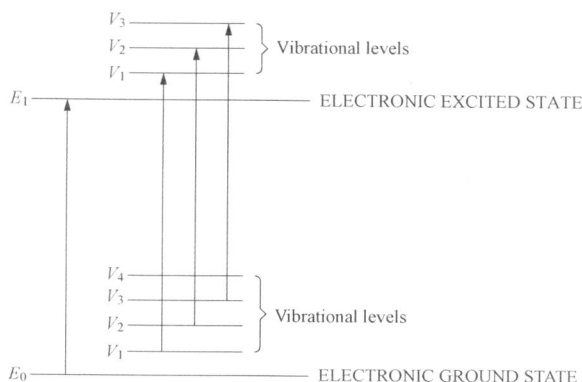

Figure 7.1   Electronic transitions with vibrational transitions superimposed (Rotational levels, which are very closely spaced within the vibrational levels, are omitted for clarity [2])

Due to the vast number of possible transitions and their closely spaced nature, the absorption lines in a molecule's UV spectrum are often too close together for a spectrophotometer to resolve individually. Instead of seeing distinct lines, the instrument records an "envelope" that covers the entire pattern of transitions. As a result, the UV absorption spectrum of a molecule usually appears as a broad band, centered around the wavelength of the primary electronic transition.

Therefore, while atoms exhibit sharp absorption lines due to quantized energy transitions between discrete levels, molecules show broad absorption bands due to the superimposed vibrational and rotational transitions on top of electronic transitions. This leads to a broad and continuous spectrum rather than distinct lines, which is a key characteristic of molecular spectroscopy in the UV range.

7.2.2   The Origin of Electric Discharge and Solar Blind Bands

In atomic spectrometry, plasma is produced by an electrical discharge within a gas-filled tube. The process involves applying a voltage between two electrodes, which accelerates free electrons and ions, leading to further ionization of the gas. This creates a plasma, where charged particles move and interact, leading to the emission of light that can be analyzed for spectrometric purposes.

The measurable current in the discharge tube is a result of the movement of these charged particles under the influence of the electric field. Detailed procedure could be explained as follows:

- The electrons gain energy from the field until they can cause ionization during the collisions.
- Electrons are expelled from the electrode (cathode) through collisions with energy-rich ions (secondary emission).
- Field emission starts as the binding energy of the electrons is surpassed (from 2.4–3.0×10^6 V/m onwards).
- Electrons or ions can be freed as a result of an increase in the temperature of the electrode (thermo-emission).

As a result of all these processes, the gas becomes partially ionized and a so-called plasma is formed.

Corona is a visual phenomenon caused by electrical discharge from high-voltage surfaces into the air. It occurs when the voltage or electric field intensity exceeds a certain threshold, leading to ionization of the surrounding air. Photons are emitted during this discharge,[3] making the corona visible and allowing for its detection and analysis in high-voltage systems, as shown in Figure 7.2. This can be an important indicator of the condition of electrical equipment and is often monitored in high-voltage applications to prevent damage or failure.

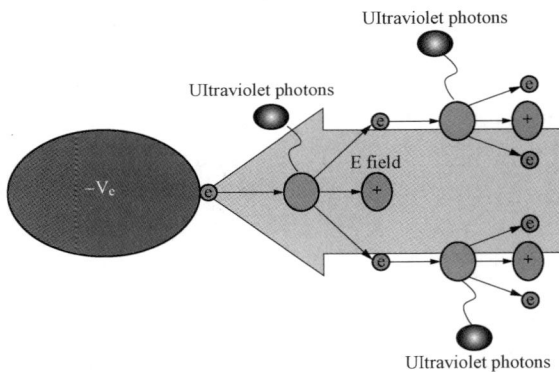

Figure 7.2 Photons are produced during corona discharges[3]

While current external insulation testing methods like high voltage testing, insulation resistance measurements, and voltage distribution measurements are useful, they are often laborintensive, risky, and require power supply interruptions, limiting their effectiveness. Traditional visual inspections are insufficient for detecting insulation defects and flashover locations in overhead lines. Therefore, the development of long-distance, non-contact detection technologies is crucial, offering a more practical and efficient approach to maintaining and monito- ring highvoltage systems.[4]

The commonly used instruments for corona discharge testing have ultrasonic detectors and infrared thermal imagers. Both ultrasonic detectors and infrared thermal imagers have their applications come with limitations. Ultrasonic detection struggles with accuracy over long distances and quantitative analysis, while infrared thermal imaging is an indirect method that detects the effects of discharge rather than the discharge itself. These limitations highlight the need for complementary detection methods or advanced technologies to improve the accuracy and effectiveness of discharge testing in electrical systems.

The characteristics of corona discharge include:

- Corona discharge is often the early sign of severe flashover phenomenon.
- Generating ion impact effect, accelerating damage of all kinds of materials in electrical equipment.
- Generating ozone, nitrogen oxide, nitric acid, etc., above chemical products, will accelerate insulation aging, and make the surface oxidation, decreasing the conductive effect of copper metal contact surface.
- The pulse electromagnetic wave will produce radio, TV, and radio frequency interference.
- Decrease the efficiency of power transmission.

Figure 7.3 provides a graphical representation of solar radiation intensity across different wavelengths, highlighting how the radiation varies within the specified range (250-2,500 nm).[5] The figure covers a broad spectrum of solar radiation, including the ultraviolet (UV), visible, and near-infrared (NIR) regions. Generally, the wavelength of ultraviolet (UV) is between 10 and 400 nanometers. Understanding the distribution of solar radiation across this spectrum is crucial for various applications, such as photovoltaic energy conversion, where different materials may absorb or reflect specific wavelengths more efficiently. This figure also highlights the weakness of solar radiation in the 200-280 nm UV range due to absorption by the Earth's ozone layer. This creates a "blind" region where UV cameras can operate with significant advantages, including low background noise, reduced false alarms, low cooling requirements, no need for scanning, and a compact, lightweight design. These features make UV cameras particularly effective for detecting UV emissions in applications where precision and reliability are essential.

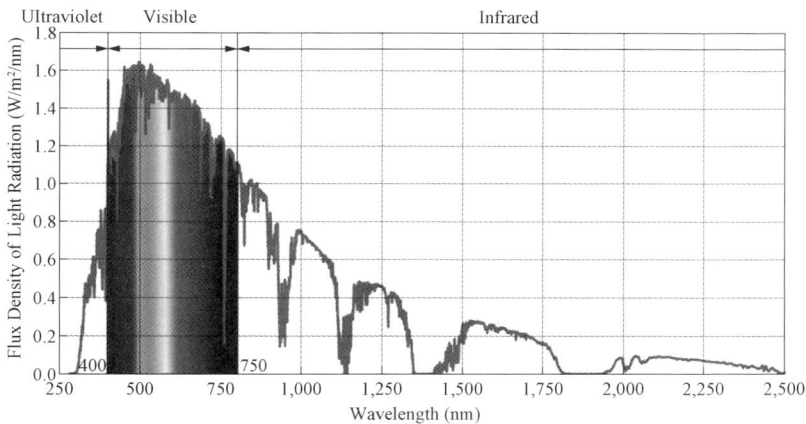

Figure 7.3   The spectrum of solar radiation between wavelength 250 and 2,500 nm

Figure 7.4 illustrates the relationship between the relative intensity of the spectrum produced by electrical discharge in the air and the corresponding wavelength. Figure 7.5 shows the curve in vertical coordinates amplified 100 times between 200 and 300 nanometres.[3] These figures demonstrate the relationship between spectrum intensity and wavelength for electrical discharges in the air. The magnification shown in Figure 7.5 emphasizes that even weak spectral emissions in this UV band can be detected, which is crucial for applications requiring precise and sensitive detection of electrical discharge phenomena.

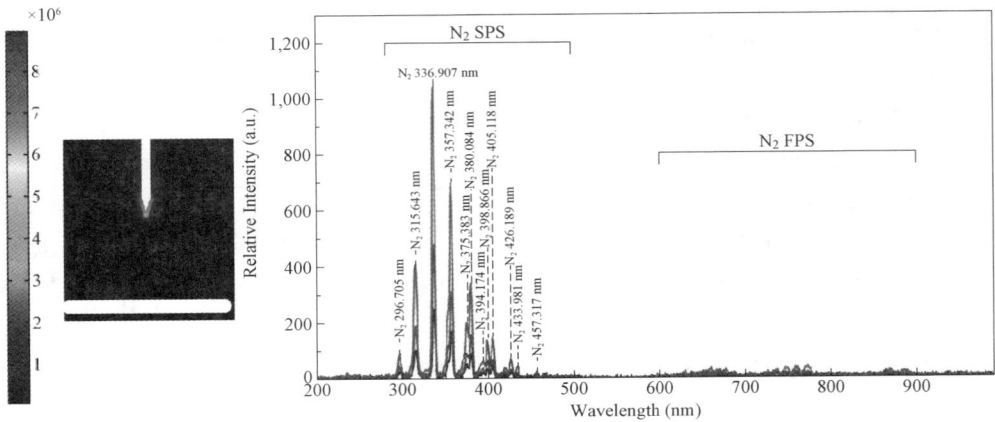

Figure 7.4　The relationship between relative intensity of spectrum due to discharge in the air and corresponding wavelength[6]

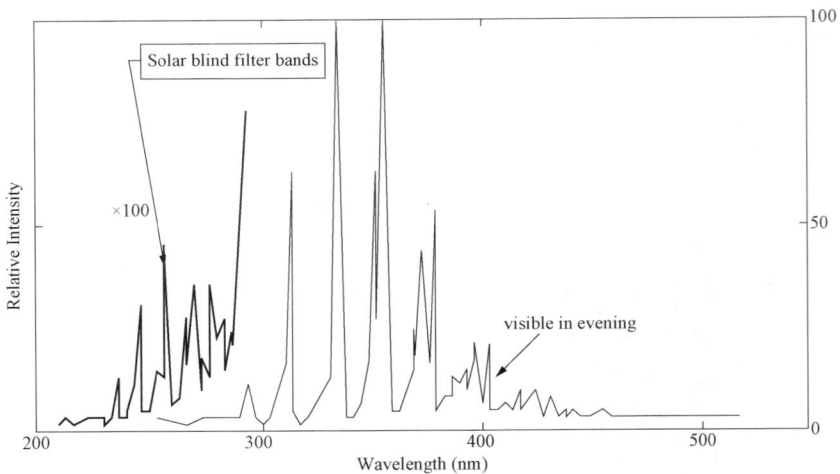

Figure 7.5　The data in vertical coordinates amplified 100 times compared with Figure 7.4 between 200 and 300 nanometres[3]

7.3 Main Components of UV Cameras

When solar radiation reaches the Earth's surface, wavelengths below 280 nm in the ultraviolet spectrum are almost entirely absorbed by the ozone layer in the atmosphere. This absorption creates a "blind area" in solar radiation, meaning that these UV wavelengths (below 280 nm) do not reach the ground and are effectively blocked. The term "solar blind" refers to this region of the spectrum where solar radiation is absent or minimal due to atmospheric absorption. In the solar blind band, the background radiation is extremely low, making it easier to extract signals from targets that emit or reflect UV light in this range. The clean background means there is minimal interference from natural sunlight, which enhances the detection and identification of specific UV signals. The solar blind band is used to detect corona discharges on high-voltage equipment. Since the background UV radiation is low, even weak corona emissions can be detected with high sensitivity. Moreover, this technology has been widely used in missile warning, ultraviolet-aided landing, and ultraviolet search and rescue.[7]

Figure 7.6 shows the internal structure and main components of a UV camera, the main components include UV solar blind filter, UV zoom and UV camera, charge-coupled devices (CCD) or intensified charge-coupled devices (ICCD).

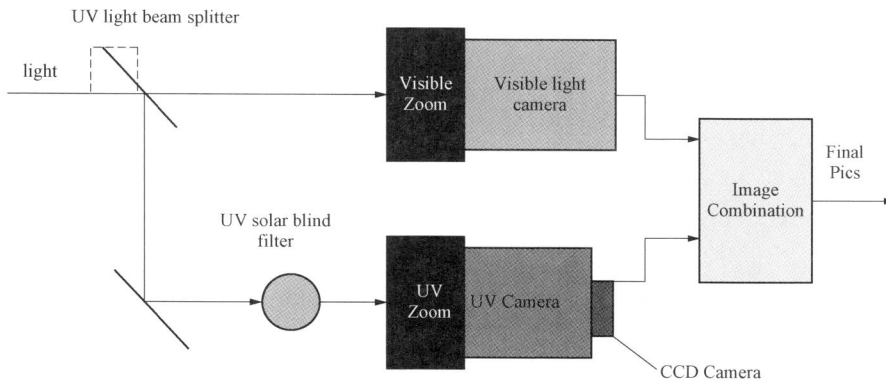

Figure 7.6   The internal structure and main components of a UV camera [3]

7.3.1 UV Filters

The filter transmission waveform between wavelength 200 and 1,000 nm of a typical solar blind filter is shown in Figure 7.7. The UV bands could be divided further as follows:
- Vacuum UV: 10-200 nm.
- UV-C short wavelength UV: 200-280 nm.
- UV-B medium wavelength UV: 280-320 nm.

- UV-A long wavelength UV: 320-400 nm.

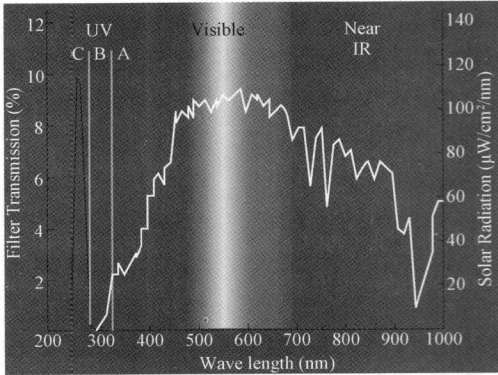

Figure 7.7   The filter transmission waveform
between wavelength 200 and 1,000 nm of a
typical solar blind filter [3]

To fabricate a UV camera, selecting the appropriate UV filter is crucial. Broadband UV filters block background visible and infrared light, but if a broadband filter is chosen, the resulting background noise will be too large, failing to meet the necessary requirements. Additionally, ordinary lenses are typically designed for a specific monochromatic light. Consequently, UV cameras utilize UV narrow-band filters. To capture a clear ultraviolet image, it is essential to filter out background light, which primarily originates from solar radiation. Filtering this background light does not interfere with the ultraviolet image of the target UV light source. Therefore, a filter that can transmit light in the UV-C band is used for UV cameras.

7.3.2   UV Lens

In ultraviolet imaging, the UV signal is processed and superimposed with a visible light image on the screen, allowing for the precise determination of the discharge location and intensity. UV lenses, developed for both X-rays and visible light, operate within a wavelength range of 200 to 385 nanometers. These lenses, equipped with an achromatic design, enable clear image formation across the UV to visible spectrum. They are used in UV cameras or image enhancement tubes for various observational purposes. Additionally, close-up lenses can detect and effectively remove faint fingerprints from materials like glass.

UV lenses feature high precision, high resolution, and light tracking aperture characteristics. In solar-blind imaging, solar-blind UV light is typically present under three conditions: (1) unnatural danger signals such as gunfire, explosions, fires, or corona discharge from high-voltage transmission lines; (2) man-made UV light sources; and (3) abnormal weather conditions like strong lightning strikes. Detection of solar-blind UV signals in a "dark room" indicates specific events, such as a missile attack. Within 15 km of the Earth's surface, this imaging is free from noise interference, allowing target detection without the need for complex image processing.

7.3.3   Photomultiplier Tubes

Photomultiplier tubes (photomultipliers or PMTs for short) are highly sensitive light detectors that are effective across the ultraviolet, visible, and near-infrared spectrums. These detectors can amplify the signal generated by incident light up to 10^8 times, allowing for the detection of individual photons. Thanks to their high gain, low noise, fast response, and large collection area, PMTs continue to be widely used in fields such as nuclear and particle physics, astronomy, medical imaging, and even motion picture film scanning (telecine). Although semiconductor devices like avalanche photodiodes have replaced PMTs in some applications, PMTs remain indispensable in others.[8]

A photomultiplier tube is composed of several key components: an input window, a photocathode, focusing electrodes, an electron multiplier, and an anode, all of which are typically enclosed in a vacuum-sealed glass tube. Figure 7.8 illustrates the schematic construction of a photomultiplier tube.[9]

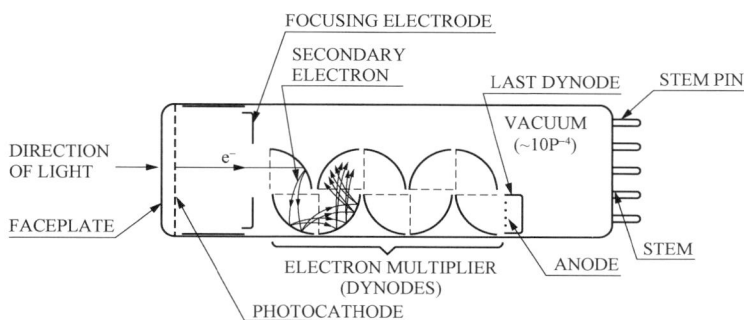

Figure 7.8   Construction of a photomultiplier tube [9]

The light which enters a photomultiplier tube is detected and produces an output signal through the following processes:

- Light passes through the input window.
- Light excites the electrons in the photocathode so that photo-electrons are emitted into the vacuum (external photoelectric effect).
- Photo-electrons are accelerated and focused by the focusing electrode onto the first dynode where they are multiplied by means of secondary electron emission. This secondary emission is repeated at each of the successive dynodes.
- The multiplied secondary electrons emitted from the last dynode are finally collected by the anode.

To efficiently collect photoelectrons and secondary electrons on a dynode and minimize the spread in electron transit time, the electrode design in a photomultiplier tube (PMT) must be optimized. This optimization is achieved through electron trajectory analysis, which is influenced by the electrode configuration, arrangement, and the applied voltage. Numerical analysis using high-speed, large-capacity computers has become a common method for optimizing this design. By dividing the area into a grid and solving the motion equations based on the potential distribution, the electron trajectory can be predicted.

When designing a PMT, particular attention is given to the electron trajectory from the photocathode to the first dynode. The shape of the photocathode, the configuration and arrangement of the focusing electrodes, and the applied supply voltage are crucial factors in ensuring that photoelectrons are efficiently focused onto the first dynode. The collection efficiency of the first dynode, defined as the ratio of electrons landing on its effective area to the total emitted photoelectrons, typically ranges from 60% to 90%. In applications requiring minimized electron transit time, the electrode design must also consider higher electric fields.

The dynode section of a PMT typically consists of several to more than ten stages of secondaryemissive electrodes, which are arranged to enhance collection efficiency and minimize electron transit time spread. The optimal configuration and arrangement of these dynodes are determined through electron trajectory analysis. Additionally, the design must prevent ion or light feedback from the latter stages, which could adversely affect the tube's performance.

Additionally, various characteristics of a photomultiplier tube, such as collection efficiency, uniformity, and electron transit time, can be assessed using computer simulations. By setting the initial conditions for photoelectrons and secondary electrons, a Monte Carlo simulation can predict these parameters, enabling a comprehensive evaluation of PMTs. Figure 7.9, Figure 7.10, Figure 7.11 and Figure 7.12 illustrate cross-sections of photomultiplier tubes with different dynode structures—circularcage, box-and-grid, and linear-focused—depicting their typical electron trajectories.

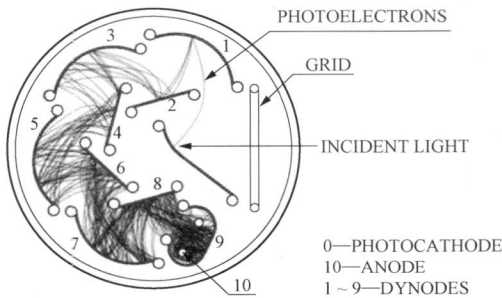

0—PHOTOCATHODE
10—ANODE
1 ~ 9—DYNODES

Figure 7.9   A photomultiplier tube with a circular cage with dynode structure[9]

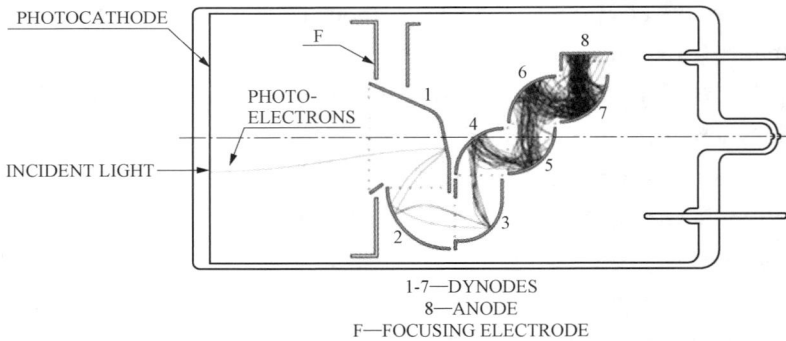

1-7—DYNODES
8—ANODE
F—FOCUSING ELECTRODE

Figure 7.10   A photomultiplier tube with box-and-grid dynode structure[9]

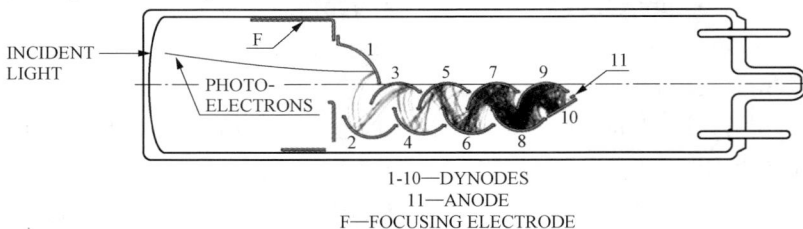

1-10—DYNODES
11—ANODE
F—FOCUSING ELECTRODE

Figure 7.11   A photomultiplier tube with linear-focused dynode structure [9]

Figure 7.12　A typical picture of a photomultiplier tube [9]

The key technology in developing UV image intensifiers lies in constructing suitable ultraviolet photocathodes. The optimal selection of materials for the light window is crucial: quartz or sapphire is required for UV light wavelengths longer than 180 nm, while lithium fluoride is used for wavelengths below 180 nm. Additionally, the photocathode materials must possess "solar blind" characteristics, making Cesium Telluride (Cs-Te) the most suitable option among available cathode materials. Cs-Te has a forbidden bandwidth of 3.5 eV and an electron affinity of less than 1.0 eV. With the appropriate UV cathode, an ultraviolet intensifier can be successfully developed.

7.3.4　Comparison among CCD, EMCCD, ICCD

The research on the detection of solar blind ultraviolet CCD, ICCD, or EMCCD modules is of great significance to the application of solar blind ultraviolet detection technology.[10]

This subsection will compare CCD vs EMCCD vs ICCD and mentions the difference between CCD, EMCCD, and ICCD.

- CCD stands for charge coupled device.
- EMCCD stands for electron multiplying CCD.
- ICCD stands for intensified CCD.

Scientific digital cameras come in four primary types based on the sensor technology employed. These technologies are CCD, EMCCD, CMOS, and ICCD, as shown in Table 7.1.

Table 7.1　Comparison of CCD vs EMCCD vs ICCD and Summarizes Difference between CCD, EMCCD, and ICCD [10]

Features	Ideal	CCD	EMCCD	ICCD
Full form	—	Charge-coupled device	Electron multiplying CCD	Intensified CCD
Quantum efficiency (%)	100	93	93	50
Robust noise	0	10	60	20
Gain	1	1	1,000	1,000
Spurious noise	0	0.05	0.05	0
Dark noise	0	0.001	0.001	0.001
Noise factor	1	1	1.41	1.6

CCD transfers each pixel's charge packet sequentially to convert its charge to a voltage. CCDs consist of an array of thousands to millions of light-sensitive elements known as pixels etched on a silicon surface. Each of the pixels is buried channel MOS capacitor. CCDs are fabricated using a p-type substrate and a buried channel is implemented by forming a thin n-type region on this surface. The size of CCDs is specified in Megapixels. A megapixel value is equivalent to the multiplication of the number of pixels in a row and the number of pixels in a column.

Let us understand the operation of the CCD sensor:

- When the sensor array is exposed to light, the number of electrons (i.e. quantum of charge) held under certain pixels will vary directly as per luminous intensity exposure of that particular pixel.
- Charge is read out by suitable electronics and converted into a digital bit pattern which can be analyzed and stored on the computer. This digital bit pattern represents the image.
- In order to record images in full color, Bayer's color filter array is bonded to the sensor substrate. This filter array is made of alternating rows of red/green, and blue/green filters and is known as an RGBG filter. A particular color filter allows photons of that color to pass through to the pixel.

EMCCD uses an electron-multiplying structure on the chip. Hence EMCCD can detect single photon events without an image intensifier. The EMCCD has essentially the same structure as CCD with the addition of a very important feature. The stored charge is transferred through the parallel registers to a linear register as before but now prior to being readout at the output node the charge is shifted through an additional register, the multiplication register in which the charge is amplified. A signal can therefore be amplified above the readout noise of the amplifier and hence an EMCCD can have a higher sensitivity and higher speed than a CCD. This is due to the amplification of the charge signal before the charge amplifier.

To overcome the limitations of CCD (Charge Coupled Device), ICCD has been developed. It exploits optical amplification provided by an image intensifier to overcome the limitations of a basic CCD sensor. ICCD comprises of image intensifier tube coupled with a traditional CCD sensor device. The light output of the tube is coupled with CCD in two ways viz. using a fiber optic coupler and lens coupled. These are the two possible types of constructions in ICCD. Lens coupled type used a lens between the output of the image intensifier and CCD instead of a fiber coupler. The major component in ICCD design is the power supply which generates DC voltages and voltage pulses for its operation.[11]

In ultraviolet imaging detection systems, the UV radiation is generally very weak, if we use UV-sensitive CCD for detecting weak UV signals directly, due to its strength being too small, the result is undetectable. To solve this problem, we need to enhance UV signal amplification, and then detection. In order to achieve enhanced UV light signal amplification, using ultraviolet light as an intensifier is more appropriate. In the ultraviolet imaging detection system, the characteristics of ultraviolet light should be developed into ultraviolet light as an intensifier has two ways. (1) the development of suitable cathode ultraviolet light, directly to the development of ultraviolet intensifier; (2) by spectral conversion technology, the use of ultraviolet light as an intensifier. Spectral conversion technology and ultraviolet light as intensifier technology have been relatively mature, so it is easy to implement and the process is simple. The former has the advantage of high resolution, while the latter is simple to implement.[12] The choice between CCD, EMCCD, and ICCD depends on the specific needs of the solar blind ultraviolet detection application. ICCDs are favored for their extreme sensitivity to faint UV signals, making them ideal for applications requiring the detection of very weak emissions. EMCCDs, with their low noise and high sensitivity, provide a balanced option where noise minimization is crucial. Standard CCDs, while less sensitive, are still valuable for general imaging applications where extreme sensitivity is not required.

7.4 Application Examples of UV Cameras in Smart Grids

UV imaging technology can be used in, but is not limited to, the following application scenarios:
- The ship navigates through fog.
- Forest fire alarm.
- Power grid security monitoring.
- Maritime search and rescue.
- Satellite navigation.
- The plane made a blind landing through fog.
- Missile approach warning.
- Identification of document security features under daylight conditions (passport, license, etc.).
- Close criminal investigation; Seek potential.

In this section, we shall focus on the power grid application of UV cameras.[13, 14, 15]

When high-voltage equipment discharges electricity, phenomena such as corona discharge, arc flash, or arc discharge can occur depending on the electric field strength. During these events, electrons in the air continuously gain and release energy, emitting ultraviolet (UV) rays. UV imaging utilizes this principle to detect UV signals, and it involves solar-blind imaging techniques. Solar radiation in the wavelength band of 190-285 nm is completely absorbed by the ozone layer as it passes through the atmosphere. The scattering by atmospheric components and absorption by surface ozone below the ozone layer further contribute to creating a natural "solar-blind" region near the ground, where naturally occurring solar signals are almost undetectable.

When partial discharge occurs, the intensity of UV radiation is directly related to the electric field intensity. The ultraviolet detection system can measure the intensity of discharge by counting the number of corona pulses within a unit of time. Table 7.2 shows the relationship between electric field intensity and UV radiation intensity.

Table 7.2   The relationship between electric field intensity and UV radiation intensity [3]

Electric field intensity	Photon number per minute	Descriptions	Countermeasures
High Intensity	≥5,000	The corrosion speed is fast and components already damaged seriously	Repairing or replacement of damaged components quickly
Middle Intensity	1,000 - 5,000	Maybe corroded and have some evidence of damage	Making a plan for maintenance or replacement
Low Intensity	<1,000	May shorten the lifespan or parts may be slightly damaged	Pay attention to corona development

Using a specific UV sensor to receive ultraviolet signals generated by the discharge of highvoltage equipment, the intensity of the discharge can be determined through statistical processing by counting the number of discharge pulses. This provides a basis for online condition monitoring of high-voltage transmission lines. Figure 7.13 (a), (b), and (c) show overlapped images combining daylight and UV photographs.

(a) Daylight photo

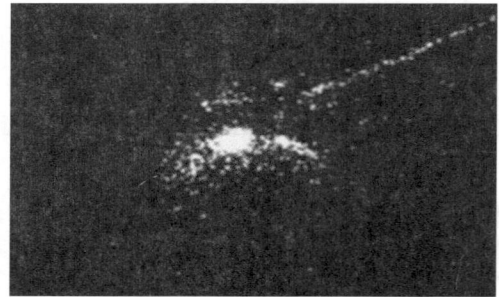

(b) UV photo

(c) Overlapped imaging photo

Figure 7.13   The overlapped imaging by daylight photo and UV photo [3]

The main electrical applications of UV imagers are concentrated in the following areas:

- Detection of deterioration in insulators and composite insulators, as well as electrical corrosion of their jackets.
- Maintenance of high-voltage substations and transmission lines.
- Detection of micro-cracks in support insulators.
- Insulation testing of suspension insulators.
- Evaluation of the layout, structure, installation, and acceptance of high-voltage electrical equipment to assess the reasonableness of the design.
- Detection of flashover marks on insulators during the operation of power equipment.
- Identification of potential overlaps between metal wires and conductive objects in highvoltage power transmission and transformation equipment.
- Detection of corona discharge at the stator rod end and slot wall of large generators.
- Searching for sources of radio interference; electrical equipment may produce strong radio interference that affects nearby communication and TV signal reception. Ultraviolet imaging technology can be used to quickly locate the source of this interference.
- Performing partial discharge tests on high-voltage electrical equipment using ultraviolet imaging technology to find or locate external discharges.

The main features of UV imaging detection include:

- Utilizing solar-blind ultraviolet imaging to detect corona, flashover, or arc discharges, completely unaffected by solar radiation.
- The instrument can be used in any weather conditions, including sunny days, rainy days, fog, etc.
- High detection efficiency; remote testing is possible, allowing for comprehensive testing of all facilities.
- The instrument has high sensitivity for detecting tiny discharges, capable of identifying as low as 1.5 pC. Since partial discharges can cause component corrosion and produce larger discharges, this helps prevent the formation of a vicious cycle.
- Easy to operate with intuitive results; the instrument can be handheld or mounted on cars, trains, or planes. The instrument can take photos or record videos to clearly identify problem areas.
- The testing method is safe and reliable, allowing for detection without taking the equipment out of service.
- Quantitative measurement of ultraviolet intensity, allowing for classification based on intensity levels.

The solar blind UV imager is widely used currently in electric power industry, especially in detecting the discharge phenomena of high voltage equipment such as corona discharge, flashover and arc discharge. However, despite the advantages of this technology in some aspects, it still has some disadvantages and shortcomings, include:

- Environmental adaptability: Although sun-blind UV imagers can work in sunlight, performance may suffer in extreme light conditions (such as strong direct sunlight or areas with high UV reflection). Even in the case of sun-blind filters, it is difficult to completely eliminate interference from the environment.

- Detection distance and viewing angle limitation: When the sun-blind ultraviolet imager detects at a distance, the detected ultraviolet signal may be weakened, resulting in the detection of distant discharge phenomena not being sensitive enough to capture weak discharge signals. Furthermore, UV imagers typically have a narrow viewing angle, requiring adjustment of the device position several times to fully cover the detection area, increasing the complexity and time cost of operation.

- Cost and maintenance: Sun-blind UV imagers are generally more expensive than other imaging equipment, especially high-precision equipment used in the electric power industry. The high cost of the device may limit the popularity of applications in small and mediumsized enterprises.

- Complicated maintenance needed: Sun-blind UV imagers require regular calibration and maintenance to ensure their performance and accuracy. This not only increases maintenance costs, but also requires a high level of skill on the part of the operator.

- Challenges of quantitative analysis: At present, the detection results of sun-blind UV im-agers are mostly qualitative, with a lack of unified quantitative standards. This makes it difficult to assess the severity and potential risk of discharge, and is not conducive to comparisons between different devices or measurements.

- Complicated data processing: Quantitative analysis of UV images requires complex image processing algorithms, especially in field environments where the accuracy and reliability of data processing can be challenging.

- Risk of false positives and omissions: In some cases, sun-blind UV imagers may produce false positives due to UV reflections or other optical phenomena in the environment, resulting in unnecessary maintenance operations.

- Underreporting problems: Some weak discharges may go undetected due to the sensitivity limitations of the equipment, which can lead to potential power system failures that are not prevented in time.

- Operational complexity: The operation of the sun-blind UV imager requires professional technical knowledge and experience, and the operator needs to be specially trained. This limits the application of the device to a certain extent; Although the sun-blind UV imager can work under various weather conditions, it may still need to be manually adjusted and set according to the actual situation in a complex power environment, increasing the complexity of operation.

Among the nine disadvantages and shortcomings mentioned above, particularly the detection distance and viewing angle limitation, as well as the challenges of quantitative analysis, some Chinese researchers are making continuous efforts to improve these aspects.

Owing to the fact that the solar blind UV imager currently only provides qualitative results, it is challenging to assess the severity and potential hazards of discharges due to the lack of standardized testing result assessments. Some Chinese researchers [16] have developed software that evaluates the number and size of spots detected in CCD, providing a quantitative basis and guidance for field UV detection. For example, they consider the two factors of observation distance and imager's gain as independent variables to establish the relationship with the apparent discharge magnitude. According to the discharge UV image features, the discharge facular regions in UV images were extracted using a digital image processing algorithm, and then the facular area parameter was defined. In the laboratory, the relationships of facular area to apparent discharge magnitude, imager's gain, and observation distance were studied respectively. Research shows that the facular area increases with the enhancement of discharge intensity, but it has a non-linear characteristic. In the gain range of 50%-80%, the relationship between facular area and gain follows an approximate exponential function, and when the gain is 50%, 60%, 70%, and 80%, respectively, the relationship between facular area and observation distance follows an approximate power function. Based on the above research, with the sample data, an adaptive neuro-fuzzy inference system (ANFIS) model was established, which successfully estimated the apparent discharge magnitude with high prediction accuracy.

7.5 Comparison of Ultraviolet and Infrared Imaging

Figure 7.14 (a) and (b) provide a comparison between ultraviolet imaging and infrared imaging. The left image (a) represents ultraviolet imaging, while the right image (b) shows infrared imaging. Both images were captured from similar poles with overhead lines and electrical equipment. The ultraviolet imaging highlights areas of electrical discharge, such as corona effects, while the infrared imaging reveals heat signatures from the equipment and lines.

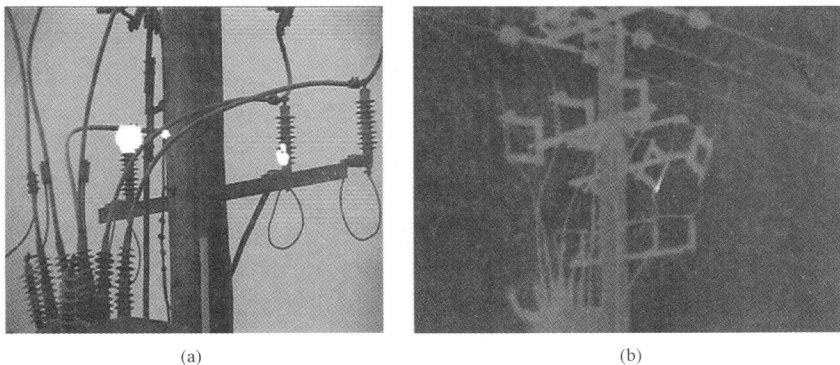

(a)   (b)

Figure 7.14   An example for comparison of ultraviolet imaging and infrared imaging [3]

It is evident that UV imaging can be used to evaluate over-excitation due to high voltage or detect defects that produce ultraviolet radiation caused by high electric field intensity. On the other hand, infrared imaging is effective for detecting overloads by identifying temperature differences or defects associated with high current.

Table 7.3 shows their detailed comparison by technical specifications.[3]

Table 7.3   **Comparison between infrared and ultraviolet Imaging[3]**

Item	Ultraviolet UV detection	Infrared IR detection
Spectrum range	230-280 nm, daylight corona or discharge detection	8-14 μm
Application	Corona produce ultraviolet radiation	Maybe corroded and have some evidence of damage
Reasons for defects	Insulator; bushing; Wire; Pollution; Wire damage; The separator relaxation; sharp edges of the parts; Insulator improperly installed; Lack of curved horns	Current type resistance defects due to hot spot; unreliable connection; charged insulator internal defects, strong arc
Related factors	Closely related to voltage	Closely related to current
On load during inspection	unnecessary	necessary
Interference by sunshine	No, sunlight blind detection	Yes
Weather conditions	High humidity, low air pressure and high temperature will benefit to partial discharge	High temperature will disturb detection, avoid raining day detection
Audible noise	Defect detection will lead to the audio noise	No signal
Interference	Will have Interference to TV or radio	No signal
Channel number	Dual channel: visible light + UV, overlapped imaging and audio	Single channel or visible + infrared
FOV	Narrow, details could be seen The early-stage defects or	Much wider than UV imager
Inspection stages	degradations can be generally detected	Late, and seriously damage of defects could be detected

7.6 Review Questions

Q1: Who is Johann Wilhelm Ritter? How did he discover the existence of ultraviolet light?

Q2: Please tell us the wavelength range of ultraviolet light?

Q3: According to ISO standard 21,348, the Ultraviolet light falls into three categories based on wavelength, would you tell us three wavelength range one by one?

Q4: The primary source of UV light is the Sun, which emits radiation across the entire UV spectrum. However, only UVA and UVB radiation reach the Earth's surface, as the ozone layer absorbs UVC. How many percentage of UVA, UVB and UVC could reach the surface of the Earth?

Q5: The ozone layer is a crucial component of Earth's stratosphere that absorbs most of the Sun's UVC radiation and a portion of UVB radiation. According to the Kyoto Protocol, seven kinds of chemical compound are harmful to the ozone layer, would you tell us the names of them?

Q6: Would you tell us the working principles of the solar blind UV camera?

Q7: What are the shortcomings of the commercial solar blind UV camera? Would you tell us your plan how to improve them?

Bibliography

[1] Küchler, Andreas. High Voltage Engineering: Fundamentals-Technology-Applications. Springer, 2017.

[2] Pavia, Donald L and Lampman, Gary M and Kriz, George S and Vyvyan, James A. Introduction to spectroscopy. Nelson Education, 2014.

[3] OFIL company, OFIL UV imager testing training handout. [Online]. Available: https://ofilsystems.com. Accessed: Nov. 2020.

[4] Hu, Yi and Liu, Kai. Inspection and Monitoring Technologies of Transmission Lines with Remote Sensing. Academic Press, 2017.

[5] Smets, Arno and Jäger, Klaus and Isabella, Olindo and Van Swaaij, René and Zeman, Miro. The physics and engineering of photovoltaic conversion, technologies and systems. Solar Energy. Cambridge: UIT, 2015.

[6] Yanfei Li. Research on discharge pattern recognition of gas insulated electrical equipment based on emission spectroscopy. Master's thesis, University of Chinese Academy of Sciences, 2021.

[7] Richards, Austin A. Alien vision: exploring the electromagnetic spectrum with imaging technology. SPIE, 2011.

[8] Wright, AG. The photomultiplier handbook. Oxford University Press, 2017.

[9] Hamamatsu Photonics, K.K. Photomultiplier tubes. basics and applications. Edition 3, 2006.

[10] RF Wireless World. CCD vs EMCCD vs ICCD | difference between CCD, EMCCD, ICCD. [Online] //https://www.rfwireless-world.com/Terminology/ CCD-vs-EMCCD-vs-ICCD.html/. Accessed: August 2024.

[11] Delin Liu. Study on the technology of solar blind ultraviolet ICCD module. Master's thesis, Southeast China University, 2017.

[12] Muhan Cui. Parametric Test and Radiometric Calibration of the Solar-Blind UV Image Intensifier and Intensified CCD. PhD thesis, Changchun Institute of Optics, University of Chinese Academy of Science, 2016.

[13] Yunpeng Liu. Research on corona onset characteristics of conductors at different altitudes based on uv imaging technology. Master's thesis, North China Electric Power University, 2008.

[14] Rijun Dai. Discharge Detection Method of High Voltage Electrical Equipment Based on Ultraviolet Signal. PhD thesis, North China Electric Power University, 2012.

[15] Shenghui Wang. Detection and Assessment of Containated Suspension Insulator Discharge Based on Ultraviolet Imaging. PhD thesis, North China Electric Power University, 2010.

[16] Wang, Shenghui and LV, Fangcheng and Liu, Yunpeng. Variation characteristic of composite insulator corona discharge ultraviolet image parameter and estimation of discharge magnitude. Proceedings of the CSEE, 33(34):233-240, 2013.

8

Electrochemical Gas Sensors for GIS Off-line Detection

8.1 Introduction

High-voltage switchgear, a crucial component in any substation (or switching station), is fundamentally a combination of switching and measuring devices such as circuit breakers (CBs), voltage transformers (VTs), and current transformers (CTs). The circuit breakers are responsible for connecting and disconnecting circuits, while the instrument transformers monitor the system and detect faults. The selection of switchgear is critical, as the security of the power supply relies heavily on the reliability of these switching and measuring devices.[1]

The reliability of circuit breakers depends on several factors, including insulation integrity, circuit-breaking capability, mechanical design, and current-carrying capacity. In modern switchgear practice, circuit breakers can be broadly classified based on the insulating medium used for arc extinction. The main types include: bulk-oil circuit breakers, air-break circuit breakers, SF_6 gas circuit breakers, small oil-volume circuit breakers, gas-blast circuit breakers, and vacuum circuit breakers. Each type has its own advantages and applications depending on the specific requirements of the power system.

Recent developments in the field of high-voltage switchgear have seen a significant shift towards "oil-free" devices. There is an increasing demand for vacuum circuit breakers and rotating-arc SF_6 circuit breakers in distribution systems. Additionally, the use of SF_6 circuit breakers has expanded, with applications now extending to system voltages up to 1,000 kV.

This chapter will focus on transmission and distribution SF_6 switchgear and explore various condition monitoring techniques used to ensure their reliability and efficiency.

8.2 Basic Principles

The insulating fluid in a circuit breaker plays a critical role in both arc extinction and providing electrical insulation. When the circuit breaker operates, the contacts separate, and an electrical arc is drawn between them. This arc must be extinguished quickly to prevent damage to the breaker and ensure safe operation. The insulating fluid helps to cool and dissipate the energy of the arc. It absorbs the heat generated and provides the necessary dielectric strength to interrupt the current flow. The fluid's ability to deionize the medium between the contacts helps in extinguishing the arc. The insulating fluid must also provide reliable electrical insulation to prevent any unwanted conduction between the separated contacts and from the contacts to the earth. The fluid should have high dielectric strength, ensuring that even at high voltages, there is no breakdown leading to flashovers or unintended current paths. [1]

The selection of insulating fluid is influenced by the type and rating of the circuit breaker. For instance, different circuit breakers like air circuit breakers (ACB), oil circuit breakers (OCB), sulfur hexafluoride (SF_6) circuit breakers, and vacuum circuit breakers (VCB) use different fluids or methods. Higher voltage circuit breakers generally require fluids with higher dielectric strength to prevent insulation failure. High current applications need fluids that can effectively extinguish large arcs, preventing damage and ensuring safety. The choice of insulating fluid or medium is therefore critical and must be carefully matched to the specific application and operational requirements of the circuit breaker.

The insulating media commonly used for circuit-breakers and their features are:

- Atmospheric air at ordinary pressure 0.1 MPa: It was used in low to medium voltage circuit breakers. Air serves as the insulating medium but isn't as effective at arc extinction as other methods.
- Oil (which produces hydrogen for arc-extinction) : In oil circuit breakers, the oil acts as both the arc extinguishing medium and the insulator. The oil decomposes under the heat of the arc, forming gas bubbles that help quench the arc.
- Sulphur hexafluoride SF_6 at pressure <0.85 MPa: Sulfur hexafluoride gas is used in high voltage circuit breakers due to its superior insulating and arc-quenching properties. It is non-flammable and has excellent dielectric strength.
- Vacuum: In vacuum circuit breakers, the contacts are separated in a vacuum, where no arc can persist for long due to the absence of any medium for ionization.

The dielectric strength of atmospheric air is considerably lower than that of other insulating media like SF_6 gas or oil, making it less effective at preventing electrical breakdowns, especially at high voltages. This necessitates larger clearances and geometrical dimensions to prevent arcing, leading to larger transmission towers, substations, and power lines. The need for larger infrastructure increases material and land costs, making air-insulated systems less economically viable. The substantial size of these installations often detracts from the visual aesthetics, raising concerns within communities and among environmentalists. Air-insulated systems often exhibit nonuniform electric fields, leading to localized high-stress points. These can result in partial discharge, corona formation, and eventual dielectric breakdown, all of which threaten system reliability. Systems relying on air insulation are highly susceptible to environmental factors like humidity, temperature, and atmospheric pressure, which can degrade their performance and increase the likelihood of insulation failure. Dust, salt, and industrial emissions further compromise air's insulating properties, necessitating more frequent maintenance and decreasing the risk of flashover events. Therefore, the inherent limitations of the atmosphere as an insulating medium, in particu-lar its low dielectric strength and sensitivity to environmental conditions, pose a major challenge in high-pressure applications. These challenges drive up costs, impact system reliability, and lead to aesthetic and enviro-nmental issues, making it necessary to explore alternative insulation solutions or employ advanced design and maintenance strategies to mitigate these issues. [2]

To address the challenges associated with atmospheric air as an insulating medium, there is a need to develop electrical devices with more compact designs. This can be accomplished by transitioning from the highly heterogeneous fields found in atmospheric air-insulated systems to the weakly heterogeneous fields present in gas-insulated systems (GIS), also known as "metalclad" systems. In these weakly inhomogeneous fields, the insulating properties of the medium are more efficiently utilized, allowing for a reduction in the size of the device while maintaining the same rated voltage. This shift not only enhances the performance and reliabi-lity of the system but also contributes to more economical and aesthetically favorable installat-ions.

The introduction of gaseous dielectrics with high dielectric strength, such as sulfur hexa-fluoride (SF_6), has significantly contributed to the reduction of geometrical dimensions in electrical installations. Since the 1960s, SF_6 has been widely adopted as a superior alternative to air and nitrogen due to its excellent insulating and arc-quenching properties. This gas has enabled the development of more compact and efficient gas-insulated systems (GIS). After the initial introduction of GIS technology, Japan installed its first complete 550 kV GIS system in 1976. The adoption of GIS rapidly expanded, and by 1986, nearly 40% of newly constructed and replacement substations in Japan were utilizing this technology. GIS systems with rated voltages up to 765 kV have been in service for over 25 years, demonstrating the long-term reliability and effectiveness of SF_6-insulated equipment. By the year 2,000, Japan had develo-ped the world's first complete SF_6 gas-insulated substation with a maximum rated voltage of 1,100 kV, marking a significant milestone in the evolution of high-voltage technology.

The use of sulfur hexafluoride (SF_6) gas in high-voltage circuit breakers led to the abbre-viation "GIS" for "Gas Insulated Switchgear". As technology advanced and complete metal-clad substations utilizing SF_6 gas were developed, the same abbreviation, GIS, was adopted by many countries to refer to "Gas Insulated Substations". Both terms are widely used, though their meanings can vary depending on the region. In addition, SF_6 gas-insulated coaxial cables are referred to as "Compressed Gas Insulated Transmission Lines" (GIL) by some, highlig-hting another application of SF_6 technology. Interestingly, the abbreviation GIS became more commonly recognized than the full terms it represented. To create a more standardized and inclusive terminology, some authors in recent literature have begun using the term "Gas Insul-ated Systems" to encompass all SF_6 gas-insulated technologies, including switchgear, substations, and transmission lines. This broader term helps streamline communication and reflects the diverse applications of SF_6 gas in modern electrical systems.

Figure 8.1 shows an example of 1,000 kV outdoor metal-clad SF_6 gas insulated substation (GIS), Figure 8.2 shows an example of 1,000 kV indoor metal-clad SF_6 gas insulated power transmission line (GIL) in the tunnel across the Yangtze river. [3]

Figure 8.1   An example of 1,000 kV outdoor metal-clad SF_6 gas insulated substation (GIS)

Figure 8.2   An example of 1,000 kV indoor metal-clad SF_6 gas insulated power transmission line (GIL) in the tunnel across the Yangtze river

8.2.1 Properties of Sulphur Hexafluoride SF_6 Gas

It is recommended that gases used in electrical insulation systems possess the following characteristics for optimal performance:[2]

- High dielectric strength: The gas should exhibit a high dielectric strength to ensure effective insulation.
- Resistance to liquefaction: Liquefaction should not occur within the operating temperature range of −40 ℃ to +60 ℃ or under design pressure conditions, as the presence of liquefied gas in a closed volume can significantly reduce the medium's insulating capability.
- Chemical stability: The gas and its decomposition products should be chemically inert and not react with other materials such as metals, solid dielectrics, or sealing components. Stability is crucial, as decomposition products might possess inferior properties, including lower dielectric strength.
- Safety: The gas and its decomposition products must be non-toxic and non-flammable to ensure the safety of personnel working with the equipment.
- Inertness and density: Given the complexity of electrical equipment assembly, the gas should ideally be inert and possess a high density to enhance insulation performance.
- Environmental impact: The gas should not contribute to the greenhouse effect, minimizing its environmental footprint.

SF_6 gas is currently the most significant gaseous dielectric used in power systems. It is employed in a variety of high-voltage applications, including SF_6 gas-insulated switchgear (such as circuit breakers and isolators), coaxial gas-filled cables (also known as gas-insulated transmission lines), power transformers, and entire substations that integrate circuit breakers, isolators, grounding switches, busbars, and instrument transformers. These components are utilized at the highest rated voltages of 1,000 kV AC or ±800 kV DC. To optimize cost-efficiency, extensive research has been conducted to identify suitable mixtures of SF_6 with other gases to evaluate their effectiveness and applicability in power systems.

Figure 8.3 and Figure 8.4 show the molecular structure of SF_6.[4]

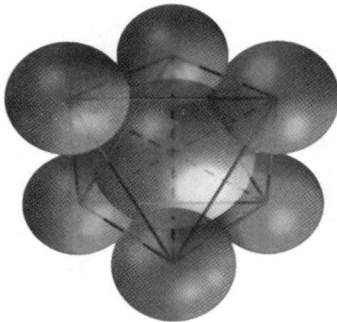

Figure 8.3 SF_6 molecular structure is octahedron
and completely symmetric nonpolarity

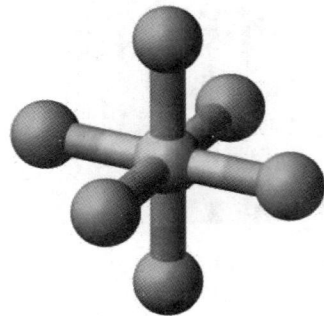

Figure 8.4   F-S-F chemical bond angles of SF_6
molecular are all 90 degree

In a SF_6 molecule, six fluorine atoms arrange themselves uniformly like an octahedron on a central sulfur atom. An excited sulfur atom can therefore form six stable covalence bonds with the strongly electronegative fluorine atoms by sharing a pair of electrons. Amongst halogens, the fluorine element and the sulfur atom both have very high coefficients of electronegativity, of the order of 4 and 2.5, respectively. This coefficient is a measure of the tendency to attract electrons of other atoms to form a dipole bond. The S-F bond length is 1.58×10^{-10} m, its molecular equivalent diameter is 4.58×10^{-10} m.

The rigid, symmetrical structure, short bond length, and high binding energy between the atoms in an SF_6 molecule confer exceptional stability, making the gas's properties resemble those of noble gases at relatively low temperatures. Thermal dissociation of highly purified SF_6 gas only begins at extremely high temperatures, typically above 1,000 K, which are only encountered in power systems during electrical arcs. Even at sustained temperatures up to around 500 K, there have been no reports of thermal decomposition or chemical reactions between SF_6 gas and other materials. Additionally, SF_6 is a non-toxic, colorless, odorless, non-flammable, non-explosive gas that is chemically inert and thermally stable. These attributes make SF_6 gas particularly well-suited for use in power system equipment.

The density of gases generally increases with their relative molecular mass. Given the relatively high molecular mass of SF_6 (146), the gas exhibits a high density. At 20 ℃ and 101.325 kPa, its density is 6.16 g/L, which is about five times that of air, leading to a strong asphyxiating effect. The high density of SF_6 results in a shorter mean free path for charge carriers. Combined with its properties of electron attachment, such as high electronegativity and high ionization energy, this contributes to the gas's high dielectric strength.

The density of a gas is fundamentally influenced by external pressure and temperature. Figure 8.5 illustrates the phase states of SF_6, liquid or gaseous, under varying pressures and temperatures for different constant gas densities. This diagram is crucial for the practical application of SF_6 gas in GIS. It is essential that within the operational pressure and temperature range, the gas does not liquefy, as this would adversely affect its electrical insulation properties.

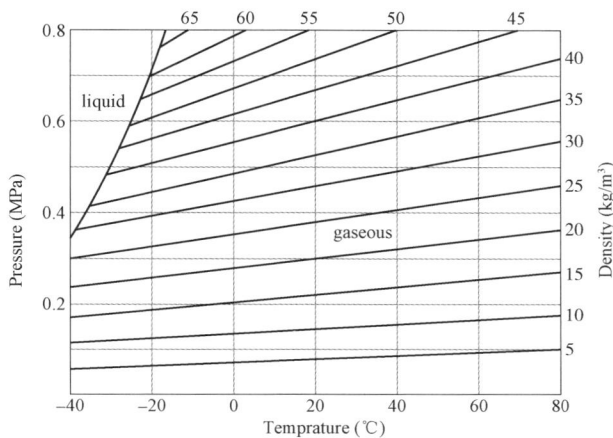

Figure 8.5   States of SF_6 gas for different constant gas densities and varying temperature

and pressure within practical application range[2]

8.2.2　Electrical Characteristics of SF₆ Gas

Figure 8.6 illustrates the breakdown voltage-pressure characteristics of SF_6, CO_2, N_2, and air in rod-plane gaps under alternating voltage. The results are presented for a 5 mm electrode gap spacing, with the relative gas pressure varied within the range of 100 to 500 kPa. The experimental data shows that the breakdown voltages of these gases increase approximately linearly with pressure for the 5 mm gap spacing. Since short rod-plane gaps resemble a uniform field, there is no discontinuity in the breakdown voltage-pressure characteristics for short electrode gaps. The breakdown voltage of SF_6 is observed to be about twice that of air, N_2, and CO_2.[5]

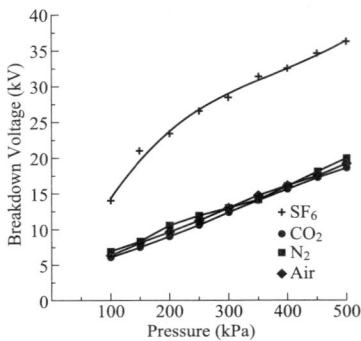

Figure 8.6　Variation of breakdown voltage with pressure in SF_6, CO_2, N_2, Air, for 5 mm gap spacing [5]

The high breakdown strength of electronegative gases like SF_6 is primarily due to their ability to capture free electrons, forming heavy negative ions. This electron attachment process significantly enhances the dielectric strength of these gases compared to others. While the ionization coefficients of electronegative gases are similar to those of other gases, their superior electron attachment properties contribute to their substantially higher electrical strength.

Under uniform electric fields, the breakdown voltage of SF_6 is generally 2.5 to 3.0 times higher than that of air at the same pressure. At a gas pressure of 0.3 MPa, the electric strength of SF_6 is comparable to that of insulating oil. If the gas pressure exceeds 0.3 MPa, the breakdown voltage of SF_6 surpasses that of insulating oil.

8.3　Decomposition of SF₆ and Its Mixtures in Gas Insulated Equipment

Analyzing decomposition byproducts is highly appealing for partial discharge detection in GIS because chemical decomposition remains unaffected by the electrical interference that is typically present in these systems. With a continuous discharge, the concentration of diagnostic gases should gradually increase to detectable levels, provided that an absorbing reagent is not used in the chambers.

The primary decomposition product of sulfur hexafluoride (SF_6) is sulfur tetrafluoride (SF_4), a highly reactive gas. This compound further reacts, usually with trace amounts of water vapor, to form more stable byproducts, such as thionyl fluoride (SOF_2) and sulfuryl fluoride (SO_2F_2). These two gases are the most common diagnostic indicators and can be detected with sensitivities down to 1 μL/L using gas chromatography and mass spectrometry. Although chemical detection tubes offer a simpler but less sensitive alternative, they are capable of detecting a relatively small discharge of 10 to 15 pC in small-volume laboratory tests, typically after several hours. However, in GIS, the diagnostic gases would be significantly diluted by the large volume of SF_6, necessitating much longer detection times. Consequently, chemical detection tubes may be too insensitive for effective partial discharge monitoring in GIS.[6]

The increasing use of SF_6 in gas-insulated substations (GIS), gas-insulated transmission lines (GIL), gas-insulated transformers (GIT), and similar equipment has raised significant concerns about the mechanisms of SF_6 decomposition and the impact of its byproducts. When exposed to conditions such as corona discharge, spark breakdown, or electric power arcs, SF_6 decomposes into lower sulfur oxyfluorides. These compounds can react with electrodes, gas impurities, or other solid dielectrics, leading to the formation of various chemically active byproducts. While SF_6 itself is chemically inert and non-toxic, its decomposition products are known to be both toxic and corrosive. The accumulation of these byproducts within the equipment has led to concerns about personnel safety, material compatibility, and the overall lifespan of the equipment.[7, 8, 9, 10]

The decomposition of SF_6 gas is significantly affected by gaseous impurities. In industrialgrade SF_6, common impurities include CF_4, N_2, O_2 (air), and H_2O. These impurities are typically introduced during the gas filling process and partly due to the desorption of moisture into the dry SF_6 after filling. Standards such as IEC 60376[11] and GB 12022[12] provide recommended values for acceptable levels of impurities and moisture content. However, in practical GIS environments, the presence of such impurities is unavoidable.

Three fundamental processes can lead to the decomposition of SF_6: electronic, thermal, and optical. In high-voltage gas-insulated equipment, electronic and thermal processes are the most significant, resulting in the formation of stable decomposition products. Both of these processes can occur in discharges, such as spark breakdown and power (heavy current) arcs. The decomposition of SF_6 during partial breakdown primarily occurs under non-thermodynamic equilibrium conditions, where the electron temperature exceeds the gas temperature.

In the presence of electrical arcs, flashovers, partial breakdowns, or corona discharges, a portion of SF_6 decomposes into lower sulfur fluorides, which can react to form various chemically active byproducts. The possible formation of SF_4, SF_2, S_2F_{10}, SO_2, SOF_2, SOF_4, SO_2F_2, SOF_{10}, $S_2O_2F_{10}$, HF, and H_2S during the degradation of SF_6 is well-documented and widely accepted. These compounds include gaseous sulfur fluorides and sulfur oxyfluorides, as well as metal fluorides formed through reactions with electrode materials and spacers. Experiments have shown that the production of certain contaminants, such as S_2F_{10}, can be influenced by the presence of water, oxygen, or even surface reactions that occur in the presence of organic insulating materials under corona, flashover, and arc conditions. Table 8.1 presents the typical byproducts and their concentrations.

Table 8.1   Typical concentration of byproducts found in a test sample of sparked SF_6

By-Product	Concentration in $\mu L/L$	Percentage (%) relative to SF_6
$SOF_2(SF_4)$	3,870	1.10
SOF_4	720	0.22
SiF_4	78	0.03
SO_2F_2	140	0.015
SO_2	—	0.002
Total	—	1.37

For the experimental conditions, E-the total discharge energy = 36 kJ, p-pressure of SF_6 = 0.133 MPa, t-time following discharge = 24 h.

The decomposition products of SF_6 due to arcs are calculated as a function of temperature using known thermochemical data for SF_6 and its decomposition products. As the temperature exceeds 1,500 K, there is a sharp decline in SF_6 concentration, accompanied by corresponding increases in SF_4 and fluorine (F) densities. The equilibrium concentrations of SF_6 and its decomposition products at temperatures between 1,000 and 3,500 K have been validated through shock tube studies, and the decomposition rates have been determined. At 4,000 K, the SF_6 molecules are nearly completely dissociated, though most of the electrons remain bound due to the formation of negative ions. Beyond 15,000 K, the particles are predominantly positive ions and free electrons. In the core of a high-current arc column, the temperature has been measured to be approximately 20,000 K.

As the arc cools, the recombination of sulfur and fluorine to form SF_6 occurs rapidly. However, in the presence of oxygen, H_2O, and metal vapor, resulting from electrode heating, the recombination process is altered, leading to the formation of various arc by-products, primarily SF_2, SF_4, and fluorine (F). The most stable of these, SF_4, diffuses out of the arc zone, where it predominantly reacts with H_2O to form lower oxyfluorides. A reaction scheme illustrating the entire family of decomposition by-products derived from SF_4 as the primary decomposition product is shown in Figure 8.7.

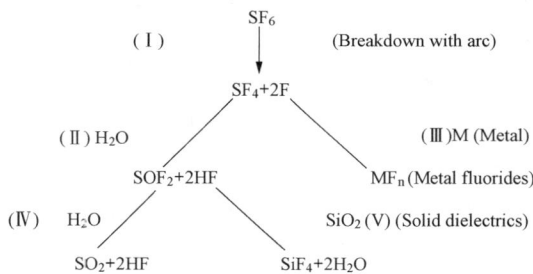

Figure 8.7   Reaction scheme for arced SF_6 leading to the formation of the long-lived decomposition products

8.4 Electrochemical Sensors for GIS

Electrochemical methods for sensing various gas molecules offer a low-power approach while maintaining strong analytical performance, including sensitivity, selectivity, and cost-effectiveness. For a wide range of applications, such as online monitoring, the simplicity, affordability, compact size, and minimal power consumption of the sensor are essential. [13] Electrochemical sensors are highly versatile, as they can detect a broad spectrum of toxic gases, such as CO, NH_3, SO_2, H_2S, NO, NO_2, as well as oxygen, making them ideal for miniaturization. An important advantage of liquid and polymer electrolyte electrochemical gas sensors is their ability to operate at room temperature, eliminating the need for a power-consuming heater and preventing disturbance to the gas sample and sensing environment by the measuring device. Room temperature operation is also a critical factor for achieving intrinsically safe performance in potentially hazardous situations.

Electrochemical sensors generally have a reported lifespan of several years; as with all sensors, the actual longevity depends on the conditions of use, but a lifespan of 5 years or more is not uncommon.

The electrochemical sensing approach, encompassing potentiometric, amperometric, and conductometric sensors, offers a compelling combination of analytical performance and cost-effectiveness for many applications. As a result, a wide range of electrochemical sensors are utilized in realworld gas detection, both in stationary and portable devices. Low-temperature operation typically employs liquid or polymer electrolytes, whereas high-temperature applications necessitate the use of solid-state materials for sensor components.

8.4.1 Fundamentals of Electrochemistry for SO_2 & H_2S Sensors

Electrochemical sensors function by measuring the properties of the electrode and/or electrolyte as they interact with the target analyte. The performance characteristics of these sensors are influenced by the entire electrochemical detection system, which includes sampling, transport within the electrolyte, electron-transfer reactions, and the diffusion of reaction products.

To achieve optimal performance, electrochemical sensors must be carefully matched to their operating environment, device design, electrode materials, and structural configurations, as well as the specific conditions of the sensing process. Bridging the gap between fundamental electrochemical behavior and observed analytical performance is a complex challenge that requires a balanced synthesis of empirical data and theoretical insight.

A study on the use of ionic liquids as electrolytes found that the presence of SO_2 in a gas environment reduced the activation energy for diffusion, which was attributed to a decrease in the viscosity of the ionic liquid as SO_2 content increased. Cyclic voltammetry of redox species like ferrocene in this ionic liquid demonstrated that the limiting current was linearly proportional to the SO_2 concentration across the entire range from 0 to 100%, highlighting a promising approach for SO_2 sensing.

Experiments have shown that a variety of gases, including H_2, O_2, CO, NO_2, NO, O_3, SO_2, H_2S, and organic vapors with electroactive functional groups such as alcohols or aldehydes, can be accurately measured using appropriately designed electrochemical gas sensors. In principle, any electroactive gas or one that can generate an electroactive species can be detected using electrochemical methods.

For the analysis of NO_2, N_2O, CO_2, and O_2, the reactions involve electrochemical reduction, while other analytes rely on electro-oxidation to produce the analytical signal. It's important to note that each sensor may have a unique design and utilize different materials and geometries for membranes, electrolytes, and electrodes, tailored to leverage the specific chemical properties of the target analyte and to withstand various operating conditions.

Electrochemical sensors typically consist of six major components (see Figure 8.8): [13] a filter, membrane (or capillary), working or sensing electrode, electrolyte, counter-electrode, and reference electrode. Each component significantly impacts the sensor's overall performance and analytical characteristics. Therefore, research efforts often focus on establishing a link between the sensor's analytical performance and its construction/design, materials, and various testing factors, such as the mass transfer of the analyte to the electrode. Selecting appropriate materials for sensor construction and optimizing sensor geometry are critical for achieving optimal operation and performance. The materials involved in specific chemical interactions, their fundamental electronic and electrochemical properties, and their stability dictate the sensor's reaction thermodynamics and kinetics. Additionally, the sensor's geometry and dimensions profoundly affect its analytical performance, including sensitivity, selectivity, response time, and signal stability. Even "minor" details in sensor design can substantially influence the accuracy, precision, background current, noise, stability, lifetime, and selectivity of the sensor. It is important to note that the optimal choice of sensor materials and geometry for a specific application is not always immediately apparent.

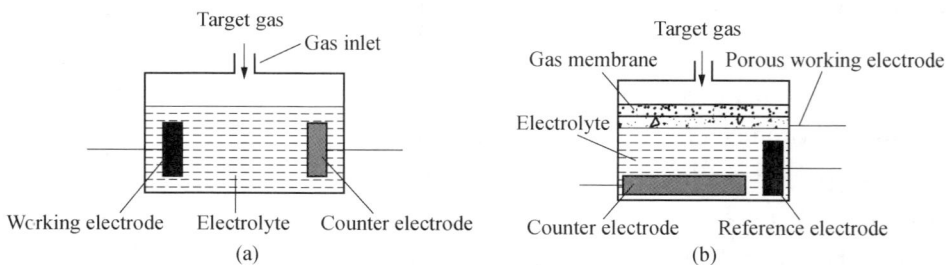

Figure 8.8　Schematic diagram of electrochemical gas sensors with (a) two and (b) three-electrode

Table 8.2 presents possible electrochemical reactions involved in gas analysis. [13] For instance, during the detection of H_2, CO, NO_2, and SO_2, the number of electrons per molecule participating in the reaction is $n = 2$, whereas for H_2S detection, $n = 8$.

Despite advances in sensor technology, aqueous solutions remain widely used as electrolytes for gas sensing. For example, acidic or halide electrolytic solutions are employed for acidic gases and various other gas-detection applications. The electrolytes commonly used in designing electrochemical sensors include H_2SO_4 and NaOH, which are among the most frequently utilized.

Table 8.2　　　**Example electrode reactions for aqueous electrolyte amperometric gas sensors**

TARGET GAS	ELECTRODE REACIION
H_2	$H_2+2OH^-=2H_2O+2e^-$
CO	$CO+H_2O=CO_2+2H^++2e^-$
O_2	$O_2+4H^++4e^-=2H_2O$
NO_2	$NO_2+2H^++2e^-=NO+H_2O$
NO	$NO+2H_2O=NO_3^{2-}+4H^++3e^-$
H_2S	$H_2S+4H_2O=SO_4^{2-}+10H^++8e^-$
SO_2	$SO_2+2H_2O=SO_4^{2-}+4H^++2e^-$

8.4.2　Market for Electrochemical Gas Sensors

Currently, many companies market electrochemical sensors, with the practical capabilities of some modern industrial electrochemical gas sensors summarized in Table 8.3. The initial sensors were developed to detect more common atmospheric pollutants, including CO, NO, NO_2, H_2S, and SO_2.

Table 8.3　　　**Example capabilities of electrochemical sensors**

Gas/vapor measured	Detection range available	Resolution
O_2	0-25% vol	0.1% vol
CO	0-500-20,000 μL/L	0.5-1 μL/L
H_2	0-100-2,000 μL/L	1 μL/L
NO	0-100-1,000 μL/L	1 μL/L
NO_2	0-20-500 μL/L	10 nL/L - 0.1 μL/L
H_2S	0-100-1,000 μL/L	0.1-1 μL/L
SO_2	0-20-5,000 μL/L	0.1-1 μL/L
Cl_2	0-20-200 μL/L	0.1-1 μL/L
O_3	0-5-20 μL/L	5-50 nL/L
HCl	0-200 μL/L	1 μL/L
PH_3	0-100 μL/L	0.1 μL/L
NH_3	0-50-1,000 μL/L	1 μL/L
MMH	0-500 μL/L	0.1 μL/L

Sensors have also been developed for specialized cases such as hydrazine, CH_4, O_3, alcohol, and medical applications for detecting and measuring N_2O, O_2, CO_2, and $CHClBrCF_3$.

Other candidates for analysis using electrochemical gas sensors include electrochemically active compounds such as acetylene, alcohols, aldehydes (e.g., acetone), phosphine, arsine, and phosgene, many of which have already been studied. These compounds are often encountered as interferences with current commercial instruments.[14, 15]

8.5 Application Examples of Electrochemical Sensors for GIS

When an electrical apparatus is operating normally, no significant decomposition byproducts are detected, regardless of whether the sensor is located inside or outside the arcing chamber. However, in the event of a fault in SF_6 electrical equipment, the energy released can generate extremely high temperatures (5,000~20,000 ℃), leading to the decomposition of SF_6 gas and insu-lating materials. The presence of impurities and moisture exacerbates this decomposition, causing the SF_6 gas and nearby solid insulation materials to produce a substantial number of byproducts. [4, 16, 17]

The decomposition products of SF_6 gas primarily include SO_2, SOF_2, HF, and SO_2F_2. In cases where solid insulation materials are involved in the malfunction, additional byproducts such as CF_4, H_2S, CO, and CO_2 are also produced.

Four types of chemicals—SO_2, H_2S, HF, and CO can be detected using electrochemical sensors, with the concentrations of SO_2 and H_2S serving as key indicators for fault estimation. The relationship between these concentration ranges and the corresponding countermeasures that engineers in substations should take can be found in Table 8.4.

Table 8.4   Q/GDW 1896—2013 Guideline for SF_6 decomposition byproducts detection

Gas	Detection	Data ($\mu L/L$)	Countermeasures
SO_2	≤1	Normal	Normal
	1-5	Attention	Shorten detecting period
	5-10	Alarm	Tracking monitoring, comprehensive diagnosis
	>10	Alarm	Comprehensive diagnosis
H_2S	≤1	Normal	Normal
	1-2	Attention	Shorten detecting period
	2-5	Alarm	Tracking monitoring, comprehensive diagnosis
	>5	Alarm	Comprehensive diagnosis

Note:

1) Arcing chamber sampling time should be within 48 hours since GIS has a normal cut-off at rated current or below rated current.

2) As auxiliary icons, CO and CF_4 should be compared with initial value (values at handover acceptance inspection on site), tracking the incremental change, if changing is significant and should make a comprehensive diagnosis.

Figure 8.9 illustrates an instrument designed to measure SO_2 and H_2S using electrochemical sensors. This device leverages the property that the target gases react chemically with the sensor, generating an electrical signal proportional to the gas concentration, which is then processed to yield concentration data. The instrument primarily detects three gases: SO_2, H_2S, and CO. The advantages of this instrument include its high detection sensitivity, rapid response speed, minimal gas consumption, and suitability for on-site rapid detection. However, it also has disadvantages: it is prone to "cross interference" (e.g., between H_2S and SO_2, and between SOF_2 and SO_2), sensor drift (both zero drift and temperature drift), and requires regular calibration (typically once a year). Additionally, it has a limited service life, with some sensors lasting only 1 to 3 years, and suffers from poor repeatability and durability. The lowest detectable limits for SO_2, H_2S, and CO are less than 0.5 μL/L, with detection ranges of 0.5 to 100 μL/L for SO_2 and H_2S, and 0.5 to 500 μL/L for CO.

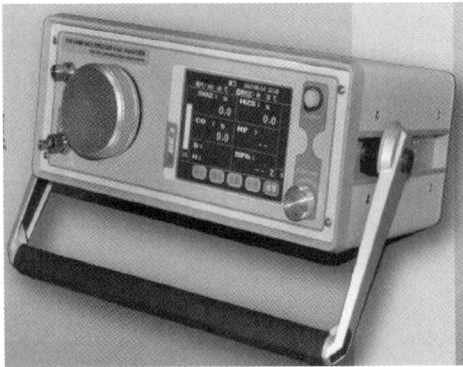

Figure 8.9   Schematic diagram of electrochemical gas sensors

Figure 8.10   Discharge occurred between conductor and ground in a 220 kV GIS Phase B, after an inspection under disassembled condition found $SO_2 > 200$ μL/L and $H_2S > 120$ μL/L[4]

Using the aforementioned electrochemical sensor instruments, along with other similar German instruments [17] for measuring SO_2 and H_2S, the State Grid of China completed a comprehensive health survey of approximately 20,000 SF_6 GIS units. Additionally, fault analysis for nearly 200 GIS units was conducted using these instruments.[4] The typical faults identified can be classified as follows:
- Discharge occurred between conductor and ground, see Figure 8.10.
- Discharge by floating potential objects, see Figure 8.11.
- Unreliable connection of conducting bars, see Figure 8.12.
- Turn to turn short circuit or the shielding layers short circuit of high voltage bushings, see Figure 8.13.
- Breaker could not be switched off, see Figure 8.14.
- Resistors and capacitors internal short circuits, see Figure 8.15.

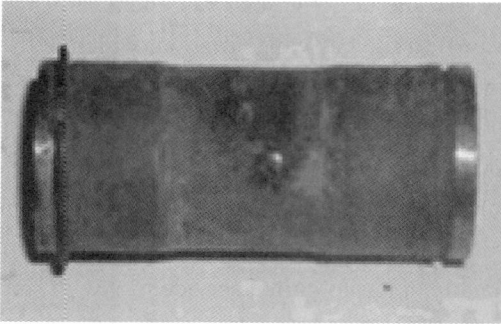

Figure 8.11   The suspension potential discharge of pull rod connecting bolt in phase A of a 220 kV circuit breaker was detected during operation in a domestic hydropower station, the content of SO_2 and H_2S is very high[4]

Figure 8.12   A discharge occurred in an equal-potential ring due to the loosening of the internal bolts and nuts of a 500 kV lightning arrester in a power plant in north China. The content of SO_2 and H_2S exceed 110 μL/L, but the amount of CO was very little [4]

Figure 8.13   A serious overheating fault of plum blossom contact connecting phase B bus of a 330 kV GIS GB1 gas chamber in XX substation was detected during operation on December 28, 2006 [4]

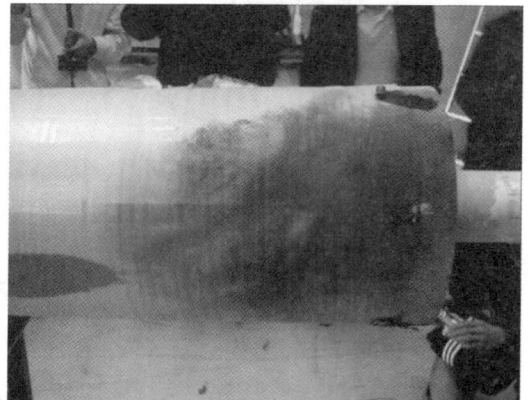

Figure 8.14   Internal spark discharge of a 500 kV current transformer in a 500 kV substation of Guangdong Province, both SO_2 and H_2S are >146 μL/L[4]

During the switching-off process of a circuit breaker, the arc is typically extinguished within one to two cycles. However, if the arc extinguishing performance is inadequate or the current does not reach zero, the arc may not be extinguished in time, leading to the burning of the arc extinguishing chamber and contacts. This can result in the decomposition of SF_6 gas and Teflon, with characteristic decomposition by-products such as SO_2, CF_4, and CO being detected.

Traditionally, several gas analysis techniques have been employed for measuring SF_6

decomposition by-products, including gas chromatography (GC), gas chromatograph-mass spectrometry (GCMS), and Fourier transform infrared spectroscopy (FTIR). These methods, however, are generally confined to laboratory settings. Other techniques, such as gas tube detectors and electrochemical sensors, are used for field monitoring of GIS, though they are limited to detecting a few diagnostic gases and can suffer from long-term accuracy issues like "zero drift." Recently, photoacoustic spectroscopy (PAS) has been recognized as a suitable technique for online dissolved gas analysis (DGA) in transformers. Due to its high sensitivity, accuracy, and remarkable stability, PAS is also considered a promising method for monitoring GIS by measuring diagnostic gases.[18]

Figure 8.15   The contact was burned when the 5013 C-phase circuit breaker was broken in a 500 kV substation in Fujian due to the reignition of the arc[4]

8.6   Review Questions

Q1: The typical technical parameters of electrochemical sensors should include sensitivity, selectivity, cost, size, power consumption and so on. Please list the technical parameters such as CO, SO_2, H_2S gas sensors which are working at room temperatures?

Q2: Electrochemical sensors may consist of six major parts (see Figure 8.8), would you tell us what are these items? Please explain the functions of each components?

Q3: Would you tell us some application examples of the electrochemical sensors in electric power apparatus?

Bibliography

[1] Ryan, Hugh M. High-voltage engineering and testing. The Institution of Engineering and Technology, 2013.

[2] Arora, Ravindra and Mosch, Wolfgang. High voltage and electrical insulation engineering, volume 69. John Wiley & Sons, 2011.

[3] Xinhua Silk Road Database. State grid puts 1,000 kV GIL utility tunnel project into service in E. China, 2024. URL https://en.imsilkroad.com/p/308570.html. Accessed: August 20, 2024.

[4] YU NaiHai. Detection technology of humidity and decomposition products of sulfur hexafluoride gas. Electric Power Research Institute, State Grid Shandong Electric Power Co., Ltd., 2015.

[5] Emel, ÖNAL. Breakdown characteristics of gases in non-uniform fields. Istanbul UniversityJournal of Electrical & Electronics Engineering, 4(2): 1177-1182, 2004.

[6] Haddad, Amy and Haddad, Manu and Warne, DF and Warne, Doug. Advances in high voltage engineering, volume 40. IET, 2004.

[7] Han, Dong and Lin, Tao and Zhang, Guoqiang and Liu, Yilu and Yu, Qiang. SF_6 gas decomposition analysis under point-to-plane 50 hz ac corona discharge. IEEE Transactions on Dielectrics and Electrical Insulation, 22(2):799-805, 2015.

[8] Lin, Tao and Han, Dong and Zhang, Guoqiang. Experimental study on SF_6 decomposition characteristics under spark discharges. In 2014 17th International Conference on Electrical Machines and Systems (ICEMS), pages 1003-1006. IEEE, 2014.

[9] Lin, Tao and Han, Dong and Zhong, Haifeng and Zhang, Guoqiang. Study on the decomposition by-products of SF_6 under the overheat faults. In 2012 China International Conference on Electricity Distribution, pages 1-4. IEEE, 2012.

[10] Zhou, Zhenrui and Han, Dong and Zhao, Mingyue and Zhang, Guoqiang. Influence of trace H_2O and O_2 on SF_6 decomposition under corona discharge and spark discharge based on oxygen isotope tracer. In 2020 IEEE Electrical Insulation Conference (EIC), pages 469-473. IEEE, 2020.

[11] IEC 60376—2018, Specification of technical grade sulphur hexafluoride (SF_6) and complementary gases to be used in its mixtures for use in electrical equipment, 2018.

[12] GB/T 12022—2014, Industrial sulfur hexafluoride (SF_6), 2014.

[13] Korotcenkov, Ghenadii. Chemical Sensors: Comprehensive Sensor Technologies Volume 5: Electrochemical and Optical Sensors, volume 5. Momentum Press, 2011.

[14] Torriero, Angel AJ. Electrochemistry in ionic liquids, volume 1. Springer, 2015.

[15] Fortuna, Luigi and Graziani, Salvatore and Rizzo, Alessandro and Xibilia, Maria Gabriella. Soft sensors for monitoring and control of industrial processes. Springer Science & Business Media, 2007.

[16] Catalog of Xiamen Jiahua Electrical Technology Co., Ltd, SF_6 decomposition products detector by electrochemical gas sensors for GIS. [Online]. Available: http://www.jiahuatech. com.cn/cpny.asp?id=1056. Accessed: Nov. 2020.

[17] DILO GmbH. Devices for the determination of the SF_6 gas quality. URL https://dilo. eu/en/SF_6-gas/SF_6-gas-measuring-devices/devices-to-determine-SF_6-quality/SF_6-multi-analyser. Accessed: August 20, 2024.

[18] Lin, Tao and Zhang, Guoqiang and Qiu, Zongjia and Guo, Runrai and Li, Kang and Han, Dong. Photoacoustic detection of SF_6 decomposition by-products with broadband infrared source. In 2014 International Conference on Power System Technology, pages 1541-1546. IEEE, 2014.

Chemiluminescence and UV Absorption Detection

9.1 Introduction

Air is one of the common insulating materials for electrical equipment in power transmission lines and substations. Partial discharge caused by metal protrusion, contamination discharge, and free metal or floating potential may cause loss of air insulation ability, resulting in damage to electrical equipment and affecting reliable power supply. Partial discharge of electrical equipment in the air can cause air to be ionized into various decomposition by-products, such as ozone (O_3) and nitrogen oxide gases, which are simplified to NO_x ($NO_x=NO_x+NO_2+N_2O_5+NO_3+N_2O$). These decomposition by-product gases greatly influence the degradation of insulating materials and surface corrosion of metallic materials. Therefore, this chapter attempts to describe the relationship between air decomposition by-products and partial discharge and conducts offline detection as well as online monitoring of air-insulated electrical equipment based on the decomposition gas method. This provides an effective detection technology and inspection method for the online continuous monitoring of power equipment.[1]

Firstly, the theoretical chemical reactions for air decomposition by-products under corona discharge were presented, and the working principles for measuring ozone and NO_x gas concentrations by ultraviolet absorption spectrometer and chemiluminescence analyzer were described in detail. Secondly, by recording the maximum and incremental values of the air decomposition by-products, the functional relationship between the local electric field intensity and the concentration of air decomposition by-products in the chamber is presented. Finally, the actual effect of the decomposition gas method in detecting partial discharge defects of air-insulated electrical equipment is verified by three real-world examples: an air-cooled hydropower generator and two units of high-voltage switchgear. These examples indicate that the decomposition gas method could be used to detect partial discharges inside semi-closed or fully enclosed containers, such as switch cabinets, hydro-generator stators, and UHV DC smoothing reactors.

The air decomposition by-products are produced after partial discharge under high voltage. By monitoring and detecting them, the experimental data are analyzed. Based on the empirical mathematical relationship between the concentration of air decomposition by-products and partial discharge, this chapter proposes different types of discharges. Certain patterns of air decomposition by-products are generated under failure conditions. The study indirectly monitors and identifies partial discharges of electrical equipment by measuring and analyzing the rate of generation of air by-product gases.

9.2　Basic Principles

9.2.1　Theoretical Foundation of Partial Discharge Phenomenon

Air is present in the atmosphere of the earth and is a mixture of many gases as well as particles of dust. It is a type of clear gas that all living bodies breathe. Its volume and shape are indefinite, including no color or smell. Due to being matter, it has definite weight and mass. Air does not exist in outer space, and its weight creates atmospheric pressure. Air is composed of 78% nitrogen, 21% oxygen, 0.9% inert gases, and other minimal amounts of gases. Water vapor is also present in the atmosphere in varying quantities, up to 2%, with an average of about 1% water vapor in humid air. The molecular gap is very high in the air compared to other gases, which makes it a very important insulator. This is one of the main reasons for using air as a dielectric material in power system equipment, especially in transmission lines, HVDC reactors, and air-insulated substations. Air is used to support or separate electrical conductors without allowing current to pass through them. Normally, electrons are strongly bonded to their atoms in atmospheric air; however, they cannot resist indefinitely high-voltage stress levels. When higher voltage is applied, any insulating material will eventually suffer from the high electrical "pressure", leading to electron flow through it.[1]

The breakdown in air (which is a kind of spark breakdown) is a transition from a non-sustaining discharge into a self-sustaining discharge. The creation of high currents during breakdown happens due to the ionization created from neutral atoms or molecules, and their flow from anode to cathode, respectively. Streamer theory and Townsend theory are the two main theories that explain the breakdown under different conditions such as temperature, pressure, electrode field configuration, nature of electrode surfaces, and the presence of initial conducting particles. According to the literature, the breakdown strength of air dielectric is about 3 kV/mm. Its exact value varies with the size and shape of the electrodes. The air pressure value also affects the breakdown phenomena in air.

There are three fundamental processes that occur in the air before its breakdown: the electron avalanche process, the streamer discharge process, and the leader process.

Electron Avalanche Process: Free electrons are always present in the air, and under the influence of an electric field, these electrons are accelerated. Collisions occur between these energized electrons and neutral atoms and molecules, increasing the ionization process. Some collisions result in photon emission, which can further ionize atoms or molecules. These ionization effects produce an ever-increasing number of electrons, a phenomenon known as the electron avalanche process.

Streamer Discharge Process: Streamer discharges can form after the electron avalanche process. In the large electric field created by the high applied voltage, accelerated electrons strike air molecules with high enough energy to knock other electrons off them, ionizing the molecules. The freed electrons continue to strike more molecules, resulting in a constant chain reaction. The electrically conductive regions in the air near the electrode are created by electron avalanches, which further generate an electric field. Additionally, an electric field is created by space charge from the electron avalanches. This field can promote the growth of new avalanches in specific regions. Consequently, the ionized region grows rapidly in that direction, forming a finger-like discharge known as a streamer.

Leader Process: The streamer discharge produced by a high voltage electric field leads to the final spark, often followed by the development of a "leader" before the breakdown itself. The leader process is an area of active research in high-voltage engineering. The leader forms as the streamer discharge progresses, creating a highly conductive channel through which a large current can flow, ultimately leading to a complete breakdown of the air gap.

As per IEC 60270 (International Electrotechnical Commission) standard, partial discharge is defined as: "localized electrical discharge that only partially bridges the insulation between conductors and which may or may not occur adjacent to a conductor. Partial discharges are in general a consequence of local electrical stress concentration in the insulation or on the surface of the insulation. Generally, such discharges appear as pulses having a duration of much less than 1μs". In general, partial discharge activity is noted in high-voltage power equipment like bushings, transformers, and many others.

Partial discharge phenomena can be divided into two basic categories: (1) External Partial Discharge: This type occurs in ambient air and is commonly known as corona discharge. These discharges are usually reversible and are generally considered harmless. (2) Internal Partial Discharge: This type occurs within insulating solids and liquid dielectrics, and can also be present in compressed gases. It is caused by undesirable features within the insulation material and can lead to more significant damage over time.

Major types of partial discharge are as follows:

- Corona discharge: This occurs due to the non-uniformity of the electric field at the sharp edges of a conductor subjected to high voltage stress. It involves gases, air, or liquid insulation. It often persists for a long duration along the bare conductor. The formation of ozone due to the corona effect is a major cause of degradation of the insulation material.

- Surface discharge: This occurs at the interfaces of dielectric materials, such as solid/gas interfaces. It commonly occurs in bushings, at points on the insulator surface, at the end of cables, or between electrodes (between high voltage terminal and ground). The existence of surface discharge is influenced by factors such as the permittivity of the dielectric materials used, properties of the insulating materials, and voltage distribution between conductors.

- Treeing channel: This is a continuous partial discharge that occurs at the sharp edges of insulating material under high-intensity fields, resulting in the deterioration of the insulating material.
- Cavity discharge: This occurs when the gas in a cavity is over-stressed due to high voltages. Solid or liquid insulating materials are involved in this type of discharge.

In this chapter, the effect of air decomposition by-products has been described under external partial discharge phenomena, especially corona and surface discharge in air. Their details are discussed in the following sections as per Figure 9.1.

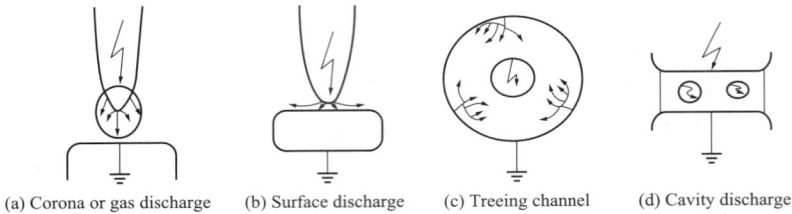

(a) Corona or gas discharge   (b) Surface discharge   (c) Treeing channel   (d) Cavity discharge

Figure 9.1   Different types of partial discharge occurring in the insulator

Corona Discharge Theory: In a uniform electric field, exceeding the dielectric strength of the surrounding media typically results in complete electrical breakdown. However, if the field distribution between electrodes is strongly non-uniform, such as in sphere-plane, coaxial cylinders geometries, or point-plane configurations, electrical discharges may be observed before a complete breakdown occurs. These discharges, or partial breakdowns, are commonly referred to as "corona." For a corona discharge to take place, certain conditions must be met, that is the electric field distribution between electrodes must be non-uniform, which prevents a complete breakdown of the gap.

The voltage applied to the corona electrode must be high enough to ensure that the electric field strength is sufficient to initiate ionization of the insulating gas medium. In practice, corona discharges are commonly found in high-voltage power transmission lines or in equipment exposed to high-voltage stress, where they are usually considered problematic. However, corona discharges can also be beneficial and are utilized in certain industrial applications, such as high-speed print-ing, electrostatic precipitators, and Geiger counters.

Coronas can be divided into impulse coronas and static field coronas. The first takes place when the voltage level surpasses the corona onset voltage for a short duration of time, while the latter occurs when the onset voltage is surpassed over a longer period of time. With a longer time period, phenomena of space charge drift and accumulation can be observed. Coronas can also be classified as positive or negative depending on corona electrode polarity.

Surface Discharge Faults: Surface discharge faults occur in low-conductivity materials such as PTFE, PMMA, ceramics, and polycarbonate, as well as on surfaces with conductive substances like water, ice, or earth. These faults are prevalent in high-voltage equipment such as bushings of outdoor switchgear and rigid insulators covered by moisture. Surface discharge breakdown can lead to significant issues in high-voltage systems. Different terms have been used in the literature for surface discharge, such as surface discharge, creepage and sliding discharge, and dielectric barrier discharge (DBD). The main meanings are as follows: (1) Surface Discharge: Refers to discharges occurring along the surface of insulating materials. (2) Creepage and Sliding Discharge: Different terms used in literature to describe surface discharge phenomena. (3) Dielectric Barrier Discharge: Occurs when one or both electrodes are covered by insulating materials.

Partial discharge (PD) can severely impact the integrity of insulating systems in high-voltage transmission cables and equipment. Once initiated, PD can lead to progressive deterioration of the insulating material, eventually causing electrical breakdown of the system. Here are some key points on the effects of partial discharge: (1) Initiation and Propagation: Once PD begins, it progressively deteriorates the insulating material, leading to eventual electrical breakdown. (2) Treeing Effect: Partially conducting discharge channels created by PD can initiate the treeing effect in solid dielectrics. (3) Irreversible Damage: Repetitive discharges cause irreversible mechanical and chemical damage to the insulating material. (4) Energy Dissipation: High-energy electrons or ions from discharges dissipate energy, causing damage to the material. (5) Gas Liberation: Chemical breakdown processes liberate gases at high pressure, contributing to the chemical transformation of the dielectric. (6) Increased Conductivity: Chemical transformations increase the conductivity of the material dielectric surrounding the voids, enhancing electrical stress in the unaffected gap region. Overall, the effect of PD can be very serious, leading to significant degradation and eventual failure of high-voltage insulation systems.

Partial discharge (PD) can dissipate energy in various forms, primarily as heat, but sometimes also as light and sound. Here are some detailed points on these aspects and the importance of monitoring PD in high-voltage electrical equipment: (1) PD primarily dissipates energy in the form of heat, causing minor thermal degradation of the insulation. The heat generated is often localized, leading to hot spots that can accelerate the deterioration of the insulating material. (2) PD can sometimes emit light, visible as a dim, glowing discharge around the insulators of overhead transmission lines. (3) PD can also produce sound, often described as a hissing noise, indicative of electrical discharge activity. Monitoring PD activities is crucial for ensuring the reliability of high-voltage electrical equipment. By observing PD, early warning signals for potential insulation failure can be detected, allowing for timely maintenance and preventing catastrophic failures. The reliability of the insulating material can be confirmed through continuous PD monitoring. Regular inspection and maintenance based on PD activity help in maintaining the long-term operational sustainability of high-voltage electrical equipment. Therefore, partial discharge not only causes minor thermal degradation but can also serve as an important indicator of potential issues in high-voltage electrical equipment. Continuous monitoring and timely maintenance based on PD activities are essential for ensuring the reliability and longevity of these systems.

9.2.2 Chemical Reactions of Air under Corona Discharge

Ozone is a colorless gas with a pungent odor and powerful oxidizing properties, formed from oxygen by electrical discharges or ultraviolet light, as shown in Figure 9.2. Its formation procedure is detailed on the website (https://ozotech.com/the-ozone-molecule-2/). Ozone (O_3) differs from normal oxygen (O_2) in that it consists of three oxygen atoms per molecule. This unstable form of oxygen is created when oxygen encounters an electric arc, such as lightning, or a particular wavelength of ultraviolet rays, like those from the sun. During an electrical storm, bolts of lightning in the atmosphere split oxygen molecules into single oxygen atoms, which quickly bond with O_2 molecules to form O_3. This is why the air smells fresh after a lightning storm, lightning acts as nature's air purifier. Similarly, the ozone layer in the lower stratosphere is generated by the sun's short-wave ultraviolet rays through molecular fission and fusion. This ozone layer, created by short UV waves, protects life on Earth from the longer, more harmful UV waves within the spectrum.[2]

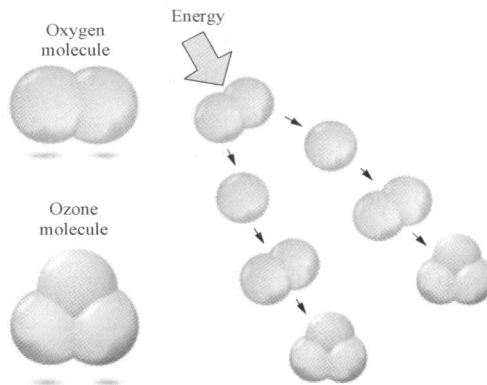

Figure 9.2 Ozone was formed from oxygen in three ways (Corona discharge or Ultraviolet (UV) radiation or Dielectric barrier discharge)

Ozone generators that utilize the electric arc method to generate ozone are known as corona discharge ozone generators. These devices pass high voltage (2,000 volts and up) through a dielectric material, which transmits electrical force without conduction, while being fed and cooled by ambient air supplied by a fan located at the rear of the unit, as shown in Figure 9.3. More detailed information can be found at the website NerdTechy.

(https://nerdtechy. com/best-ozone-generator-under-100/)

Ozone will last between 30 minutes to 4 hours for the third oxygen atom to break apart and revert back into oxygen. At lower concentrations, it typically takes 30 minutes to 1-2 hours for ozone to break down into breathable oxygen. At higher concentrations, it will take 3-4 hours for the ozone to break down into normal oxygen. Ambient air is composed of 78.08% nitrogen, 20.95% oxygen, 0.93% argon, and the remaining 0.04% consists of trace amounts of other gases. In corona discharge ozone generators, a sustained electric arc severs the O_2 molecules, resulting in the manufacture of ozone. Waves of ultraviolet (UV) light "slice" O_2 molecules in half in UV ozone generators. UV bulbs are mounted inside the generator, emitting specific wavelengths of ultraviolet light that separate the atoms of passing O_2 molecules. This process mimics how our sun creates and maintains the protective ozone layer between the troposphere and stratosphere. However, UV is much less effective at generating ozone than corona discharge, leading to a lower ozone output in UV generators.

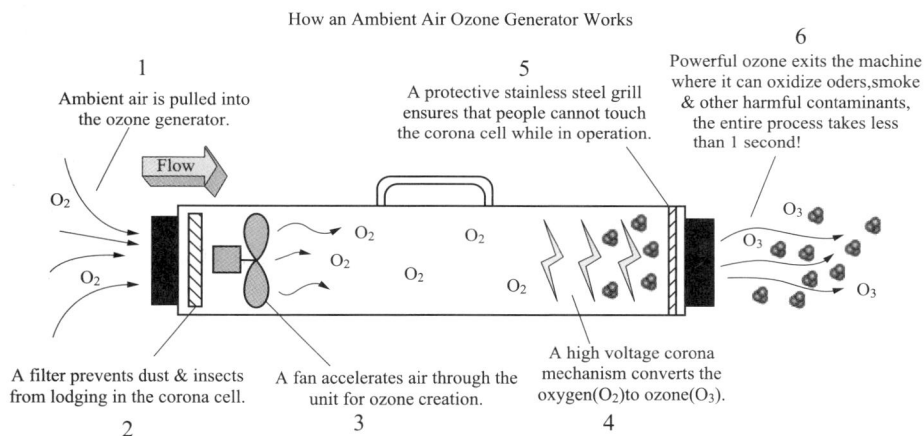

Figure 9.3   Working principle of an ozone generator

NO_x is a generic term for the mono-nitrogen oxides NO and NO_2 (nitric oxide and nitrogen dioxide). They are produced from the reaction of nitrogen, oxygen, and even hydroc-arbons during combustion, especially at high temperatures. In areas with high motor vehicle traffic, such as large cities, the amount of nitrogen oxides emitted into the atmosphere as air pollution can be significant. NO_x gases are formed whenever combustion occurs in the pres-ence of nitrogen, as in an air-breathing engine; they are also produced naturally by lightning.[3]

From experimental observation, it is known that humidity retards streamer development in the air. Water vapor acts as an electronegative gas, limiting the total charge injected into the gap by the streamer channels. Consequently, the voltage discharge required to bridge long air gaps is increased in highly divergent fields. However, under a uniform electric field, the infl-uence of moisture is practically negligible. Therefore, the extent of humidity becomes an important factor for further discussions.

Low humidity theoretical study: Atmospheric air is the primary insulation medium in electricity transmission and distribution systems. Under the influence of an electric field, free electrons and photons initiate an ionization process. This avalanche process generates a series of discharges in the air, which react with air components like oxygen and nitrogen, resulting in the formation of by-products such as ozone and NO_x gases. The key chemical reactions in dry air following a discharge are detailed in Table 9.1. [1]

Table 9.1   **Chemical Reaction for Dry Air Decomposition By-product under Corona Discharge**

Sr#	Chemical Reaction	Sr#	Chemical Reaction
R1	$e+O_2 \rightarrow O+O+e$	R16	$NO+O_3 \rightarrow NO_2+O_2$
R2	$e+O_3 \rightarrow e+O+O_2$	R17	$NO+O_3 \rightarrow NO_3+O$
R3	$O+O+O_2 \rightarrow O_2+O_2$	R18	$NO+O_3 \rightarrow 2NO_2$
R4	$O+O_2+O_2 \rightarrow O_3+O_2$	R19	$NO_2+NO_3 \rightarrow N_2O_5$
R5	$O+O_2+N_2 \rightarrow O_3+N_2$	R20	$N_2+OH \rightarrow NO_2+H$
R6	$e+O_3 \rightarrow O_2+O$	R21	$N+OH \rightarrow NO+H$
R7	$O+O_2+O_3 \rightarrow O_3+O_3$	R22	$NO_2+NO_2 \rightarrow NO+NO+O_2$
R8	$O+O_2+H_2O \rightarrow O_3+H_2O$	R23	$NO_2+NO_3 \rightarrow NO+NO_2+O_2$
R9	$O+O_3 \rightarrow O_2+O_2$	R24	$NO_2+NO_3+O_2 \rightarrow N_2O_5+O_2$
R10	$O_3+O_2 \rightarrow O+O_2+O_3$	R25	$NO_3+O \rightarrow O_2+NO_2$
R11	$O_3+O_2 \rightarrow O+O_2+O_3$	R26	$NO_3+O_2 \rightarrow O_3+NO_2$
R12	$O_3+O_3 \rightarrow O+O_2+O_3$	R27	$NO_3+NO \rightarrow NO_2+NO_2$
R13	$e+N_2 \rightarrow N+N+e$	R28	$N_2O+O \rightarrow NO+NO$
R14	$N+O_2 \rightarrow NO+O$	R29	$N_2O_5 \rightarrow NO_2+NO_3+O_2$
R15	$N+O_3 \rightarrow NO+O_2$	R30	$O+N_2O_5 \rightarrow 2NO_2+O_2$

Table 9.1 is divided into two types of chain reactions. The first type includes reactions with electron impacts, such as R1, R2, R6, and R13, while the remaining reactions occur without electron impacts. Reactions involving electron impacts are responsible for creating ozone and NO_x gases, including NO, NO_2, NO_3, and N_2O_5. It is crucial to note from this table that electrons, which represent discharge, are the primary key factor in the generation of air byproducts. Thus, their quantity and kinetic energy play a vital role in the generation rate of air decomposition by-products. Additionally, the table illustrates that ozone and NO_x also interact with each other. Therefore, a relationship exists between air by-product generation and discharge phenomena, leading to an indirect method for discharge monitoring through the measurement of these gases.

High humidity theoretical study: Before explaining the experimental results, it is important to understand the theory of air by-product generation in high humidity. It is known that discharge decomposes air into its by-products such as ozone and NO_x gases.

The presence of high humidity in the air causes a reduction in discharge capacity. Table 9.2[1] shows the chain of reactions that occur in high humidity during corona discharge. The bold font (red color) indicates the reactions that differ from those in Table 9.1. In a needle-plate experimental configuration, the presence of water vapor in the air changes the shape and propagation of the streamer, ultimately reducing the concentration of air by-products. This happens because water vapor generates new gases that interact with and reduce the concentrations of ozone and NO_x. These additional gases include hydrogen peroxide (H_2O_2), nitrous acid (HNO_2), and nitric acid (HNO_3), which are formed from reactions involving partial discharge and water vapor. The detailed chemical reactions in high humidity, as shown in Table 9.2, highlight the interaction between water vapor and the discharge process. The new reaction pathways reduce the overall production of ozone and NO_x gases by converting them into other compounds. This understanding is crucial for accurately assessing the effects of high humidity on the formation of air decomposition by-products in partial discharge phenomena. Understanding the reduced discharge capacity and the formation of new gases under high humidity conditions helps to interpret experimental results and their implications for partial discharge monitoring and diagnosis in high-voltage equipment.

Table 9.2   **Chemical Reaction During Partial Discharge in High Humidity**

Sr#	Chemical Reaction	Sr#	Chemical Reaction
R1	$e+O_2 \rightarrow O+O+e$	R16	$NO+O_3 \rightarrow NO_2+O_2$
R2	$H_2O+O \rightarrow 2OH$	R17	$NO_2+O_3 \rightarrow NO_3+O_2$
R3	$O+O_2+M \rightarrow O_3+M$	R18	$NO+NO_3 \rightarrow 2NO_2$
R4	$O_2+H \rightarrow O+OH$	R19	$NO_2+NO_3 \rightarrow N_2O_5$
R5	$OH+O_3 \rightarrow H_2O+O_2$	R20	$N_2+OH \rightarrow N_2O+H$
R6	$e+O_3 \rightarrow O_2+O$	R21	$N+OH \rightarrow NO+H$
R7	$OH+OH \rightarrow H_2O +O$	R22	$NO_2+OH+M \rightarrow HNO_3+M$ $M=N_2+O_2$
R8	$OH+OH+M \rightarrow H_2O_2+M \; M=O_2$	R23	$NO_3+ H_2O \rightarrow HNO_3+OH$
R9	$OH+OH+M \rightarrow H_2O_2+M \; M=N_2$	R24	$N_2O_5+H_2O \rightarrow HNO_2+HNO_3$
R10	$OH+OH \rightarrow HO_2+H$	R25	$NO_2+HO_2 \rightarrow HNO_2+O_2$
R11	$OH+ H_2O_2 \rightarrow H_2O+HO_2$	R26	$HNO_2+OH \rightarrow NO_2+H_2O$
R12	$HO_2+O_3 \rightarrow OH+2O_2$	R27	$HNO_3+O \rightarrow NO_3+OH$
R13	$e+N_2 \rightarrow N+N+e$	R28	$NO+OH+M \rightarrow HNO_2+M \; M=N_2$
R14	$N+O_2 \rightarrow NO+O$	R29	$NO_2+OH+M \rightarrow HNO_3+M$ $M=N_2+O_2$
R15	$N+O_3 \rightarrow NO+O_2$	R30	$O+N_2O_5 \rightarrow 2NO_2+O_2$

In dry air, ozone (O_3) and nitrogen oxides (NO_x) are present at concentrations of approximately 8 parts per billion by volume (nL/L) for O_3 and about 20 nL/L for NO_2, the main component of NO_x. The initial concentrations of oxygen (O_2) and nitrogen (N_2) are 20.95% oxygen, 78.08% nitrogen, respectively. The yields of other species are assumed to be zero. The dissociation rate coefficients of O_2, O_3, and N_2, which vary slightly under different temperatures, are obtained from reference.[4] For simplicity, it is assumed that these coefficients remain con-stant from 273 K to 313 K (0-40 ℃). The calculation method is verified by comparing the results with the earlier literature. The results show in Figure 9.4.

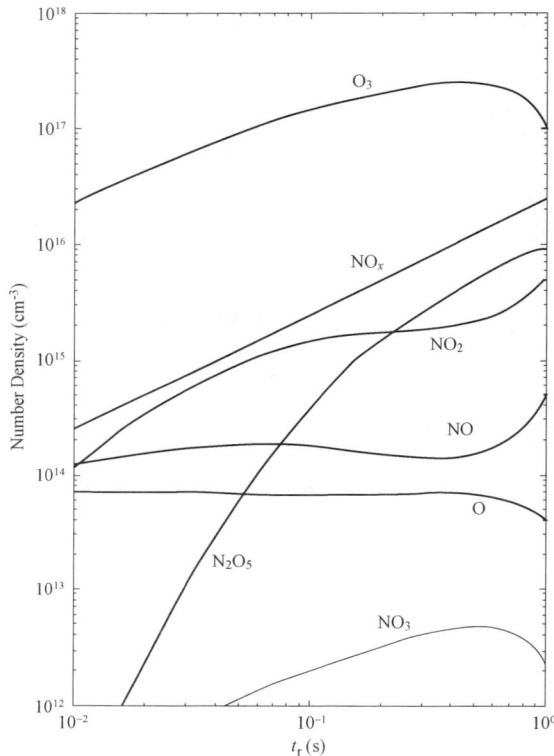

Figure 9.4　Generation of O_3 and NO_x from dry air under streamer discharge

The results of calculations of the volume density of each species are presented as solid curves in Figure 9.4. The abscissa of Figure 9.5 is the residence time t_r of the gas in the disch-arge zone, in here it means the needle tip area, and the ordinate is the species density expres-sed in volume density μL/L. As we can see from Figure 9.5 [4] as residence time tr increases the density of most species increase except O and NO. It indicates that the flow speed of gas near the needle tip is the main character that affects the concentration of species. Meanwhile, the O_3 is the main by-product. NO_x is little compare to O_3 and is less than 1% of O_3. The density of NO is bigger than NO_2 when residence time is small. When t_r is bigger than about

0.2 s, NO_2 becomes bigger. That is because as the chemical reaction continues the NO transfer to NO_2. The square wave indicates the density of O, it clearly reflects the periodicity of partial discharge (PD) under AC voltage. When there is PD happens, the density of O increases; when PD stops, the O is consumed by R2, R3, etc. The density of O decreases. The top value of O is nearly stable. This shows the generation of O is faster than its consumption in the calculated t_r. The main product of NO_x is NO and NO_2. The production of NO_3 and N_2O_5 are very small compared to NO_x especially when residence time is little and are not listed in the following table. The calculation results also show that productions of NO_x tell more messages about the reactions.

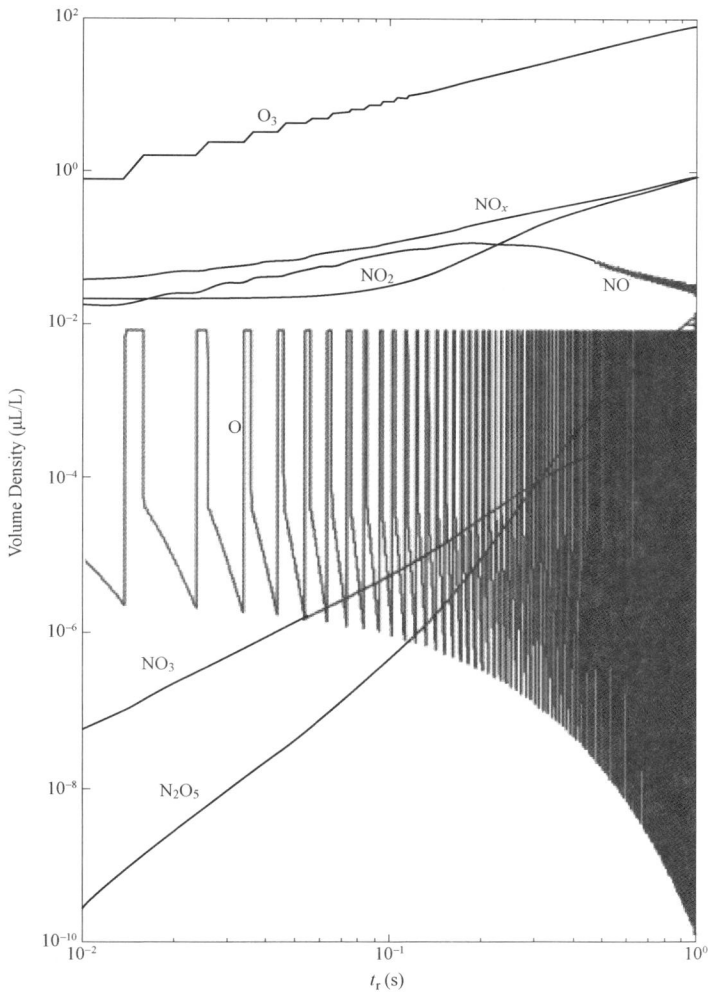

Figure 9.5   Generation of O_3 and NO_x from dry air under corona discharge

9.3 Components of Ozone and NO$_x$ Monitors

9.3.1 UV Absorption Spectrum for Ozone Monitoring

An absorption spectrum of ozone can be seen in Figure 9.6.[5] This figure also includes the absorption spectra of isoprene and water vapor, which exhibit relatively strong peak absorption intensities in the same spectral region. The spectra of the standard gases are normalized by their peak intensities. The ozone molecule has an absorption maximum at 254 nm, which coincides with the principal emission wavelength of a low-pressure mercury lamp.[2] This figure demonstrates the absorption characteristics of ozone compared to other standard gases like isoprene and water vapor. The sharp absorption peak at 254 nm for ozone is particularly notable, making it a distinct marker for identifying and measuring ozone concentrations using UV absorption spectroscopy. The alignment with the emission wavelength of low-pressure mercury lamps is advantageous for practical detection and measurement setups, such as those used in monitoring partial discharges in high-voltage equipment.

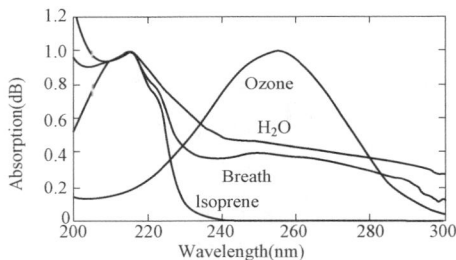

Figure 9.6　Absorption spectrum of ozone, isoprene, and water vapor

The absorption of UV light has long been utilized for the precise and accurate measurement of atmospheric ozone. Fortunately, few molecules found in significant concentrations in the atmosphere absorb at this wavelength. However, interferences, such as organic compounds containing aromatic rings, can occur in highly polluted air. These interferences can complicate the measurement process, requiring careful consideration and possibly additional calibration or correction methods to ensure the accuracy of the ozone measurements.

Figure 9.7 illustrates the working principle of an ozone monitor used for measuring ozone concentrations. Ozone measurement is based on the attenuation of light passing through a 14 cm absorption cell with quartz windows.[6, 7] A low-pressure mercury lamp, emitting light primarily at 254 nm, is placed on one side of the absorption cell. On the opposite side of the cell, a photodiode with a built-in interference filter centered at 254 nm detects the light passing through the cell. An air pump draws sample air into the instrument at a flow rate of approximately 1.0 L/min. A solenoid valve alternates the flow of air between two paths: (1) Direct Path: Air is sent directly into the absorption cell. (2) Scrubber Path: Air passes through an ozone scrubber before entering the absorption cell. The intensity of light measured by the photodiode when the air has passed through the ozone scrubber (I_0), which removes ozone. The intensity of light measured by the photodiode when the air has not passed through the scrubber, thus containing ozone (I). Ozone concentration is calculated from the measurements of I_o and I_R according to the Beer-Lambert Law:

$$C = \frac{1}{\sigma l} \ln\left(\frac{I_0}{I}\right) \tag{9.1}$$

where l is the path length (14 cm) and σ is the absorption cross section for ozone at 254 nm ($1.15 \times 10^{-17} cm^2$ molecule^{-1} or 308 atm^{-1} cm^{-1}), which is known with an accuracy of approximately 1%. Most similar commercial instruments use the same absorption cross-section (extinction coefficient).

In addition to measuring light intensity, the ozone monitor also measures the pressure and temperature within the absorption cell to ensure accurate ozone concentration readings. These measurements allow the instrument to express the ozone concentration as a mixing ratio in partsper-billion by volume (nL/L). The

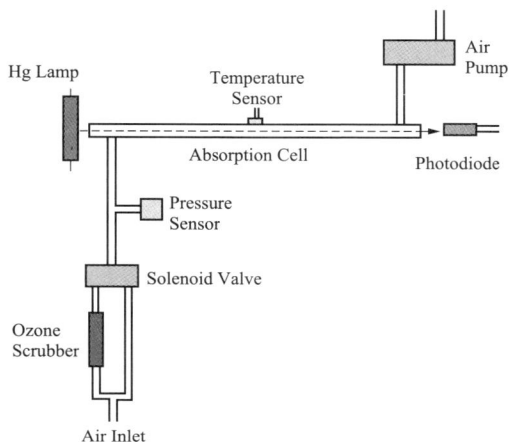

Figure 9.7   Schematic Diagram of the Ozone Monitor

ozone monitor were installed a pressure sensor to measure the pressure inside the absorption cell, with readings displayed and logged in units of mbar or torr. The ozone monitor were installed also a temperature sensor to measure the cell temperature, with readings displayed and logged in units of either degrees Celsius (℃) or Kelvin (K).

The UV absorption method for measuring ozone is inherently absolute, meaning it should not require external calibration under ideal conditions. However, due to potential non-linearity in the photodiode response and associated electronics, minor measurement errors can occur. To ensure accuracy, each ozone monitor is calibrated against a NIST-traceable standard ozone spectrophotometer in the laboratory. This process covers a wide range of ozone mixing ratios to identify any discrepancies. The instrument is compared with a standard ozone spectrophotometer to determine any offset and slope (gain or sensitivity) deviations.These calibration results are used to adjust the ozone monitor for accurate measurements. The determined corrections for offset and slope are recorded in the instrument's Birth Certificate. These parameters are programmed into the microprocessor before the instrument is shipped to the user. Users have the ability to change the calibration parameters via the front panel if necessary. It is recommended to recalibrate the instrument at least once a year to ensure continued accuracy. More frequent recalibration may be required if significant changes in temperature or chemical contamination of the absorption cell are observed. The offset can drift over time due to temperature changes or chemical contamination. The instrument is equipped with an external ozone scrubber that allows for periodic accurate offset corrections. This helps maintain the precision of the ozone measurements by compensating for any drift in the offset. By following these calibration and maintenance procedures, the ozone monitor ensures reliable and accurate measurements of ozone concentration, which is critical for various applications ranging from environmental monitoring to industrial processes.

Understanding and compensating for the effects of humidity is crucial for maintaining the accuracy of ozone measurements, particularly in varying environmental conditions. Water vapor can adsorb onto the inner walls of the detection cell, altering the cell's reflectivity. This change can affect the measurements of light intensity used to determine ozone concentration. If humidity levels differ between the two measurement phases (I and Io), an offset in the ozone measurement will occur. This offset can reach several tens of nL/L during sudden changes in ambient humidity. Over time, as the internal ozone scrubber equilibrates with water vapor, the offset will adjust accordingly. If the Ozone Monitor is zeroed using dry tank air but then used to measure ozone in humid air, a measurement offset will occur. This is due to the difference in humidity between the calibration environment (dry air) and the measurement environment (humid air). Regular calibration with consideration for ambient humidity conditions can help mitigate these offsets. Keeping track of and adjusting for environmental changes, particularly humidity, can help maintain accurate ozone measurements.

In Figure 9.7, the most important internal component of the ozone monitor is a photodiode. A photodiode is a lightweight sensor that converts light energy into electrical voltage or current. It is a type of semiconductor device with a PN junction, which consists of P (positive) and N (negative) layers. Between these layers, there is an intrinsic layer that enhances its sensitivity to light. The photodiode accepts light energy as input to generate an electric current.[8]

A photodiode is also called a photodetector, photo sensor, or light detector. The photodiode operates in reverse bias condition, meaning the p-side of the photodiode is connected to the negative terminal of the battery (or power supply) and the n-side to the positive terminal of the battery. Typical photodiode materials include Silicon, Germanium, Indium Gallium Arsenide Phosphide, and Indium Gallium Arsenide. Internally, a photodiode has optical filters, a built-in lens, and a surface area. When the surface area of the photodiode increases, it results in a longer response time. It has two terminals, as shown in Figure 9.8. The smaller terminal acts as the cathode and the longer terminal acts as the anode.

The symbol of the photodiode is similar to that of an LED but the arrows point inwards as opposed to outwards in the LED. The following image shows the symbol of a photodiode, as shown in Figure 9.9.

When light illuminates the PN junction of a photodiode, covalent bonds are ionized, generating electron-hole pairs. Photocurrents are produced due to the generation of these electron-hole pairs. When photons with energy greater than 1.1 eV hit the diode, electron-hole pairs are formed. When the photon enters the depletion region of the diode, it collides with an atom with high energy. This collision results in the release of electrons from the atomic structure, producing free electrons and holes, see Figure 9.10.

Figure 9.8   Typical outline of a photodiode [8]

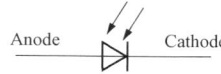

Figure 9.9   The symbol of the photodiode [8]

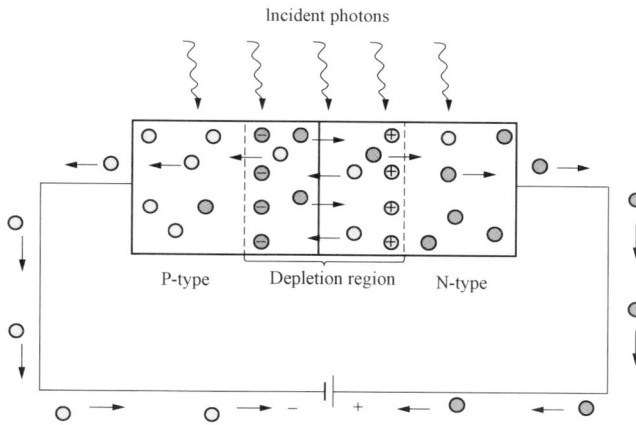

Figure 9.10   Working principle of PN junction photodiodes[8]

An electron has a negative charge, and holes have a positive charge. The depletion region has a built-in electric field, which causes the electron-hole pairs to move away from the junction. Consequently, holes move to the anode, and electrons move to the cathode, producing the photocurrent. The photon absorption intensity and photon energy are directly proportional to each other. When the energy of photons is low, the absorption will be higher. This entire process is known as the Inner Photoelectric Effect.

Photon excitation occurs via two methods: Intrinsic Excitations and Extrinsic Excitations. Intrinsic excitation happens when an electron in the valence band is excited by a photon to the conduction band.

A photodiode operates in three different modes: Photovoltaic mode, Photoconductive mode, and Avalanche diode mode.

When a photodiode operates in low-frequency and ultra-low light applications. When irradiated by light, the photodiode generates a voltage. The voltage produced is within a very small dynamic range and has a non-linear characteristic. When configured with an operational amplifier (OP-AMP) in this mode, there is minimal temperature variation, as shown in Figure

9.11. Then we say a photodiode operates in photovoltaic mode (Zero Bias Mode).

In photoconductive mode, the photodiode operates under reverse bias conditions, where the cathode is positive and the anode is negative. Applying a reverse voltage to the photodiode increases the width of the depletion layer at the p-n junction. This reduces the junction capacitance and response time, allowing the photodiode to operate faster. The increased reverse voltage causes the depletion layer to widen, which helps in quickly separating the photogenerated electron-hole pairs and thus increases the speed of response. When photons strike the photodiode, they generate electron-hole pairs in the depletion region. The reverse bias accelerates these carriers, creating a current proportional to the incident light intensity. To convert the photocurrent into a measurable voltage, a transimpedance amplifier is used. This amplifier maintains a constant voltage across the photodiode, ensuring it operates in photoconductive mode. The photodiode's current is then converted to a voltage output by the transimpedance amplifier. While photoconductive mode provides fast response times, it also introduces electronic noise. This noise can be mitigated by proper circuit design and filtering techniques, see Figure 9.12.

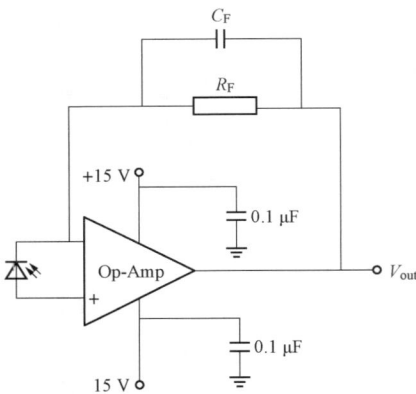

Figure 9.11   Photodiode operates in photovoltaic mode [8]

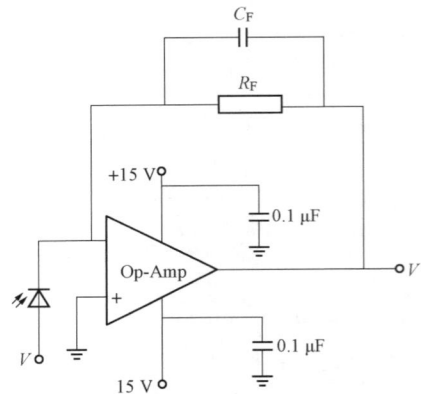

Figure 9.12   Photodiode operates in photoconductive mode [8]

In avalanche diode mode, the photodiode operates under high reverse bias conditions, resulting in the following effects: (1) High Reverse Bias: The avalanche diode is subjected to a high reverse voltage, which causes the electric field within the diode to increase substantially. This strong electric field accelerates the free electrons and holes. (2) Avalanche Breakdown: As electrons gain energy from the high electric field, they collide with the atoms in the diode material. These collisions produce additional electron-hole pairs through a process called avalanche breakdown. This cascade effect significantly multiplies the number of charge carriers. (3) Internal Gain: The avalanche breakdown results in a multiplication factor, or internal gain, where each photoproduced electron-hole pair generates multiple charge carriers. This internal gain enhances the photodiode's sensitivity and response. (4) Increased Device Response: Due to the internal gain, the avalanche photodiode can detect very weak light signals with higher sensitivity compared to standard photodiodes. This makes it suitable for applications requiring high sensitivity and detection of low light levels. (5) Operation and Noise: Although the internal gain improves the device's response, it can also introduce additional noise. Proper circuit design and noise reduction techniques are necessary to maintain accuracy in measurements, as shown in Figure 9.13.

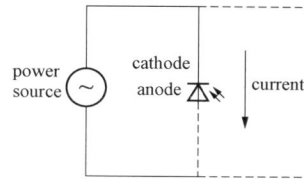

Figure 9.13　A photodiode operates in a circuit in reverse bias [8]

When photodiodes are integrated into external circuits, they require a power source to function effectively. Photodiodes generate a very small current in response to light exposure. This current alone is usually insufficient to drive external electronic devices or circuits effectively. To enhance the photodiode's performance and ensure that it can drive external devices, it is connected to an external power source. This power source, typically a battery, provides additional current to the circuit. The battery supplies the necessary power to increase the current flowing through the photodiode, which in turn helps power the connected electronic devices. By boosting the current, the battery ensures that the photodiode's output is sufficient for practical applications.

V-I Characteristics of photodiode could be found in Figure 9.14. The photodiode operates in reverse bias conditions. Reverse voltages are plotted along the *X*-axis in volts and reverse currents are plotted along *Y*-axis in microampere.

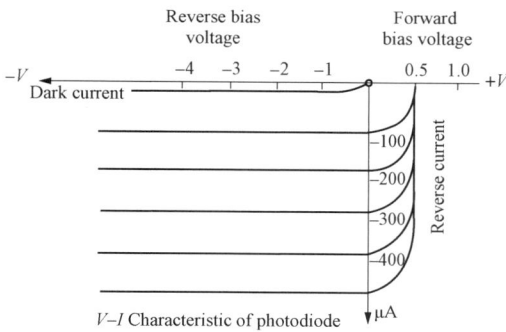

Figure 9.14　Photodiode operates in reverse bias condition [8]

Reverse current does not depend on reverse voltage. When there is no light illumination, the reverse current is nearly zero. This minimal current is referred to as the "Dark Current", which is the current flowing through the photodiode when no light is present. When light hits the photodiode, it generates electron-hole pairs, leading to an increase in reverse current. This increase is linear with respect to the intensity of the light. The *V-I* curve shows that the reverse current is independent of the reverse voltage in the normal operating range. The reverse current increases linearly with increasing light intensity, while the Dark Current remains relatively constant in the absence of light.

Photodiodes are highly effective for detecting optical signals due to their ability to generate a current directly proportional to the intensity of incident light. Photodiodes are commonly used as photodetectors in various applications, including optical communication systems, light sensors, and imaging devices. They convert optical signals into electrical signals, which can then be processed by electronic circuits. Built-in Lenses can be used to focus light onto the photodiode, increasing the amount of light collected and improving the device's sensitivity and performance. Optical filters can be used to select specific wavelengths of light, allowing the photodiode to detect signals at particular wavelengths while rejecting others. This is useful for applications requiring wavelength-selective detection. Photodiodes are often coupled with optical fibers in communication systems to ensure efficient light collection and signal transmission.

The applications of photodiode include:

- With the help of optocouplers, photodiodes provide electric isolation. When two isolated circuits are illuminated by light, optocouplers are used to couple the circuit optically. Optocouplers are faster compared to conventional devices.
- Photodiodes are used in safety electronics such as fire and smoke detectors.
- Photodiodes are used in numerous medical applications. They are used in instruments that analyze samples, detectors for computed tomography, and also used in blood gas monitors.
- Photodiodes are used in solar cell panels.
- Photodiodes are used in logic circuits.
- Photodiodes are used in the detection circuits.
- Photodiodes are used in character recognition circuits.
- Photodiodes are used for the exact measurement of the intensity of light in science and industry.
- Photodiodes are used frequently in lighting regulation and optical communication due to it is faster and more complex than normal PN junction diodes.

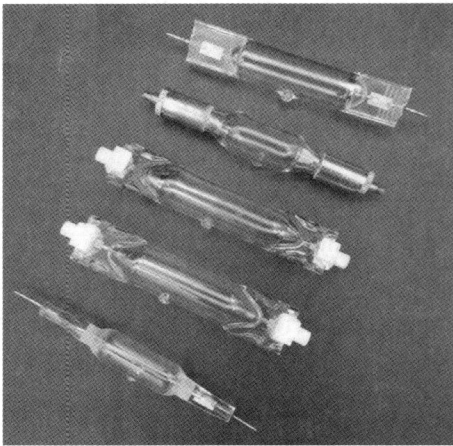

Figure 9.15   Typical outline of a
low-pressure mercury lamp

Figure 9.15 are low-pressure mercury lamps (Hg lamp in Figure 9.7). Operating at 253.7 nm, it is indeed designed to be highly effective for ozone measurement due to its close alignment with the peak absorption wavelength of ozone. Quartz glass is used in these lamps due to its high UV transmission and durability, which helps in maintaining consistent performance and longevity. The choice between ozone-free and ozone-generating lamps depends on the specific application and the need for accurate ozone measurement or control.

Low-pressure mercury vapor lamps are indeed effective sources of UV light for applications requiring short-wavelength UV, particularly for germicidal purposes or ozone measurement. Their stability and efficiency in emitting UV-C light at

254 nm make them suitable for various applications, including disinfection and ozone monitoring.

In contrast, UV LEDs offer a more targeted emission spectrum with peaks at specific wavelengths such as 365, 385, 395, and 405 nm ($\pm$5nm), as shown in Figure 9.16. Detailed information could be found at website. [9] Their monochromatic nature allows for higher irradiance at specific wavelengths compared to traditional UV lamps, but they typically lack the broad emission spectrum of mercury lamps. UV LEDs are also more energy efficient and have a longer lifespan, but they do not cover the UV-B and UV-C ranges as extensively as mercury lamps. The comparison between UV LED and UV mercury lamp spectra is crucial for choosing the appropriate UV source for a given application, especially if precise wavelength control or specific emission characteristics are required.

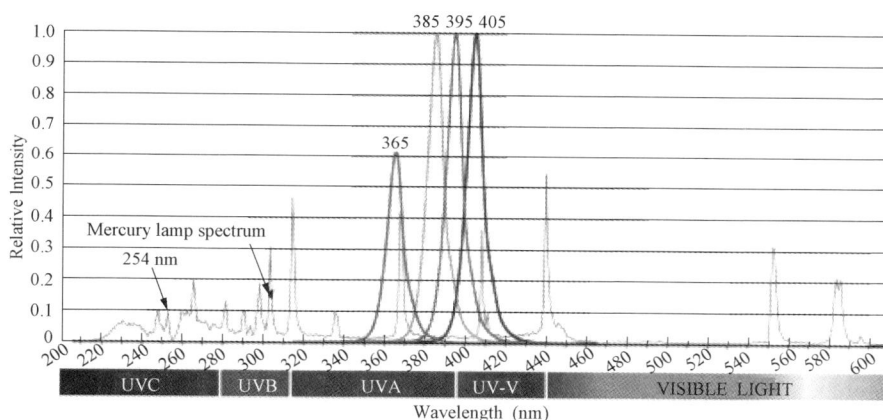

Figure 9.16   The relative intensity of low-pressure mercury lamps and several UV LEDs

9.3.2   Chemiluminescence Sensors for NO$_x$ Measurement

Chemiluminescence (also known as chemoluminescence) is the emission of light (luminescence) as the result of a chemical reaction, i.e., a chemical reaction that results in a flash or glow of light. These special exothermic reactions release energy primarily in the form of light rather than heat. The general term for this effect is chemiluminescence. Detailed information could be found at website (https://en.wikipedia.org/wiki/Chemiluminescence/).[3, 10]

Chemiluminescence sensors are devices that measure the intensity of light emitted during a chemical reaction to quantify the concentration of specific substances. In environmental monitoring, they are commonly used to detect nitrogen oxides (NO$_x$) and ozone (O$_3$) due to the predictable light emissions resulting from specific reactions involving these gases.[11, 12]

A spontaneous chemical reaction between nitric oxide and ozone (an unstable molecule formed of three oxygen atoms: O$_3$) is known to produce chemiluminescence:

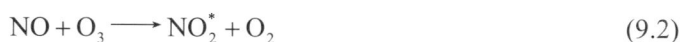

$$NO + O_3 \longrightarrow NO_2^* + O_2 \tag{9.2}$$

$$NO_2^* \longrightarrow NO_2 + light \qquad (9.3)$$

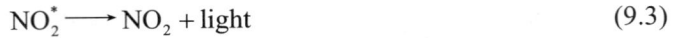

Although above chemical reaction only emitting a small amount of light (that means only a small fraction of the NO_2 molecules formed by this reaction will emit light), it is enough to measure for nitric oxide gas as a quantitative method.

In the nitric oxide (NO) gas optical analyzers, a photomultiplier tube (PMT) serves as the light-detecting sensor. The PMT generates an electrical signal proportional to the amount of light observed inside the reaction chamber. When the concentration of nitric oxide (NO) molecules in the sample gas stream increases, more light is emitted inside the reaction chamber. This increased light emission results in a stronger electrical signal produced by the photomultiplier tube, allowing for precise measurement of NO concentration.

Although the nitric oxide (NO) gas optical analyzers can be used easily to measure the concentration of nitric oxide (NO), it is not sensitive to other nitrogen oxides (NO_2, NO_3, etc., collectively known as NO_x). Normally, this selectivity is beneficial because it eliminates interference from other gases. However, other oxides of nitrogen are just as polluting as nitric oxide. Therefore, when monitoring pollution, it is important to measure a combination of these other oxides as well.

For accurately measuring all oxides of nitrogen, a chemical converter is essential. This converter changes other oxides into nitric oxide (NO) before the sample gas enters the reaction chamber.

The chemical converter can be realized in two ways: one by heating NO_x to convert them into NO molecules, and the other by using a metallic reactant.

The first method to achieve the $NO_x \rightarrow NO$ chemical conversion is to heat the sample gas to around 705 ℃. Because when the temperature is high enough, the molecular structure of NO is favored over more complicated oxides such as NO_2, resulting in the release of oxygen from the NO_2 and NO_3 molecules, converting them to NO molecules.

The second method for realization of the conversion $NO_x \rightarrow NO$ is to use a metallic reactant in the converter to remove the extra oxygen atoms from the NO_2 molecules. One such metal for this purpose is molybdenum (Mo). When Mo is heated to around 400 ℃. The chemical reaction of NO_2 converting to NO is as follows:[3, 10]

$$3NO_2 + Mo \longrightarrow MoO_3 + 3NO \qquad (9.4)$$

When other nitric oxides (such as NO_3) are converted into NO, it occurs in a similar way, leaving their excess oxygen atoms bound to molybdenum atoms and becoming nitric oxide (NO). The only difference between these reactions and the one shown for NO_2 is the proportional ratios between molecules. When the molybdenum metal is converted into the compound molybdenum trioxide over time, it becomes expendable and requires periodic replacement. The rate at which the molybdenum metal depletes inside the converter depends on the sample flow rate and the concentration of NO_2.

The pressure control of the gas sample is critically important for good measurement accuracy. If the pressure of the sampled gas inside the chemiluminescence reaction chamber varies, the amount of light emitted will change accordingly even if the relative concentration of NO_x gas remains stable. This is because higher pressures pack gas molecules closer together, resulting in more reactive molecules inside the chamber for any given percentage or μL/L concentration. For this reason, analyzers should be equipped with a pressure regulation module to ensure the gas pressure inside the measurement chamber remains constant.

9.4　Application Examples of Air Decomposition Analysis

9.4.1　The Relationship between Ozone and Partial Discharge

The detection of ozone (O_3) and nitrogen oxides (NO_x) gases are critical for various fields like atmospheric science, the pharmaceutical industry, process engineering, and water treatment, and significant research has been conducted in this area, there are still challenges with the devices used for detection. Three types of methods are available and will be described below:

- Differential optical absorption spectroscopy (DOAS).
- UV absorption method.
- Chemiluminescence.

The advantage of DOAS includes it is highly sensitive compared to other methods. However, several challenges still exist: (1) Requires a complex setup with larger sizes and volumes. (2) Cooling systems is needed which adds to the cost and size. (3) The overall system is costly due to its complexity. (4) Efficiency is compromised on cloudy days due to its reliance on a constant light source and diffraction.

The advantage of Chemiluminescence (CL) Detection includes it is a widely used method for nitrogen oxides detecting. The challenges include: (1) This method is not applicable for detecting Ozone because the energy required for chemiluminescence emission is produced through a chemical reaction, not from an exciting light beam. (2) Another is sensitivity issues, this method is not as sensitive as needed for detecting trace amounts.

The advantage of UV Absorption Method includes it is highly sensitive for ozone detection. The challenges include: (1) The primary limitation is the expense associated with the system. (2) Efforts are being made to develop an optimized UV absorption method that balances sensitivity with cost.

Therefore, above problems lead us to develop a highly sensitive low-cost detecting method which leads us to the measurement of partial discharges through air decomposition by-products. The development of a new method is very necessary which will help us to make the life of electrical equipment better and for stable power system operation.

Figure 9.17, Figure 9.18 and Figure 9.19 are the physical models for establishing the relationship between partial discharge and air decomposition byproducts.[13] Figure 9.17 shows the test circuit for measuring partial discharge between needle and plate electrodes. T is a 150 kVA single-phase non-PD transformer controlled by a regulating transformer at the low voltage side. Z is a current limited resistor. C_x represents a cylinder with needle-plane electrodes which is used to simulate the corona discharge, and it is a semi-closed cell with both gas inlet and outlet on the bottom cover. C_k is the coupling capacitor. M_{pd} is the PD inspection device, its model is TWPD-2C based on pulse current method. F is a filter and is used to strain the dust in the air. Z_m is the input impedance of PD signal acquisition portion of measuring instrument. A represents the amplification circuit. The special notation MO_3 and MNO_x represent the measuring instruments for Ozone and NO_x gases in nL/L.

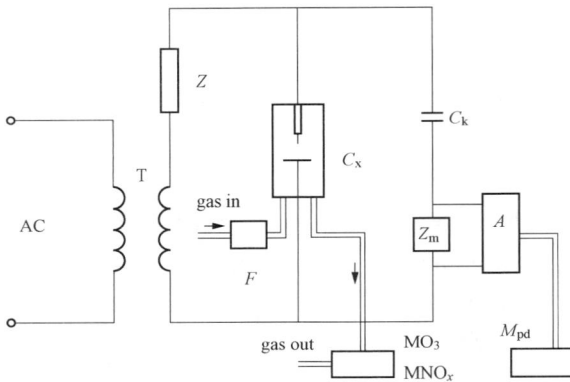

Figure 9.17   Schematic diagram of O_3 and
NO_x measurement model

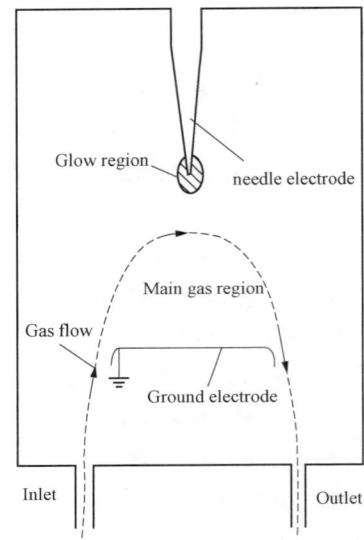

Figure 9.18   Schematic diagram of air
flow between needle and plate electrodes
in the test cell

Figure 9.20 and Figure 9.21 display images of the ozone and NO_x monitors, respectively, both manufactured by 2B Technologies. The ozone concentration is measured using model 106-L, while models 401 and 410 are used for NO_x detection. These monitors have a lower detection limit of approximately 2 nL/L and generally offer a precision better than $\pm 2\%$. The devices provide readings every 10 seconds. Both monitors are connected to the test cell through PTFE tubing. As illustrated in Figure 9.18, the system is designed for measuring decomposition gases. Thegas from the test cell is pumped to the detectors, with the flow rate remaining nearly constant throughout the testing process. Since the gas concentration is measured in real-time, there is a delay before the concentration stabilizes.

Ozone and NO_x can naturally occur at ground level, typically found in concentrations ranging from 5 to 35 nL/L for ozone and 10 to 100 nL/L for NO_x, especially in heavily polluted areas. These concentrations vary based on the specific location and environmental conditions. Human activities, particularly in urban settings, contribute significantly to the presence of both ozone and NO_x, often leading to higher levels in those regions.

In the following experiment, the influence of background ozone and NO_x concentrations is accounted for by subtracting the background gas levels. It is crucial to analyze these background values in the absence of partial discharge to ensure accurate measurements. Figure 9.22 and 9.23 present the background concentration data for ozone and NO_x recorded during the experiments. These measurements were taken over the course of one hour on the same day when no discharge occurred in the experimental setup. It is important to highlight that the background levels of ozone and NO_x can vary at different times throughout the day. For ozone, the average fluctuation in concentration is approximately 2-3 nL/L over the course of a full day. Table 9.3 provides the maximum, minimum, and average values of these background gases during the experiment.

Figure 9.19   Outline of the actual experimental setup

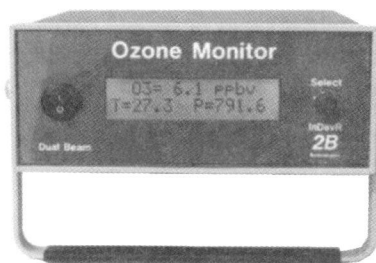

Figure 9.20   Outline of ozone monitor type 106-L of 2B Technologies

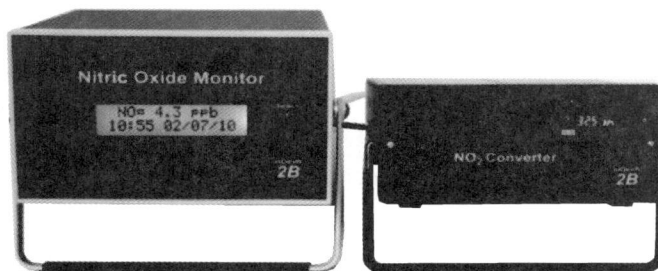

Figure 9.21   Outline of NO monitor type 401 and NO_2 converter type 410 of 2B Technologies

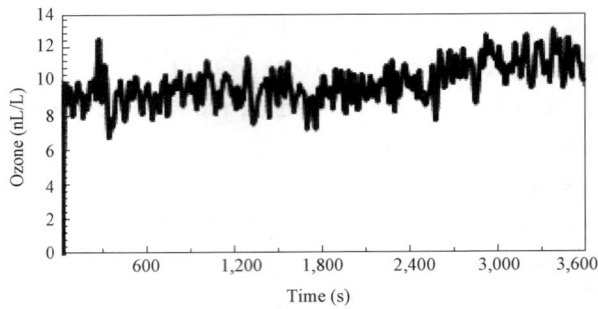

Figure 9.22   Background data of ozone without PD during the experiment

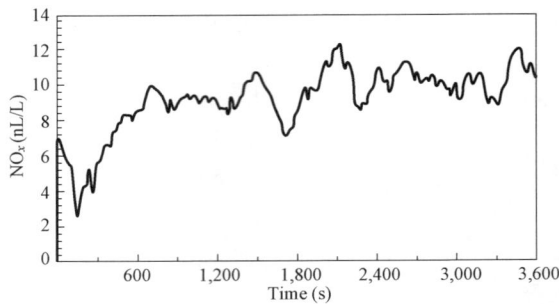

Figure 9.23   Background data of NO_x without PD during experiment

Table 9.3   **Background values of Ozone and NO_x**

Gases	Max (nL/L)	Min (nL/L)	Average (nL/L)
Ozone	13.1	8.1	9.96
NO_x	12.2	2.5	9.10

A protrusion defect is a common issue that can arise during equipment installation, manufacturing, or due to aging. These defects are particularly prevalent in transmission lines, switch-boards, and switchgears, where air serves as the insulating material. Understanding the partial discharge (PD) phenomena caused by these defects is crucial, especially when air is used as the insulating medium in both low and high-voltage equipment under various experimental conditions. In this study, the online monitoring of partial discharge resulting from a protrusion defect was conducted under two different humidity levels—low humidity (18% RH) and high humidity (44% RH). The analysis focused on measuring the air decomposition by-products, specifically ozone and NO_x gases. During all experiments, a needle with a diameter of 0.2 mm was used. The AC corona frequency was maintained at 50 Hz. All experiments were carried out on the same day, ensuring consistent temperature (25 ℃) and atmospheric pressure conditions. The inception discharge voltage, also known as corona starting voltage (CSV), was recorded along with the steady discharge voltage under the specified conditions. Humidity levels of 18% and 44% RH were generated using a humidifier and measured with a Victor-231 device, which has an accuracy of ±2% RH. The partial discharge was measured using the pulse current method, while ozone and NO_x concentrations were measured using the UV absorption method and the chemiluminescence method, respectively, as described in previous sections.

To ensure that the partial discharge (PD) was generated by the artificial defects rather than the test cell itself, the inception discharge voltage (U) of the test cell was first measured without the defects present. During the experiments involving the defects, the applied voltage was kept below this threshold (U). High voltage was then gradually applied to the test cell until the PD detection device could identify the presence of PD. This voltage, at the moment when PD was first detected, is referred to as the PD inception voltage (U_0). Subsequently, a series of voltages higher than U_0 was applied to the experimental setup to facilitate the decomposition of air. The sketch map of gas flow within the test cell during the experiment is illustrated in Figure 9.19. Both gas concentrations and PD were monitored simultaneously. As depicted in Figure 9.24 and Figure 9.25, the gas flow containing decomposition products continuously exits the test cell and reaches the measurement devices once the discharge is active. The measurement data show changes until the concentration of decomposition gases entering the detection devices stabilizes, reflecting the amount generated by the discharge. The final recorded values represent the rate of decomposition gas generation at the applied voltage.[13]

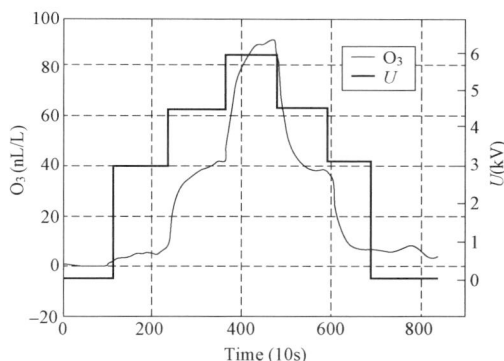

Figure 9.24 Concentration of O_3 vary with applied voltage

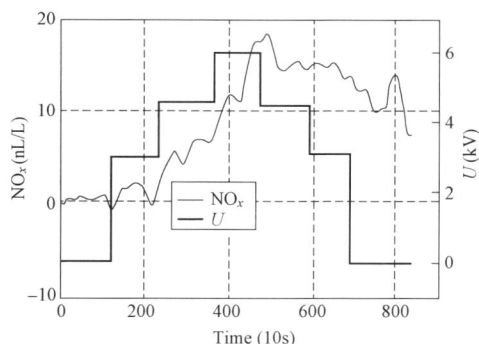

Figure 9.25 Concentration of NO_x vary with applied voltage

Figure 9.26 and Figure 9.27 illustrate the concentration of ozone and NO_x under various applied voltage levels. The ozone concentration, as measured online and depicted in Figure 9.17, represents the real-time value after subtracting the stable concentration observed at lower voltage levels. Similarly, the measuring time corresponds to the difference between the current time and the time recorded at the lower voltage. From these figures, we can observe that the lines reflect the rate of generation for both ozone and NO_x. Additionally, it is evident that higher applied voltages correspond to a steeper increase in the generation rates of these compounds.

Figure 9.28 and Figure 9.29 illustrate the effects of applied voltage and maximum partial discharge (PD) capacity under different humidity levels—low humidity (18%) and high humidity (44%). It is evident from these figures that the ozone concentration increases with both

PD capacity and applied voltage. Notably, higher applied voltages result in a greater ozone generation rate. Response time is a critical factor for condition monitoring devices, which is why the ozone measurement frequency is set at 0.1 Hz, and the testing duration is 2 hours. All results are obtained within a 10-second time frame. Additionally, Figure 9.30 displays the relationship between ozone concentration and maximum PD capacity, with an electrode gap of 30 mm.

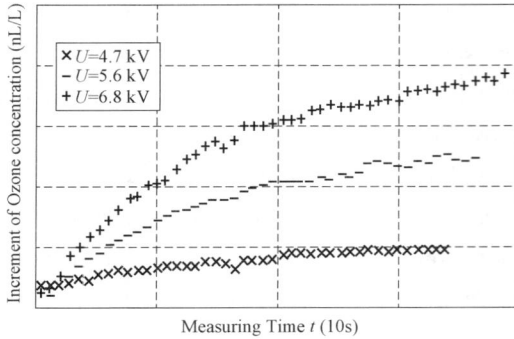

Figure 9.26   Increment of O_3 concentration go with measuring time under different applied voltage

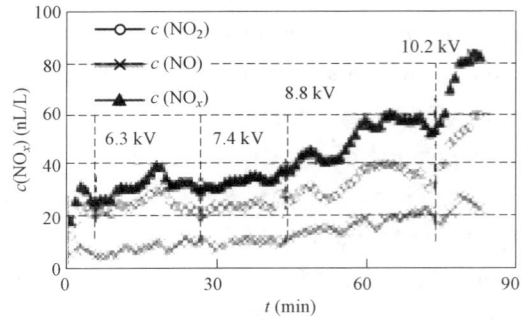

Figure 9.27   Increment of NO_x concentration under corona discharge

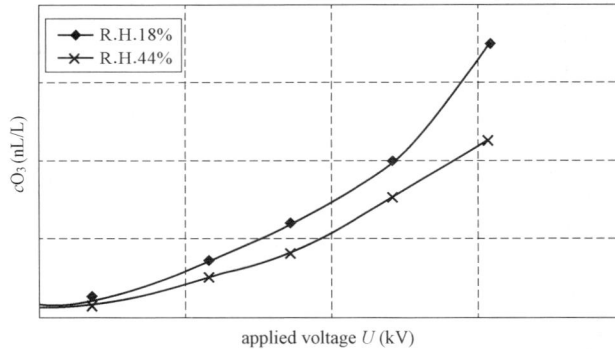

Figure 9.28   Concentration of O_3 under different conditions

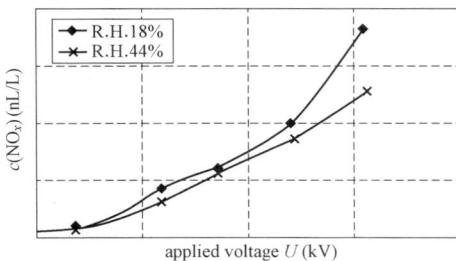

Figure 9.29   Concentration of NO_x under different conditions

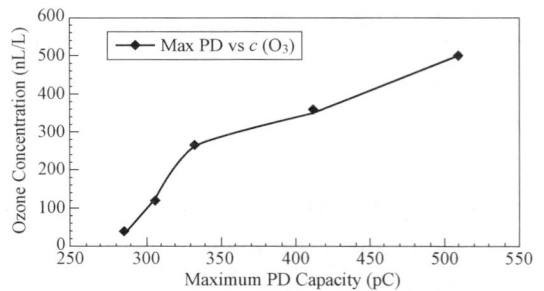

Figure 9.30   Ozone concentration against maximum PD capacity (electrodes gap: 30 mm)

The protrusion defect has been simulated using the COMSOL Multiphysics software package. A 3D model was constructed based on the experimental setup shown in Figure 9.19. The relationship between the gas concentration at the outlet of the vessel and the gas concentration inside the vessel was computed through numerical simulation. By combining this simulation data with experimental measurements of applied voltage (from which the electric field can be calculated) and the gas concentration at the outlet (measured using instruments depicted in Figure 9.20 and Figure 9.21), an empirical formula relating the electric field and gas concentration inside the vessel can be derived.

COMSOL Multiphysics is an integrated environment designed for solving systems of timedependent or stationary second-order partial differential equations (PDEs) in one, two, and three dimensions. These equations can be coupled in almost any configuration, making the software highly versatile. COMSOL Multiphysics also provides advanced tools for geometric modeling, simplifying the process of setting up simulations for many standard problems. The software includes predefined application modes that serve as templates, concealing much of the complex details involved in modeling with equations. These built-in physics modes enable users to construct models by defining relevant physical quantities—such as material properties, loads, constraints, sources, and fluxes—without the need to manually input the underlying equations. These application modes utilize the terminology and concepts familiar to the specific engineering discipline being modeled. Additionally, COMSOL Multiphysics can be accessed as a standalone tool through a flexible graphical user interface (GUI), or via script programming in the COMSOL Script language or MATLAB language, offering users extensive customization options.[1]

In this analysis, a protrusion defect was simulated using a needle-plate experimental configuration. The experimental results indicate that the large distance between the needle and plate significantly influences the rate of generation of air decomposition by-product gases. It is crucial to verify these results through numerical simulation. Studies have shown that the rate of generation of air decomposition by-product gases can be successfully modeled for complex geometries using COMSOL Multiphysics, a commercial software package that employs equation-based multiphysics modeling for various physical processes through the finite element method. In this study, the governing equations, boundary conditions, and numerical results obtained with COMSOL Multiphysics are presented and compared with the experimental data. The numerical results show good agreement with the experimental findings.

A 3D model of the protrusion defect was created in COMSOL, as shown in Figure 9.31 (a) and (b). It's important to note that simulating the concentration of air decomposition by-product gases involves the interaction between the electric field, ion motion, and the flow of electrically neutral gas molecules. This process is governed by a system of coupled nonlinear equations. For accurate modeling, it is crucial to also consider the effects of fluid dynamics and to include the complete space charge transport equation.

Figure 9.31 (a) illustrates a 3D diagram highlighting the critical regions involved in the generation of air decomposition by-product gases. When a high-intensity electric field is applied between a high-curvature corona electrode and a low-curvature collector electrode, gas molecules near the corona discharge region become ionized. These ionized gas molecules then travel toward the collector electrode, colliding with neutral air molecules along the way. These collisions impart momentum to the neutral molecules, driving the movement of gas towards the collector electrode and leading to the decomposition of air into ozone (O_3) and nitrogen oxides (NO_x). The operating voltage range for corona discharge lies between the corona onset voltage and the air gap breakdown voltage. Corona-induced airflow can be generated using both positive and negative voltages and polarities. The choice of polarity generally depends on several factors, including the electrode material, device geometry, ozone and NO_x generation constraints, among others.

Poisson's equation is employed as the governing equation for the plasma model, while the Navier-Stokes equation is used to model the fluid dynamics of air decomposition by-product gases. Detailed descriptions of the governing equations that describe the interaction between electric charges and air molecules, leading to air decomposition, as well as the boundary conditions, can be found in. [1, 14]

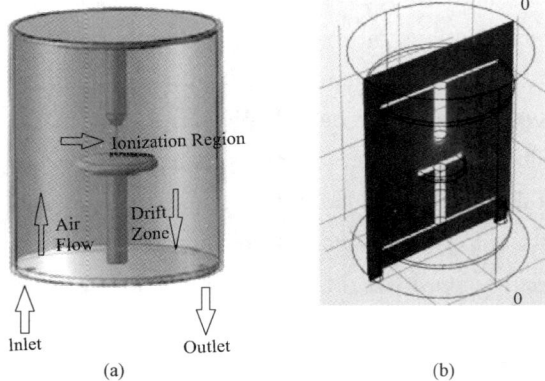

Figure 9.31   COMSOL 3D model of protrusion fault

For the electrostatics simulation, a constant positive AC voltage of 4 kV was applied to the surface of the corona needle, while the plate electrode was maintained at zero volts. In the charge transport model, a space charge surface density was applied to the corona needle surface. A zero diffusive flux condition was imposed on all boundaries, except for the surface of the corona needle.

In the fluid dynamics model, a no-slip boundary condition was applied to the surfaces of both the needle and plate electrodes. At the inlet air boundary, normal flow with pressure prescribed according to Bernoulli's equation was used. The outlet air boundary was modeled with an outletflow condition and atmospheric pressure.

The numerical simulation space was discretized into approximately 44,000 triangular elements, with the highest element density concentrated around the corona electrode and in regions with relatively high space charge density, electric field intensity, and at the outlet channel, as shown in Figure 9.32.

Figure 9.33 shows the 3D simulations of gas concentration after discharge at a 15 mm distance between needle and plate.

Figure 9.34 and Figure 9.35 illustrate the placement of the needle within the switchgear setup, as described in.[1] The ozone measuring tube is positioned 12 cm away from the corona discharge region.

While the distance of the measuring tube from the corona discharge region might influence the gas concentration measurements, this factor is not discussed in this study. All experiments were conducted under controlled conditions, with a constant temperature of 25 ℃, atmospheric pressure, and 30% relative humidity (RH) within the switchgear enclosure.

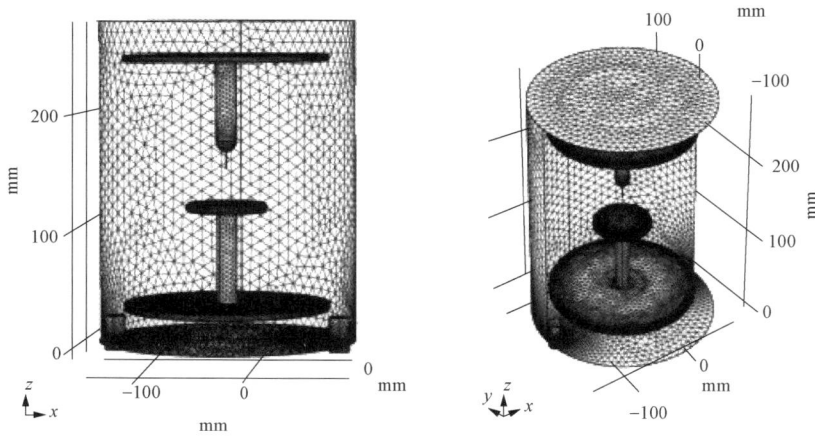

Figure 9.32　3D Mesh, domains and subdomains

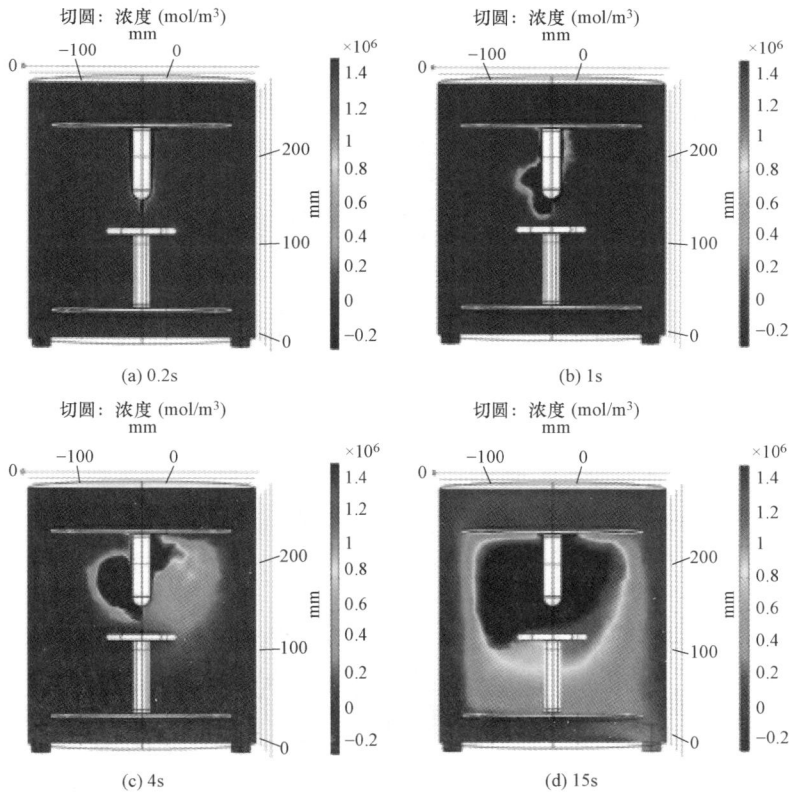

(a) 0.2s

(b) 1s

(c) 4s

(d) 15s

Figure 9.33　The simulation result of gas concentration at clearance of

15 mm between needle and plate (d=15 mm) (1)

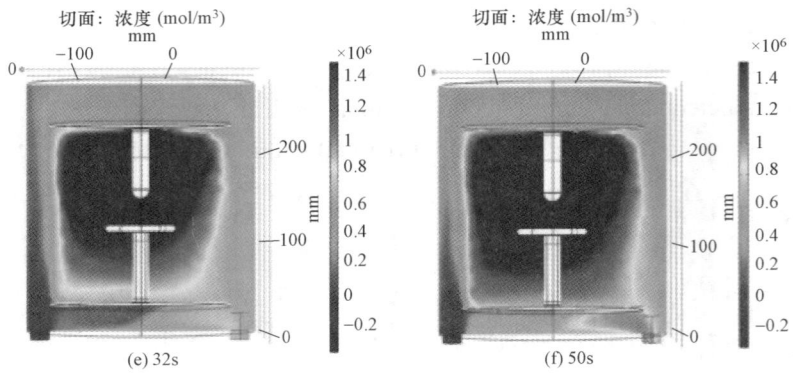

切面：浓度 (mol/m³)

(e) 32s

切面：浓度 (mol/m³)

(f) 50s

Figure 9.33　The simulation result of gas concentration at clearance of

15 mm between needle and plate (*d*=15 mm) (2)

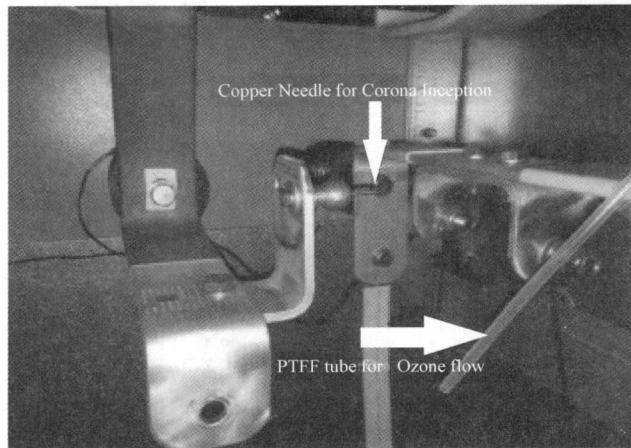

Figure 9.34　Corona discharge initiations along with Ozone flow tube

Figure 9.35　Experimental setup of a switchgear

When different voltages are applied to MV switchgear for simulation corona discharge, the results are shown in Figure 9.36. The higher voltages generate higher concentrations of ozone which can easily measure and analyzed. The experimental data were recorded for 30 min time intervals on the same day and constant conditions as discussed above. [15]

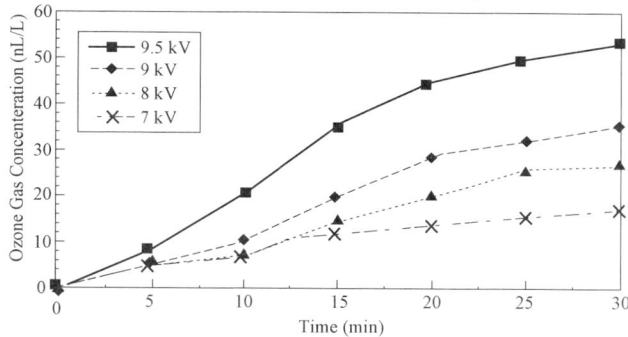

Figure 9.36   Generation of ozone after discharge in MV switchgear under different voltage levels

9.4.2   Ozone Distribution in Hydro-generators

It is well known that electrical discharges in air produce ozone. High ozone concentrations in air-cooled generators have long been associated with the presence of partial discharges (PD). This highly reactive and aggressive gas can chemically attack the semi-conductive coating, ground insulation, and exposed metallic or rubber components. However, its reactivity and inherent instability, along with physical factors such as air velocity, temperature, and humidity, can contribute to the elimination of the ozone generated by PD. [16]

L. Lepine and D. N. Nguyen from Hydro-Quebec power plants, Canada, measured the ozone gas distribution inside the stator core of an air-cooled hydro-generator and demonstrated that ozone measurements can be used to qualitatively locate partial discharge activities. Typically, indoor background ozone levels range from 5 to 20 nL/L. However, in 2002, ozone concentrations ranging from 30 to 60 nL/L were measured at the louvers exhaust of six air-cooled 48 MW generators at a Hydro-Quebec power plant, where partial discharges had already been diagnosed in one of the generators. Updated measurements in 2007 revealed that ozone levels had more than doubled compared to the 2002 values in all of these cases, indicating an increase in partial discharge activity.[17]

In 2007, engineers at Hydro-Quebec power plants used a scale model of a stator core to study the ozone distribution profile inside and at the exit of a ventilation duct in the presence of slot partial discharge activity. The discharge defect was simulated by removing the semi-conductive coating on one side of a bar over the length of two consecutive core stackings. This bar was then positioned at the front of the slot.

Figure 9.37 (a) shows a section of the stator core, composed of a full-height stacking core, along with an enlarged image of the slot discharge activity captured with long-exposure photography. Figure 9.37 (b) illustrates the linear-flow blower placed directly in front of the stacking, aligned at the level of the discharge. Figure 9.37 (c) depicts the back of the stacking, featuring the heated mass flow sensor used to measure the airflow velocity exiting the ventilation duct. Additionally, it shows the 1/8" Teflon sampling tube positioned at the ventilation duct exit, which was inserted inside the ventilation channel located between the two stackings at the defect location. Ozone measurements were conducted using an API 400A UV-based ozone monitor, connected to the other end of the sampling tube. A phenolic guide was also employed to move the tubing linearly at the back of the duct to acquire the ozone profile exiting the duct.

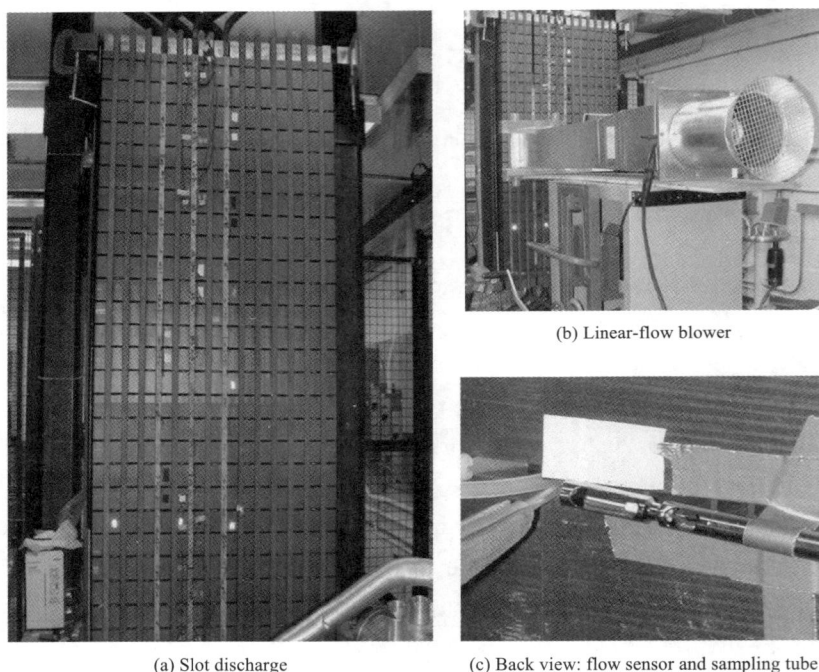

(b) Linear-flow blower

(a) Slot discharge   (c) Back view: flow sensor and sampling tube

Figure 9.37   Stator section scale model of Hydro-Quebec power plants [16]

Figure 9.38 shows the ozone concentration profile recorded inside the ventilation channel in a static mode, without ventilation, with 8 kV and 12 kV applied to the bar. As expected, the maximum concentration of ozone was recorded right next to the discharge source. Similar profiles were observed at both voltage levels, with the 12 kV experiment generating slightly more ozone.

The situation changes significantly in the presence of ventilation, as shown in Figure 9.39 and Figure 9.40, where the engineers varied the air velocity and the voltage applied to the bar, respectively. With a fixed 8 kV applied to the bar (see Figure 9.39), the ozone generated by the discharges is swept away by the ventilation towards the back of the stator, with its maximum concentration occurring approximately 5 cm away from the core. Even at the lowest air speed of 1.5 m/s, the maximum ozone concentration is two orders of magnitude lower than it is without ventilation. This maximum concentration further decreases as the air speed increases.[16]

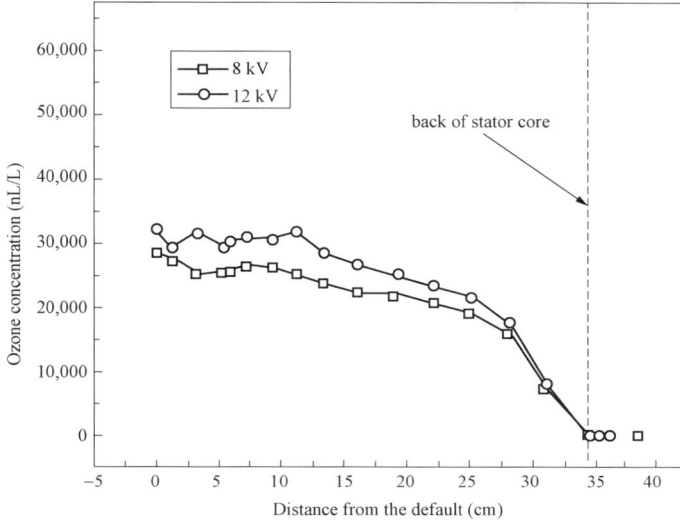

Figure 9.38　Stator section scale model of Hydro-Quebec power plants [16]

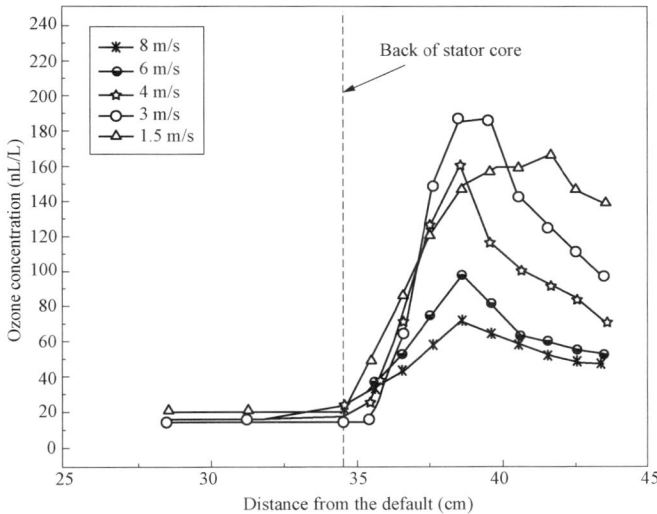

Figure 9.39　Ozone profile inside and out of the stator core at 8 kV and varying airflow [16]

When the airspeed is fixed at a typical value of 8 m/s and the voltage applied to the bar is varied, the maximum ozone concentration is again observed approximately 5 cm away from the stator core. As expected, the concentration of ozone exiting the ventilation channel increases with the voltage applied to the bar (and consequently, with the intensity of the partial discharge activity). Additionally, as shown in Figure 9.40, the ozone concentration tends to stabilize at distances greater than 12 cm outside the core and remains constant as far as could be measured.

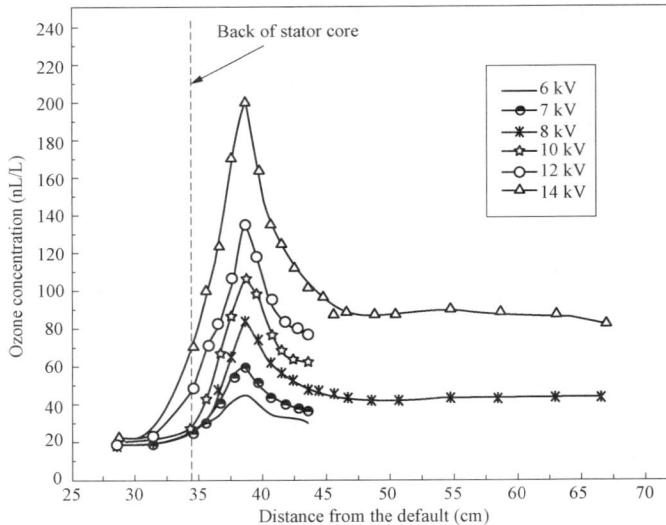

Figure 9.40   Ozone profile inside and out of the stator core with an 8 m/s air flow and
varying tension applied to the front bar [16]

These results suggest that it is possible to measure the ozone generated by slot partial discharges at certain points behind the stator core, and that the amount of ozone can be indicative of the partial discharge activity.

The Isle Maligne hydropower plant features 12 units of air-cooled 30 MW generators, as shown in Figure 9.41. The open architecture of these generators allowed for ozone measurements through openings in the outer shell, directly behind the core stacking around the machines. Measurements were conducted at the middle ventilation opening in each of the 36 sections of the stator core (see the red arrow in Figure 9.41).

Figure 9.41   Sampling location on Isle Maligne generators for ozone and PD measurement [16]

A direct relationship between the presence of partial discharges and the production of ozone inside generators can be established. Ozone is generated inside the slot in the air gap surrounding the bar, but the substantial airflow disperses it towards the back of the stator core through the ventilation ducts. The maximum ozone concentration is typically measured outside the stator core, at its rear. Ozone measurements on operational generators have demonstrated a strong correlation with diagnostic results obtained from partial discharge (PD) monitoring through capacitive couplers. Furthermore, in cases of localized faults, this technique has shown to complement PD monitoring by helping pinpoint the exact location of the fault.

Consequently, ozone monitoring could be a valuable technique for detecting slot partial discharge (PD) activity inside generators. This method is straightforward, easy to implement during normal operation, and does not require generator downtime. It is particularly useful as an initial diagnostic tool for generators that do not have capacitive couplers installed.

9.4.3 Ozone Sensors in an Medium Voltage Switchboard

Letizia DeMaria and Giuseppe Rizzi in Italy developed an ozone sensor for detecting predischarges and surface discharges in medium voltage (MV) switchboards. Their sensor utilizes the DOA method with UV light to detect ozone. They later introduced a fiber optic multisensor scheme to detect pre-discharges in medium voltage equipment, as shown in Figure 9.42. This system includes a UV light sensor to detect light emitted by pre-discharges and surface discharges, a sound sensor to measure the noise produced by discharges, and an ozone sensor to measure the concentration of ozone generated by the discharges. [18, 19, 20]

The reliability of electrical equipment in medium voltage (MV) substations is crucial for ensuring the continuity of service in an MV distribution network. Failure statistics reveal that MV substations are a major cause of outages, with MV switchboards frequently suffering from flashovers. Detecting corona or surface pre-discharge phenomena is the most effective technique for early failure warning in these electrical systems. Since these phenomena can eventually lead to breakdowns and outages, it is

Figure 9.42 Scheme of the fiber optic multisensory device

important to detect their onset and monitor their progression by observing induced effects such as light emission, acoustic noise, and ozone production. However, these signals are relatively weak, and high levels of electrical interference complicate field measurements. Additionally, the high cost of traditional sensing devices has historically limited the application of diagnostic systems for assessing the condition of these low-cost electrical components.

Recently, the feasibility of an innovative combined system has been investigated. This

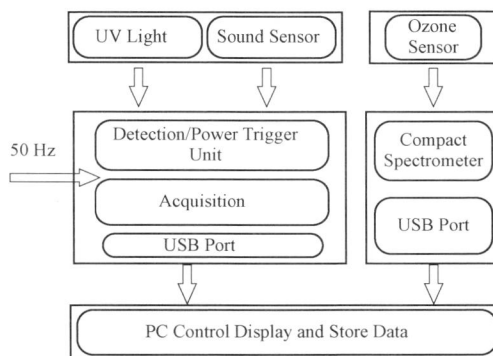

system was assembled using high-sensitivity and cost-effective sensors, which include both commercially available sensors and custom-developed prototypes.

Fiber-optic-based sensors were chosen for their advantages: they are non-invasive, unaffected by electrical disturbances, chemically inert, and relatively inexpensive.

An optical microphone and a fluorescent fiber-optic sensor were used to detect the sound pressure and light generated by pre-discharges inside the MV switchboard, respectively. Additionally, the research project aimed to develop a fiber-optic sensor for detecting ozone produced by pre-discharges. Integrating this sensor into the diagnostic prototype allows for the simultaneous detection of three different pre-discharge effects, helping to minimize false alarms.

The ozone sensor is based on a novel open-path optical design. This optical scheme ensures high sensitivity while maintaining durability in the harsh environment of the MV switchboard.

In previous work, the sensor was validated through calibration measurements conducted in a gas flow cell with varying ozone concentrations, as well as in a laboratory mock-up designed to simulate a corona discharge.

The developed sensor utilizes the differential optical absorption of UV light by ozone. Ozone strongly absorbs UV light across a broad spectral range from 200 nm to 375 nm, while it shows negligible absorption between 377.5 nm and 450 nm. The sensing probe, shown on the left in Figure 9.43, consists of a fixed-length (40 cm) cylindrical enclosure made from ozone-compatible material (PTFE). The longitudinal sides of the cylinder are open to allow the passage of ozone.

A UV fused-silica retroreflector (a circular corner cube with a 7 cm clear aperture) is securely mounted on one end of the enclosure. This retroreflector doubles the path length of the sensing probe, thereby enhancing the sensor's sensitivity to lower ozone concentrations.

On the opposite side of the enclosure, two optical fibers are connected to collimating lenses. A UV filter (peak wavelength = 300 nm, FHWD = 140 nm) is positioned in front of the collecting lens to eliminate spectral contributions from the UV source outside the two primary ozoneabsorbing (at 254 nm) and non-absorbing (at 375 nm) regions. UV-resistant fibers (300 cm long, 400-micron diameter) are used as launching and collecting fibers to prevent degradation from UV exposure.

The launching fiber is coupled to the UV light source, which is a deuterium-tungsten lamp. The collecting fiber is connected to a compact spectrometer with a fixed grating, covering a spectral range of 195-950 nm and a resolution of 0.25 nm. A variable attenuator (A) is positioned between the lamp and the launching fiber to prevent saturation of the spectrometer.

The initial optical alignment of the back-reflected beam is achieved by adjusting three screws on the top of the flange. Once the optics are properly aligned, these screws are locked to maintain the alignment.

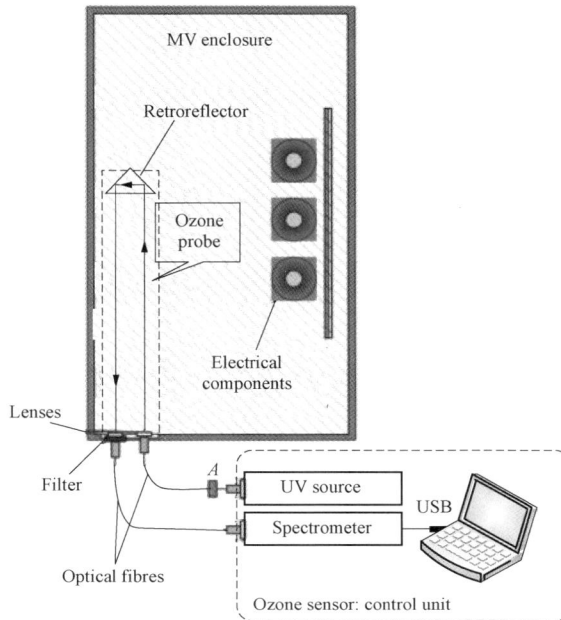

Figure 9.43   Optical scheme of the ozone sensor. The probe (on the left in the figure) mounted inside the MV switchboard is connected to the remote control unit by means of optical fibers[18]

The spectrometer is connected to a PC notebook via USB and is controlled using a proprietary software interface developed with LabVIEW.

This interface was designed as a multifunctional tool for the smart management of the entire prototype assembled for diagnosing pre-discharge phenomena. The prototype includes three types of sensors: acoustic and optical sensors for detecting the sound and light emission of the pre-discharge, respectively, and the ozone sensor. The interface facilitates easy configuration of parameters, initiation of measurements, and display and saving functions for all three sensors. Additionally, it allows simultaneous display of the time behavior of signals from each sensor, correlated to the same pre-discharge event. This combined analysis enhances the efficiency and reliability of diagnosing the initial failure of electrical components.

The program also simultaneously acquires the transmitted intensity at the non-absorbing wavelength. This allows for real-time monitoring of any intensity losses within the sensor that are not related to ozone variations.

To evaluate the feasibility of the new sensor for detecting ozone concentration under real operating conditions, a typical MV switchboard configuration was recreated in the laboratory using normalized components from the Italian distribution network (12/20 kV). Two different artificial defects were introduced sequentially to simulate pre-discharge activity. A wire (300 mm long and 1 mm diameter) was used to generate a corona discharge, while a strip of electrical semiconducting tape (70 mm long and 15 mm wide) was attached to the phase cable termination to simulate a surface discharge.

A standard Partial Discharge (PD) electrical system was used to monitor the pre-discharge activity. [14] This method measures the amplitude of the pre-discharge in terms of apparent charge (picoCoulombs, or pC).

For each type of defect, a series of tests were conducted by increasing the voltage applied to the cable termination from 0 kV to the pre-discharge inception level (U_i), as detected by the PD system, and then progressively up to 25 kV.

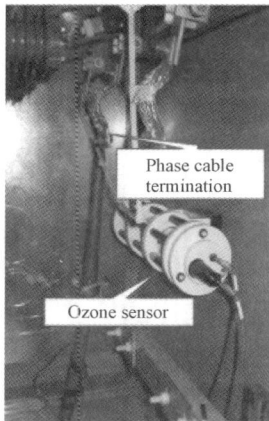

Figure 9.44   Experimental layout of ozone measurement in an MV switchboard [18]

The ozone probe was installed inside the MV switchboard and connected to the control unit located outside the enclosure, as shown schematically in Figure 9.43. The probe was positioned on a lateral wall of the enclosure, as depicted in Figure 9.44. To verify ozone concentration within the MV switchboard, a commercial ozone sensor with a typical acquisition time of one minute was also placed inside the metallic case.

Figure 9.45 illustrates the ozone concentration trend (red points) measured by the optical sensor over time during three consecutive voltage ramp tests. The data shown pertain to ozone levels measured inside the MV enclosure in the presence of a corona defect. Each acquisition lasted 2.5 seconds, with the solid blue line representing the average of over 20 data points. The vertical axis on the left shows the pre-discharge amplitude values (in pC) recorded by the standard PD system (blue dots in the figure). As depicted in Figure 9.45, similar trends were observed across the three repetitions with the corona defect, indicating consistency. This behavior aligns well with the data from the PD system, confirming the stability of the corona phenomenon. A slight delay was noted in detecting both the inception and extinction of the corona, although the delay was within a few tenths of a second. The ozone levels measured by the optical sensor ranged from 0.2 to 0.3 µL/L (on the plateau), which correlated well with the average value of 0.23 µL/L measured by the commercial ozone sensor.

As shown in Figure 9.45 the sensor started to detect ozone presence over three repetitions, when the amplitude of the corona predischarge was equal to 400 pC. This value was measured by the PD system at a voltage level equal to 1.25 U_i. This is the minimum applied voltage necessary to record an ozone activity by means of this optical sensor. This value was compared with detectable threshold values of both the optical microphone and the FL fluorescent fiber-based sensor of the diagnostic prototype, respectively, equal to 1.15 U_i and to 1.1 U_i. As evidence, the sensitivity of the ozone sensor is slightly lower than the sensitivity of optical and acoustic sensors. A lower sensitivity was obtained in presence of surface predischarge. The sensor began to appreciate the presence of ozone (0.15-0.2 µL/L) only when the predischarge activity was higher than 1,500 pC (equal to 2.5 U_i). The amplitude of the surface predischarge, measured by the PD system at inception was about 600 pC. This is probably due to the high instability of this phenomenon. To improve the sensitivity to both predischarge inception an optimization of the path length will be performed.

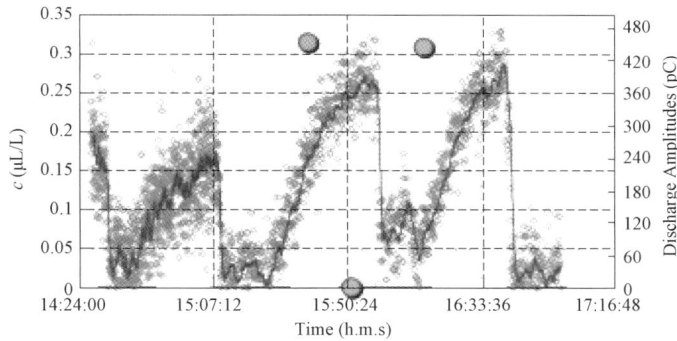

Figure 9.45   Ozone sensor response versus time measured in presence of the corona defect. The right vertical axis reports predischarge amplitudes (pC) by the standard PD electrical method[18]

As shown in Figure 9.45, the ozone sensor began detecting ozone presence during the three repetitions when the amplitude of the corona pre-discharge reached 400 pC. This value, recorded by the PD system, corresponds to a voltage level of 1.25 U_i, which is the minimum applied voltage necessary for the optical sensor to detect ozone activity. This threshold was compared with the detectable threshold values of the optical microphone and the fluorescent fiber-based sensor in the diagnostic prototype, which were 1.15 U_i and 1.1 U_i, respectively. This indicates that the sensitivity of the ozone sensor is slightly lower than that of the optical and acoustic sensors. In the case of surface pre-discharge, the ozone sensor's sensitivity was further reduced. It only detected ozone (0.15-0.2 µL/L) when the pre-discharge activity exceeded 1,500 pC (equal to 2.5 U_i). The amplitude of the surface pre-discharge, measured by the PD system at inception, was approximately 600 pC. This lower sensitivity is likely due to the high instability of the surface pre-discharge phenomenon. To enhance sensitivity for detecting both pre-discharge inception and activity, optimization of the path length of the sensor will be performed.

9.4.4   Forthcoming Applications of Ozone Sensors

Air discharge fault monitoring through the analysis of air decomposition by-product gases represents a novel methodology with significant potential for future applications. To establish criteria at the micro level, it is essential to correlate discharge phenomena with the rate of gas generation. Plasma models for the generation rate of air by-product gases are needed to study these faults comprehensively. Additionally, the impact of varying atmospheric factors, such as temperature and pressure, on the gases generated after discharge must be examined before deploying the actual prototype.

Validation of this methodology across different power system equipment is crucial. Ongoing studies aim to detect air discharge faults using air decomposition by-products in XLPE cables, wind turbine generators, HVDC reactors, air cooled generators, motors, and switchgears, as illustrated in Figure 9.46 and Figure 9.47. Furthermore, we are comparing this methodology with other existing partial discharge monitoring techniques, including optical, acoustic, and chemical methods.

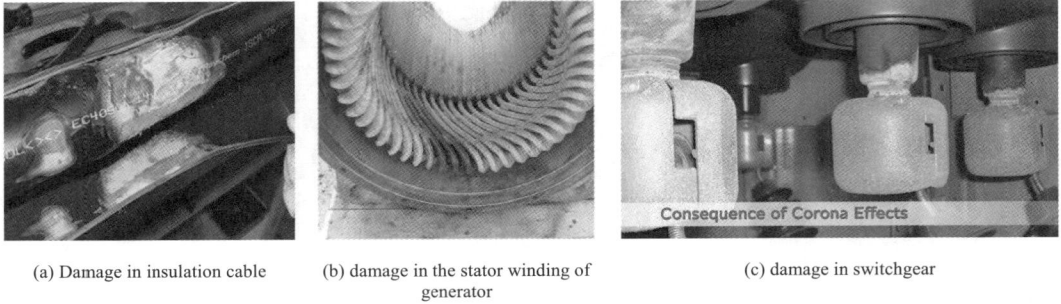

(a) Damage in insulation cable   (b) damage in the stator winding of   (c) damage in switchgear
generator

Figure 9.46   Failures due to PD in electrical power systems

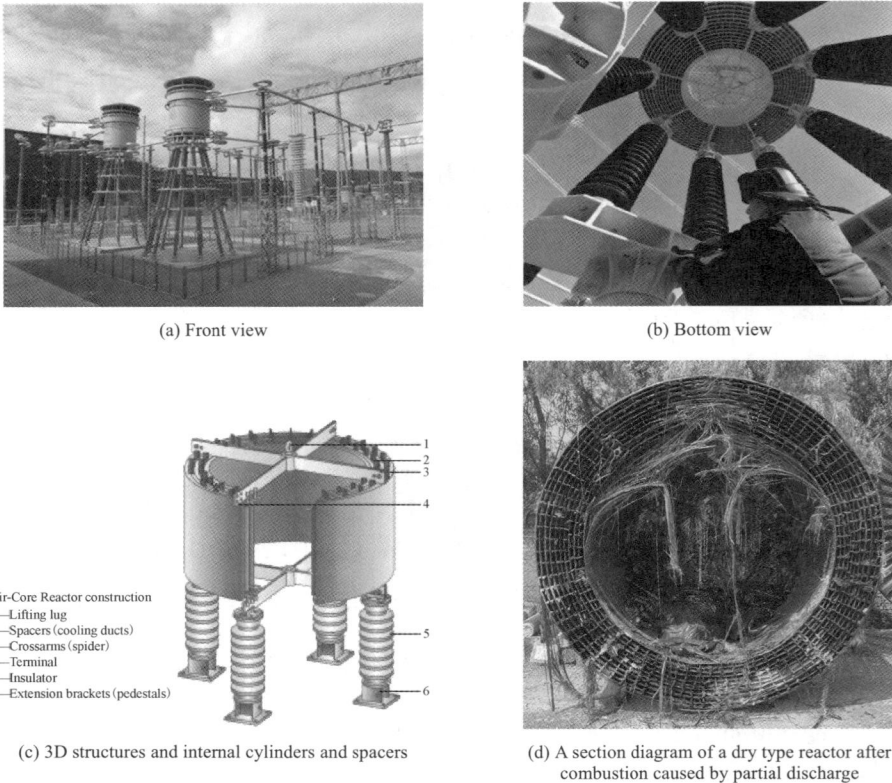

(a) Front view   (b) Bottom view

Air-Core Reactor construction
1—Lifting lug
2—Spacers (cooling ducts)
3—Crossarms (spider)
4—Terminal
5—Insulator
6—Extension brackets (pedestals)

(c) 3D structures and internal cylinders and spacers   (d) A section diagram of a dry type reactor after
combustion caused by partial discharge

Figure 9.47   Schematic diagram of a dry type shunt reactor and 3D internal structure

as well as its combustion caused by partial discharge

9.5 Fundamental Study of Emission Spectrum in GIS

9.5.1 Experimental Model of Emission Spectra Analysis in SF₆

A needle-plate electrode is used to simulate corona discharge and spark discharge. Figure 9.48 shows the schematic diagram of the needle-plate electrode. Herein, the electrode distance is 10 mm, the curvature radius of the tip of the needle electrode is about 0.3 mm, and its material is aluminum. The integration time of the spectrometer is set to 9,000 ms and 100 ms under corona discharge and spark discharge, respectively.

Schematic diagram for simulating the surface discharge is shown in Figure 9.49. Two plates of brass material are used to clamp a bisphenol-A epoxy resin cylindrical insulation block. A thin copper wire with a diameter of 0.2 mm and a length of 20 mm is attached to the side of the insulation block to simulate the surface flashover caused by metal debris attached to the insulation surface. The size of the plate electrode is shown in the left part and the geometric dimension of the insulation block is shown in the right part of Figure 9.49. Copper powder is plated on the upper and lower surfaces of the insulation block to achieve equal potential, thereby avoiding air gap discharge between the insulation block and the plate electrode. The integration time of the spectrometer is set to 100 ms.

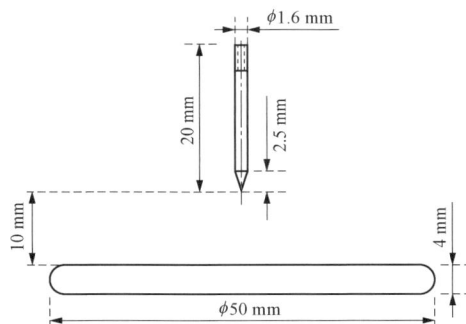

Figure 9.48 Schematic diagram of the geometry of the needle-plate electrode

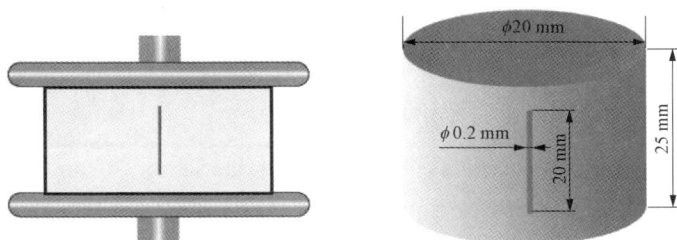

Figure 9.49 Schematic diagram of the geometry of the needle-plate electrode

SF₆ is a strongly electromotive gas. It adsorbs electrons and suppresses the accumulation of electron energy, resulting in a weak light signal emitted under corona discharge. Thus, the Savitzky Golay smoothing method is used for filtering after subtracting the background noise to improve the signal-to-noise ratio. The spectra of spark and surface discharges have high signal to noise ratios, so only background subtraction processing is performed.

9.5.2 Emission Spectra of Corona Discharge in SF$_6$

Figure 9.50 illustrates the emission spectrum distribution in the 200-650 nm range observed under corona discharge. The spectral regions around 260-347 nm [21] and 420-510 nm exhibit the most significant intensity, with the ultraviolet wavelength at 308.642 nm showing the highest spectral peak.[22]

The 260-347 nm range is generated by free radical OH radiation decomposed from trace amount of H$_2$O.[22]

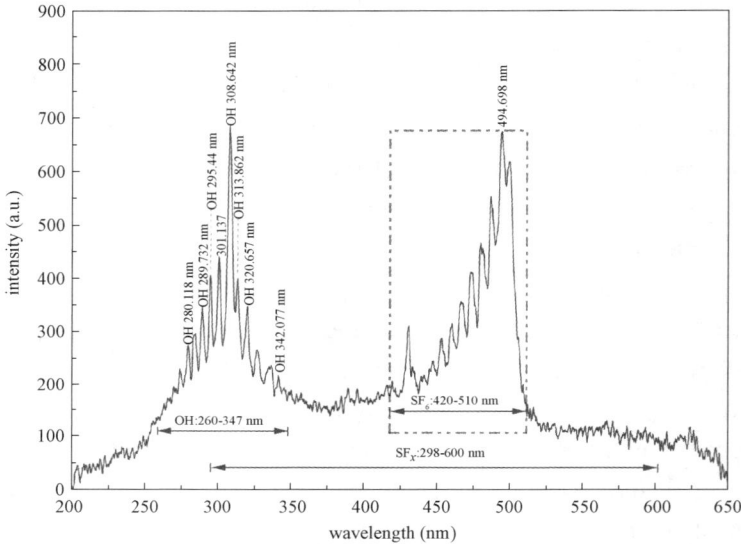

Figure 9.50   Emission spectra of SF$_6$ under corona discharge in
wavelength range 200-650 nm

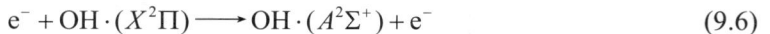

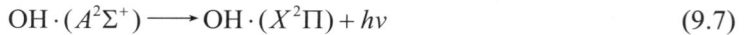

$$e^- + H_2O \longrightarrow e^- + H\cdot + OH\cdot \tag{9.5}$$

$$e^- + OH\cdot(X^2\Pi) \longrightarrow OH\cdot(A^2\Sigma^+) + e^- \tag{9.6}$$

$$OH\cdot(A^2\Sigma^+) \longrightarrow OH\cdot(X^2\Pi) + h\nu \tag{9.7}$$

Electrons continuously accumulate energy under the action of an electric field, collide with H$_2$O molecules as described in Equation (9.5), and dissociate to produce OH radicals. According to different energies, OH radicals near the needle tip exist in the form of ground state OH ($X^2\Pi$) and excited state OH ($A^2\Sigma^+$). Ground-state OH ($X^2\Pi$) may also collide again with high-energy electrons, as described in Equation (9.6), reaching the excited state OH ($A^2\Sigma^+$). The excited OH ($A^2\Sigma^+$) state has a brief lifespan, after which photons are emitted through the process outlined in Equation (9.7). These photons return to the ground state in the form of luminescence, resulting in a free radical emission band around 308.642 nm.

The spectrum of SF$_6$ molecules is in the 420-510 nm range.[23] Under the action of an electric field, electrons within a molecule are always in a state of motion. In addition, there are also relative vibrations of atomic nuclei and rotation of the entire molecule. When corona discharge occurs, due to the low input energy, SF_6 can adsorb electrons, causing energy level transitions of valence electrons inside SF_6 molecules. The transition of electronic energy levels is accompanied by changes in molecular vibrational and rotational dynamics. In this way, SF_6 molecules distributes a band of 420-510 nm.

Meanwhile, weak spectral information is also detected in the 650-1,037 nm band. As shown in Figure 9.51, the emission spectra is mainly composed of low-intensity linear spectra related to fluorine atoms gradually decomposed from SF_6 molecules under the impact of high-energy electron. [24, 25]

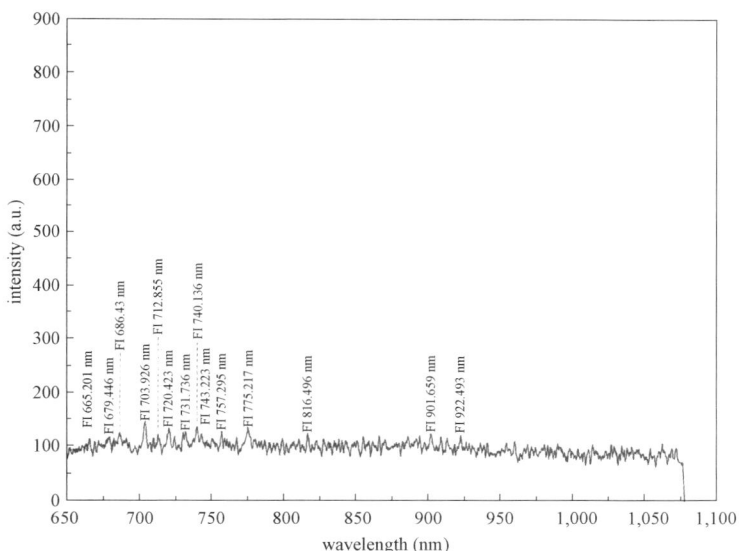

Figure 9.51　Emission spectra of SF_6 under corona discharge in the 650-1,037 nm band

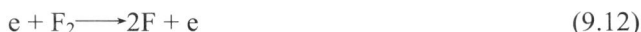

$$e + SF_6 \longrightarrow SF_5 + F + e \qquad (9.8)$$

$$e + SF_5 \longrightarrow SF_4 + F + e \qquad (9.9)$$

$$e + SF_6 \longrightarrow SF_2 + F_2 + 2F + e \qquad (9.10)$$

$$e + SF_6 \longrightarrow SF + F_2 + 3F + e \qquad (9.11)$$

$$e + F_2 \longrightarrow 2F + e \qquad (9.12)$$

The F atoms produced by decomposition are excited by the external energy in the SF_6 gap, and the outer electrons transition to different high-energy states. After a brief stay, they will return to the low-energy state and release photons, thus producing a series of F I spectral lines (Roman numeral I stands for neutral atom emission). It indicates the presence of F atoms in the glow zone near the needle electrode during corona discharge. Table 9.4 shows the corresponding energy state transitions.[26]

Table 9.4   Energy State Transition of F Atoms Observed in the Emission Spectrum

Particles	Observed wavelength (mm)	Theoretical wavelength (nm)	Low energy state			High energy state		
			Configuration	Term	J	Configuration	Term	J
FI	686.43	685.603	$2s^22p^4(^3P)3s$	^4P	5/2	$2s^22p^4(^3P)3p$	$^4D^o$	7/2
FI	703.926	703.746	$2s^22p^4(^3P)3s$	^2P	3/2	$2s^22p^4(^3P)3p$	$^2P^o$	3/2
FI	712.855	712.789	$2s^22p^4(^3P)3s$	^2P	1/2	$2s^22p^4(^3P)3p$	$^2P^o$	5/2
FI	720.423	720.236	$2s^22p^4(^3P)3s$	^2P	1/2	$2s^22p^4(^3P)3p$	$^2P^o$	3/2
FI	731.736	731.430	$2s^22p^4(^3P)3s$	^2D	3/2	$2s^22p^4(^3P)3p$	$^2F^o$	5/2
FI	740.136	739.868	$2s^22p^4(^3P)3s$	^4P	5/2	$2s^22p^4(^3P)3p$	$^2P^o$	5/2
FI	922.493	923.288	$2s^22p^4(^3P)3p$	$^4D^o$	1/2	$2s^22p^4(^3P)3d$	^2D	3/2

9.5.3 Emission Spectra of Spark Discharge in SF_6

Continuous corona discharge can deteriorate SF_6 insulation performance, resulting in spark discharge and even developing into arc discharge. Due to the discontinuity of discharge channels and the non-uniformity of plasma distribution, the state of spark discharge is unstable, and the spectral intensity has significant dispersion. For ease of comparison, the average of 100 spark discharge spectral samples is discussed here.

Figure 9.52 and Figure 9.53 show the average spectra of spark discharge. Compared with corona discharge (the integration time is 9,000 ms, which is 90 times spark discharge), strong spectral intensity exists in the full range of 200-1,037 nm under spark discharge.

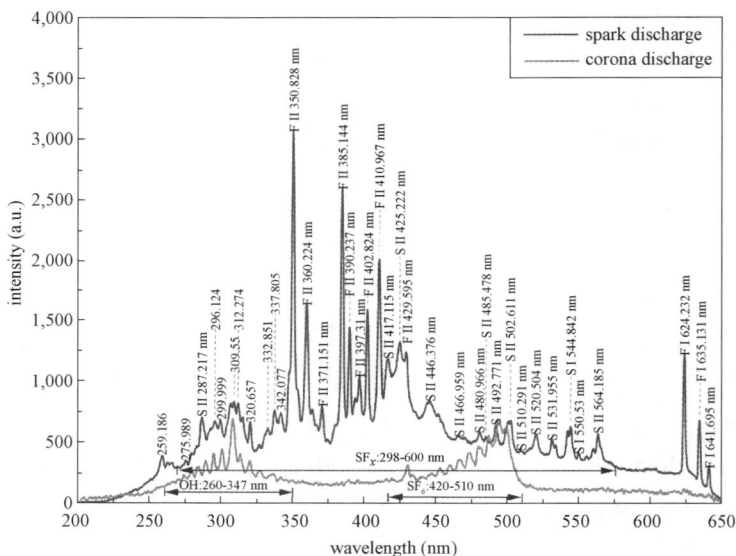

Figure 9.52   Emission spectra of SF_6 under spark discharge (200-650 nm) band

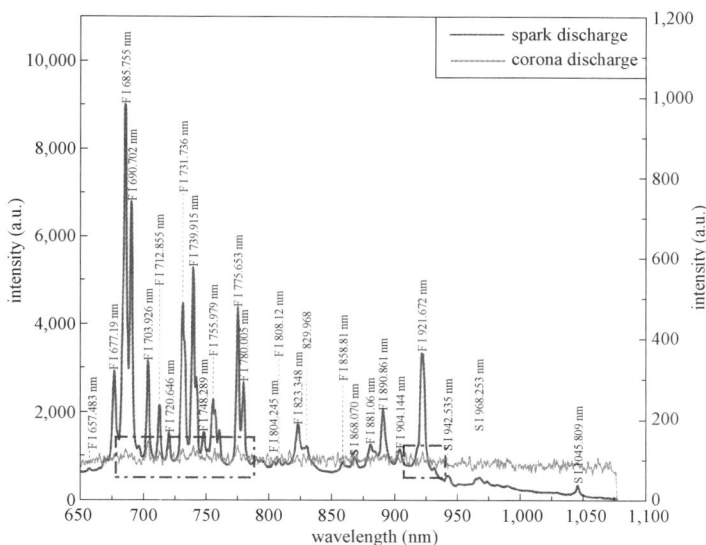

Figure 9.53　Emission spectra of SF_6 under spark discharge (650-1,037 nm band)

In the range of 200-650 nm, spark discharge has many linear spectra formed by fluoride ions and sulfur ions, namely F Ⅱ spectral lines and S Ⅱ spectral lines (Roman numeral Ⅱ stands for the emission of primary ionized ions), and the intensity of the linear spectra is more significant. Spark discharge forms highly ionized channels, which, in addition to reactions (9.8) - (9.12), mainly include the following reactions.[24, 25]

Based on Equation (9.12), electrons collide with low fluoride SF and decompose to produce S atoms:

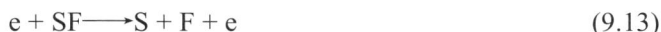

$$e + SF \longrightarrow S + F + e \tag{9.13}$$

The generated fluorine and sulfur atoms are further impacted by high-energy electrons, obtaining sufficient energy to ionize and form fluorine and sulfur ions:

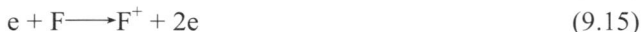

$$e + S \longrightarrow S^+ + 2e \tag{9.14}$$

$$e + F \longrightarrow F^+ + 2e \tag{9.15}$$

In the 650-1,037 nm range, the spectral lines of F atoms are strong and diverse. As a result, Equations (9.8) - (9.12) are more intense, generating more neutral F atoms. Since the material of the needle electrode is aluminum, and reference [27] reported the emission spectra of aluminum at 308.215 nm and 309.288 nm. Therefore, it is speculated that the emission spectrum containing Al atoms is masked by the band spectrum of OH radicals.

9.5.4　Emission Spectra of Surface Discharge in SF_6

The metal wire attached to the surface of the insulating material is at a floating potential. Under the action of an electric field, a large amount of induced charges accumulate at both ends. When the electric field strength near the end exceeds the insulation strength of SF_6 gas, a floating discharge will be formed. If a discharge channel through the conductor is formed, surface discharge will occur. The emission spectra of surface discharge are similar to that of spark discharge. This section still discusses its average spectrum.

When the floating potential triggers surface discharge, the spectral distribution in the 650-1,037 nm band is consistent with that of spark discharge. For the 200-650 nm range, there are local differences between the emission spectra of surface discharge and spark discharge. As shown in Figure 9.54, distinct spectral peaks appear at 324.953 nm, 327.663 nm, and 589.039 nm during surface discharge. The peaks at 324.953 nm and 327.663 nm are the sensitive lines of copper element, and 589.039 nm is the peak of copper ion.[28] Therefore, we speculate that these three peaks are generated by the radiation of copper atoms and copper ions in the thin copper wires on the surface of the insulating material. Table 9.5 lists the energy state transitions of the corresponding particles.

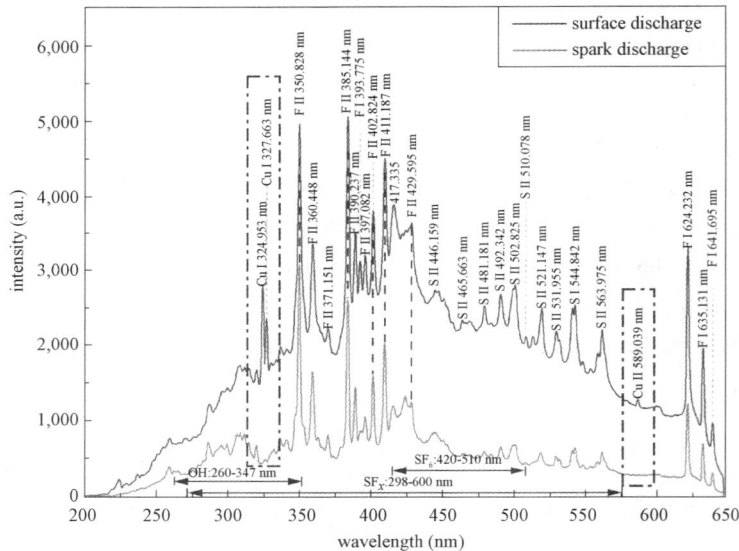

Figure 9.54   Emission spectra of SF$_6$ under surface discharge
in the 200-650 nm band

Table 9.5   **The Energy State Transitions of Copper Atoms and Copper Ions**

Particles	Observed wavelength (mm)	Theoretical wavelength (nm)	Low energy state			High energy state		
			Configuration	Term	J	Configuration	Term	J
Cu I	324.953	324.754,0	$2s^2 2p^4(^3P)3s$	^4P	5/2	$3d^{10}4s$	^2S	1/2
Cu I	327.663	327.395,7	$3d^{10}4s$	^2S	1/2	$3d^{10}4p$	$^2P^o$	1/2
Cu II	589.039	589.797,1	$3d^8(^3F)4s4p(^3P^o)$	$^3G^o$	3	$3d^9(^3D_{5/2})6s$	2[5/2]	3

9.6　Review Questions

Q1: Would you explain the working principle of UV absorption spectrum for ozone monitoring? What is the suitable absorption spectrum wavelength for an ozone gas analyzer?

Q2: Would you tell us what is Lambert-Beer law? Please give us some application examples of this law.

Q3: Please explain the internal components of an ozone gas analyzer, and please tell us the actual functions of each component.

Q4: Would you tell us the working principle of the NO_x-detecting instrument by chemiluminescence? please give us their application examples in electric power apparatus.

Bibliography

[1] Hassan Javed. Study of Air Decomposition Characteristics under Different Discharge Faults of Air Insulated Electrical Equipment. PhD thesis, University of Chinese Academy of Science, 2019.

[2] Huffman, Robert E. Atmospheric ultraviolet remote sensing. Academic Press, 1992.

[3] Garcia-Campana, Ana M. Chemiluminescence in analytical chemistry. CRC Press, 2001.

[4] Li, Kang and Javed, Hassan and Zhang, Guoqiang. Calculation of ozone and NO_x production under ac corona discharge in dry air used for faults diagnostic. In 2015 2nd International Conference on Machinery, Materials Engineering, Chemical Engineering and Biotechnology. Atlantis Press, 2015.

[5] Iwata, Takuro and Katagiri, Takashi and Matsuura, Yuji. Real-time analysis of isoprene in breath by using ultraviolet-absorption spectroscopy with a hollow optical fiber gas cell. Sensors, 2016.

[6] Ozone Monitor 2B Tech. OPERATION MANUAL, Models 106-L and OEM-106-L, MultiChannel Models 106-L-MC3 and 106-L-MC6.

[7] 2b technologies: Portable instruments for air pollution measurements. [Online] https:// twobtech.com/index. html. Accessed: Dec. 2020.

[8] Teja Ravi. What is a photodiode? working, v-i characteristics, applications. [Online] https://www.electronicshub.org/photodiode-working-characteristics-applications/. Accessed: April 2024.

[9] Qurtech. Comparison between mercury & led uv systems from a chemistry and process perspective. [Online] https://www.qurtech.com/ comparison-between-mercury-led-uv-systems-chemistry-process-perspective/.Accessed: July 2024.

[10] Chemiluminescence. [Online] https://instrumentationtools.com/chemiluminescence/. Accessed: Dec. 2020.

[11] $NO_2/NO/O_3$ Calibration Source^{TM}, 2B Tech. OPERATION MANUAL, Model 714.

[12] Briks, John and Turnipseed, Andrew and Anderden, Peter and et. al. Portable calibrator for NO based on the photolysis of NO_2 and a combined $NO_2/NO/O_3$ source for field calibrations of air pollution monitors. Atmospheric Measurement Techniques, 2020.

[13] Li, Kang and Wan, Liujie and Zhang, Guoqiang. A new detection method for corona discharge in high

voltage power modules based on decomposition gas analysis. In 2018 International Conference on Power System Technology (POWERCON), pages 3123-3128. IEEE, 2018.

[14] Bai, X-Y and Shen, L and Bai, M-D and Zhang, Z-T. High concentration ozone produced by strong ionization discharge of dielectric barrier. PHYSICS-BEIJING-, 29(10):615-619, 2000.

[15] Li, Kang and Javed, Hassan and Zhang, Guoqiang and Plesca, Adrian Traian. Analysis of air decomposition by-products under four kinds of partial discharge defects. IEEE Transactions on Dielectrics and Electrical Insulation, 24(6): 3713-3721, 2017.

[16] Lépine, Louis and Lessard-Deziel, Denise and Belec, Mario and Guddemi, Calogero and Nguyen, Duc Ngoc. Understanding ozone distribution inside stator core and measurements inside air-cooled generators to assess partial discharges problems. In Iris rotating ma-chine Conference, 2007.

[17] Millet, C and Nguyen, Duc Ngoc and Lepine, L and Bélec, M and Lessard-Deziel, D and Guddemi, Calogero. Case study-high ozone concentration in hydro generators. In 2009 IEEE Electrical Insulation Conference, pages 178-182. IEEE, 2009.

[18] De Maria, Letizia and Rizzi, Giuseppe. Ozone sensor for application in medium voltage switchboard. Journal of Sensors, 2009.

[19] De Maria, Letizia and Rizzi, Giuseppe and Serragli, Paolo and Marini, Remo and Fialdini, Lucio. Optical sensor for ozone detection in medium voltage switchboard. In IEEE SENSORS 2008 Conference, pages 1297-1300. IEEE, 2008.

[20] De Maria, Letizia and Bartalesi, Daniele. A fiber-optic multisensor system for predischarges detection on electrical equipment. IEEE Sensors Journal, 12(1): 207-212, 2011.

[21] Pearse R. W. B., Gaydon A. G. The identification of molecular spectra. London: Chapman and Hall, 1963.

[22] Anand V., Nair A., Nagendirakumar A. K. K., et al. Estimating the number density and energy distribution of electrons in a cold atmospheric plasma using optical emission spectroscopy. Journal of Vacuum Science Technology A, 36(4): 5, 2018.

[23] Casanovas A. M., Casanovas J., Dubroca V., et al. Optical-detection of corona discharges in SF_6, CF_4, and SO_2 under DC and 50 Hz AC voltages. Journal of Applied Physics, 70(3): 1220-1226, 1991.

[24] Christophorou L. G., Olthoff J. K. Electron interactions with SF_6. Journal of Physical and Chemical Reference Data, 29(3): 267-330, 2000.

[25] Yang W., Zhao S., Wen D., et al. F-atom kinetics in SF_6/Ar inductively coupled plasmas. Journal of Vacuum Science and Technology A, 34(3): 031305, 2016.

[26] Kushner M. J., Anderson H. N., Hargis P. J. Simulation of spatially dependent excitation rates and power deposition in RF discharges for plasma processing. Applied Physics Letters, 38:201-213, 1985.

[27] Lin Xin, Li Xintao, Xu Jianyuan, et al. Research on numerical computation of SF_6 breakdown voltages and spectral experiment in uniform electric fields. Proceedings of the CSEE, 1:301-309, 2016.

[28] NIST Atomic Spectra Database Lines Form. URL https://physics.nist.gov/PhysRefData/ASD/lines_form.html. Accessed: Sep. 30, 2024.

10

Electro-optic and Magneto-optic Sensors

10. 1　Introduction

A series of optical experiments has demonstrated how light interacts with matter when exposed to strong external magnetic or electric fields. Experiments that rely on magnetic fields are classified as magneto-optics, while those that rely on electric fields are classified as electro-optics.

The well-known electro-optic and magneto-optic effects are primarily categorized under these headings [1]:

Electro-optics

- Stark effect.
- Inverse Stark effect.
- Electric double refraction.
- Kerr electro-optic effect.
- Pockels electro-optic effect.

Magneto-optics

- Zeeman effect.
- Inverse Zeeman effect.
- Voigt effect.
- Cotton-Mouton effect.
- Faraday rotation effect.
- Kerr magneto-optic effect.

The Kerr electro-optic effect, Pockels electro-optic effect, and Faraday magneto-optic rotation effect are commonly employed methods for detecting electric field intensity, magnetic field intensity, or current in nearby conductors. The following descriptions will focus on these types of sensors.

The fundamental principles of electro-optic and magneto-optic effects, named after their discoverers, have been understood for over a century. The Pockels and Kerr effects describe how the optical properties of certain crystals, liquids, and gases change under the influence of an electric field, affecting the polarization of a light wave propagating along the optical axis of the medium. Similarly, according to the Faraday effect, a magnetic field alters the polarization of a passing light wave. In all these effects, the polarization plane of the light rotates within the medium, which is detected by a downstream analyzer and photodetector as the corresponding electric or magnetic field strength. These optical processes occur on a nanosecond scale, enabling electro-optic and magneto-optic sensors to achieve bandwidths ranging from zero to the GHz range.[2]

When combined with fiber optics, these sensors show significant potential for application in high-voltage environments. After on-site calibration, the voltages or currents that generate the fields can be directly measured. Over the past two decades, technical advancements in sensors based on the Pockels and Faraday effects have resolved numerous individual challenges, leading to the development of various electro-optic and magneto-optic transducers for high-voltage networks.

10.2 Basic Principles

10.2.1 Kerr Electro-optic Effect

In 1875, Kerr discovered that when a glass plate is subjected to a strong electric field, it becomes birefringent. This effect is not caused by mechanical strains in the glass, as it also occurs in many liquids and can even be observed in gases. When a liquid is placed in an electric field, it optically behaves like a uniaxial crystal, with the optic axis aligned with the field direction. When viewed perpendicularly, this setup produces interference phenomena.

Experimentally, the effect can be conveniently observed by passing light between two parallel, oppositely charged plates placed in a glass cell containing the liquid. This setup, known as a Kerr cell, is illustrated at the center of Figure 10.1.[2] A Kerr cell placed between crossed polarizer and analyzer forms a highly effective optical device. When the electric field is off, the analyzer blocks all light. When the electric field is applied, the liquid becomes birefringent, allowing light to pass through. With the cell oriented at 45°, the incident plane-polarized light from the polarizer is split into two equal components, parallel and perpendicular to the field. These components travel at different speeds, creating a phase difference, and the light emerges as elliptically polarized. The analyzer then transmits the horizontal component of this light.

The change in phase of the two vibrations in a Kerr cell is found to be proportional to the path length, i.e., the length of the electrodes l and to the square of the field strength E. The magnitude of the effect is determined by the Kerr constant K, see equation (10.1).

$$\delta = 2\pi K l E^2 \tag{10.1}$$

where δ is in radians, l is in meters, E is in volts per meter, and K is in meters per volts squared.

Nitrobenzene is one of the most suitable substances for use in a Kerr cell due to its relatively high Kerr constant.

In the Kerr effect, the phenomenon typically arises from the natural or induced anisotropy of molecules and their alignment within an electric field. This alignment causes the medium to become optically anisotropic as a whole. The Kerr effect is temperature-dependent and is the exact electrical analog of the corresponding magnetic effect.

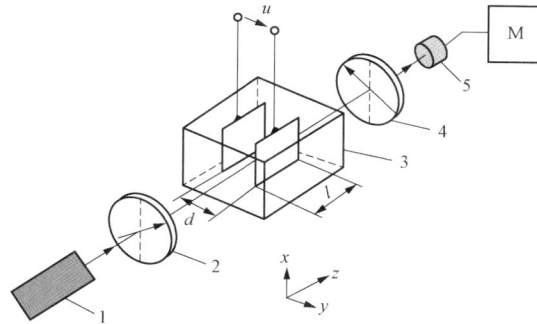

Figure 10.1　Electro-optic Kerr effect for high-voltage measurement (principle) [2]

1—laser; 2—polarizer; 3—containers filled with liquid Kerr medium; 4—analyzer;

5—photodetector, M measuring instrument, e.g., oscilloscope

Figure 10.2 illustrates the structure and a photograph of the designed electric field sensor. The sensor comprises an optical fiber collimator, K9 glass windows, a total reflection prism, and an optical fiber coupler, all housed within a custom-designed shell.[3]

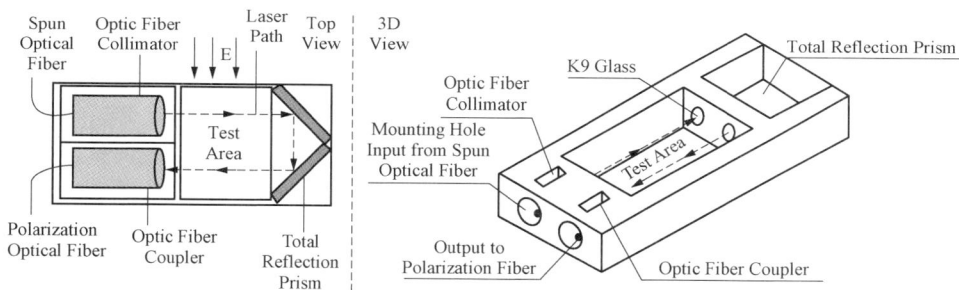

Figure 10.2　Schematic of the Kerr electro-optic sensor[3]

10.2.2　Pockels Electro-optic Effect

Various uniaxial crystals have been discovered where the induced birefringence varies linearly with the applied electric field. This effect, first studied by F. Pockels in 1893, was subsequently named after him. Recent research has developed a range of electro-optic crystals, such as ammonium dihydrogen phosphate ($NH_4H_2PO_4$) (ADP) and potassium dihydrogen phosphate (KH_2PO_4) (KDP), which exhibit significant Pockels birefringence at relatively low voltages.

A Pockels cell, commonly used as a fast light modulator or shutter, typically involves a crystal mounted with its optic axis and applied electric field aligned parallel to the beam direction [see Figure 10.3] [2]. When placed between crossed polarizers, the transmission can be modulated at frequencies exceeding 10 GHz, and the cell can function as a shutter with a response time of less than 1 ns. Since the light beam passes through the electrodes, these are often made of transparent metallic oxides like CdO, SnO, or InO, or are designed as thin metallic rings or grids.

$$\delta = \frac{2\pi}{\lambda} n_0^3 \gamma_{41} \frac{d}{l} U = kU \tag{10.2}$$

where δ is in radians, λ is the wavelength of incident light wave in vacuum, n_0 is the refractive index of crystal, γ_{41} is the electro-optic coefficient of crystal, l is the length of crystal along the light axis, and d is the thickness of crystal along applied voltage direction, in meters, U is applied voltage in volts, and k is a constant.

From Equation (10.2), it is evident that the phase difference δ has a linear relationship with the measured voltage U. Therefore, once the phase difference is determined, the corresponding voltage can be accurately measured.

Pockels cells, similar to Kerr cells, are utilized in a wide range of electro-optic systems, including their role as Q switches for generating ultrashort laser pulses. These systems have been proposed for use in wideband laser-beam communication, not only for terrestrial applications but also for interplanetary space communication.

Electric field measurement technology based on the Pockels effect utilizes non-metallic materials, allowing for easy miniaturization and significantly reducing the size of the photoelectric sensor. This technology does not interfere with the measured electric field and enables precise measurement. As a result, Pockels effect-based electric field measurement has garnered significant research interest, leading many universities and institutions to actively pursue research in this area.

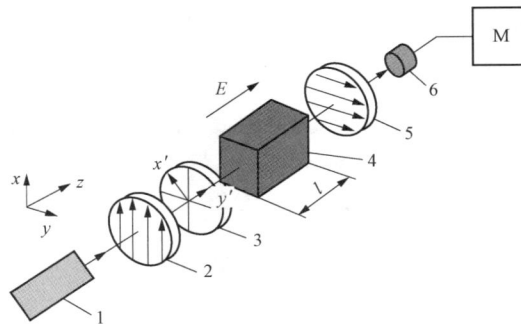

Figure 10.3   Longitudinal Pockels effect for high-voltage measurement (principle) [2]

1—laser; 2—polarizer; 3—k/4 wave plate; 4—crystal; 5—analyzer;

6—photodetector, M measuring instrument, e.g., oscilloscope or recorder

The structure diagram of the full-voltage optical voltage transformer is shown in Figure 10.4 (a). This type of voltage transformer is also known as the longitudinal modulation mode, and the voltage between the high-voltage and low-voltage electrodes is all borne by the electro-optic crystal. Figure 10.4 (b) shows the transverse modulation mode[4]. The transformer in the transverse modulation mode mainly calculates the measured high voltage through the intensity of the induced electric field.

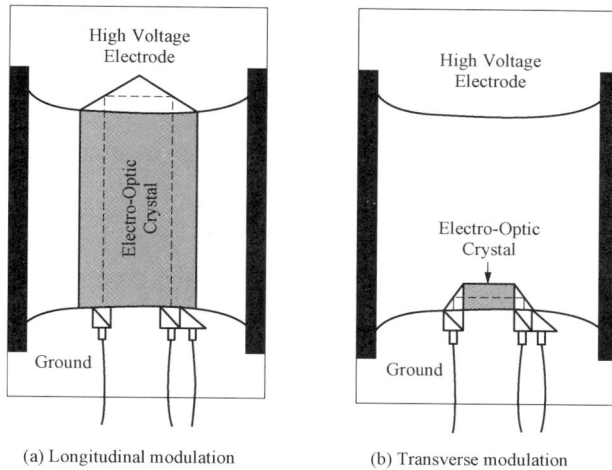

(a) Longitudinal modulation   (b) Transverse modulation

Figure 10.4   Schematic diagram of full voltage optical voltage transformer [4]

The detailed technical information of electro-optic crystal could be found in.[5] Compared to the Kerr effect, the Pockels effect exhibits the following characteristics:

- The transparent medium in a Pockels cell is typically a crystal, whereas in a Kerr cell, it is generally a liquid;
- The Pockels effect is a linear electro-optic effect, where the phase difference caused by birefringence between the ordinary (o) and extraordinary (e) light is directly proportional to the applied electric field strength (or voltage). In contrast, in the Kerr effect, the phase difference between the o and e light is proportional to the square of the applied voltage. This makes Pockels cells more suitable for applications such as light modulators and other devices.
- The voltage required for a Pockels cell is significantly lower than that for a Kerr cell—typically only 1/5-1/10, or even less—making Pockels cells more convenient to use.

10.2.3   Faraday Magneto-optical Effect

The Faraday effect, discovered by Michael Faraday in 1845, provided the first experimental evidence linking light and electromagnetism. This relationship, now understood as electromagnetic radiation, was further developed by James Clerk Maxwell in the 1860s and 1870s. The Faraday effect occurs in most optically transparent dielectric materials (including liquids) when subjected to strong magnetic fields. When a block of glass is placed in a strong magnetic field, it becomes optically active, causing the plane of vibration of plane-polarized light traveling parallel to the magnetic field to rotate. This phenomenon has been observed in various solids, liquids, and gases.

In physics, the Faraday effect, or Faraday rotation, is a magneto-optical phenomenon resulting from the interaction between light and a magnetic field in a dielectric material. The rotation of the plane of polarization is proportional to the intensity of the magnetic field component aligned with the direction of the light beam.

The Faraday effect arises from ferromagnetic resonance, where the permittivity of a material is described by a tensor. This resonance causes light waves to split into two circularly polarized rays that propagate at different speeds, a property known as circular birefringence. As the rays re-combine upon exiting the medium, the difference in propagation speed introduces a net phase offset, resulting in a rotation of the angle of linear polarization.

The Faraday effect has various applications in measuring instruments. It is used to measure optical rotatory power, modulate light amplitude, and remotely sense magnetic fields. In spintronics research, the Faraday effect is utilized to study electron spin polarization in semiconductors.

The Faraday effect involves the rotation of the plane of polarization of light as it propagates through certain transparent materials. A transmissive fiber optic sensor can measure magnetic fields and determine the current in a nearby conductor (or an encircled conductor) by placing a magneto-optical material in the light path. This material can be integrated intrinsically within the fiber or positioned between two fibers in a transmissive mode, as illustrated in Figure 10.5. [2]

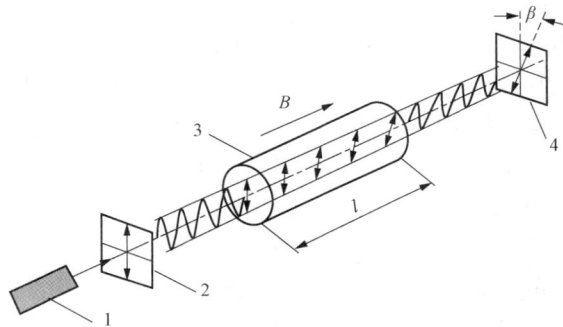

Figure 10.5 Principle of the magneto-optic Faraday effect for measuring magnetic fields or currents [2]

1—laser; 2—polarizer; 3—glass rod; 4—alyzer

Linearly polarized light will rotate β radians after traveling through a Faraday Rotator over l meters when in a magnetic field with flux density B.

$$\beta = V lB = V l\mu_0\mu_r H \qquad (10.3)$$

V represents the Verdet constant of the medium, with $\mu_r = 1$. The Verdet constant can be either positive or negative and varies based on the wavelength of the light, the medium, and the temperature. [6] Occasionally, instead of V, the product $\mu_0 V$ is provided in numerical form. The processes within the medium occur on a nanosecond timescale. A downstream analyzer converts the rotation of the polarization plane into an intensity modulation, which is then converted into a corresponding electrical signal by a photodetector, allowing for measurement by an instrument.

Faraday Rotators are ferromagnetic crystals surrounded by strong permanent magnets, which form a magneto-optic device. Linearly polarized light sent through a faraday rotator will be rotated by 45° independently of the input angle. Polarization is rotated non-reciprocally, meaning that light entering from the opposite side of the crystal will continue rotating in the same direction relative to the crystal. The rotation direction is indicated by an arrow engraved on the aluminum housing. The entrance and exit faces of the crystals are wedged approximately 1° to 2° and coated with an antireflection coating to reduce back reflections. Each Faraday rotator could be designed to operate at a specific center wavelength and at ambient room temperature (approx. 22 ℃); operation outside of this range will cause the output rotation angle and/or transmission to fluctuate outside of their specifications.

A notable characteristic of the Faraday effect is its non-reciprocal nature, meaning the rotation angle remains the same regardless of the direction of light travel. For example, if a light signal is reflected back through the magneto-optic material, the rotation angle is doubled rather than canceled out.

10.3 Applications of Electro-optic and Magneto-optic Sensors

10.3.1 Current and Voltage Combo Sensors for Substations

Fiber optic sensors for magnetic and electric fields have found significant applications in the electric power industry.[7] These sensors offer a key advantage in providing dielectric isolation between high-power equipment and personnel, enhancing safety and potentially reducing costs compared to conventional technologies. Additionally, the inherent performance characteristics of optical techniques, such as their superior theoretical bandwidth, surpass those of traditional conductive devices. Moreover, optical components are free from issues like hysteresis and saturation (dynamic range limitations) that commonly affect magnetic materials, making them highly reliable for practical use.

The compact size and lightweight nature of these devices offer significant advantages over traditional iron-core-based designs. These attributes have driven extensive research and engineering efforts aimed at developing and commercializing optical-based products. As a result, various successful sensing approaches utilizing fiber optics for both magnetic and electric field sensing have been realized.

For magnetic field sensing, which is currently a specialized area, most designs leverage the Faraday effect and rely on polarization-based techniques. Interferometric methods, involving metallic glass and various magnetostrictive coatings on fibers, have also been successfully demonstrated. In the realm of electric field and voltage sensing, polarization-based approaches have been developed using electro-optical materials to harness the Pockels and Kerr effects.

Figure 10.6 illustrates a typical setup for a high-voltage electro-optical Pockels sensor, where a Pockels crystal is positioned between two electrodes, with the polarized input light aligned with the direction of the electric field. The distance between the two electrodes matches the length of the Pockels crystal.[8] Figure 10.7 depicts a combined system of a magnetic optical current transducer (MOCT) and an electric optical voltage transducer (EOVT).[7]

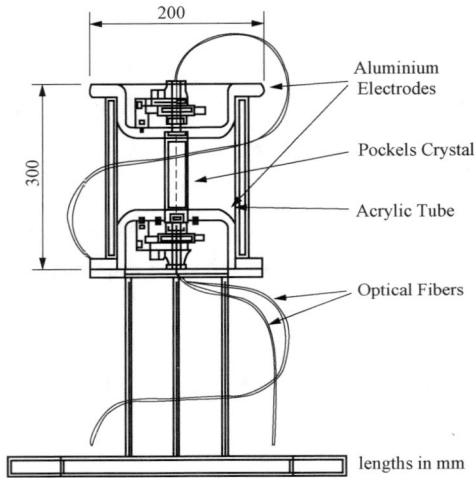

Figure 10.6   Cross section view of a high voltage electro-optical Pockels sensor [8]

The electro-optic voltage transducer (EOVT) utilizes the Pockels effect in a cylindrical BGO ($Bi_4Ge_3O_{12}$) crystal, which features cubic crystal symmetry. The full voltage is applied across the crystal's end faces, and light passes through the electro-optic material twice. At voltages in the several hundred kilovolt range, the field-induced birefringence generates a differential optical phase shift between two orthogonal polarizations, corresponding to several wave periods. The applied voltage waveform is accurately reconstructed from two output signals in quadrature. The crystal is oriented with its cylinder axis along a [1,0,0] direction, ensuring that only the field components parallel to the light path contribute to the electro-optic phase shift. Consequently, the sensor measures the line integral of the field and remains insensitive to external perturbations in the field distribution or to fields from neighboring electrical phases.

The assembly operates in an SF_6 atmosphere, enclosed within a composite insulator with silicone rubber sheds (see Figure 10.7). EOVTs designed for line voltages ranging from 115 to 550 kV have undergone type testing in accordance with IEEE standards. To account for the temperature dependence of the Pockels effect, an additional temperature sensor is integrated into the system. These transducers meet the class 0.2 accuracy requirements for metering, as defined by IEC standards, and are utilized as inputs for electronic metering and/or relaying systems.

An optical metering unit (OMU) integrates optical current and voltage transducers (MOCT and EOVT) into a lightweight and compact single-phase unit (see Figure 10.7). The OMU's reduced size and enhanced accuracy, compared to conventional oil-filled current and voltage transformers, make it particularly suitable for adding revenue metering to existing substations where space is limited. These advantages align well with the increasing need, driven by deregulation, to retrofit extra-high voltage substations with metering instrument transformers.[7]

Figure 10.8 (a) illustrates the schematic diagram of the measuring system, while (b) shows a photograph of the Pockels voltage sensor.[9] The system includes laser diode (LD) sources operating at wavelengths of 1,300 nm and 1,550 nm, an optical multiplexer, an optical demultiplexer, two O/E converters, optical fibers, and the Pockels voltage sensor. The sensor itself is composed of eight BGO crystal bars, two polarizing beam splitters (PBS), and two glass plates with transparent conductive coatings that act as electrodes. The BGO bars, each with millimeter dimensions, are

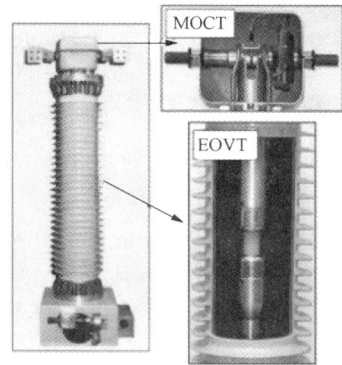

Figure 10.7　Optical metering unit (OMU) with current transducer (MOCT) and voltage transducer (EOVT) [7]

arranged in series to achieve a total length suitable for high voltage applications. Transparent electrodes are placed at both ends of the connected BGO bars, with one electrode grounded either directly or through a modulated AC power source, and the other connected to the high voltage being measured. Traditional Pockels sensors have been limited by the halfwavelength voltage; however, this new system introduces a dualwavelength laser setup, allowing the measurable voltage to be increased up to the least common multiple of the wavelengths used.

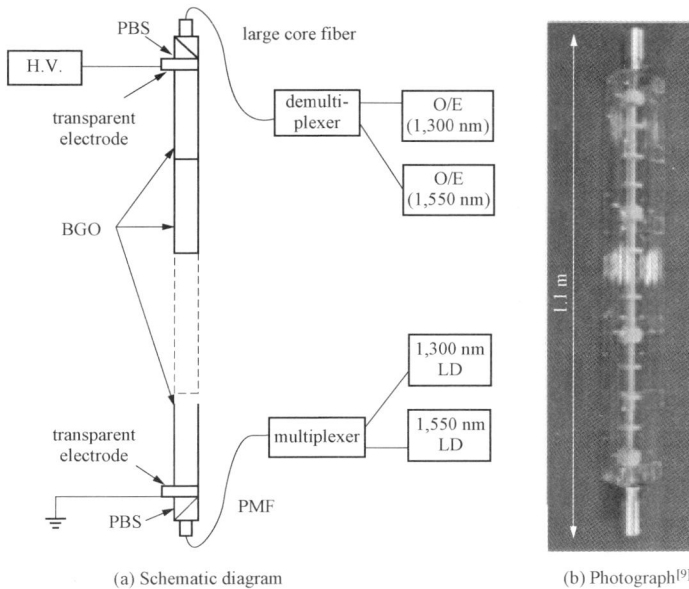

(a) Schematic diagram　　　　　(b) Photograph[9]

Figure 10.8　High Voltage Measuring System[9]

In the fiber-optic current sensor, as shown in Figure 10.9, a simple loop of optical fiber around the bus bar replaces the sophisticated head of the conventional transformer. The sensor perfectly integrates the magnetic field along the closed path described by the sensing fiber. As a result, the signal is independent of the particular magnetic field distribution and only determined by the enclosed current. All currents outside the fiber loop are of no influence. Sensor placement is therefore uncritical. The simplicity of the system reduces the time required for installation and commissioning to a few hours.[10]

Figure 10.9, Figure 10.10 and Figure 10.11 schematically depict the setup of ABB's fiber-optic current sensor.[11, 9, 12] Two light waves with orthogonal linear polarizations are transmitted from the optoelectronics module, which houses a semiconductor light source, through an interconnecting fiber to the single-ended sensing fiber. The sensing fiber wraps around the bus bar in an integer number of loops, N, where a single loop is typically sufficient for measuring high DC currents. At the entrance of the sensing fiber, a fiber-optic phase retarder converts the orthogonal linear waves into left-circularly and right-circularly polarized light. These circular waves travel through the sensing fiber coil, reflect at the fiber's end, and then retrace their optical path back to the coil entrance, where they are reconverted into orthogonal linearly polarized waves. Due to the reflection, the polarization directions of the returning waves are swapped relative to the forward-propagating waves. If a DC current, I, is flowing the two returning light waves have accumulated a phase difference given by $\Delta\Phi F = 4VNI$. This phase difference is directly proportional to the line integral of the magnetic field along the sensing fiber, providing a precise measurement of the current.

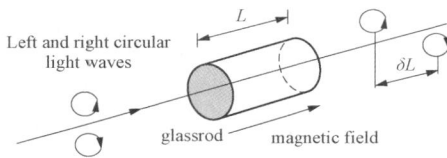

Figure 10.9 Circularly polarized light was used by Faraday rotation sensors of ABB [11]

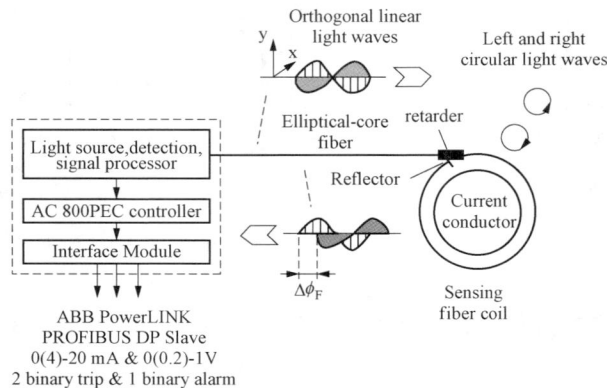

Figure 10.10 Fiber-optic current sensor developed by ABB [11]

The returning waves are brought to interference in the optoelectronics module, where the signal processor converts the optical phase difference into a digital signal. Operating the sensing coil in reflection provides several advantages, including simplicity and strong resistance to mechanical disturbances such as shock and vibration. Non-reciprocal Faraday optical phase shifts are doubled as the waves travel forward and backward, while reciprocal phase shifts caused by mechanical disturbances cancel each other out. This design ensures that the sensor signal remains largely unaffected by external mechanical influences. Figure 10.12 presents three representative fiber optic current and voltage transformers developed by NxtPhase, Canada.

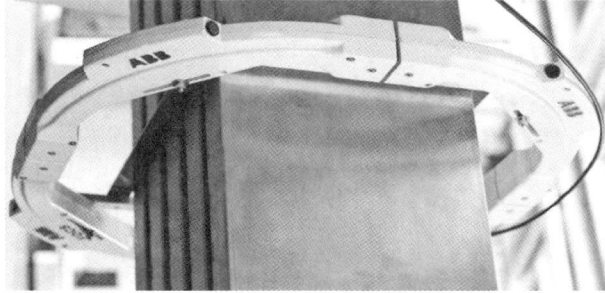

Figure 10.11　±500 kA fiber-optic current sensor developed by ABB [13]

(a) A 362 kV class fiber optic current transformer and a 362 kV class fiber optic voltage transformer

(b) A three-phase 550 kV class combined fiber optic current/ voltage transformer [12]

Figure 10.12　Two examples of optical current transformer and optical voltage transformer

10.3.2　Research and Development of ±800 kV UHVDC Fiber Optical Current Transducer

For a long time, fiber optic current transducers and their core optoelectronic components have been largely dependent on imports from developed countries. These devices often enco-unter challenges such as extremely low ambient temperatures, external vibrations, and strong electromagnetic interference, resulting in a high failure rate. Such failures frequently lead to DC blocking and operational stoppages, significantly constraining the reliable operation of DC power systems.

Fiber optic current transformers, which are oil-free, gas-free, safe, environmentally friendly, and offer fast response times along with a wide dynamic range, have become the preferred solution for current measurement in power systems. They are particularly crucial in HVDC power transmission projects, where their reliability directly influences the safe and stable operation of the system.

A research team from State Grid of China has made significant breakthroughs, including advancements in polarization optics analysis methods for fiber optic current transducers, disturbance mechanisms, wave-packet integrated sensing fiber development, and reliability testing methods for core optoelectronic components. Additionally, they have developed intelligent monitoring and early warning technologies for operational status (https://5gai.cctv.com/2023/01/09/AR-TI5UF0fc56F6f2bBIhh1O7230109.shtml).

The culmination of these efforts is the successful development of the first fully domestic-made, high-reliability fiber optic current transformer, designed to withstand temperature variations, vibrations, and other environmental challenges. This transformer has been successfully integrated into the ±800 kV Qingnan converter station, marking a significant milestone in China's capability to independently tackle the challenges of optical current measurement.

Figure 10.13　The ±800 kV UHVDC fiber optic current transformer developed by Comcore Optical Intelligent Technology Co., Ltd. was successfully put into operation at the ±800 kV Lingzhou converter station in August 2016

Figure 10.13 shows the ±800 kV UHVDC fiber optic current sensor developed by Comcore Optical Intelligent Technology Co. Ltd. Detailed information could be found at website (https://https://www.comcore.com/news-detail11-en.html/).

10.4　Review Questions

Q1: Please explain the working principle of the Kerr effect. (see Section 10.2.1)

Q2: Please explain the working principle of the Pockels effect. (see Section 10.2.2)

Q3: Please explain the working principle of the Faraday rotation effect. (see Section 10.2.3)

Q4: The Faraday magneto-optic effect and the Kerr or Pockels effects can be utilized to measure electrical current and electric fields, respectively. To design an optical fiber sensor that integrates both phenomena, one could propose an arrangement where the Faraday effect is used to measure the magnetic field associated with the electrical current, while the Kerr or Pockels effects are used to measure the electric field. The major difficulty in implementing such a combined sensor would be achieving effective isolation and accurate measurement of both the magnetic and electric fields without interference between the two effects. This challenge involves ensuring that the optical responses from the Faraday and Kerr/Pockels effects do not overlap or influence each other, which could lead to measurement inaccuracies. Additionally, maintaining sensor performance and stability under high-voltage conditions and ensuring the durability of the optical fiber under such extreme conditions would be significant technical hurdles. Why this kind of combined sensors have not been realized in ultra high-voltage direct current power transmission project? What would be the major difficulty to overcome? (see Section 10.3.1 and 10.3.2)

Bibliography

[1] Jenkins, Francis A and White, Harvey E. Fundamentals of optics. Tata McGraw-Hill Education, 2018.

[2] Schon, Klaus. High voltage measurement techniques. Springer, 2019.

[3] Gao, Chunjia and Qi, Bo and Gao, Yuan and Zhu, Zongwang and Li, Chengrong. Kerr electro-optic sensor for electric field in large-scale oil-pressboard insulation structure. IEEE Transactions on Instrumentation and Measurement, 68(10): 3626-3634, 2018.

[4] Han, Rui. Research on a Battery-less and Contact-less Optic Overvoltage Sensor and Measurement System Based on Pockels Effect. PhD thesis, Chongqing University, 2017.

[5] Montemezzani, G and Günter, P and others. Electro-optic and photorefractive properties of $(Bi_4Ge_3O_{12})$ crystals in the ultraviolet spectral range. JOSA B, 9(7): 1110-1117, 1992.

[6] Shen, Yan and Chen, Tong and Yu, Wen-bin and Ge, Jin-ming and Han, Yue and Duan, Fang-wei. A direct current measurement method based on terbium gallium garnet crystal and a double correlation detection algorithm. Sensors, 19(13):2997, 2019.

[7] Bohnert, K and Gabus, P and Brändle, H and Khan, Aftab. Fiber-optic current and voltage sensors for high-voltage substations. In Proceedings of the 16th International Conference on Optical Fiber Sensors, pages 752-754, 2003.

[8] da Silva, Luiz Pinheiro C and Santos, Josemir Coelho and Côrtes, AL and Hidaka, K. Optical high voltage measurement transformer using white light interferometry. Annals of Optics-Xxv Enfmc, pages 204-207, 2002.

[9] Bohnert, Klaus and Gabus, Philippe and Brändle, Hubert and Guggenbach, Peter. Fiberoptic dc current sensor for the electro-winning industry. In 17th International Conference on Optical Fibre Sensors, volume 5855, pages 210-213. International Society for Optics and Photonics, 2005.

[10] Kumada, Akiko and Hidaka, Kunihiko. Directly high-voltage measuring system based on pockels effect.

IEEE transactions on power delivery, 28(3):1306-1313, 2013.

[11] M. Wiestner and M., Wendler and K., Bohnert and Baden. Technology leap in dc current measurement opens new opportunities for aluminium smelters. URL https://search.abb. com/library/Download.aspx? DocumentID=3BHS209777E01&DocumentPartId=. Accessed: August 30, 2024.

[12] Rahmatian, Farnoosh and Ortega, Abraham. Applications of optical current and voltage sensors in high-voltage systems. In 2006 IEEE/PES Transmission & Distribution Conference and Exposition: Latin America, pages 1-4. IEEE, 2006.

[13] Focs - fiber-optic current sensor make light work of dc current measurement. URL https://search.abb.com/ library/Download.aspx?DocumentID=3BHS362996E01& DocumentPartId=. Accessed: August 30, 2024.

11

Michelson and Fabry-Perot Interferometer

11.1 Introduction

Interference of light waves is a phenomenon commonly observed in everyday life, such as the colorful patterns on an oil slick or a thin soap film. When white light is used, only a few colored fringes are visible. As the film's thickness increases, the optical path difference between the interfering waves also increases, causing the color changes to become less distinct and eventually fade away. However, with monochromatic light, interference fringes remain visible even with significant optical path differences.[1]

Given the narrow wavelength of visible light (around half a micrometer for green light), even minute changes in the optical path difference can result in noticeable variations in the intensity of an interference pattern. This sensitivity makes optical interferometry a highly precise measurement technique. While it has been a laboratory method for nearly a century, recent advancements have significantly broadened its scope and accuracy, making optical interferometry practical for a wide array of measurements.

The invention of the laser stands as the most significant advancement in modern interferometry. Lasers have overcome many limitations of conventional light sources, enabling a host of new interferometric techniques. Additionally, the use of single-mode optical fibers to create analogs of traditional interferometers has opened up new applications. Another transformative development has been the growing integration of photodetectors and digital electronics for signal processing. Interferometric measurements have gained further prominence with the redefinition of the meter, now based on the speed of light.

Current applications of optical interferometry include precise measurements of distances, displacements, and vibrations; testing of optical systems; analysis of gas flows and plasmas; surface topography studies; and measurements of temperature, pressure, and electric and magnetic fields. Additionally, interferometry is used for rotation sensing, high-resolution spectroscopy, and laser frequency measurements. Emerging applications under exploration include high-speed all-optical logic and the detection of gravitational waves. It is highly likely that many more applications will be discovered in the near future.

11.2 Basic Principles

11.2.1 Category of Interferometers

There are numerous types of interferometers, each designed for specific applications.

Some of the well-known interferometers include:

(1) Rayleigh Interferometer: Used mainly for measuring small angles, refractive index changes, and surface irregularities;

(2) Michelson Interferometer: Widely used in experiments involving the measurement of wavelengths, refractive index, and precise distance measurements. It is also the basis for the famous Michelson-Morley experiment and LIGO for detecting gravitational waves;

(3) Mach-Zehnder Interferometer: Often used in fluid dynamics and gas flow studies, as well as in optical communications and quantum mechanics experiments;

(4) Sagnac Interferometer: Primarily utilized in gyroscopes (e.g., ring laser gyroscopes and fiber-optic gyroscopes) to detect rotation;

(5) Fabry-Perot Interferometer: Used for high-resolution spectroscopy and the measurement of the wavelength of light, often found in lasers and optical cavities;

(6) Twyman-Green Interferometer: A variation of the Michelson Interferometer, commonly used for testing optical components like lenses and mirrors;

(7) Fizeau Interferometer: Utilized for measuring the surface quality of optical components and thin film thicknesses;

(8) Linnik Interferometer: Used for surface topography measurements with high precision;

(9) Jamin Interferometer: Applied in measuring refractive indices and studying gases under various conditions;

(10) Lloyd's Mirror Interferometer: Used in thin film studies and determining wavelength differences;

(11) Gabor Interferometer (Holographic Interferometer): Used in holography for recording and reconstructing the phase of light waves;

(12) Shearing Interferometer: Typically used for testing the quality of wavefronts in optical systems;

(13) Pound-Drever-Hall Interferometer: Used for stabilizing the frequency of a laser;

(14) White-Light Interferometer: Applied in optical coherence tomography and surface metrology;

(15) Babcock Interferometer: Used in astronomy for measuring the separation of close binary stars.

This list is not exhaustive, as new designs and adaptations of interferometers are developed for specialized applications in science and engineering. The choice of interferometer depends on the specific measurement or application requirement.

In the electric power industry, the Michelson Interferometer, Fabry-Perot Interferometer, Mach-Zehnder Interferometer, and Sagnac Interferometer have already attracted significant attention. Their effectiveness has been verified through long-term practical application.

11.2.2 Working Principle of Michelson Interferometer

The Michelson interferometer is historically significant because it was used by Albert A. Michelson and Edward W. Morley in 1887 to provide experimental evidence against the theory of the luminous aether. Michelson later used an interferometer to measure the length of a standard meter based on the wavelength of an atomic spectral line. In 1907, he became the first American citizen to receive the Nobel Prize in Physics "for his precision optical instruments and the spectroscopic and metrological investigations carried out with their aid". Improved versions of the Michelson interferometer are still widely used in research and technology today, with applications including Fourier transform spectroscopy, precision testing of optical components, and fiber optic communications.[1]

In the Michelson interferometer, the beam from the source is divided at a semi-reflecting coating on the surface of a plane-parallel glass plate G_1, as illustrated in Figure 11.1 (a), (b), and (c). This coating reflects 50% of the incident light and transmits the remaining 50%, effectively splitting the incident beam into two separate beams. One beam is transmitted toward the movable mirror (M_1), while the other is reflected toward the fixed mirror (M_2). Both mirrors then reflect the light back toward the beam-splitter. Half of the light returning from M_1 is reflected by the beam-splitter to the viewing screen, while half of the light from M_2 is transmitted through the beam-splitter to the viewing screen.

The beam splitter G_1 is used to recombine the beams reflected back from the two mirrors. To obtain interference fringes with a white-light source, the optical paths for both beams must be equal for all wavelengths. Therefore, both arms of the interferometer must contain the same thickness of glass with identical dispersion properties. However, one beam passes through the beam splitter G_1 three times, while the other passes through it only once. To compensate for this difference, a compensating plate G_2 (which is identical to the beam splitter G_1 but without the semi-reflecting coating) is introduced into the path of the second beam.

Figure 11.1(b) shows the three-dimensional structure of the Michelson interferometer.

As shown in Figure 11.1(c), the reflection at the beam splitter G_1 creates a virtual image M_1' of the mirror M_1. The interference pattern observed is analogous to that produced in a layer of air bounded by M_1 and M_2. The characteristics of this pattern depend on the properties of the light source and the separation between M_1 and M_2.

In this process, the original beam of light is split, and portions of the resulting beams are recombined. Since both beams originate from the same source, their phases are highly correlated. When a lens is placed between the laser source and the beam splitter, the light rays spread out, resulting in an interference pattern of dark and bright rings, or fringes, on the viewing screen (see Figure 11.2)[2].

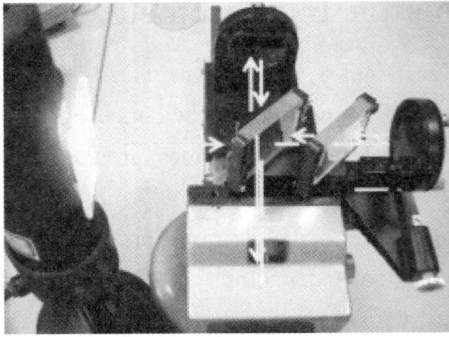

(a) Actual physical model

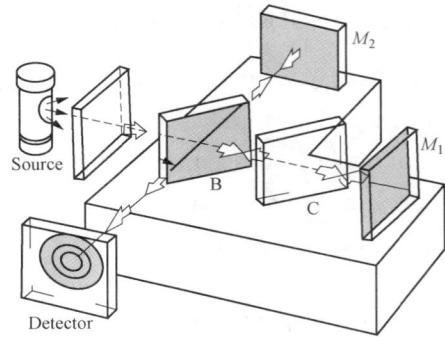

(b) Transverse modulation

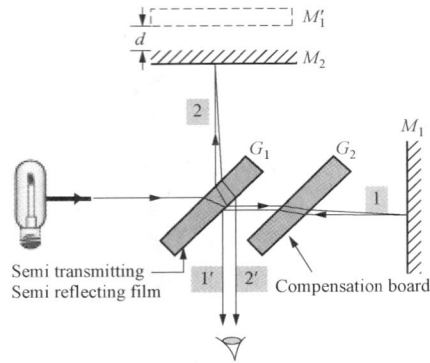

(c) Transverse modulation

Figure 11.1　Actual physical model and 3D outline of Michelson Interferometer

Figure 11.2　The Michelson interference pattern (left side) and Fabry-Perot interference
pattern (right side) of dark and bright rings (fringes)

By moving M_1, the path length of one of the beams can be adjusted. Since the beam traverses the path between M_1 and the beam splitter twice, moving M_1 1/4 wavelength closer to the beam splitter will reduce the optical path of that beam by 1/2 wavelength. As a result, the interference pattern will change, with the radii of the maxima decreasing so that they now occupy the positions of the former minima. If M_1 is moved an additional 1/4 wavelength closer to the beam splitter, the radii of the maxima will again decrease, causing the maxima and minima to trade positions once more. However, this new arrangement will be indistinguishable from the original pattern.

By slowly moving the mirror a measured distance d, and counting m, the number of times the fringe pattern is restored to its original state, the wavelength of the light λ can be calculated using the formula below:

$$\lambda = \frac{2d}{m} \tag{11.1}$$

If the wavelength of the light is known, the same procedure can be used to measure d.

The Michelson interferometer can be used to measure deformations as small as a few nanometers. With careful design and fabrication of the internal components, even higher precision measurements can be achieved.

11.2.3 Working Principle of Fabry-Perot Interferometer

The Fabry-Perot interferometer (FPI), sometimes called the Fabry-Perot etalon, is shown in Figure 11.3. [3, 4]

Fiber Fabry-Perot interferometers are highly sensitive to perturbations that affect the optical path length between two mirrors. The sensing area can be very compact, functioning as a "point" sensor sensor in certain applications. Fiber optic FPIs appear to be ideal for many smart structural sensing applications, particularly where the sensor needs to be embedded in composite materials or metals. Additionally, these multifunctional measurement devices are well-suited for spatial division, time division, frequency division, and coherence multiplexing techniques, which help reduce the cost of multi-point monitoring.

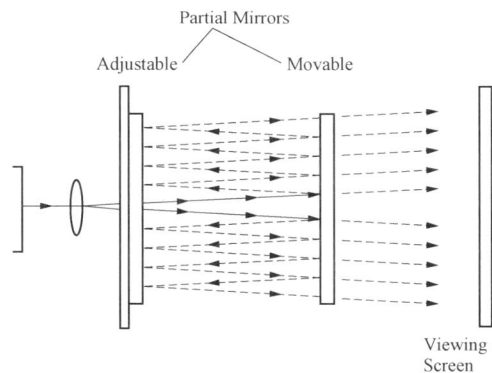

Figure 11.3　Fabry-Perot Interferometer[3]

In the Fabry-Perot Interferometer, two partially reflective mirrors are aligned parallel to each other, forming a reflective cavity. Figure 11.3 [3] illustrates two rays of light entering this cavity and reflecting back and forth inside. At each reflection, a portion of the beam is transmitted, causing each incident ray to split into a series of transmitted rays. Since these rays originate from the same incident beam, they maintain a constant phase relationship, assuming a sufficiently coherent light source is used. The phase relationship between the transmitted rays depends on the angle of entry into the cavity and the distance between the two mirrors. This results in a circular fringe pattern, similar to that produced by the Michelson interferometer, but with thinner, brighter, and more widely spaced fringes. The sharpness of the

Fabry-Perot fringes makes it a valuable tool in high-resolution spectroscopy. Like in the Michelson Interferometer, as the movable mirror is adjusted closer to or farther from the fixed mirror, the fringe pattern shifts. When the mirror movement corresponds to 1/2 of the wavelength of the light source, the new fringe pattern becomes identical to the original.

The principles of fiber Fabry-Perot interferometer (FFPI) sensors are primarily explained using Equation (11.2). When a perturbation is applied to the sensor, it alters the optical path difference (OPD) of the interferometer, thereby affecting the phase difference φ. By measuring the shift in phase or the wavelength spectrum, the applied sensing parameter can be quantitatively determined. Additionally, the free spectral range (FSR), which represents the spacing between adjacent interference peaks in the spectrum, is also influenced by changes in the OPD.

$$\varphi = \frac{2\pi}{\lambda} l \qquad (11.2)$$

where l is the optical path difference (OPD), mainly relying on the characteristics of the cavity and λ is the free-space optical wavelength. The free-space optical wavelength refers to the wavelength of light as it propagates through a vacuum or air, where the refractive index is very close to 1.0 .

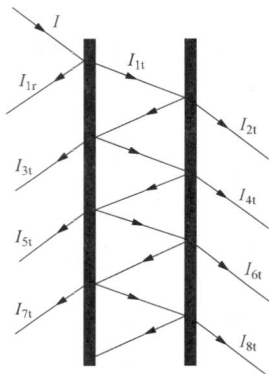

Figure 11.4　Two-beam Fabry-Perot interference schematic[5]

The principle of two-beam interference relies on the coherent addition of reflections from two surfaces with low reflectance. A common structure is formed by the Fresnel reflections from the ends of two fibers. As shown in Figure 11.4, [5] assuming the intensity of the incident light is I, and the reflectance of the fiber end is $r \approx 4\%$, the transmission coefficient t is approximately 96%. The reflective and transmitted intensities of the nth reflected beam are denoted as I_{nr} and I_{nt}, respectively. The corresponding light intensities are provided in Table 11.1. The reflected intensities are

I_{1r}, I_{3t}, I_{5t}, and so on.

Table 11.1　Reflective and Transmitted Intensities from a FPI with Low Fresnel Reflection of 4% [5]

Intensity	Equation	Value
I_{1r}	$r \cdot I$	4%I
I_{2t}	$t^2 \cdot I$	92.16%I
I_{3t}	$t^2 r \cdot I$	3.69%I
I_{4t}	$t^2 r^2 \cdot I$	0.15%I
I_{5t}	$t^2 r^3 \cdot I$	0.006%I
I_{6t}	$t^2 r^4 \cdot I$	2.4×10^{-6}%I

It is evident that after the third reflection, the intensity I_{3r} becomes significantly lower than that of the first two reflections, making its influence negligible. Therefore, it is sufficient to consider only the first two reflected beams for most practical applications.

The interference results discussed are approximate and based on reflecting planes with low reflectance, such as the fiber end with a reflectance of about 4%. This approximation is valid when considering only two beams. However, if the reflectance of the reflecting plane increases, the results will differ. For instance, with an incident light intensity of 100% and a reflectance of 90%, the intensity of the first transmitted beam would be 1%, the second 0.81%, the third 0.66%, and so on. Although the intensities of these transmitted beams decrease, the differences between adjacent beams remain relatively small. Therefore, in such cases, multibeam interference must be considered.

Consider the light shown successively reflected between the fiber ends of the FP cavity A and B in Figure 11.5. The incident ray i of unit amplitude stands for the propagating direction of the incident light. [5]

Figure 11.6 shows the Fabry-Perot interference pattern of dark and bright rings at different phase angles. Comparing Figure 11.6 (e) with Figure 11.2 (left side), it is obvious that the fringes of Fabry-Perot interference pattern are much sharper than others. That means the Fabry-Perot interferometer has much higher precision, and it could be used to measure a tiny deformation of less than several picometers.

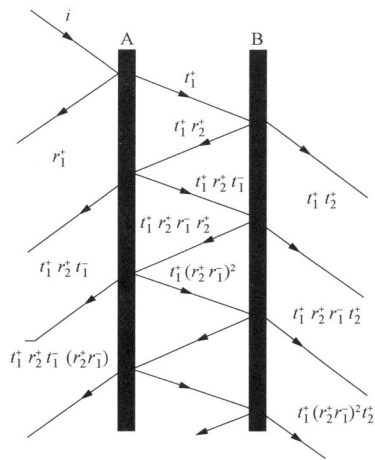

Figure 11.5　Multi beam Fabry-Perot interference schematic [5]

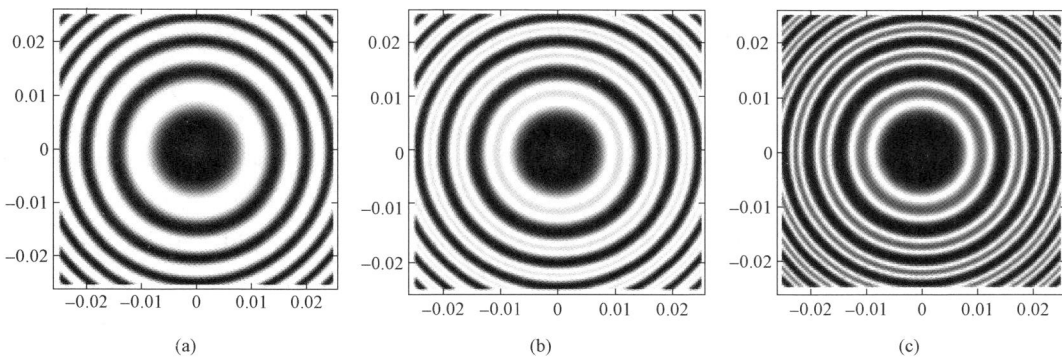

Figure 11.6　The Fabry-Perot interference pattern of dark and bright rings (fringes) (1)

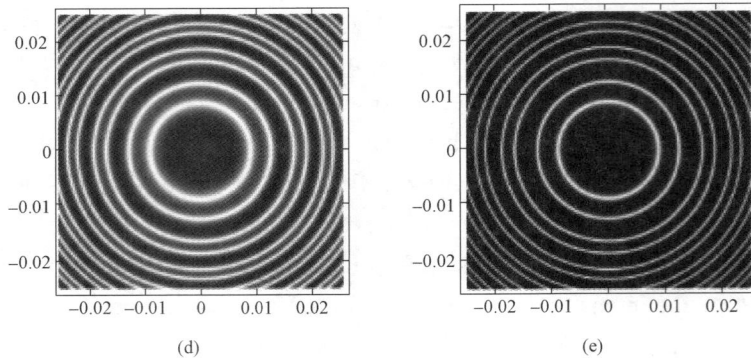

Figure 11.6   The Fabry-Perot interference pattern of dark and bright rings (fringes) (2)

11.3   Applications of Michelson and Fabry-Perot Interferometers

11.3.1   Trace Gas Detection by Michelson Interferometers

The demand for trace gas detection is steadily increasing across various fields, including biology, medicine, industrial process control, and the monitoring of pollutants or toxic gases. Among the many detection methods available, spectroscopy stands out due to its versatility, wide range of applications, and non-destructive nature, making it one of the most essential tools in materials research. In most spectroscopic techniques, the absorbed radiation is typically measured indirectly by comparing transmitted, reflected, or scattered photons with the incident light. Conventional detectors have evolved to operate near their theoretical performance limits, so extending the optical path length remains one of the few viable ways to improve sensitivity. However, this approach often results in larger or more complicated instruments and can introduce nonlinearities, particularly when analyzing wet gases. A promising solution to these challenges is the use of zero-background methods, such as photoacoustic (PA) spectroscopy, where the sensitivity is primarily limited by the microphones used.

Photoacoustic spectroscopy (PAS) is a highly sensitive, zero-background method for trace gas analysis. A typical gas measurement setup is illustrated in Figure 11.7. As the infrared (IR) beam passes through the sample cell, a portion of the gas molecules absorbs the radiation and is excited to higher energy states. The relaxation of these states through non-radiative pathways generates heat, leading to an increase in both the temperature and pressure of the gas. This modulation of IR radiation creates temperature and pressure variations, which form sound waves typically detected by a capacitive microphone. Selectivity is achieved either by using a laser wavelength that corresponds to a known absorption line of the target gas or by employing an optical filter with a broadband source. Extremely high sensitivity, even down to the sub-nL/L range, can be achieved with powerful laser sources and by leveraging the acoustic resonances of the cell. However, the ultimate sensitivity of the detector is limited by the microphone used. Replacing the condenser microphone with an interferometric cantilever sensor has shown significant improvements in performance.

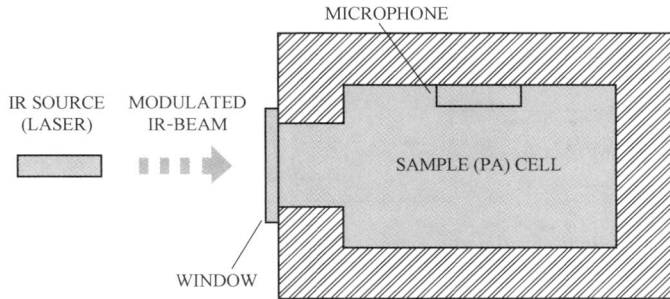

Figure 11.7   Conventional photoacoustic measurement setup [6]

The sensor in the cantilever microphone is a rectangular, flexible bar, as illustrated in Figure 11.8. Typical dimensions for its width w and length l are a few millimeters, while the thickness t ranges from 5 to 10 μm. It is separated from the thicker frame on three sides by a narrow gap (Δ=3-5 μm), functioning like a door between the photoacoustic cell and a larger balance cell (see Figure 11.9). The cantilever's displacement is measured using a compact laser interferometer, providing interferometric accuracy with a linear dynamic range of several nanometers. [6]

Figure 11.10 shows a photoacoustic measurement system, where two chambers (Volume V and Volume V_0) are both filled with the sample gas to be measured. The highly sensitive silicon cantilever is located between the two chambers (like a door on a wall). The absorption cell is a brass tube, which is polished inside. The cantilever and its frame act like a mirror reflecting the IR beam back. Figure 11.11 and Figure 11.12 show the assembly of a Michelson Interferometer as a microphone in 3D and 2D outline of Gasera instrument. [7]

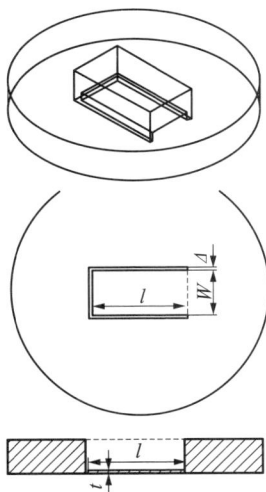

Figure 11.8   Structure of the silicon cantilever pressure sensor element [7]

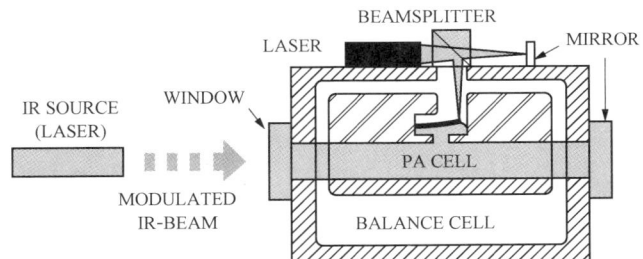

Figure 11.9   Cantilever microphone by Michelson Interferometers [6]

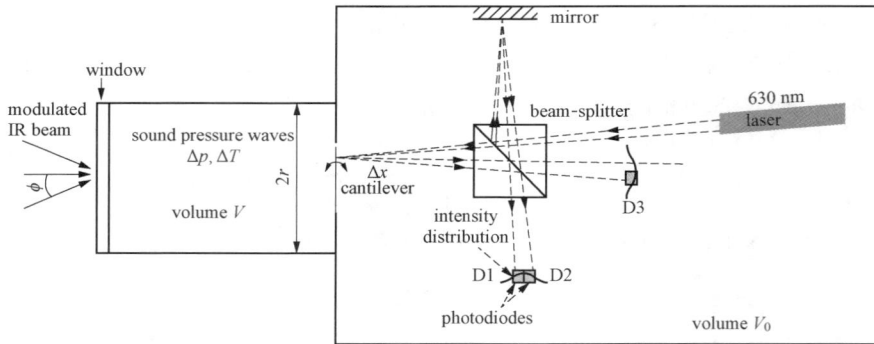

Figure 11.10   Setup for the measurements platform by Michelson Interferometer [7]

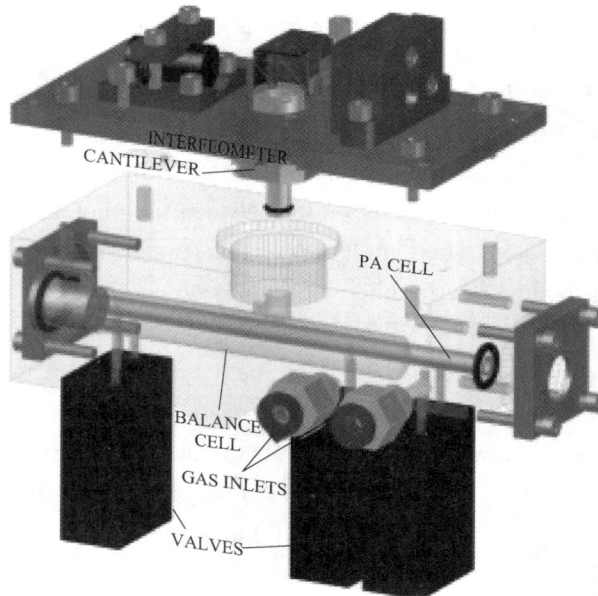

Figure 11.11   A Michelson Interferometer as acoustic wave measurements
microphone in 3D outline of Gasera instrument [6]

Figure 11.13 shows an updated version of the Michelson Interferometer in which the moving mirror and fixed mirror are located side by side, and the angle mirror has a function to split the laser beam from one to two, as so to achieve interference on CMOS array. [9, 10, 11]

11.3.2   Partial Discharge Detection by Michelson Interferometers

To prevent in-service failures in Gas Insulated Switchgear (GIS), partial discharge (PD) measurement has become a crucial method for identifying defects.

Traditionally, PD detection methods include pulse current detection, acoustic emission (AE) detection, and ultra-high frequency (UHF) detection. Both pulse current and UHF methods offer better sensitivity, but they are susceptible to electromagnetic interference (EMI), which can lead to detection failures in environments with strong EMI. As a result, AE detection is an effective alternative, especially when combined with electrical methods to enhance detection accuracy.

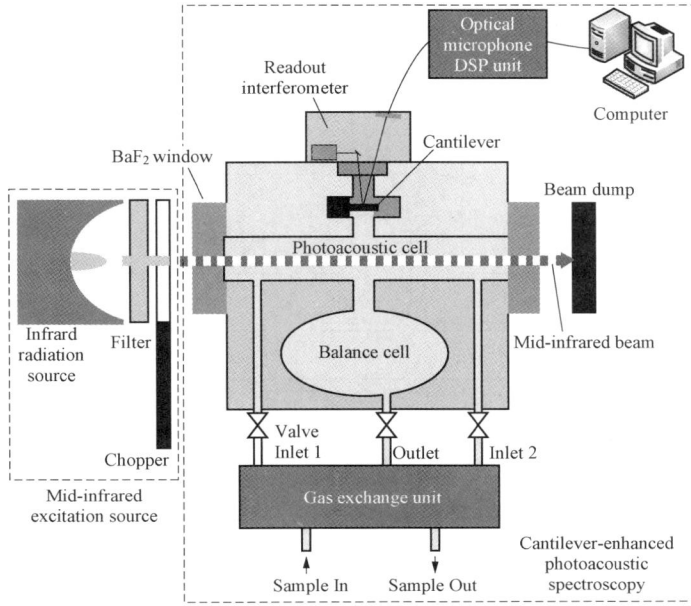

Figure 11.12   A Michelson Interferometer as acoustic wave measurements microphone in 2D outline of Gasera instrument [8]

The conventional AE detection element is a lead zirconate titanate (PZT) piezoelectric transducer, which can pinpoint defect locations using multiple sensors positioned in different locations. However, improving the sensitivity of PZT sensors to detect smaller defects remains challenging. In recent years, significant efforts have been made to develop optical interferometer acoustic sensors for PD detection due to their higher sensitivity, typically around 10 dB greater than conventional PZT sensors.

Optical sensors have demonstrated superior sensitivity in acoustic detection of PD activity compared to traditional PZT sensors. However, most optical interf-

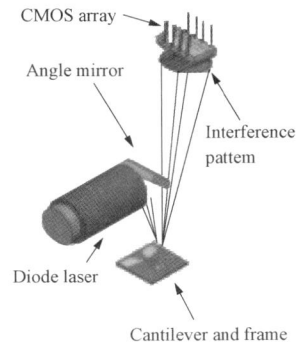

Figure 11.13   The updated version of the Michelson Interferometer in which moving and fixed mirrors are located side by side [9]

erometer investigations have been conducted on small-scale models for PD detection. In real-world applications, particularly in GIS, system stability is critical. The stability of the interferometric sensing system is strongly influenced by the initial phase, and environmental noise can adversely impact this phase, thereby affecting system reliability. Ensuring stability is crucial for the practical use of optical interferometric ultrasonic sensing systems in the field. An optical fiber ultrasonic sensing system, based on the Michelson interferometer, designed for PD detection in actual 126 kV GIS is illustrated in Figure 11.14. [12]

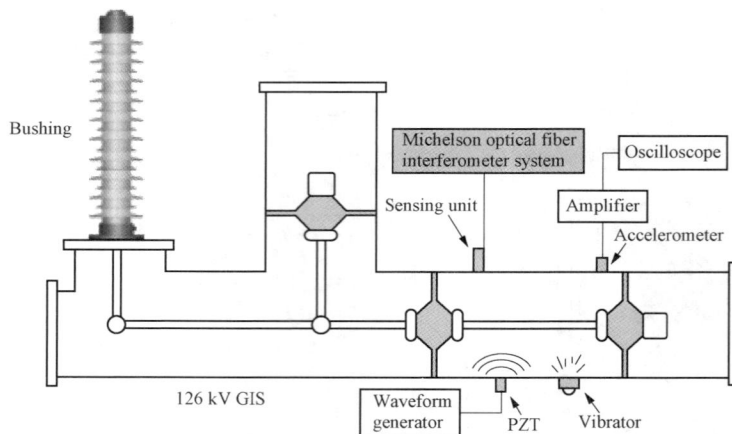

Figure 11.14　The schematic diagram of the actual 126 kV GIS experiment
setup for a partial discharge test [13]

The Michelson optical fiber interferometer, integrated with a phase feedback system, functioned as a high-pass filter with a designed cutoff frequency of 1.8 kHz. Finally, the antiinterference performance of the optical fiber interferometer was tested. The incorporation of the phase feedback control system ensured stability, making the optical interferometric acoustic sensing system suitable for on-site applications.

Another experimental study for partial discharge (PD) detection was conducted in a power transformer tank to evaluate the performance of the Michelson ultrasonic sensing system. [13] The setup, illustrated in Figure 11.15(a), involved a transformer tank filled with oil (dimensions: 2m×1m×1m) and a needle-plane defect [Figure 11.15(b)] installed near a winding to induce partial discharges. The cylindrical winding, measuring 500 mm in height and 400 mm in diameter, is depicted in Figure 11.15(c).

The proposed optical fiber ultrasonic (OFU) sensor was suspended in the oil near the inner wall of the tank. A PZT sensor (Physical Acoustics Corp. R15α) was also insulated in the oil, positioned near the OFU sensor (labeled as 1# PZT). Both the OFU sensor and the 1# PZT sensor were placed approximately 1 meter from the PD defect. Another identical PZT sensor (2# PZT) was mounted on the surface of the transformer tank, following the typical installation method used in the field.

Since amplifiers are commonly used to enhance the sensitivity of PZT ultrasonic sensors for PD detection in transformers, amplifiers were also employed to boost the PZT sensor signals. Both the amplified PZT signals and the Michelson interferometer signals were recorded simultaneously by an oscilloscope. An AC high-voltage transformer was used to apply voltage to the PD defect. The voltage was gradually increased until partial discharges were detected, and the discharge amount was measured using a PD instrument. [13]

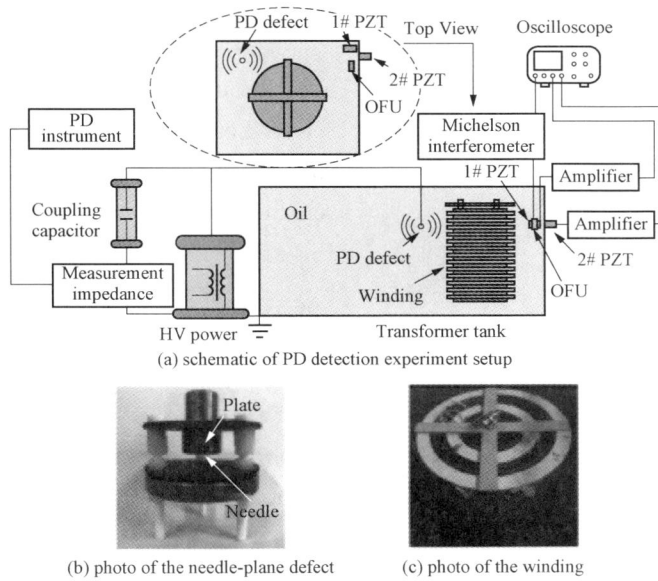

(a) schematic of PD detection experiment setup

(b) photo of the needle-plane defect   (c) photo of the winding

Figure 11.15　The setup of partial discharge detection experiment of an oil-immersed power transformer[13]

In the frequency range of 80 kHz to 200 kHz, whether the PZT sensor is placed in oil or mounted on the tank, the average response sensitivity of the proposed sensing system is higher than that of the conventional PZT system. When the distance between the sensor head and the ultrasonic source is 300 mm in oil, the average detection limit of the Michelson ultrasonic sensing system is approximately 0.26 Pa, which is about 18.6% of the PZT system's detection limit. Additionally, experimental results indicate that the detectable partial discharge inception voltage for the optical system is 21.5% lower than that of the PZT system. The improved sensitivity of the Michelson ultrasonic sensing system makes it a promising method for detecting small defects in power transformers.

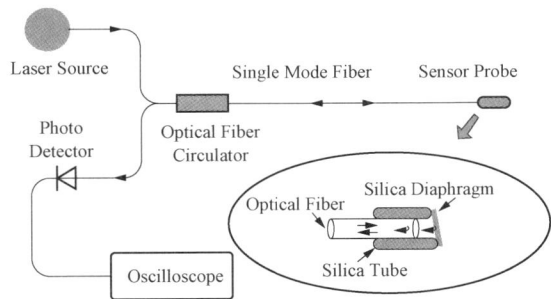

Figure 11.16　Structure of the F-P fiber-optic acoustic detection system in a power transformer model [14]

11.3.3　Partial Discharge Detection by Fabry-Perot Interferometer

As shown in Figure 11.16, the partial discharge (PD) detection system using a Fabry-Perot Interferometer (FPI) consists of a sensor head, laser source, optical fiber circulator, and photodetector, all connected via single-mode fiber. The sensor head is made up of optical fiber, a silica tube, and a silica diaphragm, with an Fabry-Perot (F-P) cavity between the diaphragm and the fiber, as illustrated in Figure 11.17. [14]

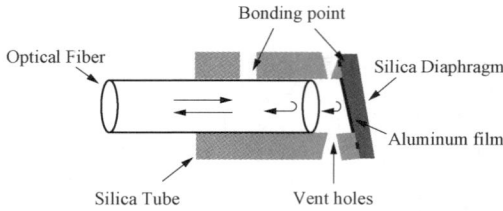

Figure 11.17　Structure of the proposed F-P fiber-optic sensorl [14]

The light from the laser source propagates through the circulator and reaches the sensor head. At the fiber's end face, the light first reflects due to the glass-air interface. The remaining light travels across the air within the F-P cavity and is then reflected a second time by the inner surface of the diaphragm. The two reflected beams travel back through the same fiber, where they interfere. The resulting light intensity, after interference, depends on the cavity length.

Upon absorbing acoustic energy, the silica diaphragm deforms, which causes a change in light intensity at the photodetector. By demodulating the light intensity at the receiving end, the cavity length changes, reflecting the diaphragm's deformation, can be measured. Thus, the acoustic signal generated by PD can be detected, as its arrival time is the key factor causing the diaphragm's deformation.

Figure 11.18 and Figure 11.19 show the F-P fiber optic sensor test setup, closely resembling real conditions for online PD detection in power transformers. The sensor's receiving end, including the laser signal processing system, photodetector, and amplifier, must be positioned far from the winding model, connected by fiber optic links to prevent potential electromag- netic interference (EMI) from the power transformer.

Figure 11.18　Test setup and the placement of PD source at the adjacent turns in the high voltage winding as an example [14]

Figure 11.20 shows the installation of four F-P fiber optic sensors and two PZT sensors in a 220 kV power transformer. This arrangement helps identify the location of partial discharge during routine testing before disassembly and transportation of the transformer. [15] A similar study can also be found in. [16]

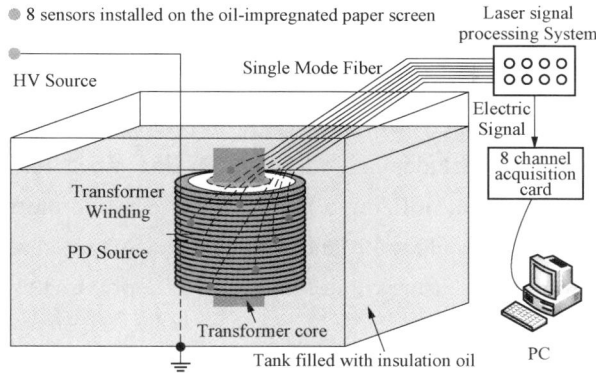

Figure 11.19   Schematic diagram of the test setup and the placement of PD source [14]

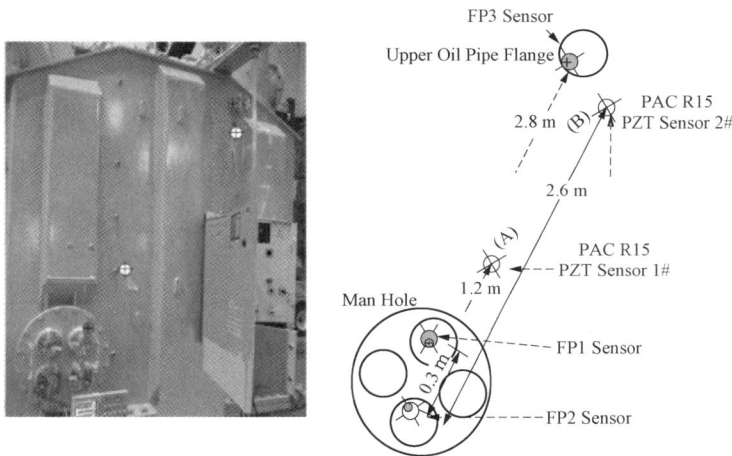

Figure 11.20   Arrangement of the F-P optic fiber sensors and PZT sensors in

a 220 kV prototype power transformer [15]

11.3.4 Temperature Sensing by Fabry-Perot Interferometers

FPI fiber-optic sensors are highly sensitive to thermal radiation, a characteristic that has long been utilized for measuring thermal variations and detecting the direction of temperature changes. The widest temperature range reported for these sensors spans from −200 to 1,050 ℃, as demonstrated by a fiber FPI temperature sensor with internal mirrors. It has been confirmed that these sensors can reliably operate at temperatures up to 1,000 ℃. Such high sensitivity allows these temperature sensors to detect even small variations, such as those produced by the human body.

Additionally, coherence-multiplexed remote FPI fiber-optic sensors are employed as point sensors for measuring remote temperature changes. These sensors are suitable for critical applications such as monitoring transformers, aiding in cancer treatments where tumors are exposed to microwave radiation, and evaluating structural materials where conventional sensors are unsuitable.

Fabry-Perot interferometers have been adapted in various ways for temperature sensing. For example, they can be combined with Michelson interferometers to create hybrid temperature sensors. One example includes an all-silica, in-line fiber Fabry-Perot etalon with feedback-controlled cavity length, utilizing a piezoelectric ceramic unimorph actuator for the simultaneous measurement of acceleration and temperature. A thin-core fiber is also used to construct a hightemperature FPI sensor with a sensitivity of approximately 18.3 pm/℃, capable of sensing temperatures up to 850 ℃.

11.3.5 Mechanical Vibration Sensing by Fabry-Perot Interferometers

FPI vibration sensors are widely used in various applications, including geological surveys, diagnostics of large civil structures (frequencies below 10 Hz), inertial navigation, consumer electronics, oil and gas field exploration, and earthquake monitoring (frequencies below 20 Hz). A low-finesse FPI vibration sensor with a reported sensitivity of 9 mrad/Hz$^{1/2}$ has also been inves-tigated. For measuring both periodic and non-periodic vibrations, dual-cavity EFPI (Extrinsic Fabry-Perot Interferometer) systems are employed. Additionally, non-contact self- calibrating FPI vibration displacement sensor systems, all-fiber FPI sensors for low-frequency vibration measurements, and various other FPI vibration sensors have been developed in earlier studies.

11.3.6 Acoustic Wave Sensing by Fabry-Perot Interferometers

The acoustic sensitivity of an FPI sensor varies with fiber length. A diaphragm-based optical fiber acoustic (DOFIA) sensor was developed to measure the attenuation properties of acoustic waves in water, showing a detection range of approximately 0°-30° with an acoustic wave attenuation coefficient of 0.062,6/cm. Additionally, a high-sensitivity, diaphragm-based interferometric fiber optical micro-electromechanical system (MEMS) sensor was fabricated for online detection of acoustic waves. A polymer diaphragm-based EFPI (Extrinsic Fabry-Perot Interferometer) fiber acoustic sensor system demonstrated an acoustic sensitivity of 31 mV/Pa • A, with a maximum detected frequency of 80 MHz. Furthermore, an FP acoustic sensor using a multilayer graphene diaphragm exhibited acoustic pressure sensitivity of 1,100 nm/kPa and a minimum detectable pressure equivalent to a noise level of approximately 60 μPa/Hz$^{1/2}$ at a 10 kHz input frequency.

11.3.7 Ultrasound Sensing by Fabry-Perot Interferometers

Ultrasound has become a widely used method for detecting various material defects, monitoring structural health, and overseeing processes. Piezoelectric devices are commonly employed for ultrasound sensing but often face issues with non-uniform frequency response due to poor acoustic impedance matching with liquids. The development of extrinsic optical-fiber ultrasound sensors has provided improved sensitivity, with one such sensor achieving 61 mV/MPa. Another approach involves a Fabry-Perot (FP) polymer film sensing interferometer based on optical ultrasound sensing, offering excellent detection sensitivity below 10 kPa.

Additionally, FPI (Fabry-Perot Interferometer) fiber optic ultrasound sensors hold potential for use in fiber optic smart structure applications. An in-line FPI made from hollow-core photonic crystal fiber (PCF) demonstrates wavelength pressure sensitivity approximately twice that of Fiber Bragg Grating (FBG) sensors, at around 7.29×10^{-3} nm/MPa. Another notable development is the in-line silica capillary tube all-silica fiber-optic Fabry-Perot (ILSCT-ASFP) interferometric sensor, which has shown an acoustic sensitivity of 65.4 mV/MPa across a measurement bandwidth of 2.5 MHz.

11.3.8 Pressure Sensing by Fabry-Perot Interferometers

Fabricating high-performance Fabry-Perot (FPI) pressure sensors with minimal measurement errors is a challenging task. An FPI fiber-optic sensor has demonstrated effective performance for static pressures ranging from 15 to 1,000 psi by measuring pressure-induced deflections of a membrane. An Extrinsic Fabry-Perot Interferometer (EFPI) pressure sensor achieved a sensitivity of 2.75×10^{-8} 1/kPa. Another ultra-high sensitivity FPI pressure sensor has shown a sensitivity greater than 1,000 nm/kPa. Additionally, a miniature FPI sensor, integrated at the tip of a Fiber Bragg Grating (FBG), exhibited a sensitivity of 0.010,6 μm/psi over a pressure range of 1.9 to 7.9 psi.

11.4 Review Questions

Q1: Please explain the working principle of the Michelson interferometer. (see Section 11.2.2)

Q2: Please explain the working principle of the Fabry-Perot interferometer. (see Section 11.2.3)

Q3: Please give an application example of a Fabry-Perot interferometer in electrical apparatus. (see Section 11.3.3)

Q4: What are the differences between the Michelson interferometer and the Fabry-Perot Interferometer?

Q5: Why it has a balance cell in the Cantilever microphone by Michelson interferometer? What will happen if balance cells do not exist? (see Section 11.3.1)

Bibliography

[1] Hariharan, Parameswaran. Basics of interferometry. Elsevier, 2010.

[2] United Scientific Supplies, INC. Michelson and fabry-pérot interferometer. URL https://www.novatech-usa. com/pdf/MFPI01%20Michelson%20&%20Fabry-Perot% 20Interferometer%20Datasheet.pdf. Accessed: August 30, 2024.

[3] PASCO Scientific. Precision interferometer. URL https://cdn.pasco.com/product_ document/Precision-Interferometer-Manual-OS-9255A.pdf. Accessed: August 30, 2024.

[4] Shizhuo Yin, Paul B. Ruffin, Francis T. S. Yu. Fiber optic sensors, Second Edition. CRC Press, Taylor & Francis Group, 2008.

[5] Yun-Jiang Rao, Zeng-Liang Ran and Yuan Gong. Fiber-Optic Fabry-Perot Sensors: An Introduction. CRC Press, Taylor & Francis Group, 2017.

[6] Koskinen, V and Fonsen, J and Roth, K and Kauppinen, J. Progress in cantilever enhanced photoacoustic spectroscopy. Vibrational spectroscopy, 48(1):16-21, 2008.

[7] Kauppinen, Jyrki and Wilcken, Klaus and Kauppinen, Ismo and Koskinen, Vesa. High sensitivity in gas analysis with photoacoustic detection. Microchemical journal, 76(1-2): 151-159, 2004.

[8] Fengxiang Ma, Zhenghai Liao, Yue Zhao, Zongjia Qiu, Liujie Wan, Kang Li, Guoqiang Zhang. Detection of trace C_2H_2 in N_2 buffer gas with cantilever-enhanced photoacoustic spectrometer. Optik, 232:166525, 2021.

[9] Gasera user manual - PA201, 2012.

[10] Hirschmann, CB and Lehtinen, J and Uotila, J and Ojala, Satu and Keiski, RL. Sub-ppb detection of formaldehyde with cantilever enhanced photoacoustic spectroscopy using quantum cascade laser source. Applied Physics B, 111(4):603-610, 2013.

[11] Hirschmann, Christian B and Sinisalo, Sauli and Uotila, Juho and Ojala, Satu and Keiski, Riitta L. Trace gas detection of benzene, toluene, p-, m-and o-xylene with a compact measurement system using cantilever enhanced photoacoustic spectroscopy and optical parametric oscillator. Vibrational Spectroscopy, 68:170-176, 2013.

[12] H. Zhou, M. Zhang, G. Ma, C. Li, B. Cui, Y. Yin and Y. Wu. The anti-interference method of michelson optical fiber interferometer for gis partial discharge ultrasonic detection. In 2019 IEEE Conference on Electrical Insulation and Dielectric Phenomena (CEIDP), pages 283-286, 2019.

[13] j. Zhou, Hong-Yang and Ma, Guo-Ming and Zhang, Meng and Zhang, Han-Chi and Li, Cheng-Rong. A high sensitivity optical fiber interferometer sensor for acoustic emission detection of partial discharge in power transformer. 21(1):24-32, 2019.

[14] Gao, Chaofei and Yu, Lei and Xu, Yue and Wang, Wei and Wang, Shijie and Wang, Peng. Partial discharge localization inside transformer windings via fiber-optic acoustic sensor array. IEEE Transactions on Power Delivery, 34(4):1251-1260, 2018.

[15] Guo Shaopeng. Research on some Key Technologies of Fiber Fabry-Perot Sensor Applied for Partial

Discharge Detection in Transformers. PhD thesis, The University of Chinese Academy of Sciences, 2015.

[16] Si, Wenrong and Fu, Chenzhao and Li, Delin and Li, Haoyong and Yuan, Peng and Yu, Yiting. Directional sensitivity of a mems-based fiber-optic extrinsic fabry-perot ultrasonic sensor for partial discharge detection. Sensors, 18(6): 1975, 2018.

12

Acoustic and Ultrasonic Imaging

12.1 Introduction

Sound waves can propagate through air, solids, and liquids. The measurement and analysis of these waves are vital across various industries, including production testing, machine and engine performance evaluation, and process control.

Sensors such as microphones, accelerometers, and interferometers detect sound waves, providing crucial insights into the behavior and movement of objects. Acoustic localization is another important application of sound waves. Active techniques generate sound to analyze echo responses, while passive methods involve listening to naturally occurring sounds or vibrations to determine the direction and location of a source. SONAR, one of the most well-known examples of acoustic localization, employs hydrophones to monitor waveforms traveling through water. Beyond underwater applications, acoustic localization is widely utilized in fields like materials science and geology.

Using the latest acoustic spectrum digitizers and arbitrary waveform generators makes excellent tools for use in the development, testing, and operation of acoustic systems. Small and compact they offer a wide range of bandwidths and sampling rates so that they can best match the measurement needs.

Typical acoustic applications include acoustic ranging, acoustic location, SONAR, seismology, acoustic emission, vibration analysis, engine testing, process control, ocean acoustic tomography, and bio-acoustics.

For various electric apparatus in smart grid, acoustic detection technology is applied in the following areas:
- Acoustic fingerprint recording and fault identification;
- Noise source localization, vibration and noise analysis by sound spectrum analysis;
- Ultrasonic partial discharge localization and detection.

12.2 Basic Principles

12.2.1 Audio Spectrum

An audio spectrum (or sound spectrum) represents the various frequencies that combine to form a sound. Most sounds are made up of a complex mix of vibrations. For instance, you might hear the rustling of the wind, the rumble of passing traffic, or background music containing both high and low notes. Some sounds, like those produced by drums or cymbals, may not have a specific pitch or tone but still contribute to the overall mix of frequencies. [1]

A sound spectrum represents a snapshot of a sound, typically a short sample, by showing the amount of vibration at each specific frequency. It is usually displayed as a graph, where either power or pressure is plotted against frequency. Power or pressure is typically measured in decibels, while frequency is measured in hertz (Hz), representing vibrations per second, or in kilohertz (kHz), representing thousands of vibrations per second. You can think of the sound spectrum like a recipe: combining different amounts of various frequencies to form the complete, complex sound.

Today, sound spectra (the plural of spectrum) are typically measured using the following devices:

- A microphone that measures the sound pressure over a certain time interval.
- An analog-digital converter that converts this to a series of numbers (representing the microphone voltage) as a function of time, and.
- A computer that performs a calculation upon these numbers.

Your computer may already have the necessary hardware (such as a sound card) to analyze sound spectra. Many sound analysis or editing software packages can take a short sample of a sound recording, calculate its spectrum using a digital Fourier transform (DFT), and display it in real-time (i.e., after a brief delay). With these tools, you can experiment by singing sustained notes into a microphone (or playing an instrument) and observing the results. When you change the loudness, the amplitude of the spectral components increases. If you change the pitch, the frequency of all components rises. By definition, altering a sound without changing its loudness or pitch affects its timbre. Timbre refers to the sum of differences between two sounds that share the same pitch and loudness. One key factor in determining timbre is the relative amplitude of different spectral components. For instance, singing "ah" and "ee" at the same pitch and loudness will produce noticeably different spectra. Figure 12.1 illustrates the audio spectrum, spanning from zero to several hundred MHz. The infrasound frequency range is between 0 and 20 Hz, while the human hearing range extends from 20 Hz to 20 kHz. Ultrasound refers to acoustic sound pressure waves with frequencies exceeding the upper limit of human hearing. Ultrasound devices typically operate within a frequency range of 20 kHz to several gigahertz. Table 12.1 summarizes the characteristics of various common ultrasonic applications. [3]

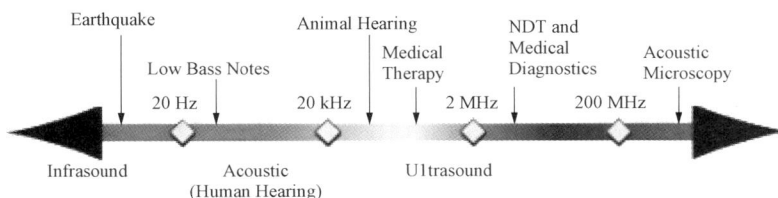

Figure 12.1   Audio spectrum could be divided into Infrasound, human hearing sound, and ultrasound [2]

Table 12.1   **Characteristics of common ultrasonic applications along with**
the recommended spectrum frequency range [3]

Application	Frequency Range	Dynamic Range
Nondestructive Testing (NDT)	0.1 to 100 MHz	100 to 120 dB
Medical Imaging	1 to 18 MHz	60 to 80 dB
Ultrasonic Welding Inspection	2 to 10 MHz	100 to 110 dB
Ultrasonic Cleaning	20 to 100 kHz	Not specified
Ultrasonic Welding	20 to 100 KHz	Not specified
Flowmeters	0.5 to 10 MHz	60 to 90 dB
Sonochemistry	20 to 500 kHz	Not specified
Range Finding	40 to 70 kHz	Not specified
Food Processing	20 to 100 kHz	Not specified
Seismology and Geological Exploration	1 to 10 Hz	60 to 100 dB
High-Intensity Focused Ultrasound	0.8 to 3 MHz	up to 100 MPa
Time of Flight Diffraction	1 to 15 MHz	Not specified

Figure 12.2 shows the human hearing audio spectrum range spans from 20 Hz to 20,000 Hz and can be effectively broken down into eight different frequency bands, with each band having a different impact on the total sound. [4]

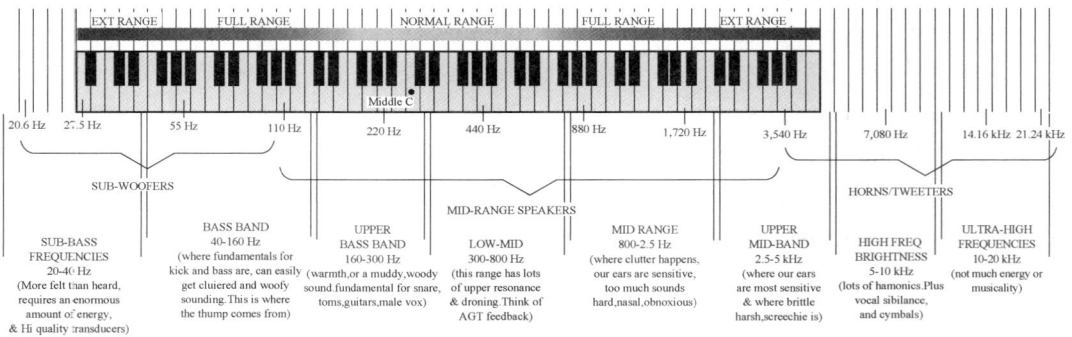

Figure 12.2   Human hearing sound could be divided into eight sub-frequency bands [4]

The eight frequency bands of the human hearing audio spectrum are:
- Sub-bass: 20-60 Hz.
- Bass band: 60-250 Hz.
- Upper bass band: 250-500 Hz.
- Low midrange: 500-2 kHz.
- Midrange: 2-4 kHz.
- Upper midrange: 4-6 kHz.
- High frequency: 6-10 kHz.
- Ultra-high frequency: 10-20 kHz.

Sub-20 Hz sounds fall into the category of infrasound, which requires specialized equipment to detect. Infrasound refers to low-frequency sound waves that are both mechanical vibrations and acoustic oscillations, typically below the range perceptible by the human ear (approximately 0 Hz to 16-20 Hz). At these lower frequencies, human hearing is no longer sensitive enough to perceive them directly. Numerous studies have investigated the production, transmission, and human sensitivity to both infrasound and ultrasound. Natural sources of infrasound are often consistent and include wind, ocean waves, and the Earth's natural seismic activity. On the other hand, artificial sources of infrasound are becoming increasingly common with modern technological and industrial developments.

The sensitivity of human hearing to both the sound spectrum and sound intensity varies between individuals and is influenced by factors such as age and attention. Generally, human hearing spans from approximately 20 Hz to 20 kHz. The frequency range between 20 Hz and 40 Hz serves as a transition zone between infrasound and audible sound. Frequencies above 20 kHz and below 20 Hz are typically imperceptible to the ear, except in cases where infrasound is of high intensity. In such instances, other parts of the body, such as the rib cage, abdomen, skin, eyeballs, muscles, skeleton, and skull, may resonate or perceive the sound as a vibratory sensation due to the induced energy.

It has long been known that high-intensity noise damages hearing, and for decades, efforts have been made to detect and measure the potential effects of exposure to inaudible vibrations. NASA has taken a particular interest in this field because its pilots and astronauts are subjected to extreme levels of vibration and noise. During experiments conducted for the Apollo missions, volunteers were exposed to very high levels of infrasound (120 to 140 decibels) under medical supervision, with no harmful health effects observed. Interestingly, high levels of infrasound are found to be more tolerable than the same sound intensity within the normal audible frequency range.

Animal studies indicate that infrasound can have physiological effects, but only after prolonged exposure to very high levels. Some humans, especially those living near large wind turbines, have reported symptoms like fatigue, depression, stress, irritability, exhaustion, headaches, issues with alertness and balance, and nausea. The audibility threshold, the minimum sound level the human ear can detect, rises as sound frequency decreases, meaning lower-frequency sounds need to be louder to be perceived. These reactions are believed to be caused by internal organs, such as those in the digestive, cardiovascular, and respiratory systems, or the eyeball responding to specific infrasound frequencies, though such effects occur only at much higher levels than those generated by wind turbines.

Today's fixed or mobile sensor networks enable the detection, measurement, and tracking of low-frequency acoustic waves, primarily infrasound, with frequencies as low as a few hertz, over thousands of kilometers. These waves propagate through the ground, sea, or Earth's atmosphere, and computer analysis helps isolate and identify them. This technology can detect and locate both natural phenomena, such as tsunamis, volcanic eruptions, and meteorite entries, as well as artificial events like sonic booms from supersonic aircraft or explosions, including nuclear tests or accidents. Each of these emissions has a distinct "signature" that allows for precise identification.

12.2.2 Working Principle of Microphones

Microphones (often abbreviated as mic or mike) and speakers are commonly used audio equipment. They were used not only in public meetings and conferences but also in everyday devices like cell phone. A microphone and a speaker function in opposite ways: a microphone converts sound vibrations into an electrical signal (voltage/current), while a speaker converts electrical signals into sound vibrations by moving its diaphragm to create vibrations in the air. In a microphone, the diaphragm moves in response to sound pressure, and this movement is converted into a proportional electrical voltage using various types of transducers.

Once sound is in electrical form, it can be amplified, mixed, and recorded. The acoustic waveform can be transformed into an electrical waveform of the same shape, where amplitude becomes voltage (V) and the motion of air particles corresponds to electrical current (I). Electrons, in this case, take on the role of air particles. The cycles of compression and rarefaction in the acoustic wave create an alternating current (AC), leading to changes in the direction of electron flow. The movement of electrons through a conductor faces resistance (R), and the relationship between V, I, and R is governed by Ohm's law: $V = IR$. Their relationship to power (W) is given by: $W = I^2R = V^2/R$. In AC systems, resistance is replaced by impedance, which also accounts for reactance and is frequency-dependent.

Electromagnetic transducers convert mechanical motion (like an acoustic wave) into an electrical signal. This process generates an electrical current (and voltage) in two scenarios: (i) when a stationary conductor is in a changing magnetic field, or (ii) when a conductor moves within a static magnetic field. This is explained by Faraday's law of electromagnetic induction. The direction of movement, when perpendicular to the lines of magnetic flux, dictates the direction of current flow in the conductor (such as a wire). Back-and-forth movements result in an alternating current (AC), which is related in both frequency and amplitude to the motion of the wire.

A microphone (mic) is a transducer that converts acoustic sound energy into electrical energy. Its primary function is to transform sound waves into electrical audio signals, which can then be processed, amplified, or recorded for further use.

We can classify the microphones based on construction/directivity. [5] On the basis of construction/type of transducer used, the categories can be classified as follows:

- Condenser Microphone (also capicitor/electrostatic microphone).
- Dynamic microphone.
- Ribbon microphone.

- Carbon microphone.
- Piezoelectric microphone.
- Fiber optic microphone.
- Laser microphone.
- MEMS (Micro electrical mechanical sysstem).

On the basis of pick up or directionality properties (Figure 12.3), the category can be classified as follow:[6]

- Omni-directional.
- Unidirectional or cardioid.
- Bidirectional or supercardioid.

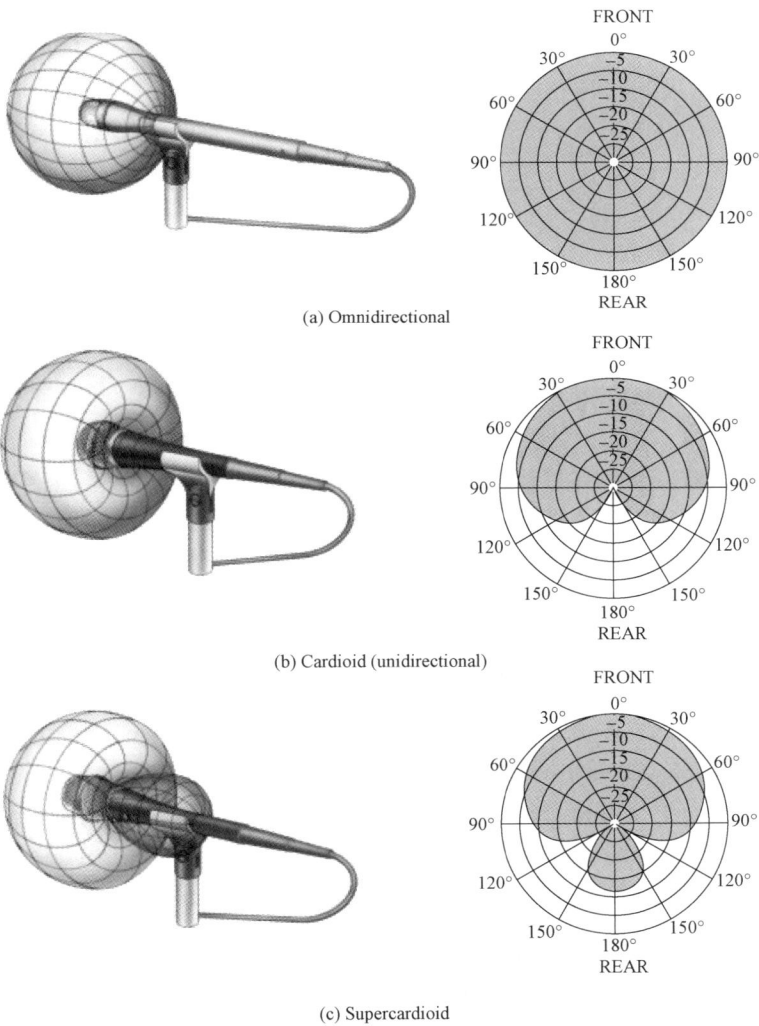

(a) Omnidirectional

(b) Cardioid (unidirectional)

(c) Supercardioid

Figure 12.3   The sensitivity of a microphone to sound depends on the direction or angle of the incoming sound [6]

Dynamic   Condenser   Ribbon

Figure 12.4   Outline of three main types
of microphones (Dynamic microphone,
Condenser microphone and Ribbon
microphone) [7]

In this chapter, we will primarily discuss the working principles of dynamic microphones, ribbon microphones, and condenser microphones (as illustrated in Figure 12.4 [7]). Additionally, we will provide a brief overview of MEMS (Micro Electrical-Mechanical System) microphones, carbon microphones, fiber optic microphones, laser microphones and piezoelectric microphones. [8]

12.2.2.1   Dynamic Microphone

A dynamic microphone operates on the principle of electromagnetic induction and is often referred to as a moving coil microphone. In this design, a small coil is attached to a diaphragm and suspended within the magnetic field of a magnet, as illustrated in Figure 12.5.[9] When sound waves strike the diaphragm, it vibrates, causing the attached coil to move within the magnetic field. This movement generates an electromotive force (EMF) across the coil's terminals. The current produced in the coil is proportional to the sound waves.

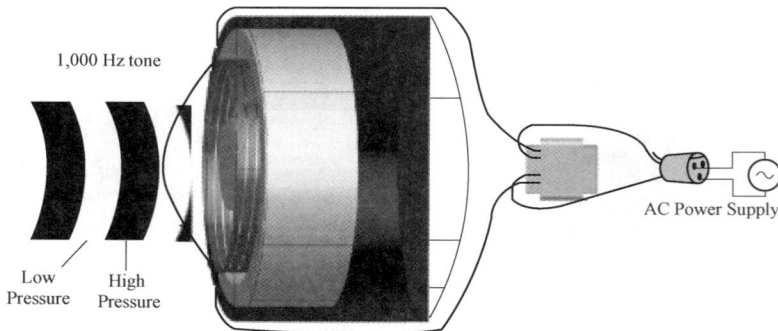

Figure 12.5   Working principle of a dynamic microphone [9]

Figure 12.6 illustrates the internal components of a dynamic microphone. [10] It consists of a diaphragm suspended in front of a magnet, with a coil of wire attached to the diaphragm. The coil is positioned within the magnetic gaps. As the diaphragm vibrates in response to sound waves, it causes the coil to move within the magnetic field, generating an alternating current (AC). However, the mass of the coil-diaphragm structure limits its responsiveness at higher frequencies, where the response tends to decrease. A resonant peak typically occurs around 5 kHz, which makes this type of microphone popular with vocalists.

12.2.2.2   Ribbon Microphone

A ribbon microphone operates using a corrugated metal ribbon suspended within a magnetic field, as shown in Figure 12.7. [11] When sound waves strike the ribbon, it vibrates, causing changes in the magnetic flux across the ribbon, which in turn induces an electrical current. This current drives the speaker. When passed through a coil attached to the microphone's diaphragm, the diaphragm vibrates and generates sound. Advanced materials developed through nanotechnology are now used to create ribbons that are both lightweight and strong, enhancing the microphone's sensitivity. Unlike conventional microphones that respond to pressure alone, the ribbon microphone detects pressure gradients, making it sensitive to sound from both sides.

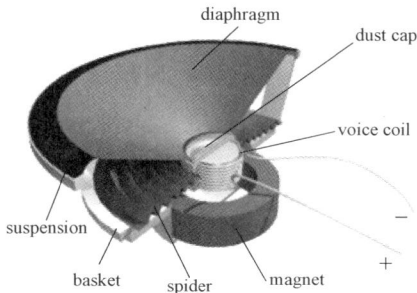

Figure 12.6   Internal structure and main components of a dynamic microphone [10]

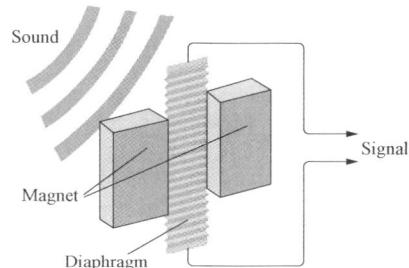

Figure 12.7   Internal structure and main components of the ribbon microphone[11]

12.2.2.3 Condenser Microphone

Condenser microphones, invented at Bell Labs in 1916, are also known as capacitor or electrostatic microphones. They consist of two very thin parallel plates, one positively charged and the other negatively charged. Figure 12.8 illustrates a condenser microphone. [8] The diaphragm, typically 1 to 10 micrometers thick, serves as one of these plates. A micrometer (or micron) is one-millionth of a meter, or one-thousandth of a millimeter. Positioned close to this diaphragm is another metallic plate, often with holes. These two plates act as electrodes and are kept at opposite polarities by supplying DC voltage, allowing the system to function as a capacitor. To maintain proper operation, the plates are insulated from each other.

When sound waves hit the diaphragm, it vibrates, causing changes in the capacitance between the two plates. Capacitance is directly proportional to the potential difference and inversely proportional to the separation between the plates. Any movement of the diaphragm changes this separation, thereby altering the capacitance. Although capacitance can also depend on the medium between the plates, this factor remains constant here and is therefore disregarded. The resistance and capacitance are selected so that any variation in capacitance immediately affects the voltage across a series resistance. These changes in sound waves lead to changes in capacitance, which are translated into voltage variations. The resulting voltage is then amplified to make the signal usable.

In condenser microphones, the diaphragm serves as the front plate, vibrating in response to sound waves. The charge (Q) remains fixed, so changes in the distance (d) between the plates result in corresponding changes in voltage (V). The light diaphragm enables a flat frequency response, with the resonance peak occurring above 12 kHz. Condenser microphones have a much higher output than dynamic microphones, making them more resistant to noise.

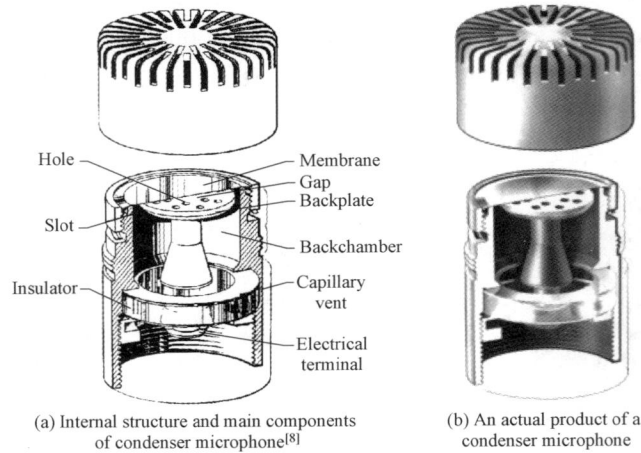

(a) Internal structure and main components
of condenser microphone[8]

(b) An actual product of a
condenser microphone

Figure 12.8   A condenser microphone

12.2.2.4   MEMS (Micro Electrical-Mechanical System) Microphone

The MEMS (Microelectromechanical System) microphone, also known as a microphone chip or silicon microphone, is a compact device comprising two key components: the sensor and the integrated circuit. These components are housed together in a small package. The pressure-sensitive diaphragm is directly etched into a silicon chip using advanced MEMS fabrication techniques, often accompanied by an integrated preamplifier. Despite its small size, just a few millimeters, it contains intricate parts, including the detection mechanism and electronic circuitry, which convert the sensed sound into electrical signals for further processing. This combination of miniaturization and integration allows MEMS microphones to be highly efficient and versatile, finding use in devices like smartphones, hearing aids, and other portable electronics. see Figure 12.9. [12]

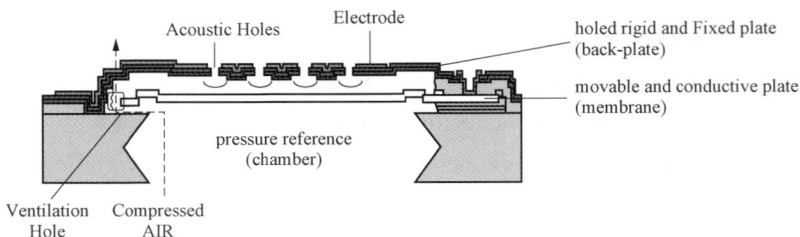

Figure 12.9   MEMS transducer mechanical specifications [12]

The MEMS sensor functions as a variable silicon capacitor, composed of two plates: one fixed and one movable (depicted in the figure below, with the green plate representing the fixed one and the grey plate the movable one). The fixed plate is coated with an electrode to enhance conductivity and has acoustic holes to allow sound waves to pass through. The movable plate, attached only on one side, can oscillate in response to sound pressure. A ventilation hole ensures that air compressed in the back chamber is released, allowing the membrane to return to its original position after displacement. The chamber design influences both the movement of the membrane and the overall acoustic performance, impacting aspects such as frequency response and signal-to-noise ratio (SNR). Essentially, the MEMS microphone operates as a variable capacitor, with sound waves causing changes in capacitance between the fixed back plate and the movable membrane, as illustrated in Figure 12.10. [12]

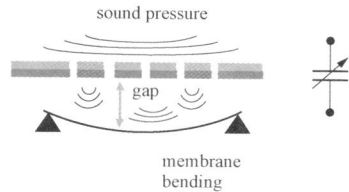

Figure 12.10   Microphone MEMS sensor is a variable capacitor between a fixed plate and a movable plate (membrane) [12]

Table 12.2 compares the technical specifications of three types of microphones.[13] The B&K 4,134 is a condenser microphone, and Electret refers to another type of condenser microphone. It is clear that MEMS microphones have the smallest size and offer reasonable sensitivity, making them ideal for applications where compact dimensions are critical, as seen in Figure 12.11. [12]

Table 12.2   Technical specifications comparison among B&K 4,134 and Electret as well as MEMS microphones [13]

	B&K 4,134	Electret	MEMS
Volume	>150 mm^3	100 mm^3	25 mm^3
Support Electronics	High	Medium	Low
Power	Low	Low	Low
Voltage	200 V	3 V	1-3 V
Mass	High	Low	Low
Cost	Very High	Low	Low
Sensitivity	High	Medium	Medium
Range	$<$100kHz	$<$20kHz	$<$50kHz
Temperature	Low	Low	Medium

Figure 12.11   Outline of three kinds of microphones (B&K 4,134, Electret and MEMS microphones) [13]

12.2.2.5　Carbon Microphone

The carbon microphone, also known as a carbon button microphone (or sometimes just a button microphone), uses a capsule or button containing carbon granules pressed between two metal plates, see Figure 12.12 [14] and Figure 12.13. [15]

Figure 12.12　Carbon microphones were used in telephones like this vintage British GPO 300 series telephone [14]

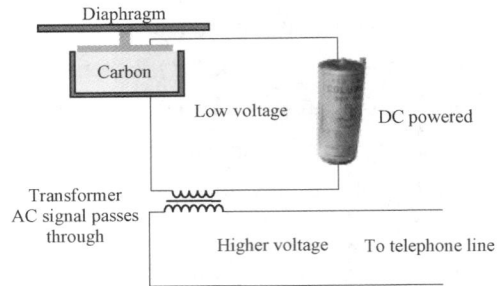

Figure 12.13　Working principle of the carbon microphone [15]

The core principle behind a carbon microphone is that the electrical resistance of carbon granules decreases when they are compressed. This happens because the pressure brings the granules into closer contact, improving conductivity. A carbon microphone consists of a small container filled with carbon granules, topped by a thin metal diaphragm. A battery is used to establish a current flow through the microphone. When sound waves hit the diaphragm, it vibrates and exerts varying pressure on the carbon granules. These pressure changes result in corresponding changes in resistance, which modulate the current flowing through the microphone. This varying current can be processed using a transformer, capacitor, or amplifier for use in telephones or audio devices.

However, carbon microphones have a limited frequency response and tend to generate significant electrical noise, often producing a crackling sound. This noise could sometimes be reduced by physically shaking the microphone or giving it a sharp tap to redistribute the carbon granules, helping restore a more consistent current flow.

12.2.2.6　Fiber Optic Microphones

Fiber optic microphones are immune to electrical, magnetic, electrostatic, and radioactive interference, offering complete EMI/RFI immunity. This makes them particularly useful in environments where traditional microphones may fail or pose risks, such as inside industrial turbines or near magnetic resonance imaging (MRI) machines. These microphones are highly durable, withstanding environmental changes in heat and moisture, and offer flexibility in terms of directionality and impedance matching for sound engineers.

One of the key advantages of fiber optic microphones is that the distance between the microphone's light source and its photodetector can extend over several kilometers without the need for a preamplifier or other electrical components. This makes them ideal for long-range industrial and surveillance acoustic monitoring. They are also suitable for direct acoustic measurements in hostile environments such as turbojets or rocket engines, where sensors must endure extreme heat and vibrations. As a result, fiber optic microphones are well-suited for applications like computational fluid dynamics (CFD) code validation, structural acoustic testing, and jet noise reduction efforts. In such conditions, fiber optic microphones using a Michelson or Fabry-Perot interferometer design are particularly effective.

Two common optical methods for detecting ultrasound waves are interferometry and refractometry.

Interferometry-based detection works by identifying changes in optical interference patterns triggered by ultrasound. These acoustic waves modify the interference conditions by directly interacting with an optical beam, vibrating a reflector, or altering the resonance frequency of a resonator. The perturbations in the interference pattern, whether through changes in the optical phase, wavelength, or the mean free path, are detected by a photodiode or a wavelength meter. This results in intensity or frequency shifts at the interferometer output, revealing critical information about the ultrasound signals.

Both the Michelson and Fabry-Perot interferometers are interferometry-based detection methods, which were extensively covered in Chapter 11 of this book. In this section, we will briefly introduce the refractometry-based detection method.

Refractometry-based detection relies on the photoelastic effect, which states that acoustic waves interacting with a medium induce mechanical stress, thereby causing changes in the refractive index (RI) of the medium. These changes in RI are proportional to the mechanical pressure exerted by the sound waves. In this method, a laser beam (called the interrogating or probe beam) measures RI variations in a single medium or at the interface between two adjacent media. These variations, caused by propagating ultrasound waves, result in changes to the beam's intensity, deflection angle, or phase, which are recorded at an optical detector. This data provides insight into the interrogated ultrasound signals.

Figure 12.14 and Figure 12.15 illustrate an membrane-free optical microphone by XARION Laser Acoustics, Wien, Austria.[16, 17]

An optical microphone is based on the unidirectional interaction of two wave phenomena: the acoustic sound wave influences the optical wave, but the optical wave does not disturb the acoustic field.

The optical microphone operates on the principle of interferometry. It consists of a control unit containing a laser that directs light through an optical fiber to a sensor head. This sensor head comprises two parallel, semi-reflective mirrors. The difference in the laser's wave velocity, which varies with air density inside the hollow sensor head, creates interference in the reflected light wave, providing direct and proportional information about the sound pressure level.

Figure 12.14　Complete system of the fiber optic microphone developed by XARION Laser Acoustics, including the sensor head, optical fiber cable, and control/detection unit [16]

As shown in Figure 12.15, a laser emitting at a telecom wavelength of 1,550 nm is connected via an optical fiber to a miniaturized laser interferometer (also known as an etalon), consisting of two parallel, millimeter-sized, semi-transparent mirrors, fixed together using spacer elements. Within the etalon, a standing optical wave forms. When the half-wavelength of the laser matches a multiple of the fixed etalon length, constructive interference occurs, resulting in maximum light intensity. If this condition is unmet, the intensity decreases.

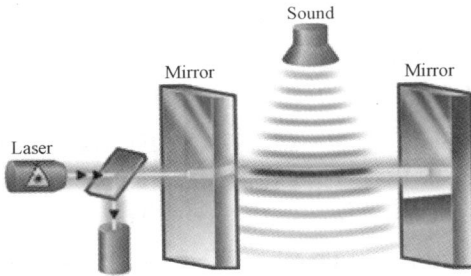

Figure 12.15　Diagram of the membrane-free optical microphone setup, featuring a laser source and a pair of mirrors (etalon), as developed by XARION Laser Acoustics [16]

When a sound wave enters the etalon, air pressure fluctuates, which alters the wavelength of the laser light, causing variations in the interference pattern and thus the light intensity. By selecting the optimal point between constructive and destructive interference, the pressure modulation is linearly transformed into an intensity modulation that is easily detected by a photodiode.

In acoustics, the alternating component of the pressure is of interest rather than static pressure. Therefore, the laser emission wavelength is stabilized within the etalon using a slow control circuit, compensating for slowly varying factors such as environmental pressure, temperature, or laser drift.

In contrast to state-of-the-art capacitive microphones, where a membrane deforms under acoustic sound pressure—resulting in a sensitivity drop at higher frequencies due to inertia—the membrane-free optical microphone involves no mechanical movement. Consequently, the detectable frequency range extends into the very high-frequency range. This technology enables resonance-free, linear signal detection across a broad bandwidth (10 Hz to 1 MHz), covering not only the human audible range but also ultrasound transmitted through the air. This feature is particularly valuable for investigating various sound generation mechanisms near high-voltage conductors.

12.2.2.7 Laser Microphone

Laser microphone is often portrayed in movies as spy gadgets, see Figure 12.16, because they can be used to pick up sound at a distance from the microphone equipment, see Figure 12.17.

Optoacoustics Ltd has demonstrated an optical microphone in which a light beam shines on a diaphragm with a modulated beam reflecting onto a photodiode. [18] Details are shown in Figure 12.17.[8] The device can be made quite small and has no electronic components at the diaphragm. [19, 20]

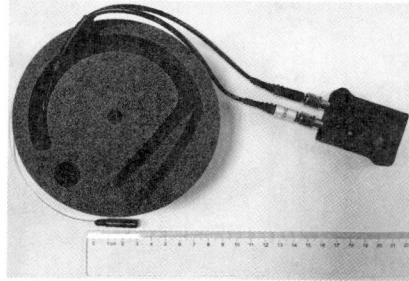

Figure 12.16　A typical product of laser microphones developed by Optoacoustics (OPTIMIC™ Special Model 1,200) [18]

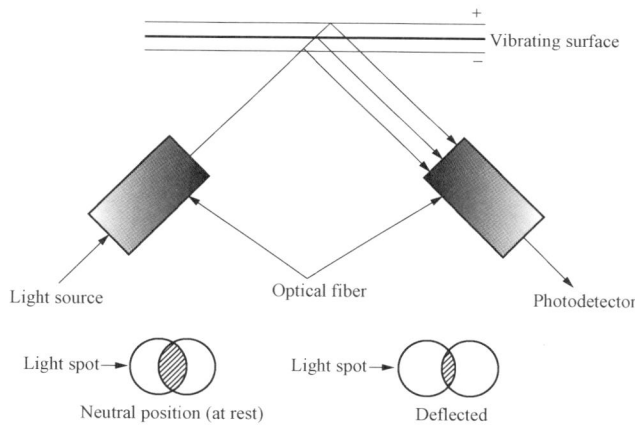

Figure 12.17　Working principle of laser microphone [8]

Figure 12.17 illustrates the internal structure of a laser microphone, which includes a surface designed to accommodate vibration signals. The optical fiber input and output terminals are positioned at an angle relative to one another on this surface. The surface allows sound waves to interact with it, while incoming sound waves, entering through the top vibrating surface, impact the upper part of the membrane. The reflected light intensity entering the output optical fiber terminal fluctuates in response to changes in the sound wave. By measuring the variations in the received light intensity, the sound wave can be accurately detected.

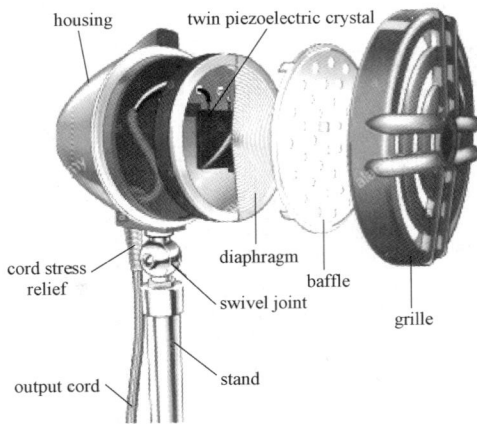

Figure 12.18 A Piezoelectric microphone [21]

12. 2. 2. 8 Piezoelectric Microphone

The piezoelectric microphone operates based on the principle of piezoelectricity—the ability of certain materials to generate an electrical voltage when subjected to pressure or mechanical stress. This allows vibrations to be converted into an electrical signal, as illustrated in Figure 12.18. [21]

Piezoelectric sensors and actuators form an important subset of piezoelectric devices. Today, these sensors and actuators are widely used across a variety of applications, including process measurement technology, nondestructive testing, medical devices, and consumer electronics.

12. 2. 3 Piezoelectric Effect and Piezoelectric Materials

Piezoelectric materials possess the unique ability to convert electrical energy into mechanical energy and vice versa. The term "piezo" originates from the ancient Greek word "piezin", meaning "to press" or "to squeeze".

Though often unnoticed, piezoelectric materials are widely integrated into everyday technology. From mobile phones and automotive electronics to medical devices and industrial systems, piezoelectric components are indispensable. For instance, ultrasound imaging, used to capture the image of an unborn baby, relies on piezoelectric technology. Even the parking sensors in cars utilize piezoelectric elements.

One of the most commonly used piezoelectric ceramics is PZT (lead zirconate titanate), with the chemical formula $Pb[Zr_{(x)}Ti_{(1-x)}]O_3$. When fired, PZT adopts a perovskite crystal structure, where each unit consists of a small tetravalent metal ion surrounded by a lattice of large divalent metal ions. In PZT, titanium or zirconium typically serves as the small tetravalent metal ion, while lead occupies the role of the large divalent ion. Under conditions that create tetragonal or rhombohedral symmetry, PZT crystals exhibit dipole moments.

PZT and other piezoelectric materials demonstrate a remarkable set of properties. Fundamentally, when a piezoelectric material undergoes mechanical stress, it generates an electric charge, a phenomenon known as the piezoelectric effect. Conversely, applying an electric field to a piezoelectric material induces mechanical deformation, referred to as the inverse piezoelectric effect, see Figure 12.19. [22, 23]

The process of manufacturing PZT (lead zirconate titanate) powders involves six key unit operations (see Figure 12.20). [24] The first step is the evaluation, selection, and sourcing of high-purity raw materials. In addition to purity, selection criteria also include material activity and limits on specific harmful impurities.

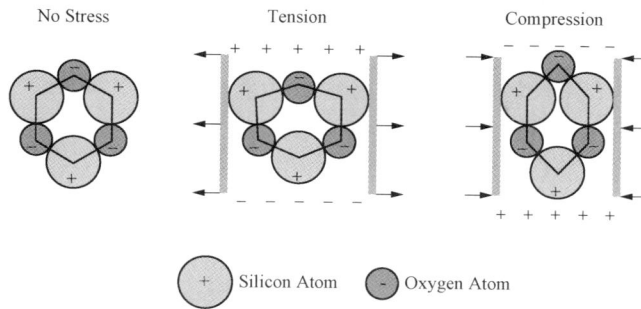

Figure 12.19   Piezoelectric effect in quartz [22, 23]

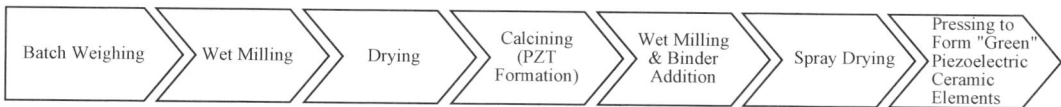

Figure 12.20   The process of manufacturing PZT powders consists of six distinct unit operations [24]

Once the materials are selected and approved, they are precisely weighed according to the specific formulation and transferred to wet mills. The ingredients are then wet-milled in their correct proportions to achieve a uniform particle size distribution. Controlling the particle size distribution is critical to ensure the desired material activity during calcination.

After wet milling, the product is dried and prepared for calcination. It is essential to use highpurity crucibles during the calcination process to avoid chemical contamination. The calcination takes place in air at approximately 1,000 ℃, where the desired PZT phase is formed.

One important consideration is that lead oxide, the primary component of PZT, is hazardous and has a relatively high vapor pressure at calcination temperatures. Excessive evaporation of lead during firing can alter the composition of the material, so proprietary measures are taken to minimize lead loss.

Following calcination, the PZT powder is returned to the mill for homogenization and preparation for the addition of an organic binder. The binder-containing slurry is then fed into a spray dryer, where water is evaporated. The success of spray drying depends on experienced operators regulating the temperatures and slurry volume to maintain the correct solid content and ensure uniform drying.

The goal of spray drying is to produce free-flowing PZT powder in the form of bindercontaining hollow spheres with a narrow particle size distribution. The morphology of the PZT material is critical for ensuring consistent filling of die cavities during dry pressing in piezoelectric ceramic manufacturing. Uniform PZT spheres with the appropriate particle size distribution enable air escapement throughout compaction, resulting in green ceramic shapes free from lamination defects.

PZT is the most widely used piezoelectric ceramic today. Piezo ceramics are preferred for their physical strength, chemical inertness, cost-effectiveness, and customizability for specific applications. PZT is especially valued for its superior sensitivity and higher operating temperature compared to other piezo ceramics.

PZT ceramics are used in a variety of applications. Soft (sensor) PZT powders are typically chosen for applications requiring high coupling and charge sensitivity, such as flow or level sensors, ultrasonic nondestructive testing, or inspections in automotive, structural, and aerospace industries. These materials are characterized by a high dielectric constant, strong coupling, high charge sensitivity, fine grain structure, high Curie point, and clean frequency response.

Hard (high-power) PZT powders, on the other hand, are used in applications requiring highpower capabilities, such as ultrasonic cleaning devices and sonar systems. These powders are noted for their high piezoelectric charge constant, higher mechanical quality factor (which reduces mechanical loss and lowers operating temperature), low dissipation factor (for cooler, more efficient operation), high dielectric stability, and low mechanical loss under demanding conditions.

12.3 Applications of Acoustic and Ultrasonic Sensors

Acoustic and ultrasonic sensors play a critical role in modern electrical power engineering, offering advanced solutions for monitoring, fault detection, and system diagnostics. These sensors are used in a wide range of applications due to their ability to detect mechanical vibrations, pressure variations, and sound waves, providing valuable insights into the health and performance of power system equipment.

- Partial Discharge Detection.
- Transformer Monitoring.
- Gas-Insulated Switchgear (GIS) Diagnostics.
- Cable Fault Location.
- Wind Turbine Monitoring.
- Circuit Breaker Monitoring.

Acoustic and ultrasonic sensors are extensively used for detecting partial discharges (PD) in high-voltage equipment such as transformers, cables, and switchgears. By capturing ultrasonic waves generated by electrical discharges, these sensors help identify insulation degradation or faults, enabling preventive maintenance and reducing the risk of equipment failure.

Transformers are vital components in power systems, and acoustic sensors are employed to monitor their internal conditions. Sensors can detect acoustic emissions caused by winding vibrations, core noise, or even arcing, which can indicate mechanical issues or electrical faults within the transformer.

Ultrasonic sensors are commonly used in gas-insulated switchgear to detect the presence of electrical discharges or internal faults. These sensors provide non-invasive monitoring of GIS, helping prevent catastrophic failures by identifying issues early.

In underground power cables, acoustic sensors can assist in locating faults by detecting sound waves generated by short circuits or other electrical anomalies. This helps in reducing downtime by pinpointing the exact location of the fault for rapid repairs.

Acoustic and ultrasonic sensors are used in wind turbines to monitor mechanical vibrations in critical components such as gearboxes and bearings. Early detection of abnormal vibrations allows for timely maintenance, extending the operational lifespan of turbines and minimizing unplanned outages.

Ultrasonic sensors are employed to monitor the performance of circuit breakers. By detecting acoustic signals generated during the operation of breakers, the sensors can assess mechanical health and operational efficiency, enabling predictive maintenance and reducing wear-related failures.

Acoustic and ultrasonic sensors are indispensable in electrical power engineering, providing non-invasive, real-time diagnostics for critical components. Their ability to detect early-stage faults, mechanical wear, and insulation degradation enhances system reliability, improves safety, and reduces maintenance costs across the power grid.

Among various electrical apparatus, such as power transformers and GIS, common faults and failures include electrical partial discharges, compressed gas leaks, vacuum system leaks, loose bolts and nuts and welding line cracks. These issues are not only costly to repair but also lead to energy inefficiency, resulting in unanticipated expenses for companies and potential disruptions in production and uptime.

An acoustic camera either in the audio frequency band or in the ultrasonic band is an effective way to detect these issues as part of a complete asset management plan. This easy-to-use technology typically allows professionals to complete their inspections 10 times faster than with traditional methods.

One of the first features to consider is the camera's frequency range. You might think that you need the widest range possible in order to pick up the widest range of sounds. However, in reality, the most effective frequency range for detecting a compressed air leak is between 20 kHz and 30 kHz. This is because using the 20-30 kHz range aids in distinguishing compressed air leaks from the background noise in a factory. The amplitude of machinery noise usually peaks below 10 kHz and trends down to zero at 60 kHz, whereas air leaks peak between 20 kHz and 30 kHz. Since there is a greater difference between the air leak noise and the background noise between 20 and 30 kHz, compared with higher frequencies, it's easier to detect the air leak in this frequency range. Both the compressed air and machinery noise follow the same downward amplitude trend in the 30-60 kHz frequency range, making it difficult to distinguish between them. Therefore, working within the 20-30 kHz range is more effective.

For users looking for partial discharge from a safe distance, the 10~30 kHz range is optimal. This is because higher frequency ranges travel shorter distances. In order to detect partial discharge from high voltage equipment in an outdoor setting, your camera needs to be attuned to lower frequency, In order to detect partial discharge from high voltage equipment in an outdoor setting, your camera needs to be attuned to lower frequency, far more distance sounds. [25]

In the pursuit of quieter noises, the microphone number, the more, the better. Acoustic imaging cameras typically employ dozens of microelectric-mechanical system (MEMS) microphones to collect and characterize sound. While MEMS are small, use little power, and are very stable, they also generate their own noise that interferes with an individual microphone's ability to pick up very quiet sounds. The solution is to increase the number of microphones in use; simply doubling the number of microphones improves the signal-to-noise ratio enough to remove three decibels of unwanted sounds. For example, one microphone might create enough self-noise to make it impossible for the system to pick up a compressed air leak generating a 16.5 kHz signal. An acoustic camera with 32 microphones would be able to detect that leak, but the signal-to-noise ratio is still too poor to hear anything quieter. In contrast, a camera with 124 microphones can pick up both the 16.5 kHz leak and one that is 18.5 kHz, making it easy to detect, locate, and quantify the small leak.

Adding just the right number of microphones to an acoustic imaging camera can also improve the chances of picking up very quiet noises from a long distance. This is especially important when inspecting high-voltage systems, which require a safe distance from the energized equipment. The force of a sound signal drops significantly as one moves further away from its source. The solution is to increase the number of microphones: quadrupling the number of microphones essentially doubles the sound detection range.

The placement of microphones on acoustic cameras factors into how the camera determines the direction and location of sounds. The camera collects data from each microphone, measures the timing and phase differences in the signals, and calculates the source location. These microphones need to be grouped closely together to ensure they collect enough data on sound waves to correctly determine from what direction they originated.

Just like frequency, there is a sweet spot for how many microphones an acoustic imager hosts. A potential downside of too many microphones is each requires processing power to convert audio data signals into images—so adding too many has diminishing returns. Some manufacturers balance this by reducing the resolution of the acoustic image pixels, or "sound" pixels, but this will affect the camera's overall performance. It's important to have enough sound pixels to detect corona and partial discharge reliably from a distance and pinpoint its exact source.

The final features to consider are the computing power and analytics provided by the acoustic imaging camera and any companion software. An inspector can classify leak severity, perform leak cost analysis, and partial discharge pattern analysis in real-time during a survey.[26]

12.3.1 Causes of Noise in Power Transformers

Power transformers, being among the most complicated and critical components in elect-rical infrastructure, are prone to generating noise and vibration that can provide insights into their health. This discussion aims to explore the origins of transformer noise, its contributing internal components, and how voiceprint analysis can be used to predict transformer condit-ions.

To understand the reason and where the noise and vibration of a power transformer come from, we shall analyze the noise contribution by its internal components.

Transformer noise primarily arises from its internal components, and understanding its sources is key to effective diagnostics. Unlike the broadband noise generated by cooling fans or pumps, transformer noise is typically tonal, consisting of even harmonics of the power frequency. The primary source of this tonal noise is the transformer core. The low-frequency, tonal nature of the core noise or "buzzing" is particularly challenging to mitigate, as low freq-uencies travel farther with less attenuation. Additionally, tonal noise stands out more promin-ently to the human ear compared to broadband noise, even in environments with significant background noise. This combination of wide propagation and acute perception makes trans-former tonal noise a significant issue, especially in nearby communities. As a result, noise ordinances often impose stricter limits or penalties for tonal noise.

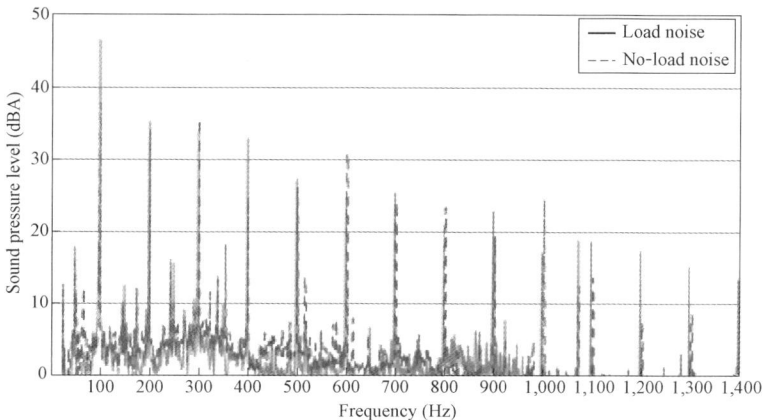

Figure 12.21   Measurement result of a power transformer noise using sound
intensity changing with frequencies [27]

While the core is the dominant source of noise, the load noise, generated by electrom-agnetic forces in the windings, can also contribute significantly, particularly in low-noise-level transformers. Cooling systems (fans and pumps) tend to dominate the very low and very high ends of the frequency spectrum, whereas core noise is most prominent in the mid-frequency range between 100 Hz and 1,000 Hz. Understanding these noise characteristics enables better prediction and assessment of transformer health through voiceprint analysis. [27] Figure 12.21 presents the measured noise characteristic of an actual power transformer, showing variations on sound intensity with frequency under both full-load and no-load.

The sound-generating mechanisms in power transformers can be characterized as follows:

Core Noise: When a strip of iron is magnetized, it undergoes a slight dimensional change (typically a few parts per million). This dimensional change, known as magnetostriction, occurs at twice the power frequency because it is independent of the magnetic flux direction. Due to the nonlinear nature of the magnetostriction curve, higher even-order harmonics appear in core vibrations, especially at higher flux density levels (above 1.4 T). Factors such as flux density, core material, geometry, and the waveform of the excitation voltage affect both the magnitude and frequency of transformer core noise. Additionally, mechanical resonance in the transformer's structure, core, and tank walls can significantly amplify vibrations, thereby increasing acoustic noise.

Load Noise: Load noise results from vibrations in the tank walls, magnetic shields, and windings caused by electromagnetic forces generated by load currents. These forces are proportional to the square of the load currents and primarily cause axial and radial vibrations in the windings. Poorly designed magnetic shielding can also contribute to load noise. A robust design with laminated magnetic shields, firmly anchored to the tank walls, can mitigate this. Load noise typically occurs at twice the power frequency, and its magnitude can be influenced by resonance in the tank walls or saturation during overloads. Radial vibrations usually have a minor effect on winding noise, except in large coils. However, axial vibrations from compressive electromagnetic forces can become a dominant noise source in inadequately supported windings. If the natural mechanical frequency of the winding clamping system resonates with these forces, load noise can intensify significantly, requiring damping. Harmonics in load current and voltage, especially in rectifier transformers, can further increase vibrations and noise by producing harmonics at multiples of the fundamental frequency. Although load noise has traditionally contributed moderately to overall transformer noise, it can become significant in transformers designed with low induction levels and improved core designs aimed at reducing core noise.

In many cases, the load noise is only slightly lower than core noise, sometimes by just a few decibels.

Fan and Pump Sound: Transformers produce substantial heat due to losses in the core, coils, and other metallic components. This heat is dissipated by fans blowing air over radiators or coolers. The noise from cooling fans is typically broadband and can be more prominent in smaller transformers or those operating at lower core induction levels. Key factors influencing fan noise include tip speed, blade design, the number of fans, and the configuration of radiators.

Figure 12.22 summarize the vibration and noise transferring path within a power transformer. [28]

12.3.2 Voiceprints and Their Recognition

Traditional diagnostic methods rely primarily on contact-type sensors, which are often installed on high-voltage electrical equipment in environments with strong electromagnetic interference. Such interference can significantly affect the accuracy of test results. Additionally, the installation and maintenance of contact-type sensors are challenging, and any malfunction in these sensors can potentially disrupt the power supply. Therefore,

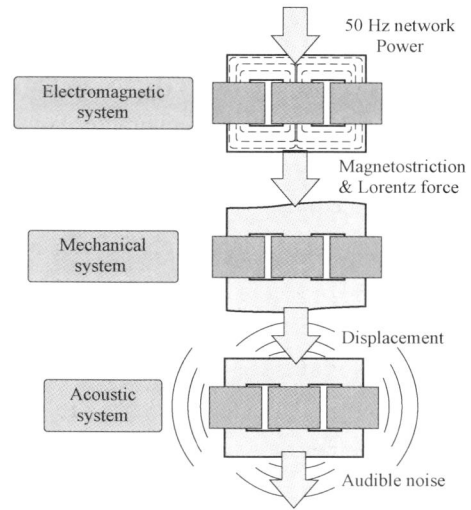

Figure 12.22   The vibration and noise transferring path within a power transformer

the need for more accurate, non-contact monitoring methods, alongside the development of online fault diagnosis and decision-aiding systems, has become crucial. [29]

Among the various fault diagnosis methods, acoustic signals stand out for their richness in conveying transformer fault information. Using sound and vibration signals to monitor the operational conditions and diagnose faults in electrical equipment—such as transformers, gas-insulated substations (GIS), and circuit breakers—offers distinct advantages.

In this subsection, we introduce the concept of voiceprints and their application in predicting the health of electrical apparatus. A voiceprint represents the acoustic spectrum of sound, visualized through electroacoustic instruments. It is a unique biological characteristic comprised of more than 100 parameters, including wavelength, frequency, and intensity. Voiceprints are known for their stability, measurability, and uniqueness.

The process of human speech production involves a complex interaction between the brain's speech center and various vocal organs. Since the size and shape of vocal organs—such as the tongue, teeth, larynx, lungs, and nasal cavity—differ from person to person, the voiceprint map of any two individuals will also differ.

While each person's voice retains both relative stability and variability, these variations can be influenced by factors like physiology, pathology, psychology, mimicry, disguise, and environmental disturbances. However, due to the unique structure of each person's vocal organs, it is generally possible to distinguish between individuals or determine whether two voiceprints belong to the same person.

Although voiceprints are not as visually intuitive as images, they can be analyzed and displayed through waveforms and spectrograms for practical analysis.

The size and shape of the vocal organs involved in speech production vary significantly from person to person, resulting in unique voiceprint patterns for each individual. These differences are primarily reflected in the following aspects:

- Resonance mode characteristics: Pharyngeal resonance, nasal resonance, and oral resonance vary across individuals.
- Voice purity characteristics: The clarity of a person's voice differs and can generally be categorized into three levels: high purity (bright), low purity (hoarse), and medium purity.
- Average pitch characteristics: This refers to whether a person's voice has a higher or lower average pitch.
- Range characteristics: This refers to the range between high and low pitches, affecting whether the voice sounds full or dry.

In a spectrogram, different voices exhibit varying distributions of formants. Voiceprint recognition works by comparing the sound of a speaker's phonemes across two speech segments to determine if they belong to the same person, enabling the ability to "identify a person by their voice."

Voiceprint identification typically relies on the technical accuracy of voiceprint recognition algorithms.

In electric power industry, voiceprint recognition has already been applied in practice. For instance, in vacuum circuit breakers, there are two distinct acoustic emissions. One originates from the arc electrodes during the switching on or off of the circuit, and the other comes from discharge between the electrodes and the shielding case when the vacuum has deteriorated.

By monitoring these two acoustic emissions, it is possible to perform online diagnostics of vacuum circuit breakers. Both analytical and experimental results demonstrate that the acoustic fingerprint method is effective for detecting vacuum degradation.

Acoustic waves exhibit specific characteristic parameters, which have a clear correlation with vacuum levels. Initially, the acoustic emission waveform and frequency characteristics are recorded, and subsequent measurements are compared to these baseline values to identify vacuum degradation.

12.3.3 Acoustic Imaging of Mechanical Vibration and Noise

Acoustic imaging cameras are now commercially available from many domestic and international instrument manufacturers. Below, we introduce some major options.

An acoustic camera is a device used to locate and characterize sound sources. It consists of an array of microphones—often referred to as a microphone array—that simultaneously capture sound to create a spatial representation of the sound sources. By measuring sound waves within a given area and comparing signal phase differences, the camera determines the location and amplitude of the sound sources. The results are displayed as an image, where color and brightness represent the intensity of the sound. These images are often overlaid with regular camera photos, providing a combined visual and acoustic representation.

By integrating acoustics, electronics, and information processing technologies, acoustic cameras convert sound into visible images. This makes it easier for users to intuitively understand the sound field, sound waves, and sound sources, facilitating the identification of noise causes and locations. See Figure 12.23 as an example.[30]

For different measurement requirements, there are varying array types and sizes available for microphone arrangements. All arrays come with an integrated fixed-focus camera, available as a USB or ethernet version as well as in different sensor sizes, resolutions, and frame rates. The included high-end tripod allows measurement set-ups in almost every measurement environment, see Figure 12.24 .[31]

Figure 12.23   Acoustic Camera -
Bionic S-112 Microphone Array [30]

Figure 12.24   Microphone arrays could be used for
beamforming, holography, and intensity measurements [31]

The right side of Figure 12.24 shows two acoustic microphone arrays in the form of ring arrays. These ring arrays typically have 32, 48, or 72 microphone channels, making them suitable for various 2D beamforming applications. They are versatile and can be used for both far-field and near-field measurements, in both indoor and outdoor environments.

Beamforming technology is a noise source identification method based on microphone array measurements. The fundamental principle is to arrange a set of microphones at different positions in space to form a microphone array, which captures sound signals. After applying appropriate delay and summation processing, information such as the sound source location, direction, and intensity is extracted. This provides details about the number, direction, and amplitude of the acoustic radiation sources. Originally used in radar, sonar, and wireless communication, beamforming is a long-distance sound source localization technology ideal for steady-state sound sources. It offers fast testing and higher resolution for sound source identification over long distances.

The star array, shown in Figure 12.25, [33] is particularly well-suited for far-field measurements using the beamforming method. [32] For 3D measurements, microphone arrays receive signals from all directions, making them ideal for interior applications. Beamforming results can be mapped onto scanned 3D point clouds or 3D CAD models of the measured object.

Acoustic imaging technology has already been applied in the electric power industry for noise source identification at a ±800 kV HVDC converter station in Loudi, Hunan Province, P. R. China. [34] Figure 12.26 illustrates a 30-channel acoustic holographic array with a diameter of 4m, composed of 5 test arms, and six 1/4-inch high-precision microphone probes are arranged on each test arm. The acquisition device uses B&K's LAN-XI acquisition instrument, and the terminal display and data processing are completed by a computer.

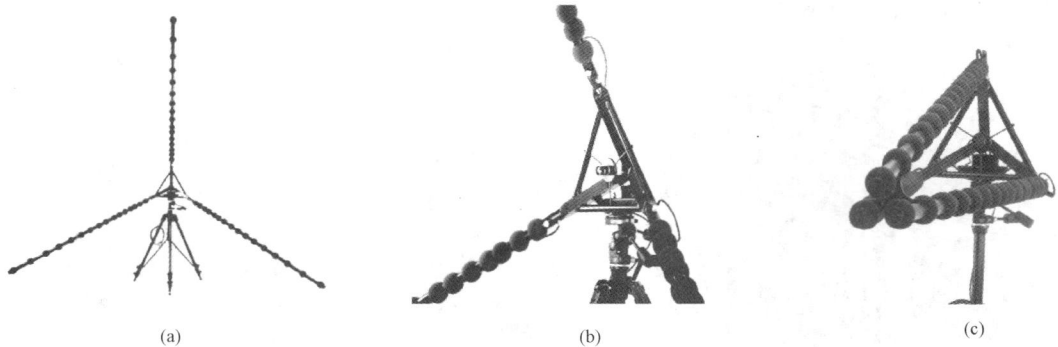

(a)   (b)   (c)

Figure 12.25   The Star microphone arrays could be used for beamforming,
holography, and intensity measurements [32]

Figure 12.26   The five-arm acoustic camera installed in a ±800 kV HVDC converter
station for noise source identification by far-field measurements [34]

The primary cause of noise in a HVDC converter station is the significant amount of harmonic current present in the DC transmission system. As this harmonic current flows through the filter system, it generates considerable noise in equipment such as capacitors, reactors, and transformers. Since the filter system occupies a large portion of the converter station, the noise produced has a substantial impact on the surrounding environment. Besides the main equipment like capacitors, reactors, and transformers, auxiliary cooling systems, including the converter heat dissipation system, valve cooling system, and external cooling water system, also contribute significantly to the overall noise levels at the converter station.

The ± 800 kV HVDC converter station includes the following mail components:

- 2 incoming ± 800 kV DC transmission lines.
- 7 outgoing 500 kV AC transmission lines.
- 24 units of converter transformers.
- 2 units of 500 kV station transformers.
- 2 units of 35 kV station transformers.
- 10 sets of AC filters.
- 9 sets of parallel capacitor banks.
- 2 sets of low impedance reactors.
- 12 high-end converter transformers.
- 12 low-end converter transformers.
- 2 groups of DC filters.
- 4 groups of smoothing reactors.

The converter station is equipped with 4 valve halls, each having its own independent closedloop valve cooling system. Each valve cooling system consists of an internal valve cooling system and an external valve cooling system, which includes a closed cooling tower, spray pump group, and external cooling water treatment system.

The B&K 3660-C-000 noise source identification system is employed, utilizing beam-forming acoustic imaging technology to locate and identify noise sources in converter stations. Figure 12.27, Figure 12.28, Figure 12.29, and Figure 12.30 illustrate the results of these measurements. [34]

After thoroughly investigating all potential noise sources in the ± 800 kV HVDC converter station, the highest noise intensities were identified in the converter transformers, reactor banks in the AC filter field, capacitor banks in the AC filter field, smoothing reactors in the DC filter field, and the water cooling systems of the valve halls. Among these, the converter transformers and reactor banks in the AC filter field ranked as the top two contributors.

Figure 12.27 and Figure 12.28 present the acoustic imaging results for the converter transformer area. The primary noise from the converter transformer is largely shielded by the surrounding wall. Acoustic imaging measurements reveal that the main noise source in this area has an acoustic intensity of approximately 76 dB, with the highest level observed at the fan array of the converter cooling system, located outside the converter. The figure also indicate that the fan array noise is predominantly low-frequency, concentrated in the band below 200 Hz.

Figure 12.27   The acoustic imaging test of the converter transformer [34]

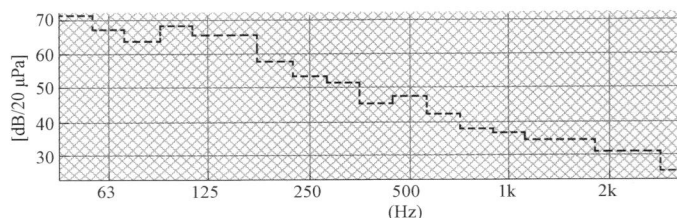

Figure 12.28   The sound spectrogram analysis by the acoustic camera for a
± 800 kV HVDC converter transformer [34]

Figure 12.29   Acoustic imaging test results of reactor banks in the AC filter field [34]

The AC filter field comprises a filter reactor bank and a filter capacitor bank. The reactor utilizes a dry-type air-core design, where the reactor coil generates periodic magnetic vibrations due to alternating electromagnetic fields, resulting in noise. Figure 12.29 presents the acoustic imaging of the reactor bank, and Figure 12.30 shows the corresponding acoustic imaging of the capacitor bank. Test results indicate that the primary noise source in the reactor bank area is the dry-type air-core reactor itself, exhibiting a prominent noise peak within the 562-708 Hz frequency range. For the capacitor bank, the main noise source is the capacitor units, though improper installation of supporting components may also contribute to vibrations and noise. In addition to the peak observed in the 89-112 Hz range, a similar peak between 562-708 Hz is noted, consistent with the reactor test results. Acoustic imaging measurements further reveal that the main noise source in this area has an acoustic intensity of approximately 73 dB.

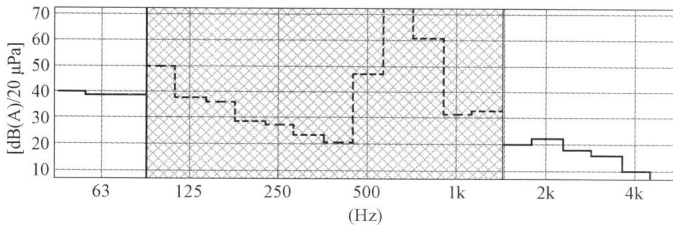

Figure 12.30　The sound spectrogram analysis by the acoustic camera for
reactor banks in the AC filter field [34]

HVDC converter transformers can experience noise issues due to DC bias, which elevates noise levels, often characterized by tonal components at multiples of the converter frequency and its harmonics. The main noise sources associated with DC bias include magnetostriction in the transformer cores and electromagnetic forces induced by the DC component in the windings.

To measure and analyze transformer noise, spectrum analyzers are used to examine the frequency spectrum. Of particular interest are tonal components and harmonics related to the converter frequency (commonly 50 Hz or 60 Hz) and their harmonics (100 Hz, 150 Hz, etc.). Microphones or acoustic cameras positioned near the transformer can detect and quantify audible noise levels, including both broadband noise and tonal components. Additionally, accelerometers are employed to measure mechanical vibrations caused by magnetostrictive forces, providing valuable data on the intensity and frequency distribution of vibrations due to DC bias.

Continuous monitoring of noise levels and spectral characteristics is essential for assessing transformer condition and identifying potential issues early. Comparative analysis over time helps detect trends or changes in noise patterns, which may indicate degradation or operational shifts affecting DC bias.

Effectively managing DC bias-related noise in HVDC converter transformers is critical for ensuring reliable operation and extending the lifespan of power transmission systems.

12.3.4　Partial Discharge Ultrasonic Detection in Enclosed Spaces

Partial discharge (PD) events generate both electromagnetic waves and acoustic pressure waves. The acoustic signals typically span a frequency range from a few kHz to several hundred kHz. One key advantage of detecting these acoustic emissions (AE) is their immunity to electromagnetic interference. To minimize the influence of mechanical vibrations from pumps, fans, and acoustic noise from transformer iron cores, caused by magnetostriction and the Barkhausen effect, the ultrasonic frequency range, typically between 40 kHz and a few hundred kHz, is selected for capturing AE signals. Ultrasonic PD detection was initially employed to localize airborne noise from corona discharges, such as identifying discharges at shielding electrodes in high-voltage test facilities or faulty discharges in broken cap-and-pin insulators. Figure 12.31 shows a hand-held battery-powered ultrasonic PD detector designed for this purpose. [35, 36]

However, using only a single ultrasonic transducer for localization is time-consuming, especially for intermittent PD events. A more efficient approach, now commonly used, is triangulation.

In a homogeneous medium, the distances x_1, x_2, and x_3 between the PD source and the AE transducers are proportional to the time-of-flight values t_1, t_2, and t_3, derived from oscilloscope records. The point where these trajectories cross, as shown in Figure 12.32, marks the location of the PD source. [35]

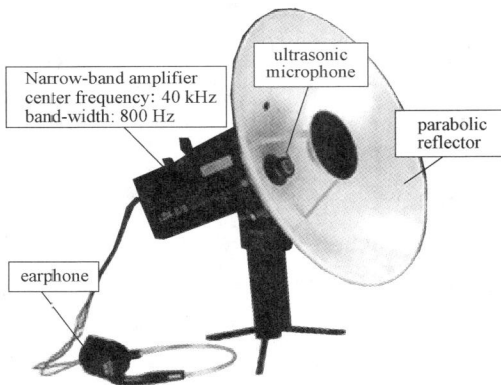

Figure 12.31   Photograph of an ultrasonic
PD detector. Courtesy of Doble Lemke [35]

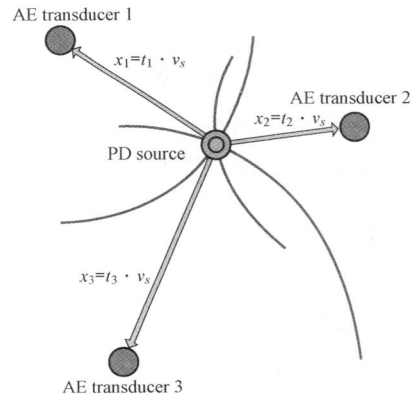

Figure 12.32   Principle of triangulation used
for localization of the PD site [35]

To improve localization accuracy, it is standard practice to combine ultrasonic techniques with electrical PD measurements. This allows the oscilloscope to be triggered by an electrical signal when the PD event occurs, as illustrated in Figure 12.33 (a). [35] Since the time delay of the electrical signal is less than a microsecond, it is negligible compared to the travel time of the acoustic signal, which travels only about 1.2 mm per microsecond in oil. In noisy conditions, the signal-to-noise ratio can be significantly improved using an averaging mode, as shown in Figure 12.33 (b). The combined acoustic-electrical method further verifies that the detected signal is indeed a PD event and not interference from other acoustic noise.

In laboratory settings, the electrical signal used to trigger the oscilloscope is often captured via a coupling capacitor or bushing tap. However, in noisy on-site conditions, the VHF/UHF technique is preferable for enhancing the signal-to-noise ratio. It is important to note that triangulation, as shown in Figure 12.32, only yields reliable results when acoustic waves propagate through a continuum where their velocity remains constant. In complex HV equipment, such as power transformers, acoustic wave velocity is influenced by various materials, including copper, steel, wood, pressboard, and insulating oil.

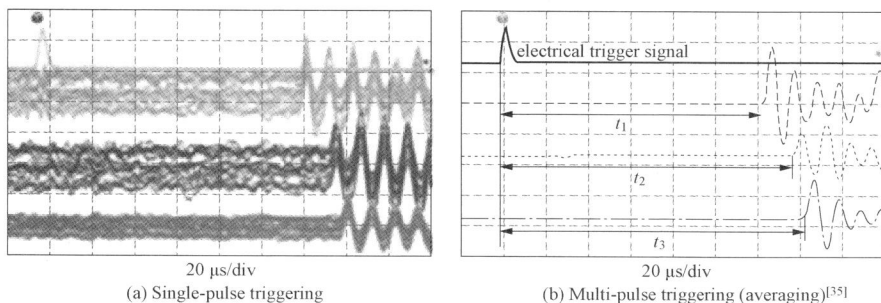

20 μs/div
(a) Single-pulse triggering

20 μs/div
(b) Multi-pulse triggering (averaging)[35]

Figure 12.33   Principle of time-of-flight measurement, where the acoustic signal is received by three ultrasonic transduces attached to the tank of a 110 kV instrument transformer. Here the oscilloscope was triggered by an electrical PD signal

Two types of acoustic waves—longitudinal (pressure) and transverse (shear) — must be considered, as each travels at different speeds. The shortest path is not always the fastest, as shown in Figure 12.34, due to the differing velocities of acoustic waves. For example, acoustic waves travel at approximately 1.25 mm/μs in oil and 5.1 mm/μs in steel. To solve the complex equations for wave velocity in real high-voltage equipment, advanced computer systems with sophisticated software are now available.

A typical AE transducer is shown in Figure 12.35. To facilitate easier installation on the side wall of electric apparatus, a permanent magnet cap, as illustrated in Figure 12.36, is used. The application photo in Figure 12.37 demonstrates that a signal amplifier is also required to

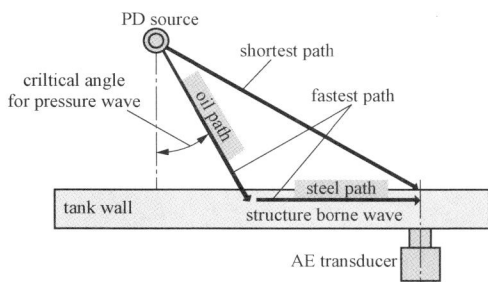

Figure 12.34   The shortest and fastest paths between a PD source in oil and an acoustic sensor placed on a power transformer tank [35]

Figure 12.35   The outline of AE-Transducer model R15α from Physical Acoustics Corporation [37]

process the acoustic waves detected by the AE transducer. Additionally, Figure 12.38 presents a typical acoustic intensity versus frequency plot for the AE transducer. In this example, the AE transducer, model $R15\alpha$, was obtained from Physical Acoustics Corporation.

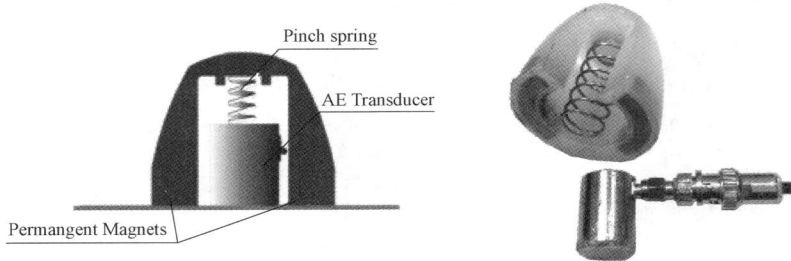

Figure 12.36   The 2D internal structure and actual picture of an AE-Transducer cap

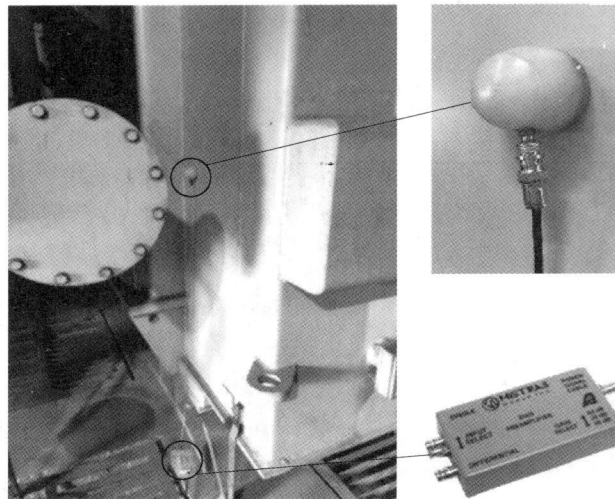

Figure 12.37   An AE-Transducer was installed on the side wall of
a power transformer including a preamplifier

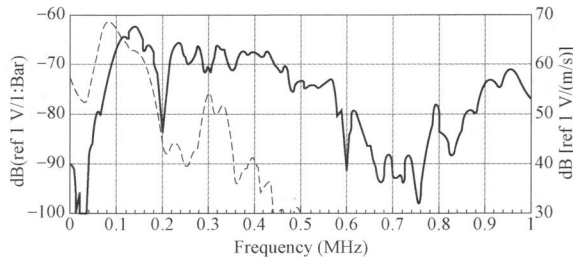

Figure 12.38   The typical acoustic intensity versus frequency plot for the AE
transducer from Physical Acoustics Corporation (model $R15\alpha$) [37]

The equipment needed for acoustic PD detection in high voltage equipment typically includes an array of AE transducers, a signal transmission unit (either via cabling or fiber optic link), and an acquisition system (such as a digital oscilloscope or computer-based system) for signal processing, visualization, and data storage. The following types of ultrasonic transducers are commonly used:

- Piezo-electric transducers.
- Structure-borne sound resonance transducers.
- Accelerometers.
- Condenser microphones, and.
- Electro-optic transducers.

To ensure effective signal transmission, the transducer surface is usually coated with hard epoxy resin to match the acoustic impedance of the high voltage apparatus' metallic enclosure. This coating also provides necessary insulation between the transducer and the test object. Special care should be taken with the coupling method, as acoustic waves may reflect at the interface between the transducer and the high voltage equipment enclosure. Acoustic coupling gel or grease should be applied to reduce reflections.

It is often beneficial to integrate a pre-amplifier into the ultrasonic transducer to enhance the signal-to-noise ratio. As mentioned earlier, mechanical vibrations from pumps, fans, and transformer core noises caused by magnetostriction and the Barkhausen effect are not correlated with acoustic waves emitted from PD sources and can be effectively filtered out. Narrowband amplifiers operating at center frequencies around 40 kHz may also improve testing results. However, under these conditions, pulse responses often exhibit oscillations with envelopes extending over hundreds of microseconds, potentially causing superposition of subsequent acoustic signals, as seen in Figure 12.39. [35] This is one reason why the acoustic PD detection method cannot accurately quantify PD magnitude. Additionally, ultrasonic signals are subject to strong attenuation and dispersion when passing through various insulation structures, which act as low-pass filters, attenuating the signal in proportion to the square of its frequency.

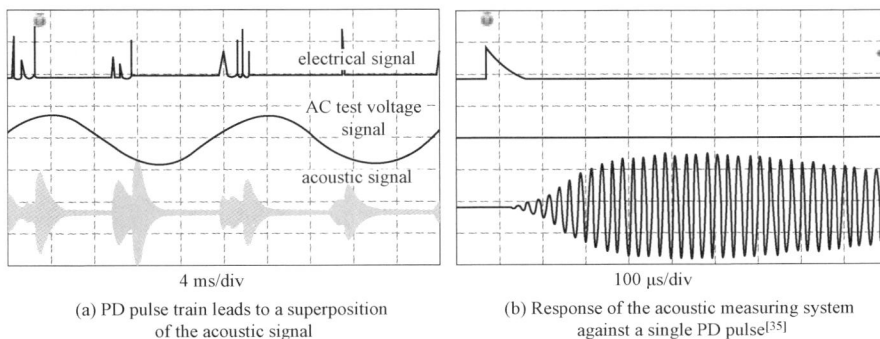

4 ms/div

(a) PD pulse train leads to a superposition
of the acoustic signal

100 μs/div

(b) Response of the acoustic measuring system
against a single PD pulse[35]

Figure 12.39   PD pulse response of a narrow-band ultrasonic measuring system having
a center frequency of 42 kHz and a bandwidth of 800 Hz

According to standards and guidelines for ultrasonic detection and positioning of partial discharge in oil-immersed power transformers and reactors, the following methods should be followed:

During the detection process, care should be taken with the electrical environment of the transformer tank, ensuring that sensors and signal cables are kept at a safe distance from hazardous areas. Additionally, testing should not be conducted during adverse weather conditions such as rain, snow, or strong wind.

Sensors should be directly placed on the side wall of the transformer tank, avoiding reinforcement belts, control boxes, or other obstructions. The sensor should also not be installed above the oil level. Before installation, the sensor surface must be cleaned and coated with silicone grease to ensure proper adhesion. The sensor should be secured using magnetic absorption or binding. Signal cables must be routed carefully to avoid interference caused by sensor vibration or cable movement during detection. Once the sensors are in place, vibration signals can be generated by lightly tapping the oil tank wall to verify that the sensors are functioning correctly.

During ultrasonic detection, the impact of interference signals such as tank vibrations, noise from oil pumps and fans, and electromagnetic interference must be considered. Attenuation and obstruction of ultrasonic signals within the transformer, due to factors like magnetic or electrical shielding inside the oil tank, should also be taken into account. The continuous monitoring time at each measuring point should be no less than 10 minutes. For low-frequency or irregular signals, extending the monitoring time may be necessary to ensure accurate detection.

12.3.5 Ultrasonic Camera Applications in Substations

In electric power systems, maintaining adequate and properly serviced insulation is essential for ensuring equipment longevity. However, factors such as manufacturing defects, improper installation practices, and vibrational wear can compromise insulation integrity. This can result in unwanted partial discharges (PD), including corona, tracking, and arcing. These PD events progressively degrade the insulation, worsening the defect over time. If partial discharges are not promptly identified and addressed, they can lead to insulation failure, resulting in short circuits and potentially complete equipment failure.

Insulation failure is not only costly in terms of repairs but also leads to significant facility downtime. To prevent this, it is essential to regularly inspect insulation components like insulators, windings, switches, and fuse elements in XLPE cables, transformers, switchgear, and disconnect units to detect and address critical defects early.

Ultrasonic camera devices provide a safe and accurate means of detecting insulation damage, even in cases where no heat signature is present. While not all insulation defects cause temperature increases, they all produce noise in the ultrasonic spectrum. Tracking or treeing occurs when electric current leaks between two points separated by insulating material, often due to contaminants like dirt, carbon particles, or moisture. Arcing refers to visible plasma discharge between electrodes, caused by ionization of gases in the air by electrical current. Arcing can result from intermittent short circuits to ground, such as lightning strikes, and tracking often leads to arcing if left unchecked. Corona is an electrical discharge caused by ionization of air around a conductor, influenced by conductor geometry and contamination. Corona discharges on utility lines lead to energy loss but can be mitigated with better insulation, the use of corona rings, and ensuring high-voltage electrodes lack sharp edges.

Each type of discharge, arcing, tracking, and corona, produces distinct sound characteristics in the ultrasonic spectrum. Experienced technicians can differentiate between them using onscreen data from handheld ultrasonic testers. These discharges also produce ionization, where neutral atoms and molecules lose or gain electrons to become ions. The byproducts of ionization, such as ozone (O_3) and nitric acid (HNO_3), are corrosive to many dielectric materials and certain metals. Additionally, partial discharges often cause radio frequency interference.

Ultrasonic cameras operate in the 20-100 kHz range and utilize "heterodyning" to convert ultrasound signals into audible frequencies for technicians. These devices typically come with high-quality over-ear headphones to attenuate ambient noise from plant operations. The detectors display signal intensity in decibels (dB), and some advanced models also provide graphical representations of sound waves in both frequency and time domains. Ultrasonic cameras range from affordable entry-level models to sophisticated multifunctional devices with touchscreens and selectable frequency tuning. Complementary software is available to capture and analyze sound files, aiding in identifying PD types and severity.

Thanks to various optional accessories and probes, ultrasonic cameras are well-suited for electrical inspections. For example, they can detect PD over short distances using basic airborne sound probes, effectively scanning panel seams or ventilation louvers. For certain equipment types, 40 kHz permanently mounted partial discharge sensors (PDS) can be installed, and an adapter cable for ultrasonic cameras can connect to these sensors. PDS units may be standalone or integrated into an infrared window or visual inspection system, allowing safe and accurate detection without compromising inspector safety. For longer distances and overhead substation applications, parabolic sensors with a range of up to 45 meters (150 feet) are available.

In addition to airborne ultrasound, loose insulators, windings, or vibrating mechanical components can transmit ultrasonic waves through the equipment's structure waves that are inaudible to the human ear. However, using a highly sensitive structure-borne probe with an ultrasonic camera allows these sound waves to be detected and converted into audible frequencies. This enables the identification of electromechanical defects, allowing for timely corrective actions. Conducting partial discharge testing on electrical assets is a critical part of a comprehensive Condition-Based Maintenance program. Defects like arcing, tracking/treeing, corona, and loose mechanical components can be detected using both airborne and structure-borne probes. Detecting and addressing these issues early ensures the longevity and reliability of electrical assets.

Figure 12.40, Figure 12.41 and Figure 12.42 show examples of how ultrasonic cameras can identify partial discharge defects in high-voltage bushings and insulators or incoming overhead lines. [38, 39, 40, 41]

Figure 12.40　Quickly and safely locate partial discharge within a high-voltage electrical system [38]

Figure 12.41　System quickly provides PD type classification and PD pattern for further analysis [38]

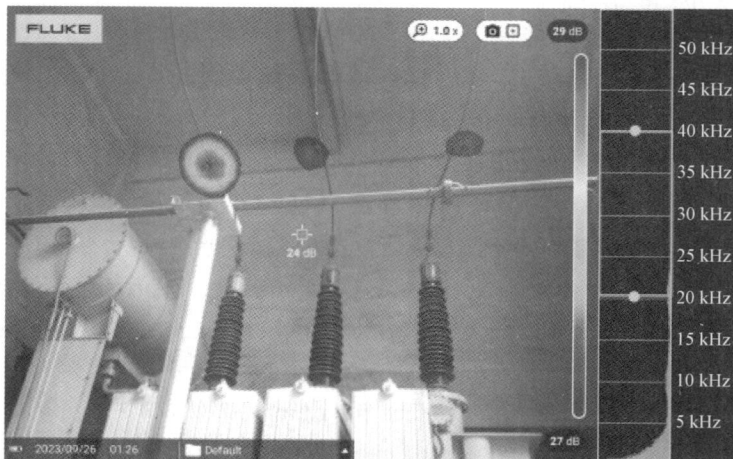

Figure 12.42　Partial discharge acoustic imaging on the high voltage incoming overhead transmission line of a 110 kV power transformer [39]

12.4 Review Questions

Q1: Please give the audio frequency band of infrasound, human hearing sound, and ultrasound. (see Section 12.2.1)

Q2: Please tell us the working principle of eight kinds of microphones. (see Section 12.2.2)

Q3: Please explain the properties of piezoelectric materials, which kind of piezoel- ectric material is the most popular used?

Q4: Please give an example of ultrasonic arrays to identify the location of partial discharge. (see Section 12.3.2)

Q5: Ultrasonic signal propagation between the partial discharge source and the ultrasonic transducer is along the shortest distance or along the fastest curve way? Please explain the reasons.(seeSection12.3.4)

Q6: Please tell us the reasons and locations of the noise and vibration coming from by power apparatus, and how to predict health conditions according to voiceprint. (see Section 12.3.1)

Bibliography

[1] Joe Wolfe. What is sound spectrum. URL https://newt.phys.unsw.edu.au/jw/sound. spectrum. html#:~: text=A%20sound%20spectrum%20displays%20the%20different%20frequencies%20present,made%20up %20of%20a%20complicated%20mixture%20of% 20vibrations. Accessed: August 30, 2024.

[2]OLYMPUS. Thickness gauge tutorial: Theory of operation. [Online]. Available: https://www.olympus-ims. com/en/ndt-tutorials/thickness-gauge/introduction/ operation/. Accessed: Sep. 2024.

[3] © Spectrum GmbH, Germany. Using Spectrum Digitizers in Ultrasonic Ap-plications. URL https://spectrum-instrumentation.com/dl/an_digitizers_in_ ultrasonic_applications.pdf. Accessed: August 30, 2024.

[4]The frequency spectrum, instrument ranges, and EQ tips. [Online]. Available: http://www. guitarbuilding. org/wp-content/uploads/2014/06/Instrument-Sound-EQ-Chart.pdf. Accessed: Aug. 2024.

[5] Classification of Microphones. URL https://www.cemca.org/ckfinder/userfiles/ files/7_Lesson-06_MICR-OPHONES.pdf. Accessed: August 30, 2024.

[6] SHURE®. Microphone Techniques for Recording. A Shure Educational Publication, 2014.

[7] Three Types of Microphones. URL https://thereviewmail.com/is-a-condenser-mic-good-for-vocals/. Accessed: August 30, 2024.

[8] John, Eargle. The microphone book, 2005.

[9] Alex Milne. How Dynamic Microphones Create Audio Signal. URL https://www.rfvenue. com/blog/2015/ 01/05/how-dynamic-microphones-create-audio-signal. Accessed: August 30, 2024.

[10] Musical Magnets. URL http://howtmm.com/week10/#/3/1. Accessed: August 30, 2024.

[11] Daren Banarsë. How ribbon microphones work. URL https://soundref.com/ different-types-of-micropho-nes/. Accessed: August 30, 2024.

[12] ST Microelectronics. AN4426 application note: Tutorial for MEMS microphones. [Online]. Available: https://www.st.com/resource/zh/application_note/dm00103199-tutorial-for-mems-microphones-stmicroel ectronics.pdf. Accessed: Sep. 2024.

[13] Walter C. Babel III, Qamar A. Shams and James F. Bockman. Qualitative Analysis of MEMS Micropho-nes. URL https://slidetodoc.com/qualitative-analysis-of-mems-microphones-16-th-annual/.Accessed: Sep. 30, 2024.

[14] SHURE®. The History Of Carbon Microphones And Artifacts From The Shure Archives. URL https://www. shure.com/en-US/insights/the-history-of-carbon-microphones-and-artifacts-from-the-shure-archives. Accessed: Sep. 30, 2024.

[15] The Edison Tech Center. Microphones. URL http://edisontechcenter.org/ microphones.html. Accessed: Sep. 30, 2024.

[16] U. Schichler OVE, W. Troppauer, B. Fischer, T. Heine, K. Reich OVE, M. Leonhardsberger OVE, O. Oberzaucher. Development of an innovative measurement system for audible noise monitoring of OHL. Elektrotechnik & Informationstechnik, 135:556-562, 2018. URL https://doi.org/10.1007/s00502-018-0670-z.

[17] Balthasar Fischer. Optical microphone hears ultrasound. NATURE PHOTONICS, 10:1-3, 2016. URL https://xarion.com/ploxmedia/_1_/9dce499b194faed208e3a728dff67e9b/XARION_NaturePhotonics_2016. pdf.

[18] Optoacoustics Ltd (OPTIMIC™). Sound Solutions from Light Technology. URL https://erii.org/wp-content/ uploads/97.pdf. Accessed: August 30, 2024.

[19] Laser listening device - Spectra laser microphone M+. [Online]. Available: https://www. detective-store. com/laser-listening-device-spectra-laser-microphone-m-458. html. Accessed: Sep. 2024.

[20] Georg Wissmeyer, Miguel A. Pleitez, Amir Rosenthal and Vasilis Ntziachristos. Looking at sound: optoacoustics with all-optical ultrasound detection. Light: Science & Applications, 7:Article number: 53, 2018. doi: 10.1038/s41377-018-0036-7. URL https://www.nature. com/articles/s41377-018-0036-7.

[21] Britannica. Crystal microphone - Electroacoustic device. [Online]. Available: https://www. britannica.com/ technology/crystal-microphone. Accessed: Feb. 2021.

[22] Paolo Visconti, Laura Bagordo, Ramiro Velázquez, Donato Cafagna and Roberto De Fazio. Available technologies and commercial devices to harvest energy by human trampling in smart flooring systems: A review. Energies, 15:432, 2022. URL https://doi.org/10. 3390/en15020432.

[23] Rupitsch, Stefan J. Piezoelectric sensors and actuators. Springer, 2019.

[24] APC International Ltd. What Is PZT? URL https://www.americanpiezo.com/ knowledge-center/piezo-theory/pzt/. Accessed: Sep. 30, 2024.

[25] McGrail, A and Risino, A and Auckland, DW and Varlow, BR. Use of a medical ultrasonic scanner for the inspection of high voltage insulation. IEEE Electrical Insulation Magazine, 9(6):5-10, 1993.

[26] FLIR. Save time and money: Complete inspections 10 times faster with acoustic imaging. [Online]. Available: https://www.flir.com/discover/instruments/acoustic-imaging/ 7-things-to-look-for-in-an-acoustic-

imager/. Accessed: Sep. 2024.

[27] Michal Kozupa, Grzegorz Kmita, Roberto Zannol and Gianluca Bustreo. Vibroacoustic analyses for noise mitigation in transformers - transformer vibroacoustic analyses. ABB REVIEW, pages 22-27, 2018. doi: 10.15199/48.2019.12.38.

[28] Daniel MARCSA. Noise and vibration analysis of a distribution transformer. PRZEGLĄD ELEKTROTE-CHNICZNY, pages 172-175, December 2019. doi: 10.15199/48.2019.12.38.

[29] Hugh M. Ryan. High voltage engineering and testing, 3rd Edition, volume: Power and Energy Series 66. The Institution of Engineering and Technology, 2013.

[30] Acoem Ecotech Industries Private Limited. The Bionic S-112 microphone array. URL https://www.india-mart.com/proddetail/ acoustic-camera-bionic-s-112-microphone-array-23043539812.html. Accessed: Sep. 30, 2024.

[31] ambergo®. Acoustic Cameras. URL https://ambergo.pt/en/produtos/manutencao-en/ acoustic-cameras-2/.Accessed: Sep. 30, 2024.

[32] ©2024 Polytec GmbH. 2D and 3D Microphone Arrays. URL https://www.polytec.com/us/acoustics/products/ 2d-microphone-arrays. Accessed: Sep. 30, 2024.

[33] Polytec. Star48-ac-pro. URL https://www.polytec.com/us/acoustics/products/ 2d-microphone-arrays/star48-ac-pro. Accessed: Sep. 30, 2024.

[34] Tao Huang, Xingyao Wang, Kai She, Wei Cai, and Cao Hao. Noise source identification for a ±800 kV converter station based on beam-forming acoustic imaging technology. IOP Conf. Series: Earth and Environmental Science, 692:020045, 2020. doi: 10.1088/1755-1315/692/ 2/020045.

[35] Hauschild, Wolfgang and Lemke, Eberhard. High-voltage test and measuring techniques, volume 1. Springer, 2014.

[36] Schon, Klaus. High voltage measurement techniques. Springer, 2019.

[37] Physical Acoustics Corporation. r15α sensor. URL https://www.physicalacoustics.com/ content/literatu-re/sensors/Model_R15a.pdf. Accessed: Sep. 30, 2024.

[38] TELEYNE FLIR. Flir acoustic cameras. URL https://www.flir.com/discover/instruments/acoustic-imag-ing/quickly-locate-partial-discharge-pd-with-acoustic-imaging/. Accessed: Sep. 30, 2024.

[39] FLUKE. Fluke acoustic cameras. URL https://www.fluke.com/en-us/product/ industrial-imaging/precis-ion-acoustic-imager-ii910. Accessed: Sep. 30, 2024.

[40] Chizhi Huang. Research on propagation characteristics of partial discharge ultrasonic signals in gis and design of detection system. Master's thesis, Hunan Ubiversity, 2016.

[41] Gong Zheng. Research and development of ultrasonic phased array detection and imaging system. Master's thesis, Nanjing University of Aeronautics and Astronautics, 2018.